LEVERAGING ARTIFICIAL INTELLIGENCE (AI) COMPETENCIES FOR NEXT-GENERATION CYBERSECURITY SOLUTIONS

LEVERAGING ARTIFICIAL INTELLIGENCE (AI) COMPETENCIES FOR NEXT-GENERATION CYBERSECURITY SOLUTIONS

Edited by

Pethuru Raj, PhD

B. Sundaravadivazhagan, PhD

V. Kavitha, PhD

AAP | APPLE ACADEMIC PRESS

First edition published 2025

Apple Academic Press Inc.
1265 Goldenrod Circle, NE,
Palm Bay, FL 32905 USA

760 Laurentian Drive, Unit 19,
Burlington, ON L7N 0A4, CANADA

CRC Press
2385 NW Executive Center Drive,
Suite 320, Boca Raton FL 33431

4 Park Square, Milton Park,
Abingdon, Oxon, OX14 4RN UK

Library and Archives Canada Cataloguing in Publication

CIP data on file with Canada Library and Archives

Library of Congress Cataloging-in-Publication Data

CIP data on file with US Library of Congress

ISBN: 978-1-77491-634-6 (hbk)
ISBN: 978-1-77491-635-3 (pbk)
ISBN: 978-1-00347-809-6 (ebk)

About the Editors

Pethuru Raj, PhD
Chief Architect and Vice President, Reliance Jio Platforms Ltd., Bangalore, Karnataka, India

Pethuru Raj, PhD, is working as a Chief Architect and Vice President at Reliance Jio Platforms Ltd. (JPL) Bangalore, India. Previously, he worked at the IBM Global Cloud Center of Excellence (CoE), Wipro Consulting Services (WCS), and Robert Bosch Corporate Research (CR). In total, he has gained more than 20 years of IT industry experience and eight years of research experience. He finished his CSIR-sponsored PhD degree at Anna University, Chennai, and continued with the UGC-sponsored postdoctoral research in the Department of Computer Science and Automation, Indian Institute of Science (IISc), Bangalore. Thereafter, he was granted a couple of international research fellowships (JSPS and JST) to work as a research scientist for 3.5 years at two leading Japanese universities. He focuses on some of the emerging technologies such as the Internet of Things (IoT), optimization of artificial intelligence (AI) models, big, fast, and streaming analytics, blockchain, digital twins, cloud-native computing, edge and serverless computing, reliability engineering, microservices architecture (MSA), quantum computing, event-driven architecture (EDA), 5G/6G, etc.

B. Sundaravadivazhagan, PhD
Professor, Department of Information Technology, University of Technology and Applied Sciences, AL Mussanah, Sultanate of Oman

B. Sundaravadivazhagan, PhD, is a Professor in the Department of Information Technology at the University of Technology and Applied Science-AL Mussanah in Oman. He earned a PhD in computer science from Anna University, Chennai, India. He has professional memberships in ISACA and ISTE and is a senior member of IEEE. His academic and research background spans more than 21 years at many institutions, including working on two financed research projects for the Ministry of Higher Education Research Innovation in Oman via TRC. His research interests include the Internet of Things, artificial intelligence and machine learning, deep learning, cloud computing, and cybersecurity. He has published more than 30 technical articles in international journals.

V. Kavitha, PhD
Associate Professor, Department of Computer Science with Cognitive Systems, School of Computing, Sri Ramakrishna College of Arts and Science, Coimbatore, Tamil Nadu, India

V. Kavitha, PhD, is currently working as an Assistant Professor in the Department of Computer Science with Cognitive Systems, Sri Ramakrishna College of Arts and Science, Coimbatore, Tamil Nadu, India. She earned her PhD in 2014 in the broad area of data mining. She has over 19 years of teaching and research experience. She has been supervising research scholars and postgraduate and undergraduate students in the recent areas of cyber security. She has published more than 75 research papers in international as well as national journals, international conferences, and book chapters. She has delivered technical talks and done research in the areas of big data analysis and cyber security. cyber security, IoT Security, artificial intelligence, machine learning, and deep learning.

Contents

Contributors

B. Aysha Banu
Assistant Professor, Department of Information Technology, Mohamed Sathak Engineering College, Tamil Nadu, India

T. Bhuvaneswari
Assistant Professor, Department of Computer Applications, Queen Mary's College, Chennai, Tamil Nadu, India

M. Jaithoon Bibi
Assistant Professor, Department of Computer Science with Cognitive Systems, Sri Ramakrishna College of Arts and Science, Nava, Coimbatore, Tamil Nadu, India

Robin Cyriac
Department of Information Technology, University of Technology and Applied Sciences, Al Mussanna, Oman

M. Renuka Devi
Associate Professor, School of Information Science and Engineering, Presidency University, Bangalore, India

S. Gavaskar
Assistant Professor, Department of Computer Applications, Bharathiar University, Coimbatore, Tamil Nadu, India

K. Geetha
Assistant Professor, Department of Computer Science, Bharathiar University, Coimbatore, Tamil Nadu, India

S. Geethanjali
Research Scholar, Department of Computer Science, IDE, University of Madras, Chennai, Tamil Nadu, India

K. Gowri
Department of Computer Science with Cognitive Systems, Sri Ramakrishna College of Arts and Science, Coimbatore, Tamil Nadu, India

M. Jayakumar
Assistant Professor, Department of Computer Applications, Dr.N.G.P Arts and Science College, Coimbatore, Tamil Nadu, India

D. Jocil
Research Scholar, Department of IT, Bharathiar University, Coimbatore, Tamil Nadu, India

R. Justin
University of Technology and Applied Sciences, Sultanate of Oman

A. Kalaivani
Assistant Professor, Department of Computer Technology, Nallamuthu Gounder Mahalingam College, Pollachi, Tamil Nadu, India

D. Kalaivani
Associate Professor and Head, Department of Computer Technology, Dr. SNS Rajalakshmi College of Arts and Science, Coimbatore, Tamil Nadu, India

Karthikeyan
Professor, Department of CSE (AI&ML), Vardhaman College of Engineering, Hyderabad, Telangana, India

S. Karthiyayini
Assistant Professor, Department of Information Technology, Mohamed Sathak Engineering College, Tamil Nadu, India

V. Kavitha
Department of Computer Science with Cognitive Systems, Sri Ramakrishna College of Arts and Science, Coimbatore, Tamil Nadu, India

V. Krishnapriya
Associate Professor, Department of Computer Science with Cognitive Systems,
Sri Ramakrishna College of Arts and Science, Nava, Coimbatore, Tamil Nadu, India

M. Kundalakesi
Assistant Professor, Department of Computer Applications, Sri Krishna Arts and Science College, Coimbatore, Tamil Nadu, India

S. R. Lavanya
Assistant Professor, Department of Computer Science, KPR College of Arts Science and Research, Coimbatore, Tamil Nadu, India

S. Leelalakshmi
Assistant Professor, Department of Computer Science, Queen Mary's College, Chennai, Tamil Nadu, India

B. Mahalakshmi
Assistant Professor, School of Computer Studies, Rathnavel Subramaniam College of Arts and Science, Coimbatore, Tamil Nadu, India

K. Maheswari
Department of Computer Science, Anna Adarsh College for Women, Chennai, Tamil Nadu, India

S. Mayukha
Research Scholar, Department of IT, Bharathiar University, Coimbatore, Tamil Nadu, India

N. Nivetha
Assistant Professor, Department of Computer Science, Adithya College of Arts and Science, Coimbatore, Tamil Nadu, India

A. Parameswari
Department of Computer Science, Anna Adarsh College for Women, Chennai, Tamil Nadu, India

G. Pradeepa
University of Technology and Applied Sciences, Sultanate of Oman

M. Praneesh
Assistant Professor, PG and Research Department of Computer Science, Sri Ramakrishna College of Arts and Science, Coimbatore, Tamil Nadu, India

A. Saleem Raja
University of Technology and Applied Sciences, Sultanate of Oman

S. Sasikala
Associate Professor, Department of Computer Science, IDE, University of Madras, Chennai, Tamil Nadu, India

D. Sasirekha
Department of Computer Science, Anna Adarsh College for Women, Chennai, Tamil Nadu, India

A. Dhanu Saswanth
Department of Cyber Security, University of Texas at Dallas, USA

V. Sindhu
School of Computer Applications, Lovely Professional University, Punjab, India

D. Sowmyadevi
Assistant Professor, Department of Computer Science, PSG College of Arts and Science, Coimbatore, Tamil Nadu, India

S. Srividhya
Associate Professor, Department of Computer Science, KPR College of Arts Science and Research, Coimbatore, Tamil Nadu, India

S. Suganyadevi
Assistant Professor, Department of Computer Science, Sri Krishna Arts and Science College, Coimbatore, Tamil Nadu, India

M. R. Sundarakumar
School of Computing Science and Engineering, Galgotias University, Greater Noida, Utter Pradesh, India

B. Sundaravadivazhagan
Department of University of Technology and Applied Sciences, Al Mussanah, Oman

R. Swathika
Research Scholar, Department of Computer Science, Bharathiar University, Coimbatore, Tamil Nadu, India

K. Tamilselvi
Assistant Professor, Department of Computer Science, Bharathi Women's College, Chennai, Tamil Nadu, India

Jayarani Thomas
Independent Researcher

R. Vadivel
Associate Professor, Department of IT, Bharathiar University, Coimbatore, Tamil Nadu, India

P. S. Vijayalakshmi
Research Scholar, Department of Computer Science, SNMV College of Arts and Science, Coimbatore-50 ,Tamil Nadu, India

Hannah Vijaykumar
Department of Computer Science, Anna Adarsh College for Women, Chennai, Tamil Nadu, India

R. Vinodini
Little Angel Institute, Karur, Tamil Nadu, India

Preface

The cyberattack surface in modern enterprises is expanding rapidly. This is literally complicating the cybersecurity posture of any growing enterprise. Due to the massive volume of security-related data being generated by business workloads, IT infrastructure services, security products, network components, and monitoring/observability solutions, human-centric cybersecurity solutions are beset with a number of practical challenges and concerns. Therefore, security experts recommend AI-based automated and advanced cybersecurity mechanisms. AI-backed cybersecurity posture involves less intervention, interpretation, and involvement of human security professionals. Considering the need for real-time decisions and actions, AI-based security solutions have momentum.

Machine and deep learning (ML/DL) models are being established and deployed to receive and analyze all kinds of log and security-related event data instantaneously to supply actionable insights and to activate the appropriate countermeasures in time. Therefore, any cybersecurity attack can be nipped in the budding stage.

AI technologies and tools are inherently capable of processing big data quickly and pinpointing useful patterns, rules, associations, and other insights hidden in data heaps. Cyber threats, such as malware menaces, illegal attempts to steal information, and phishing attacks, can be easily identified and articulated in time to plunge into corrective actions immediately. Machine and deep learning models are intrinsically empowered to learn from data streams continuously to enhance prediction accuracy. New types of cyberattacks can be identified, and necessary actions can be taken to avoid any kind of business disruption.

The advantages of AI in cybersecurity are:

Hackers and cyber-attackers are brilliantly working out various ways and means to steal confidential customer and corporate information to gain financial benefits. With the arrival and accumulation of connected and multifaceted input/output (I/O) devices in our personal, social, and professional environments, the quantity and quality of cybersecurity threats go up dramatically. Considering the growing complexity and severity of cyberattacks, AI is being prescribed as the right medicine for expertly surmounting cybersecurity problems. AI can maintain the pace to destroy any bad moves of cybercriminals with alacrity. In a nutshell, AI can automate threat detection and do a better job when compared with traditional software solutions.

In this book, we primarily focus on various kinds of cybersecurity problems and detail them to educate our esteemed readers. Then, how AI algorithms and models come in handy in eliminating those attacks is illustrated through many chapters. How AI is elevating its competency in detecting new threats is also dealt with in detail. Other chapters focus on the following topics.

It is well-known that by using sophisticated machine and deep learning algorithms, it is possible for advanced AI systems to proactively and preemptively detect malware and ransomware attacks and to recognize unique patterns therein. The natural language processing (NLP) capability can go through articles, news, blogs, white papers, etc., to assist in cyber-attack prediction. Anomaly or outlier detection helps here. AI can also contribute to attack prediction and to formulating a workable and winnable prescription strategy.

AI can also battle against bots. Bots are dangerous as they can take over accounts with stolen credentials. Bots can do bogus account creation as well.

AI can do breach risk prediction. Any IT environment is blessed with the IT asset inventory, which is a detailed record of all devices, users, and applications with different levels of access to various systems. By considering the asset inventory and threat exposure, AI systems can predict which applications can be compromised so that remedial measures can be taken quickly. This is to reinforce the much-needed cyber resilience.

AI guarantees endpoint protection. That is, endpoints are secured through the crucial role of AI. Antivirus products and virtual private network (VPN) solutions help in minimizing remote malware and ransomware attacks which are signature-based. AI-inspired endpoint protection guarantees proactive protection against threats rather than waiting for signature updates.

Thus, AI is being prescribed as the futuristic and flexible technology for tackling the growing menace of cybercriminals. This book is for clearly demystifying the unique contributions of AI algorithms, models, frameworks, and libraries in nullifying cyberattacks.

CHAPTER 1

Network Security Risks and Solutions Through Automated Toolkits in Underwater Sensor Network: A Survey

D. JOCIL and R. VADIVEL

Department of IT, Bharathiar University, Coimbatore, Tamil Nadu, India

ABSTRACT

The Underwater Wireless Sensor Network (UWSN) plays a vital role in various underwater applications, including environmental monitoring, disaster warning, gas leak detection, mine detection, assisted navigation, oceanic data collection, surveillance, and military use, among others. UWSN consists of cluster heads (CHs), autonomous underwater vehicles (AUVs), and sensors that are deployed to facilitate data communication in the underwater environment through wireless connections. Additionally, UWSN communication involves several components such as a base station, sensor nodes, buoys, satellites, and cluster heads. This chapter discusses the security risks associated with UWSNs, such as limited bandwidth, media access control, resource utilization, power constraints, Doppler spread, and other environmental vulnerabilities. To address these risks, it is crucial to develop secure communication techniques that protect against internal and external attacks, while ensuring reliable data delivery. Researchers are focusing on secure routing protocols to mitigate misbehaviors, external attacks, and internal attacks by employing lightweight authentication techniques with minimal overhead. Furthermore, the reliability of data delivery is enhanced through error correction techniques and data consistency authorizations. This chapter provides insights into the applications, components, security threats, requirements, existing solutions, and proposed solutions to overcome the challenges of UWSNs.

Leveraging Artificial Intelligence (AI) Competencies for Next-Generation Cybersecurity Solutions.
Pethuru Raj, B. Sundaravadivazhagan, & V. Kavitha (Eds.)

1.1 INTRODUCTION

Underwater data collection, disaster prevention, underwater resource detection, human rescue, and military operations are just a few of the applications that utilize UWSNs, also known as underwater wireless sensor networks. UWSNs consist of devices such as sensing nodes and independent Underwater Autonomous Vehicles (UAVs) that are installed and connected to base stations and satellites to enable communication in the underwater environment through wireless connections [1]. Limited power, high propagation delay, limited bandwidth, variable speed, and path loss [2] can lead to malicious attacks and disrupt network communication in UWSNs, causing congestion, heavy traffic, data alteration or modification, and more. Research is currently focused on less than 10% of the Earth's atmosphere, which is predominantly covered by water [3]. This is why researchers and the academic community are paying attention to UWSN studies. The underwater scenario is vastly different from the terrestrial environment due to specific characteristics of water, such as signal attenuation, salinity, temperature, low acoustic speed (≈1,500 m/s), ambient noise, and multi-path propagation [4]. UASN, short for Underwater Acoustic Sensor Networks, utilizes a popular routing technique known as a reliable cluster-based secure protocol, which offers advantages such as quality connectivity, extended lifetime, minimal power consumption, reliability, and improved measurements. However, the selection of cluster members and cluster heads presents challenges due to conflicting structures and limitations in UASN [5]. Researchers are implementing cryptographic techniques to ensure secure node-to-node communication in UASNs.

1.2 UWSN ARCHITECTURE

The UWSN system has important components such as a base station, surface buoys, acoustic links, cluster heads (CH), and sensor nodes, as stated in Figure 1.1. The architecture of the sensing nodes enables them to sense input values, compute and transfer the gathered information. Various cluster connecting points are used to transmit the gathered information to the buoys located on the water's surface. The water surface buoys serve as hubs, while the base station on land remotely controls the underwater sensor nodes. It provides guidance to the buoys in the acoustic wireless communication network on the water's surface. Figure 1.1 illustrates the basic system architecture for Underwater Acoustic Communication Networks.

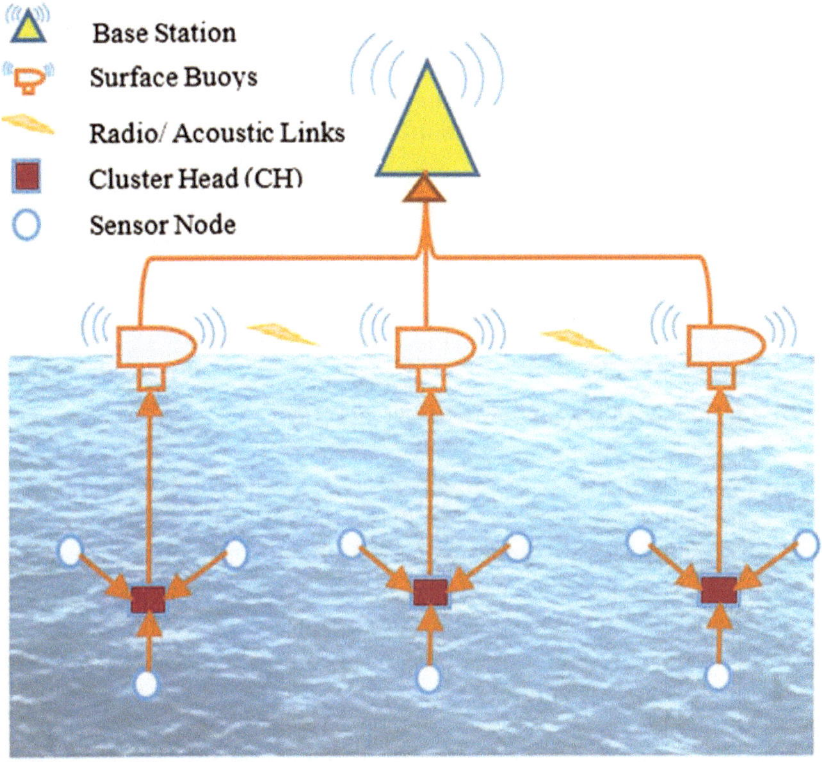

FIGURE 1.1 System architecture of UASN.

1.3 UNDERWATER ACOUSTIC CHANNELS: CHARACTERISTICS

Underwater acoustic channel characteristics are described below:

1.3.1 PHYSICAL BEHAVIORS

Water produces heat and propagates sound 4 times more efficiently than air. Heat is produced by water up to 25%. Light penetration, buoyancy, density, temperature, pressure, thermal conductivity, and acoustic characteristics are the physical behaviors of the underwater environment. The ambient pressure caused by water density can generate and intensify depth. Objects weigh less in water due to the buoyant force opposing gravity. As water depth increases, sunlight and visibility decrease.

1.3.2 VARYING AND SLOW BROADCAST SPEED

The magnitude of underwater sound waves' broadcast speed is approximately five times lower than the speed of light. Other factors that affect the speed include depth, temperature, and salinity [6], which result in the alteration of diffusion speed. The slow broadcast speed leads to a high Doppler effect in transportable UWANs. As a result, the strength of the Doppler effect directly correlates with the ratio of the frequency of data transmission to the relative speed of the transmitter and receiver. This causes significant frequency shift and distortion due to motion induction [7].

1.3.3 LIMITED AND JAM-PACKED NETWORK

A restricted bandwidth of maximum kHz is possible for the underwater sound communication network, and it is also susceptible to triangulation and localization processes. Underwater performance, in terms of bandwidth, is hampered by frequency-selective signal-heat translation, as well as the losses that result from the expansion of transmitting power over a vast area. The limitation of long-range transmission increases due to the incremental distance of signal propagation.

1.3.4 INSTABLE NETWORK

Multipath or transient propagation creates signal points from the starting point, and they can reach the final point with a phase shift through different paths. This phenomenon can be observed in sound signals initiated from the surface, ocean floor, moving objects, and so on. Unbalanced networks often result in higher Inter Symbol Interference (ISI) and interfere with succeeding symbols [8].

1.3.5 COMMUNICATION SYSTEM BANDWIDTH

The transmission system bandwidth is classified as narrowband, wideband, and ultra-wideband depending on the bandwidth of the communication. The narrow bands are referred to as the communication system's bandwidth that is less than 1% of the center frequency of the signal. The wideband refers to the bandwidth that lies between 1% and 20%. Bandwidths outside these ranges are called Ultra-Wide Band (UWB) [9].

1.4 PARTICULARITIES AND CONSTRAINTS

1.4.1 NARROW PROPERTIES

Sensor nodes in the underwater environment are limited in certain situations and have very specific hardware characteristics, such as limited battery capacity, memory space, and lack of autonomous external devices. Due to their location in both deep and surface water, recharging and replacing the sensor nodes' batteries is a challenging task. Additionally, due to the longer distance and signal attenuation, as well as the more complex signal processing in receivers, the power consumption of underwater sound network communication is higher compared to terrestrial network radio communication. This power consumption issue restricts processing speed and memory space in order to reduce power consumption and extend the network's lifetime. Therefore, current research and UWSN instruments prioritize power conservation while maintaining efficiency and security.

1.4.2 UNDEPENDABLE COMMUNICATION NETWORK

Since the communication medium's environment and the physical characteristics of the environment are hostile, the communication network in the underwater acoustic channel undergoes temporal changes. Unreliable communication networks in underwater communication networks create opportunities for malicious attacks and often result in damage to the communication system, causing data loss in UWSNs.

1.4.2.1 LONG AND FLIGHT TIME OF PACKETS (PROPAGATION DELAY)

The definition of propagation delay is the amount of time it takes for a packet to travel through a communication link at the speed of light. The airborne radio wave moves at a speed of 3 x 10^8 m/s, but the underwater acoustic wave travels at a rate of roughly 1.5 x 10^3 m/s, which is 5 orders of magnitude lower. The equation below can be used to compute other elements that affect speed, such as salinity, depth, and temperature [10].

$$d = \frac{s}{v} \tag{1}$$

Where 's' stands for the distance in meters between the transmitter and receiver, 'd' for the propagation delay in seconds, and 'v' for the acoustic speed in meters per second. Underwater, temporal synchronization and localization are challenging to accomplish due to the lengthy propagation delay and fluctuating sound speed.

1.4.2.2 DATA RESTRICTIONS AND LIMITED LATENCY TRANSMISSION

The transmission range and depth have a significant impact on the usable bandwidth of underwater acoustic channels, which is constrained [10]. Between 500 Hz and 10 kHz, it is possible to communicate over considerable distances in deep water, between 10 and 100 kHz over medium distances in shallow water, and short distances in deep water between 100 and 500 kHz. With a reduction in transmission range, especially at distances under 100 meters, the available bandwidth becomes substantially greater. There is a trade-off between data rate and channel bandwidth. Because underwater audio communication can only occur over a limited range, the speed of data transfer in an underwater environment is significantly lower compared to radio frequencies in the air. A wider communication range is more crucial in many UWSN applications, such as AUV management, than a faster transfer rate. The communication range is also affected by the data rate. The maximum data rate for long-distance communication is approximately 10 kb/s. For medium-range transmission, the maximum data rate is around 50 kb/s. Short-range transmission can achieve data rates exceeding 100 kbps.

1.4.2.3 ECOLOGICAL NOISES

Another element that significantly affects interaction in underwater communication channels is ecological noise. Here are some illustrations of ecological noises: self-noises emanating from ships, noises brought on by waves or other water motions, turbulence, heat transfer, and animal disturbances in the ocean [11]. When the time, place, or depth changes, the ambient disturbances may have a significant impact on sonar functionality and produce significant fluctuations. These impacts could affect the network's stability.

1.4.3 LOSS OF BROADCAST TRANSMISSION

Propagation loss and attenuation loss are both types of transmission loss in broadcasting.

- **Propagation Loss (PL):** It is the disruption of transmission data from the origin node to the target node. During propagation, the sound wave front occupies a larger area, so the wave energy per unit surface becomes less [12].
  The formula below can be used to define propagation loss.

$$PL = p \times log\, logn \tag{2}$$

  where; 'p' denotes the loss model; and 'n' stands for the meter range.
- **Attenuation Loss (AL):** During propagation, the energy of the sound wave is transformed into different modes and absorbed by the transmission medium. Additionally, this process depends on the frequency.

1.4.4 DOPPLER EFFECT AND MULTI-PATH

The multipath effects in UWSN are more significant than in WSN. A wave point from the origin node is transferred to the target node through two or more routes, which is a phenomenon known as the multipath effect. Additionally, under certain circumstances, the destination node may experience interference from two or more arriving signals. Due to the uniform nature of the medium in the submerged environment, representations of the surface and lowermost may be ignored. However, in a shallow aquatic setting, the communication range is greater than the water depth. Additionally, factors such as acoustic pace, temperature, salt content, and turbidity vary depending on the depth of the water. As a result, there are numerous layers that run the length of the shallow water habitat, from top to bottom. Multiple entries of identical signals may be detected at the receiver after the acoustic waveform has been transmitted and returned at numerous levels, creating a considerable multipath effect.

1.4.5 TRANSMISSION ERROR

Numerous elements, including water temperature, the slow speed of acoustic waves, transmission loss, ambient noise, the Doppler effect, and the multipath effect, have a significant impact on the underwater acoustic channel. All of these variables could result in variability in delay and bit errors, leading to an increased rate of bit errors and a higher likelihood of packet loss in UWSNs. Additionally, the underwater optical channel is accessible to all terminals within the transmission range and exhibits the characteristics of an open channel. In this scenario, an attacker can actively interfere with network functions such as localization, time synchronization, and routing, as well as passively intercept and analyze acoustic signals. Designing efficient protocol security to protect UWSNs from eavesdropping and other malicious attacks is therefore extremely challenging. The data transfer speed can be calculated using the round trip time (RTT) in the network communication process. Sending an Acknowledgment (AK) to the source node is considered a crucial aspect of a successful communication process in the network.

1.4.6 DYNAMIC TOPOLOGY FOR NETWORK COMMUNICATION

Due to the expenses and difficulties involved in the deployment itself, underwater sensor nodes are thought to be distributed less densely than those on land. Additionally, due to water flow, 95% of nodes for underwater sensors are portable. According to actual findings, items moving underwater typically travel at a speed of 2 to 3 knots, or 3 to 6 km/h, resulting in a highly changeable network structure [13]. In Underwater Sensor Networks (UWSNs), a cluster-based communication system is frequently employed. Autonomous Underwater Vehicles (AUVs) are also utilized in various ways in the underwater environment to increase surveillance and communication capabilities. The AUVs can regularly enter and exit clusters or channels, resulting in an extremely dynamic topology. As depicted in Figure 1.2, the AUV joined the third cluster as a member node in order to communicate with a coastal station or on-land sink by sending packets through Cluster-3. As seen in Figure 1.3, the AUV left the third cluster and joined Cluster-1 as a member node because it had moved out of Cluster-3's communication range. The movement of the AUV led to a disparity in the system topology. The network topology changed as a result of the AUV's movement.

1.4.7 INSECURE FUNCTIONING ENVIRONMENT

The underwater working environment is insecure for certain specific application fields, such as security monitoring, surveillance targeting, data transmission, and evading attacks by eavesdroppers, among others. Underwater nodes are deployed and operate in hostile areas within the ocean, making them highly vulnerable to malicious attacks and threats. Attackers or eavesdroppers can exploit nodes to steal data during the transmission process, leading to communication disruption. This can result in severe and persistent damage. Ensuring secure packet switching during the communication process poses a significant challenge for Underwater Wireless Sensor Networks (UWSNs).

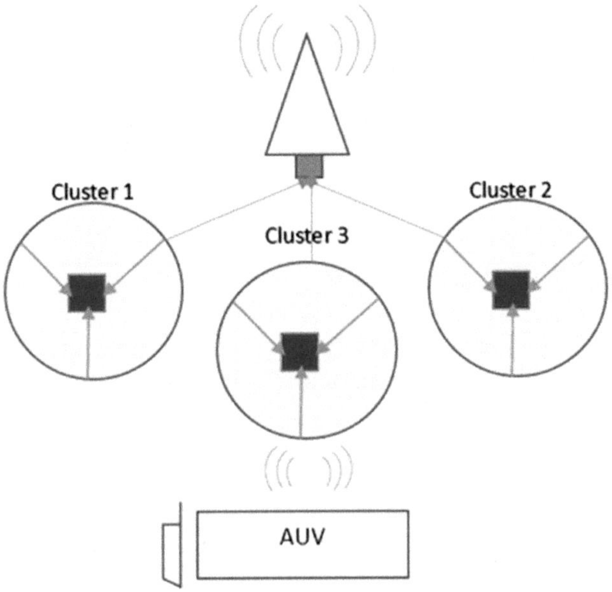

FIGURE 1.2 AUV joins Cluster 3.

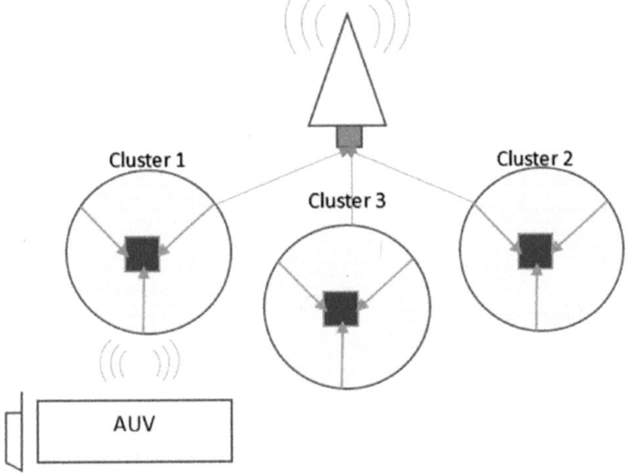

FIGURE 1.3 AUV joins Cluster 1.

1.4.8 PHYSICAL VULNERABILITY

UWSN nodes are intelligently designed to be small and waterproof. As previously mentioned, these sensor nodes are deployed in harsh and uninhabited sea zones. In the event of an attack, they are at risk of physical damage, becoming non-functional, and being vulnerable to marine life. It is not possible to shield every node from potential physical harm. Any physical damage to the sensor nodes and autonomous underwater vehicles (AUVs) leads to unnecessary transmission, frequent data transmission, data loss, path loss, and energy loss in the network.

1.5 APPLICATIONS OF UWSNS

UWSN technology finds its applications in many areas grouped into three categories: Scientific, Industrial, and Security applications as shown in Figure 1.4.

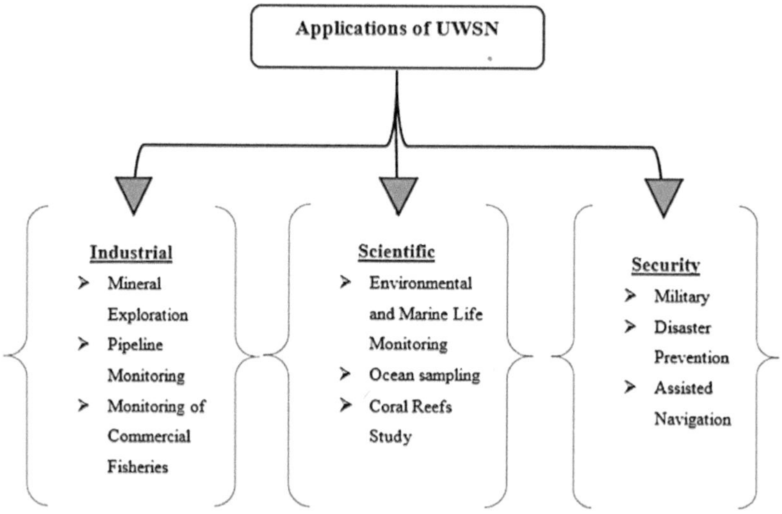

FIGURE 1.4 Applications of UWSN.

1.5.1 SCIENTIFIC APPLICATIONS AND RESEARCH

Over 71% of the Earth's surface is covered by water, as mentioned in the introduction section. These underwater parts of the Earth remain largely unexplored and contain vast resources and data waiting to be discovered. The applications of UWSNs in scientific research focus on environmental monitoring, ocean sampling, and monitoring of biological activities on the ocean floor [14].

1.5.1.1 ENVIRONMENTAL AND MARINE LIFE MONITORING

Environmental and marine life monitoring involves the observation and assessment of the underwater environment, various objects of interest, and marine life. The tracking of ocean water quality primarily focuses on evaluating the chemical and biological waste deposited on the ocean floor [15]. Additionally, measurements of water oxygen levels, temperature, and pressure are taken, and the data is transmitted to onshore monitoring stations for further analysis. By monitoring water quality, we ensure that aquatic life is not adversely affected by human activities [16]. Any deviations from the normal operating conditions of the ocean are promptly detected, and the potential causes are identified before any harm is done to human or aquatic life [16]. The objective of monitoring aquatic life, including fish, mammals, and microorganisms, is to observe, study, and gain a deeper understanding of their behavior, thereby expanding our current scientific knowledge base [16].

1.5.1.2 OCEAN SAMPLING

The aim of an ocean monitoring application, as outlined in Ref. [17], is to monitor a significant coastal area by deploying underwater sensor vehicles at various sites to explore ocean occurrences. The collected data is automatically transmitted to the shore for further observations. Each vehicle covers approximately 500 square meters during each deployment in an underwater environment [18].

1.5.1.3 CORAL REEFS STUDY

Coral reefs are underwater ecosystems composed of microorganisms that inhabit water. They are studied to assess the impact of human activity on them. In Ref. [19], the author describes a coral reef application that utilizes big data, the Internet of Things (IoT), and sensor network technologies to investigate the influence of ocean salinity, thermal readings, humidity, and pressure on coral bleaching and marine ecosystems. A combination of multiple agents and communication can be employed for long-term maritime environmental monitoring. For data collection, the research team [20] utilized highly mobile artificial fish, artificial mussels, and autonomous surface vehicles (ASVs) in trials conducted at Biograd Na Moru, Croatia.

1.5.2 INDUSTRIAL USE

1.5.2.1 MINERAL EXPLORATION

UWSN technology is used to monitor and control underwater rigs to ensure effective operation with minimum pollution to the surrounding environment. Underwater sensor technology has been deployed to identify, locate, and map these underwater deposits. AUVs mounted with optical and acoustic modules are used for such applications, facilitating significant area coverage during exploration [21].

1.5.2.2 PIPELINE MONITORING

Applications for monitoring undersea oil and gas pipelines can be monitored by UWSN. This is done when extracting oil and gas from the ocean floor and transferring them to the shore through pipelines. The authors of [22] have developed a prototype for measuring oil and gas in undersea pipes. The system was designed to provide statistical data on the condition of pipelines spanning large areas.

1.5.2.3 MONITORING OF COMMERCIAL FISHERIES

One of the most in-demand industries that generates strong economic assets is fish aquaculture. To keep an eye on the fish's habitat, however, a tight monitoring system is necessary. The creators of [23] have developed a remote user-accessible Zigbee-based UWSN system for surveillance in massive fish farms. Additionally, the system can monitor fish production using data such as oxygen concentration, temperature, water volume, and humidity. Wireless cameras are also utilized and connected to technology and the Internet for remote monitoring underwater, anytime and anywhere.

1.5.3 SECURITY APPLICATIONS

1.5.3.1 MILITARY AND DEFENSE

UWSNs are deployed to guide and assist military applications as well. The UWSN system architecture uses different sensor nodes and various protocols to detect and monitor different aspects of military applications. The AUVs are integrated with various sensors such as imaging sonars, metal detectors, cameras, etc. These sensors are used to monitor and assist in finding security ports, submarines, underwater mines, and conduct surveillance monitoring. In military applications, several underwater sensors are used to monitor and regulate ports and harbors, detect sea mines, defend boundaries from uninvited battleships or submarines, and conduct tactically distributed surveillance [24].

1.5.3.2 DISASTER PREVENTION

Furthermore, the mobile underwater sensor network, one of UWSN's innovative technologies, is employed to transmit natural disaster early warnings such as seismic activity on the seafloor. UWSNs are used to monitor the conditions of the ocean floor, and if any seismic activities are detected, the sensors send a warning to the shore stations. This information can then be relayed to those near the coast, advising them to move to higher ground [25].

1.5.3.3 ASSISTED NAVIGATION

The underwater situation is very challenging, unfamiliar, disorienting, unpredictable, and increasing in depth. There is a significant need for navigational assistance in such an environment to navigate ships, boats, and even for swimming and research purposes. Due to the differences in communication networks below and above the water's surface, many navigational aids are not applicable. Consequently, underwater assistive navigation technologies, systems, and apps are necessary to identify, guide, and navigate. In this regard, UWSN can be utilized to provide applications and assisted navigation systems.

1.6 UWSN SECURITY RISKS AND THREATS

There are many potential security threats and challenges that remain unresolved in Underwater Sensor Networks. Numerous research articles emphasize the need to deploy new approaches and techniques to enhance network communication in UWSNs. Despite being susceptible to various attacks on the OSI network, the insecure working environment of underwater sensor networks and underwater sensor channels leads to several limitations. These include transmission errors, physical vulnerability, transmission loss, dynamic network topology, limited bandwidth, low data rates, long and variable propagation delay, ambient noise, multipath, and Doppler effect, among others [26].

There are two types of attacks in UWSN: passive attacks and active attacks shown in Figure 1.5.

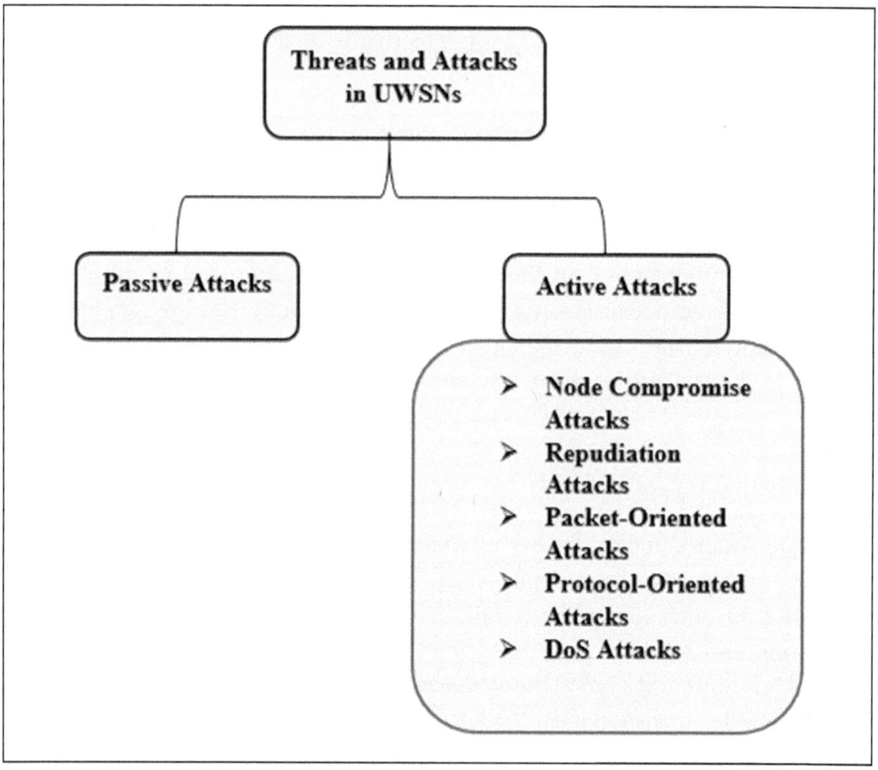

FIGURE 1.5 Types of attacks.

1.6.1 PASSIVE ATTACKS

These attacks can be produced by unauthorized eavesdropping on the communication route. The eavesdropper monitors the communication transfers and gathers the information without the knowledge of the sender and receiver, as depicted in Figure 1.6. In this attack, the attacker cannot alter, modify, or damage the transmitted data, but eavesdropping, interference, message distortion, imitation, and secret information leak are the major issues of passive attacks. Hydrophones and underwater microphones are used by eavesdroppers to capture transmitted packets in the network communication station. The attacker can take advantage of packet interchange detection, finding the data shifting points, influencing the node station, and packet movements easily since the nature of acoustic channels is open and easy to attack. The use of encryption techniques, which make it challenging for eavesdroppers to obtain any critical information or data during the communication process in UWSN, is an effective way to prevent those assaults [27].

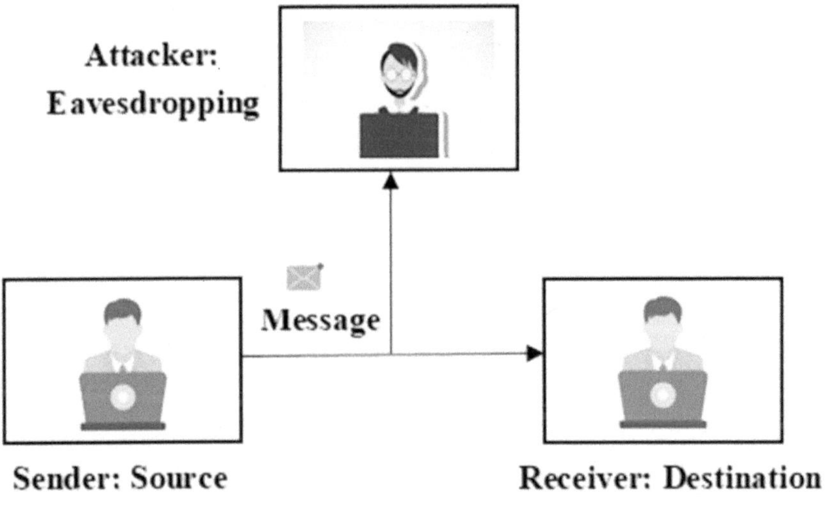

FIGURE 1.6 Passive attack.

1.6.2 ACTIVE ATTACKS

The objective of these attacks is to capture network data and insert, modify, destroy, or remove data on the communication process over network circumstances shown in Figure 1.7. It is considered to be a very complex attack as it performs all necessary operations against sensitive and authentic data. An intruder can send fake information to the target node and steal sensitive data via the communication channel. The threat launched from an insider node may cause substantial damage to the network. According to prior research, insider node attacks are tougher to prevent and can result in more damage than outside attacks [28]. Trust management techniques, authentication protocols, encryption techniques, and trusted node-to-node cooperative routing techniques can be employed to detect and prevent these attacks [29] (Table 1.1).

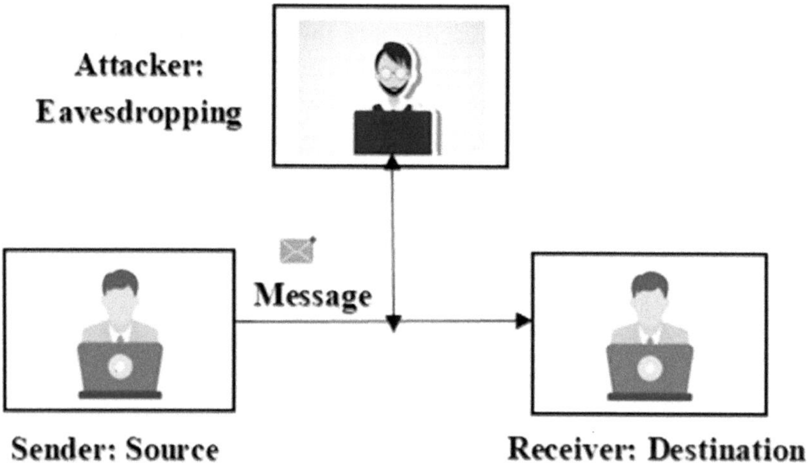

FIGURE 1.7 Active attack.

1.6.2.1 COMPROMISED NODE ATTACKS

The sensor nodes in UWASNs are deployed in harsh and hostile sea environments. They are incapable of guaranteeing the protection of all the nodes positioned on a large-scale network. Attackers can potentially capture, crack, and alter information from memory locations. The compromised nodes are then used to infiltrate the network as legitimate nodes, allowing attackers to monitor, steal information, and disrupt network communication, causing severe damage.

1.6.2.2 REPUDIATION ATTACKS

The involvement of malicious nodes denying specific actions or communication with other nodes can occur in repudiation attacks. In this attack, the

TABLE 1.1 The major concerns affected by active attacks on the physical layer, data link layer, network layer, and transport layers.

Attack Scheduled	Occurrence	Mode of occurrence	Concern Occurred
Physical Layer	Seize of nodes	Sensor node control is seized by an attacker using a physical attack.	Creates erroneous information to genuine users by making the hijacked nodes unlimited requests on behalf of the attacker.
	Jamming DoS	Continuous jamming can be passed over a network from source to destination nodes which prevents nodes from communicating.	Jamming its transmission of information with the same frequency may occur.
Data Link Layer	Denial of Sleep	Repetitive communication on the network or collision threats are the source of this attack.	Reduction of energy for battery-powered devices.
	Jamming	Nonstop Request to Send (RTS) received from hackers to make congestion.	Denial of system access by valid nodes.
	Exhaustion	The internal program code of the attacker performed with hijacked nodes in order to make exhaustion in the communication.	Hosting a malicious node for network busy and energy drainage.
	Unfairness	A minor frame technique is used to speed up the process, however, it creates communication risk.	Using a communication platform a hacker reduces the performance.
Network Layer	Selective Forwarding	The close-by nodes will post various directions to the destination node. It deliberately transmits and drops some packets to avoid detection.	Modification of packet of source nodes.
	Misdirection Attack	An attacker changes or misdirects the network communication during the transmission process in the network.	Redirection of packets to the invalid path or route modification causes data loss or data modification.

TABLE 1.1 *(Continued)*

Attack Scheduled	Occurrence	Mode of occurrence	Concern Occurred
	Black Hole Attack	Broadcasting false routing data with the shortest path toward the receiver.	Sensor node damages.
	Sybil Attack	The attacker hijacks legitimate node IDs.	Distance-based or location-based directing structures are severely harmed.
Transport Layer	Synchronization Flooding Attack	New user requests are created indefinitely by intruders until each connection requires the resources to be consumed.	Since the significant flood was created, the network is unable to distinguish between real and false traffic.
	Desynchronization Attack	Transfer forged packets with forged sequence series or signal controls for endpoint desynchronization.	A malicious user interrupts existing networks among nodes.

Active attacks are classified as follows according to their aim.

malicious user can change the authoring information using incorrect input credentials. In these attacks, the network lacks the capacity to trace the malicious actions and identify the hacker.

1.6.2.3 PACKET-ORIENTED ATTACKS

In this attack, the packets from the source node are unable to reach the destination node successfully. This attack disrupts the functionality of network communication. These attacks are embedded in routing protocols and involve operations such as packet replication, rushing attacks, routing table poisoning, and routing table overflow. These malicious actions enable attackers to capture packets, analyze them, and modify or drop them. The encryption and decryption technique is often employed to safeguard against routing attacks. A trusted third party introduced by the cryptographic technique is used for authentication purposes and ensures secure data transmission. However, due to the high computational cost of encryption, it not only increases the size of communication packets but also consumes more energy.

1.6.3 PROTOCOL-ORIENTED ATTACKS

Several routing-based protocols have been developed and designed for the computing protocols of UWSNs in existing research. However, considering security aspects, there is a need to implement additional security for the UWSN system network. This is to prevent unauthorized access through brute force attacks such as wormhole, hello flooding, Sybil, blackhole, sinkhole, and exhaustion.

1.6.4 DoS ATTACKS

The network's legitimate nodes are unable to access resources due to a denial of service (DoS) attack, which can be carried out in various ways. The objective of the attacker is to disrupt network services by preventing the legitimate nodes from accessing them. These attacks are critical and difficult to identify. Due to the low-cost and hard-to-detect nature of DoS attacks, the attacker can cause significant harm by interrupting network protocols and reducing the availability of the entire communication network (Table 1.2).

1.7 UWSN VERSUS CYBER ATTACKS

1.7.1 PROMINENT UWSN CYBER ATTACKS

The field of cyber security has seen rapid advancements over the past 10 years, frequently making headlines as the risks increase and hackers attempt to evade law enforcement. Hackers are targeting businesses, government agencies, and healthcare organizations worldwide, exploiting security vulnerabilities and sometimes demanding large sums of money as ransom. They exploit these vulnerabilities to steal sensitive information or encrypt files. Globally, every company now faces an average of 925 cyber attacks per week, a 50% increase compared to the previous year. In 2021, businesses experienced 50% more attacks per week compared to 2020, with access control attacks (20%) and Ransomware (11%) being the most common methods used against Asian organizations, followed closely by data fraud (10%) [32].

TABLE 1.2 The Comparison of Passive and Active Attacks Conceived from [30]

Comparison Based	Active Attacks	Passive Attacks
Basic	The attacker can cut off the connection and convert the information that affects the underwater acoustic channel resources.	In a passive attack, the attackers read, analyze, and convert the transmitted information.
Information Modification	Occurs	Does not take place. It can't modify any information. But can steal sensitive data on communication.
Nodes harmful	Always harm the nodes.	Can not cause any damage.
Attack Awareness	When an attack occurs the entity gets informed	The entity does not get informed.
The attacker performance	Network communication is caught by controlling the portion of the network.	Only observation takes place during transmission.
Emphasis is on	Detection	Prevention

Black holes, gray holes, flooding, wormhole attacks, repudiation attacks, node compromise attacks, packet-oriented attacks, protocol-oriented attacks, DoS attacks, active attacks, passive attacks, node compromise attacks, repudiation attacks, and repudiation attacks are some of the cybersecurity attacks launched against the underwater network platform. These attacks are performed in the physical layer, data link layer, network layer, and transport layer on UWSN platforms as mentioned in Table 1.1. These malicious attacks intrude on the UWSN communication and create malware, phishing, eavesdropping, DoSs, password attacks, etc.

1.7.1.1 PHISHING

By disseminating faked data packets with the cheapest or fastest route to the target, an attacker simulates a black hole attack. This link, which actually passes via the adversary, would be chosen by the victim routers as the best option. As a result, the attacker has access to all of the network's activity. Additionally, the attacker has complete freedom to examine, alter, or even drop information. Detectors that are installed far from the base station are vulnerable to this kind of assault. A more cunning attack can drop crucial packets at a specific moment or with a specific proportion, making it harder to catch.

1.7.1.2 EAVESDROPPING

Eavesdropping is a type of passive DoS attack that involves listening in on auditory communications between endpoints and intercepting data transmitted over the connection. If the transmissions are not encrypted, intruders can easily gain access to private and sensitive information. Consequently, it can be difficult to determine if eavesdropping is actually occurring, as the attacker does not disrupt normal transmission.

1.7.1.3 DENIAL OF SERVICES

A Cyberattack known as a DoS assault seeks to prohibit authorized users from accessing system resources by obstructing or eradicating network traffic. The most frequent and well-known kind of DoS attack is a distributed DDoS assault. A botnet is created through a DDoS attack using a number of compromised machines. The botnet bombards the target server with traffic, depleting its computing or communication capacity and resulting in a DoS. The IoT system is made up of small devices like tablets, printers, cameras, and home gateways, which can be used by attackers. There may be more than just distant DDoS attacks for the IoT. More connections can be used by the attacker to gain physical access to the system or to communicate wirelessly directly with it. An illustration of this is the use of jammers to send out greater wireless signals into the airwaves to interfere with nearby wireless transmission.

1.7.1.4 JAMMING

In a jamming attack, one or more rogue nodes continuously send out unwanted or useless signals, interfering with the ability of other genuine nodes to communicate normally in the area. What's even worse is that the rogue nodes have the capability to bring down the entire system by overwhelming just a few key endpoints, such as the base station, root node, or cluster-head node. These specialized nodes are responsible for data collection, packet transmission, and network supervision.

1.7.1.5 UNFAIRNESS

Rather than completely prohibiting legitimate nodes from connecting to the channel, the attacker utilizes this less powerful denial-of-service attack to weaken the signal. To reduce time, consider adopting a small-frame approach. This technique sacrifices efficiency in order to decrease the impact. Furthermore, it is more susceptible to unfairness. For example, an attacker may choose to retransmit at a faster rate instead of sporadically delaying.

1.7.1.6 SELECTIVE FORWARDING

In this attack, rogue nodes mimic the behavior of normal nodes while arbitrarily dropping packets and failing to relay incoming packets. There is a possibility that the neighboring node will select an alternative route to send the packet to the target node. Therefore, the attacker selectively sends some packets and drops others to avoid detection. To avoid arousing suspicion, the attacker can confidently forward the remaining packets while suppressing and modifying a packet that arrives from a select few chosen nodes.

1.7.2 PRIMARY CYBER SECURITY SOLUTIONS FOR UWSNS

The security challenges of UWSNs are conceptually segregated into distinct secure modules in accordance with the Open System Interconnection (OSI) network to provide a comprehensive solution.

1.7.2.1 SIGNATURE-BASED INTRUSION DETECTION

A database can be utilized by a signature-based intrusion detection system (IDS) to store attack signatures associated with unwanted traffic or valid signatures that correspond to regular traffic. In real-time, the IDS compares the information in incoming datagram's with the previously saved signatures [33]. The drawback of this approach is that IDSs can only partially and dependably detect malicious data entering a channel when relevant signatures are absent.

1.7.2.2 ANOMALY-BASED INTRUSION DETECTION

By employing this technique, a model of what might be the norm is created. The models may take the form of statistical methodologies [34], mathematical models [35], or rule-based systems [36]. Changes from the ordinary are viewed as assaults. When compared to signature-based identification, these solutions have the benefit of not relying on signature patterns, which frees them from organizational efforts to collect signatures.

1.7.2.3 KEY MANAGEMENT

Cryptographic methods and key management are essential for preserving secrecy, authenticity, and non-repudiation. Insecure networks, such as the undersea acoustic channel, can be used to transport or store sensitive data using encryption to prevent reading or alteration by unauthorized users. The cryptographic and credential controlling procedures for wireless networks are unsuitable for underwater networks due to the unique characteristics and limitations of UWSNs outlined in Section 4. Traditional public key cryptography (PKC) methods require the certificate authority (CA) to oversee the distribution of certificates and keys, resulting in significant computation and communication costs. For UWSNs with limited resources, it is not economical [37, 38].

1.7.2.4 TRUST MANAGEMENT

A cryptographic protocol's trust evaluation system is a crucial addition to security defense and offers significant advantages for intrusion detection. The investigation into UWSNs' trust-management techniques encounters difficulties as a result of the peculiarities and limitations of UWSNs outlined in Section 4 [39]. There are now three different types of trust management systems in use: hierarchical schemes, distributed schemes, and centralized schemes.

1.7.2.5 ROUTING SECURITY

Routing security utilizes fundamental transportation and connection security mechanisms to protect both individual nodes and routing protocols. Additionally, in order to establish the network topology, nodes need to communicate with each other within a communication system. Two aspects of routing security are secure routing and highly secure forwarding. Secure data transfer requires safeguarding data packets from manipulation, dropping, or alteration by any unauthorized party, while secure routing necessitates node cooperation to provide accurate routing information, ensuring efficient network connectivity. Numerous research papers have been published in recent years to address routing security in UWSNs.

1.7.3 AI-INSPIRED CYBER SECURITY SOLUTIONS FOR SECURING UWSNS

In UWSN, there is a need for an IDS based on AI that can identify a wide range of potential threats.

The model taxonomy of cyberattacks in Wireless Sensor Networks (WSNs) in Smart Underwater Environments includes a customized dataset for a wireless smart underwater channel that can be categorized into four types of DoS attacks [40]. The data used for training and testing the detection model is a crucial component of this process. In this study, the LEACH (Low Energy Adaptive Clustering Hierarchy) approach has been employed as a Cluster Hierarchy Protocol with Low Energy Awareness. The dataset consists of 374,668 entries that can be divided into four different types of DoS attacks: gray hole, blackhole, scheduling, and flooding. The Multilayer Perceptron (MLP) artificial neural network (ANN) classifier structure, with multiple hidden layers, is used to train the dataset. Vector-based forwarding provides a scalable routing technique for deploying wireless underwater sensor networks. The NS2-based simulation model has been proposed as a method for capturing network traffic and generating the dataset. The dataset has been trained using an ANN to classify it into various DoS attacks. By utilizing the acquired WSN dataset and the MLP ANN, it is possible to achieve highly accurate classification of DoS attacks.

To investigate the impact of reducing eavesdropping risks on the lifetime of underwater acoustic sensor networks (UWSNs), [41] rescue operations are carried out on the network. Two optimization models are proposed: the first aims to minimize the network's vulnerability to eavesdropping, while the second aims to maximize network longevity while maintaining a certain level of eavesdropping risk. The results show that network lifetimes are significantly shorter when eavesdropping risks are minimized compared to when they are not considered. Additionally, it is demonstrated that loosening safeguards against eavesdropping hazards greatly increases the lifetime of UWSNs.

In Reference [42], a novel AI-based method is implemented to identify, categorize, and localize underwater occurrences. The usefulness of this novel AI-based method for monitoring underwater activity inside water reservoirs is demonstrated and evaluated in a real-world situation. The system is based on the hydrophone network, which is one of the sensor networks, and the model provides details on the technological and cognitive AI design of the system. The challenges of setting up and using the hydrophone network in a water reservoir are explored, including the complexities of a fluctuating environment caused by traffic, tourists, and changing water levels. The AI approach utilizes a time delay estimator for sound localization, along with an auto-encoder for unsupervised learning of latent encodings for classification and anomaly detection. We provide detailed information on AI-based technology solutions for underwater event localization and recognition. The implementation of a modular AI pipeline is described, including guidance on how to address the following important tasks:

- The supervised learning technique-based classification of known events.
- Unknown signal recognition and security assessment using an unsupervised learning methodology.

A protected authentication and data aggregation system is described to improve QoS in a scalable and robust cluster-based underwater network [43]. The suggested technique consists of two modules: secure authentication for Cluster heads (CHs) and safeguarded data aggregation. In the first module, CHs authenticate with gateways (GWs) to ensure that the CH servicing each cluster is a reliable node and has not been compromised. This ensures secure operation of the cluster. In the second module, before transferring data to the CH, each sensor symmetrically encrypts it. The encrypted data is then securely consolidated and sent to the base station. By utilizing the timestamp value, the base station can analyze the data and identify any tampering. Any compromised data is discarded. The performance of the suggested scheme is evaluated in terms of end-to-end latency, average energy consumption, and data reliability ratio in comparison to existing methods in UWSN.

1.8 UWSN SECURITY REQUIREMENTS

An attacker can find numerous ways to penetrate the network for malicious attacks. The UWSN infrastructure must be protected with appropriate protocols to prevent these malicious attackers. Confidentiality, integrity, availability, authentication, freshness, and non-repudiation are the major security measures that researchers and the academic world focus on to secure UWSN from unwanted critical attacks.

1.8.1 CONFIDENTIALITY

Exchanging data packets between nodes on UWASN plays a crucial role in guaranteeing confidentiality. Communication between source nodes and destination nodes must be secured with all necessary measures against threats and attacks, and a formal acknowledgment (AK) must be sent to the source node to indicate successful communication. The exchanged packets cannot be accessed by an unauthorized party. Privacy should be ensured for the user's information, preventing MAC, routing protocols, etc. In Ref. [44], entropy-based ant colony optimization (E-ACO) is presented for UWSN's secure route selection. It provides a higher Packet Delivery Ratio (0.89) with 500 nodes compared to existing approaches such as FRR-MFO, SAPDA, IACO, and SAPDA. The PDR directly represents the overall network efficiency using Eq. (3).

$$packet\ delivery\ ratio = \frac{number\ of\ packages\ received}{number\ of\ packages\ transmitted} \quad (3)$$

1.8.2 INTEGRITY

Integrity is the protection of data received from modification, corruption, or removal by unauthorized nodes during transmission through a malicious attack. The Inter-Cluster Authentication Technique can be employed to safeguard packets during delivery from unauthorized users using cryptographic techniques. The Trusted Third Party (TTP) plays a crucial role in reviewing all critical transaction communications between the sources and the destination, as mentioned below.

Let T_{ttp} be the trusted third party (TTP)

Once the trusted CHs are selected, it sends a request message (Req) to register with T_{ttp}. On receiving the request, the TTP generates a secret key (K_{sec}) and broadcasts it to the respective CHs.

$$\text{TTP } Ksec \rightarrow \text{CHs} \quad (4)$$

CHs then creates an authentication key (K_{auth}) with the help of the secret key obtained from the TTP.

CH signs the aggregated data (AD) with this authentication key and transmits it to other CHs.

$$\text{CH: } K_{auth}\ (\text{AD}) \rightarrow \text{CHs} \quad (5)$$

If a CH receives an AD packet from a source CH, it authenticates the source CH using its authentication key.

Similarly, the sink node authenticates the CH once it receives the AD from the CH.

1.8.3 AVAILABILITY

Availability guarantees that data is available to an authorized user on-demand. It ensures that the UWSN network is sufficiently healthy and provides services even if the system is attacked. Proper redundancy tactics, disaster recovery, proper monitoring, server clustering, and continuity of operations planning can provide availability for UWSNs. Detection and prevention from DoS is a significant approach to avoid high loss in case of natural disasters [45]. DoS includes the following methods of attack.

1.8.3.1 BLACK HOLES ATTACK

These attacks occur within a network and happen when traffic is dropped or discarded without the knowledge of the source nodes. In this attack, source nodes do not receive acknowledgment (Ak) from target nodes. As a result, the source is not informed about the packets sent. Monitoring lost traffic can help identify and prevent black hole attacks in a UWSN topology [46].

1.8.3.2 FLOODING

This attack occurs when there is a high volume of network traffic. The network becomes congested, leading to a flooding attack on UWSNs. For instance, an attacker can initiate multiple handshakes on a TCP connection system. Once the handshake is complete, the attacker will not respond to any communication from the source location. Consequently, the source node does not receive any acknowledgment from a third party or the target node. As the sending node waits for an acknowledgment from the destination or a trusted third party, it allocates only limited resources to maintaining the connection. Eventually, when the server becomes overwhelmed, legitimate users are unable to log in.

1.8.3.3 GRAY HOLE

These attacks are considered a separate category of DoS and node capture attacks. A malicious attacker selects a highly sensitive sensor node to modify, manipulate, or destroy sensitive data within the network. In this attack, nodes selectively drop or discard packets that need to be forwarded along the path [47].

1.8.3.4 WORMHOLE ATTACKS

This type of attack is another special type of DoS attack that misleads the communication networks in the network during transmission. By building tunnels through which packets are sent to themselves, malevolent nodes thwart network routing protocols [48]. In Ref. [49], detection and mitigation of DoS and DDoS algorithms are proposed and used to identify and prevent DoS attacks at early stages. This algorithm was developed to adapt to unnatural and controlled work environments. In this algorithm, the source

IP of all incoming packets and the payload of the packets are compared, and if the input values are the same, the packets are dropped and DoS is detected. Finally, SourceIP was added to BlackIPList.

1.8.4 AUTHENTICITY

Authentication must be provided for incoming packets to ensure that they are sent from a valid neighbor node, allowing them to be exchanged in a protected manner. Receiving nodes need to identify the source of the data in order to filter out malicious attacks. Node authentication schemes, cooperative authentication schemes, and trusted secure identity authentication schemes are popular techniques for establishing trusted node-to-node (N2N) cooperative authentication.

1.8.4.1 ENCRYPTION-BASED AUTHENTICATION

In Reference [50], a limited computation overhead authentication method centered on Vandermonde matrices is investigated. To reduce computing overhead, matrix addition is employed instead of matrix multiplication in the recognition system. When a node connects with another, the technique uses additional orthogonal vectors to construct the key. The identities of the nodes sharing the key are then validated and a zero-knowledge verification protocol is used, where only the true declaration is sent from the proof system to the verification. The Vander monde vectors (VVs) used to generate the polynomial for creating the secret key are randomly chosen by the access point, which then transmits the configurations to each node. The creation of a common or public key occurs after a connection is established between two endpoints. Common network protocol security standards include IEEE 802.1X, which provides a basic outline for identity and permission for the secrecy of IEEE 802.1AE, IEEE LANs, and Expandable Authentication Protocol (EAP), which offers an authorization framework that supports multiple authentication techniques.

1.8.4.2 REPUTATION-BASED AUTHENTICATION

In Ref. [51], a reputation model is used by the Attack-Resistant Trust Model based on multidimensional trust Metrics (ARTMM). It considers both the node's portability and the properties of underwater sound networks. This model is made up of the following components: link trust, node trust, and data trust.

- **Link Trust:** On the basis of link utilization and link quality, it is rated. Packet-loss rate (Pl) and packet-error rate (Pe), which are both dependent on channel-error rates, are used to measure the quality of a link (1–Pe). The percentage of the link that was utilized during the time period is called link utilization.
- **Node Trust:** The honesty of the node and residual node energy are used to evaluate the trustworthiness of the node. The quantity of nominal and realistic communications through the network is used to assess the authenticity and integrity of the network.
- **Data Trust(T_{data}):** It relies on the node's integrity and the strength of the connection. The mean is taken as the most trustworthy variable with the highest confidence value for the dataset for calculating Tdata. The following fuzzy sets are used to further define a fuzzy membership function trust value based on the interrelationship property of the three trust variables: completely untrust with a trust value of (0, 0.25), untrust with a trust value of (0.25, 0.5), uncertainty with a trust value of (0.5, 0.75), and trust with a trust value of (0.75, 1).

1.8.5 NON-REPUDIATION

Denial means that a node cannot deny that it has performed certain actions, including sending or receiving data. Non-repudiation can be achieved through a digital signature. Existing studies examine a security framework for UASN. Each node has a shared group key with the other nodes in the network and a separate key for the sink, which is used to validate and encrypt all communications sent to and from the sink. Additionally, brief digital signatures like Quartz, ZSS, and BLS are allocated to provide identification for origins with non-repudiation, specifically at the application layer.

1.9 OPEN RESEARCH CHALLENGES

1.9.1 UNSECURE UWSN ENVIRONMENT

Nodes in underwater architecture are usually installed in opaque and hostile surroundings, which creates room for antagonists to capture and damage the communication nodes. It is practically impossible to implement physical counterbalancing to protect all of them [52]. The secret contained in compromised nodes can often be leaked, which should be addressed by logically removing those compromised nodes through rekeying the entire network. Adversaries, such as attackers or eavesdroppers, can exploit UWAN for an advantage in the unprotected opaque underwater environment. A secure protocol for physical removal is essential to avoid network jams and to identify and secure compromised nodes.

1.9.2 IMPERFECT BANDWIDTH

Many researchers have studied simulation analysis and experiment-based network communication. There are several important issues that need to be addressed in the future, with a primary focus on the Medium Access Control (MAC) protocol at the data link layer. The main objective of MAC is to coordinate the access of UWSN network nodes to the shared channel, while also ensuring the reliable transmission of valid data to the intended destination. Another research challenge is to send appropriate acknowledgments to the source node within a suitable time frame.

1.9.3 VARIANCE IN DELAY AND PROPAGATION SPEED

An incorrect assessment of Round Trip Time (RTT) values and the challenging definition of waiting time in the MAC protocol are caused by variations in delay. However, recent studies on the MAC protocol did not consider delay variations in their results. The underwater environment is unique due to factors such as salinity, temperature, and depth, which affect the dynamics of propagation speed. These effects result in severe Doppler Effect, unstable frequency, and distortion induced by motion.

1.9.4 UNRELIABLE DATA COMMUNICATION

The main objective of an underwater network is reliable data transfer. However, traditional positioning and localization systems are unable to achieve successful communication in the underwater working environment. As a result, the node locations and network topology are disrupted by underwater conditions, leading to unreliable data transmission. Additionally, underwater clatter further hampers communication, affecting the reliability of data transfer.

1.9.5 COMMUNICATION CHANNEL AND TRANSMISSION RANGE

The communication standard frequency for underwater acoustic networks varies depending on the communication ranges. One distinct feature of the underwater environment is the capacity of signal absorption, which is influenced by the water depth. To mitigate the effects of signal absorption, it is advisable to minimize the frequency. However, this poses a challenge as longer transmission ranges increase the possibility of disruption and high data collisions.

1.9.6 COOPERATION OF HETEROGENEOUS NODES

Satisfactory combination and data transfer between sensor nodes and underwater vehicles are necessary actions to sustain wide-ranging application developments. It becomes a critical task for heterogeneous node collaboration in a functioning system since communal standards and interfaces to support communication are lacking. Moreover, the different sorts of resources and heterogeneity that support portable and static nodes in specific application implementations also face an additional challenge.

1.9.7 BIG DATA INTEGRATION

Integrating Big Data and UWSNs is one of the hot developments that are tough in real-time analytics, accuracy, communication, and visualization. Due to the different data formats, protocols, and service constraints of diverse applications, customizing another system is a crucial and challenging task. It is difficult to reconfigure the underwater vehicles after deployment. Since UAVs are implemented by different constructors, performing cooperative communication among UWSN technologies is a challenging task and not flexible.

1.10 CONCLUSION AND FUTURE WORK

This chapter provides an overview of security in UWASNs, including the specific architecture of these networks, potential security threats and attacks, requirements for successful UWSNs implementation, and ongoing research challenges. The table presents a comparison of attacks on OSI layers, distinguishing between active and passive attacks. The chapter also discusses the necessary techniques to address security issues. It is evident that several challenges are increasingly difficult to solve. The following works are considered important for future research in overcoming these challenges in UWSNs.

- New theoretical representations are an essential need in both computational and analytical progress.
- The extensive utilization of field experiments and test beds is crucial for advancing UWASNs. Field experiments and test beds should be employed to conduct a more precise performance analysis and characterization of the network system. Such efforts can contribute to the evolution of UWASN's communication in the next generation.
- Pre-installed security scheme sensor nodes are necessary in future work, and it is necessary to reconfigure the security scheme from time to time in case of attacks on the network. The design and implementation of such reconfiguration in the network system is a challenging and critical issue.

These research issues are open for future investigations and proper UWSN communications can be made.

KEYWORDS

- **existing and proposed models**
- **round trip time**
- **security risks**
- **trusted third party.**
- **underwater autonomous vehicles**
- **UWSN characteristics**

REFERENCES

1. Irfan, A., Taj, R., Asim, Z., Inayat, K., Inam, U., Habib, H., & Omar, C., (2021). Analysis of security attacks and taxonomy in underwater wireless sensor networks. *Wireless*

Communications and Mobile Computing, 2021, 15. Article ID 1444024. https://doi.org/10.1155/2021/1444024.

2. Al Guqhaiman, A., Akanbi, O., Aljaedi, A., & Chow, C. E., (2020). Lightweight multi-factor authentication for underwater wireless sensor networks. In: *2020 International Conference on Computational Science and Computational Intelligence (CSCI)* (pp. 188–194). doi: 10.1109/CSCI51800.2020.00039.
3. Ali, T., Irfan, M., Shaf, A., Saeed, A. A., Sajid, A., Awais, M., & Aamir, M., (2020). A secure communication in IoT enabled underwater and wireless sensor network for smart cities. *Sensors, 20*, 4309. https://doi.org/10.3390/s20154309.
4. Karim, S., et al., (2021). GCORP: Geographic and cooperative opportunistic routing protocol for underwater sensor networks. *IEEE Access,9*, 27650–27667. doi: 10.1109/ACCESS.2021.3058600.
5. Vani, K., & Sunilkumar, S. M., (2021). Trusted node selection in clusters for underwater wireless acoustic sensor networks using fuzzy logic. *Physical Communication, 47*, 101388. ISSN 1874-4907. https://doi.org/10.1016/j.phycom.2021.101388.
6. Etter, P. C., (1996). *Underwater Acoustic Modeling, Principles, Techniques and Applications* (2nd edn.). London, U.K.: E & FN Spon.
7. Stojanovic, M., (2008). Underwater acoustic communications: Design considerations on the physical layer. In: *Proc. Annu. Conf. Wireless Demand Netw. Syst. Services (WONS)* (pp. 1–10). Garmisch-Partenkirchen, Germany.
8. Jiang, S. M., (2018). State-of-the-art medium access control (MAC) protocols for underwater acoustic networks: A survey based on a MAC reference model. *IEEE Commun. Surveys Tuts.,20*(1), 96–131. 1st Quart.
9. Van, W. P. A., & Otnes, R., (2013). Ultra-wideband underwater acoustic communication channels. *IEEE J. Ocean. Eng., 38*(4), 678–688.
10. Yang, G., Dai, L., & Wei, Z., (2018). Challenges threats, security issues and new trends of underwater wireless sensor networks. *Sensors, 18*(11), 3907. https://doi.org/10.3390/s18113907.
11. Burrowes, G. E., & Khan, J. Y., (2010). Investigation of a short-range underwater acoustic communication channel for MAC protocol design. In: *Proceedings of the 2010 4th IEEE International Conference on Signal Processing and Communication Systems (ICSPCS)* (pp. 1–8). Gold Coast, Australia.
12. Stojanovic, M., (2007). On the relationship between capacity and distance in an underwater acoustic communication channel. *ACM SIGMOBILE Mob. Comput. Commun. Rev., 11*, 34–43.
13. Jun-Hong, C., Kong, J., Gerla, M., & Zhou, S., (2005). *Challenges: Building Scalable and Distributed Underwater Wireless Sensor Networks (UWSNs) for Aquatic Applications.*
14. Alsulami, M., Elfouly, R., & Ammar, R., (2022). Underwater wireless sensor networks: A review, In: *Proceedings of the 11th International Conference on Sensor Networks (SENSORNETS 2022)* (pp. 202–214). ISBN: 978-989-758-551-7; ISSN: 2184-4380. doi: 10.5220/0010970700003118.
15. Collins, A., (2013). Application of subsea wireless technology to environmental monitoring. In: *2013 OCEANS* (pp. 1–5). San Diego.
16. Kiranmayi, M., & Kathirvel, A., (2015). Underwater wireless sensor networks: Applications, challenges, and design issues of the network layer-a review. *International Journal of Emerging Trends in Engineering Research, 3*(1), 5–11.

17. Luo, H., Wu, K., Ruby, R., Hong, F., Guo, Z., & Ni, L. M., (2017). Simulation and experimentation platforms for underwater acoustic sensor networks: Advancements and challenges. *ACM Comput. Surv. (CSUR), 50*, 28.
18. Fattah, S., Gani, A., Ahmedy, I., Idris, M. Y. I., & Targio, H. I. A., (2020). A survey on underwater wireless sensor networks: Requirements, taxonomy, recent advances, and open research challenges. *Sensors, 20*(18), 5393.
19. Lin, W., Li, D., Tan, Y., Chen, J., & Sun, T., (2008). Architecture of underwater acoustic sensor networks: A survey. In: *Proceedings of the First International Conference on Intelligent Networks and Intelligent Systems* (pp. 155–159). Wuhan, China.
20. Loncar, I., Babic, A., Arbanas, B., Vasiljevic, G., Petrovic, T., Bogdan, S., & Miškovic, N., (2019). A heterogeneous robotic swarm for long-term monitoring of marine environments. *Appl. Sci., 9*, 1388.
21. Maeda, F., Takagawa, S., Asada, A., Maki, T., Sasaki, Y., & Yamamoto, F., (2011). Next R & D of measurement method of offshore minerals using the low-frequency acoustic mills-cross transmission and receiving system multibeam sub-bottom profiling method and "Natsushima" cruise experiment. In: *2011 IEEE Symposium on Underwater Technology and Workshop on Scientific Use of Submarine Cables and Related Technologies* (pp. 1–5). Doi: 10.1109/UT.2011.5774143.
22. Saeed, H., Ali, S., Rashid, S., Qaisar, S., & Felemban, E., (2014). Reliable monitoring of oil and gas pipelines using wireless sensor network (WSN) – REMONG. In: *Proceedings of the 2014 9th International Conference on System of Systems Engineering (SOSE)* (pp. 230–235). Adelade, Australia.
23. Abdul-Rahaim, L. A., & Ali, A. M. A., (2015). Remote wireless automation and monitoring of large farm using wireless sensors networks and internet. *Int. J. Comput. Sci. Eng. Technol., 6*, 118–137.
24. Antonelli, G., Caffaz, A., Casalino, G., Volpi, N. C., De Jong, I. B., De Palma, D., Duarte, H., Grimsdale, J., Indiveri, G., Jesus, S., et al., (2016). The widely scalable mobile underwater sonar technology (WiMUST) H2020 project: First year status. In: *Proceedings of the OCEANS 2016-Shanghai* (pp. 1–8). Shanghai, China.
25. Zwanzig, C., (2018). Ship-based exploration of submarine gas hydrates and environmental monitoring with multibeam echo sounders. In: *2018 OCEANS–MTS/IEEE Kobe Techno-Oceans (OTO)* (pp. 1–8).
26. Shahapur, S., & Khanai, R., (2016). Localization, routing, and its security in UWSN–A survey. *International Conference on Electrical, Electronics, and Optimization Techniques, ICEEOT 2016* (pp. 1001–1006).
27. Ateniese, G., Capossele, A., Gjanci, P., Petrioli, C., & Spaccini, D., (2015). *Sec FUN: Security Framework for Underwater Acoustic Sensor Networks* (pp. 1–9). OCEANS 2015–Genova.
28. Ahmed, M. R., Aseeri, M., Kaiser, M. S., Zenia, N. Z., & Chowdhury, Z. I., (2015). A novel algorithm for malicious attack detection in UWSN. In: *2nd International Conference on Electrical Engineering and Information and Communication Technology, ICEEICT 2015* (pp. 21–23).
29. Jiang, S., (2019). On securing underwater acoustic networks: A survey. *IEEE Communications Surveys and Tutorials, 21*(1), 729–752.
30. Suresh, W., Nitin, Ra., Parma, N. A., & Ashish, K., (2020). Attacks in underwater sensor network. *International Journal of Interdisciplinary Innovative Research & Development (IJIIRD)* (Vol. 5, No. 1). ISSN: 2456-236X.

31. https://spanning.com/blog/cyberattacks-2021-phishing-ransomware-data-breach-statistics (accessed on 27 January 2024).
32. https://www.business-standard.com/article/international/cyber-attacks-india-among-top-3-most-affected-nations-in-asia-in-2021-122022400945_1.html (accessed on 27 January 2024).
33. Depren, O., Topallar, M., Anarim, E., & Ciliz, M. K., (2005). An intelligent intrusion detection system (IDS) for anomaly and misuse detection in computer networks. *Expert Syst. Appl., 29*(4), 713722.
34. Garcia-Teodoro, P., Diaz-Verdejo, J., Maciá-Fernández, G., & Vázquez, E., (2009). Anomaly-based network intrusion detection: Techniques, systems, and challenges. *Comput. Secur.,28*(12), 1828.
35. Filonov, P., Lavrentyev, A., & Vorontsov, A., (2016). *Multivariate Industrial Time Series with Cyber-Attack Simulation: Fault Detection Using an LSTM-Based Predictive Data Model*. arXiv:1612.06676. [Online]. Available: https://arxiv.org/abs/1612.06676.
36. Dufeld, N., Haffner, P., Krishnamurthy, B., & Ringberg, H. A., (2016). *Systems and Methods for Rule-Based Anomaly Detection on IP Network Flow*. U.S. Patent 9258217.
37. Teguig, E. H., Touati, Y., & Ali-Cherif, A., (2017). ECC based-approach for keys authentication and security in WSN. In: *Proceedings of the 2017 9th IEEE-GCC Conference and Exhibition (GCCCE)* (pp. 1–4). Manama, Bahrain.
38. Simplicio, M. A. Jr., Silva, M. V. M., Alves, R. C. A., & Shibata, T. K., (2017). Lightweight escrow-less authenticated key agreement for the internet of things. *Comput. Commun., 98*, 43–51.
39. Goyal, N., Dave, M., & Verma, A. K., (2017). Trust model for cluster head validation in underwater wireless sensor networks. *Underw. Technol., 34*, 107–114.
40. Ahmad, B., Jian, W., Enam, R. N., *et al., (*2021*)*. Classification of DoS attacks in smart underwater wireless sensor network. *Wireless Pers Commun.,116*, 1055–1069. https://doi.org/10.1007/s11277-019-06765-5.
41. Ozmen, A., Yildiz, H. U., & Tavli, B., (2020). Impact of minimizing the eavesdropping risks on lifetime of underwater acoustic sensor networks. In: *2020 28th Telecommunications Forum (TELFOR)* (pp. 1–4). doi: 10.1109/TELFOR51502.2020.9306557.
42. Jörg, S., Philip, W., Severin, L., Maria-Elena, A., Martin, Z., Wolfgang, K., Martin, S., et al., (2021). Underwater acoustic networks for security risk assessment in public drinking water reservoirs. *ACM Trans. Sensor Netw., 1*(1), 24. https://doi.org/10.48550/arXiv.2107.13977.
43. Goyal, N., Dave, M., & Verma, A. K., (2020). SAPDA: Secure authentication with protected data aggregation scheme for improving QoS in scalable and survivable UWSNs. *Wireless Pers Commun.,113*, 1–15 https://doi.org/10.1007/s11277-020-07175-8.
44. Premkumar, D. S., & Mukeshkrishnan, M. B., (2022). Secured route selection using E-ACO in underwater wireless sensor networks. *Intelligent Automation and Soft Computing, 32*(2), 963–978. doi: 10.32604/iasc.2022.022126.
45. Kajwadkar, S., & Jain, V. K., (2018). A novel algorithm for DoS and DDoS attack detection in internet of things. In: *2018 Conference on Information and Communication Technology (CICT)* (pp. 1–4). doi: 10.1109/INFOCOMTECH.2018.8722397.
46. Zala, D., Thummar, D., & Chandavarkar, B. R., (2021). Mitigating blackhole attack of underwater sensor networks. In: *2021 12th International Conference on Computing Communication and Networking Technologies* (ICCCNT) (pp. 1–8). doi: 10.1109/ICCCNT51525.2021.9579473.

47. Dongare, S., & Mangrulkar, R., (2016). Optimal cluster head selection-based energy efficient technique for defending against gray hole and black hole attacks in wireless sensor networks. *Procedia Computer Science, 78*, 423–430.10.1016/j.procs.2016.02.084.
48. Lu, X., & Yonghua, Z., (2012). *Modeling the Wormhole Attack in Underwater Sensor Network*, 1–4.10.1109/WiCOM.2012.6478576.
49. Vanitha, K. S., Uma, S. V., & Mahidhar, S. K., (2017). Distributed denial of service: Attack techniques and mitigation. In: *2017 International Conference on Circuits, Controls, and Communications (CCUBE)* (pp. 226–231).
50. Yuan, C., Chen, W., Zhu, Y., Li, D., & Tan, J., (2015). *A Low Computational Complexity Authentication Scheme in Underwater Wireless Sensor Network*, 116–123.10.1109/MSN.2015.26.
51. Han, G., Jiang, J., Shu, L., & Guizani, M., (2015). An attack-resistant trust model based on multidimensional trust metrics in underwater acoustic sensor network. In: *IEEE Transactions on Mobile Computing, 14*(12), 2447–2459. doi: 10.1109/TMC.2015.2402120.
52. Kong, J., Ji, Z., Wang, W., Gerla, M., Bagrodia, R., & Bhargava, B., (2005). Low-cost attacks against packet delivery, localization, and time synchronization services in underwater sensor networks. *WiSe–Proceedings of the 2005 ACM Workshop on Wireless Security* (pp. 87–96). 10.1145/1080793.1080808.

CHAPTER 2

IoT Security: Challenges, Best Practices, and Service Platforms

JAYARANI THOMAS[1] and ROBIN CYRIAC[2]

[1]*Independent Researcher, Springfield Lakes, Queensland, Australia*

[2]*Department of Information Technology, University of Technology and Applied Sciences, Al Mussanna, Oman*

ABSTRACT

Using standard Internet protocols to enable communication between people and things is known as the Internet of Things (IoT). It is widely acknowledged that IoT systems require security, and several standardization initiatives have already been launched to offer security. However, security challenges continue not just because some use cases remain unaddressed, but also because many IoT systems and devices have been created and put into use with extremely limited security capabilities. To safeguard IoT systems, it is all the more essential to have protected device provisioning, a secured connection between IoT devices and the cloud, and secured data protection throughout their processing and storage in the cloud. This chapter examines the many security issues that might arise while deploying an IoT system, the approaches and best practices that should be followed to establish a secure IoT infrastructure, and the most effective IoT service platforms.

2.1 INTRODUCTION

The term "Internet of Things" (IoT) refers to the inter-networking of extremely diverse entities that use a variety of communication protocols,

Leveraging Artificial Intelligence (AI) Competencies for Next-Generation Cybersecurity Solutions.
Pethuru Raj, B. Sundaravadivazhagan, & V. Kavitha (Eds.)

including human-to-human, human-to-thing, and thing-to-thing. In a vision to create a new world where every physical object has a radio-frequency identification (RFID) tag with a globally unique identity, the Auto-ID center coined the term "IoT" in the year 1999. This would enable data searching about the objects through the Internet in addition to real-time object monitoring. Since then, a vast range of technologies, products, and protocols have been incorporated into the concept of the IoT. Computing devices that comprehend and respond to their surroundings are the elements that make up the IoT. They are frequently called smart gadgets or smart objects.

IoT applications are built on the existence of IPv6 and CoAP, which make it possible to link IoT systems to the Internet. Additionally, they aid in the introduction of envisioned situations, such as building automation, into actual settings and LANs. According to a report from the International Data Corporation (IDC), there will be about 41.6 billion interlinked IoT devices by 2025, including machines, sensors, and cameras. This surge in the number of connected devices is accelerating. IoT devices, which do not include mobile devices, make up more than 30% of all network-connected endpoints in a typical organization [16].

This chapter can be roughly divided into three sections. The first section covers the various security issues that impact an IoT system, including its different security threats and vulnerabilities. The second section explores several methods and best practices recommended for a secure IoT architecture. Lastly, we familiarize ourselves with two service platforms that can potentially expedite and simplify the implementation of secure IoT systems.

2.2 PURPOSE AND SCOPE

A crucial necessity for every communication system is security. However, for a number of reasons, security is an even more crucial necessity in operational IoT implementations. Firstly, hacked IoT systems not only put a user's privacy and security at risk, but they also have the potential to cause physical harm. This is because IoT systems often include sensors, actuators, and other connected devices in the user's physical surroundings. If these devices are hacked, they can negatively impact the user. A weak IoT system provides attackers with a small window of time to alter a device's functionality, which not only damages the manufacturer's brand image but also exposes valuable information, such as proprietary algorithms, to the public. Since hacked IoT systems can be used illegally on any scale, the consequences of hacking an IoT system extend beyond a single device or network. They can be utilized, for example, to launch a Distributed Denial of Service (DDoS) attack, which would reduce the accessibility of services in the IoT network. Many IoT systems rely on standard IP protocols for simpler system integration. However, this also makes them vulnerable to attacks against standard IP protocols, which are generally applicable to IoT systems as well. This highlights the need for additional security implementation-related regulations in IoT.

2.3 SECURITY CHALLENGES IN IOT

Despite the fact that IoT creates novel potential across all sectors, from big data analytics (BDA) to industries, it has also given rise to fresh, challenging security issues [2]. IoT solutions expand the attack surface for new types of attacks since they connect a huge number of potentially susceptible devices and transcend many network borders. One should pay close attention to how to implement these services across the whole system, which includes devices, networks, apps, and the cloud, in addition to concentrating on device-specific security services in order to ensure security in IoT systems.

- The CIA triad of information security —Confidentiality, Integrity, and Availability — must be maintained by ensuring that data is securely acquired, stored, analyzed, transmitted, and eventually destroyed upon retirement.
- Find and fix any vulnerabilities and potential risks in IoT components.

2.4 SECURITY THREATS IN IOT

A loss of productivity, visibility, functions, and capabilities may occur from threats to an IoT infrastructure. Threats may cause financial loss or harm to a business's brand. Several threat models already exist for characterizing dangers to an IT infrastructure, and they are being modified to also account for IoT risks. Security professionals created these threat models for the analysis of IT and IoT ecosystems. Businesses and organizations may create or modify these or other models to fit their specific contexts. Among the typical ones are: [15]

- **The STRIDE threat model from Microsoft:** Microsoft's taxonomy of threats and the corresponding mitigation techniques is called STRIDE. Spoofing, Tampering, Repudiation, Information Disclosure, Denial of Service (DoS), and Elevation of Privilege are acronyms for STRIDE [4].
- **IoT Vulnerabilities Project OWASP:** The Open Web Application Security Project (OWASP) has identified IoT vulnerabilities and disclosed them in the order in which they occurred within specific timeframes [7]. This model provides an overview of the relevant attack surface for each of the main IoT vulnerabilities.
- **Threat Taxonomy of the European Union Agency for Network and Information Security (ENISA):** This taxonomy provides a comprehensive and multi-level characterization of dangers. The threat model offers a library of dangers related to various physical and digital attack routes.

Various security threats in IoT are illustrated in Figure 2.1 and discussed in the following subsection [1].

FIGURE 2.1 Security Threats in IoT.

2.4.1 PRIVACY THREAT

The privacy of people nearby may be in danger if the usage and location of a personal device are tracked. For instance, based on the data gathered and delivered by specific IoT gadgets, an attacker might infer knowledge about private and sensitive data. The distribution of such information to parties interested in marketing and targeted advertising is therefore possible [1]. Another hazard to privacy is when intelligence services illegally spy on and intercept traffic going to and from a node in IoT.

2.4.2 CLONING OF THINGS

During the production process, a thing's physical attributes, firmware/software, or security settings can be easily copied by an unreliable manufacturer. The software of deployed items may potentially be hacked, allowing for software copying or alterations. Such a copied item might be purchased for a lower price and still fulfill its intended purpose. A cloned device can be utilized to launch an attack or gain control over a genuine device. Additionally, an unreliable manufacturer may modify the functionality of a cloned item, thereby damaging the reputation of the original product. Cloned objects may potentially possess additional capabilities, such as backdoors, which make them more susceptible to attacks.

2.4.3 VULNERABLE SOFTWARE

IoT devices rely on programs that may have serious vulnerabilities or poor design decisions. This exposes the systems to a number of attacks depending on the severity of the flaws, such as buffer overflows or a lack of authentication. One of the most significant security dangers is the Mirai botnet [1]. It is a large-scale DDoS attack, which was brought on by devices with widely known or easily guessable configuration passwords.

2.4.4 MAN-IN-THE-MIDDLE (MITM) ATTACK

Both the deployment and processing phases may be subject to man-in-the-middle attacks, particularly if the operational keying components, security parameters, or configuration settings are transmitted over a wireless medium and the cryptographic algorithms employed are insufficient to safeguard the system or the network. This will result in the communication channel losing its credibility and secrecy, as well as the trustworthiness of any instructions or other traffic sent through it. The ability to eavesdrop is an example of an MITM attack. If keys are compromised because of protocol flaws or are not renewed or updated properly, an adversary may be able to eavesdrop.

2.4.5 FIRMWARE ATTACKS

A connected entity's firmware or software may be upgraded while it is in operation or maintenance to add new capabilities or modify current features. By maliciously updating the device's firmware, an attacker could be able to take advantage of such software upgrades and change the way it operates. For instance, an attacker may insert malicious code into the firmware to make the device regularly transmit its energy use to a data repository for analysis. With this knowledge, an attacker can exploit the situation to gain access to an unoccupied building or residence. A similar scenario arises when devices have outdated or improperly maintained software, which can contain vulnerabilities that hackers may exploit to attack and compromise the device.

2.4.6 EXTRACTION OF PRIVATE INFORMATION

An attacker could quickly gain physical access to unprotected IoT devices (such as sensors, actuators, etc.) and then tactfully extract private information, including secret keys, data from sensors like a user's health status, configuration information, or proprietary algorithms. Attackers can utilize traffic analysis to derive significant information even when the data coming from an IoT device is encrypted, which could jeopardize the owner's or user's privacy.

2.4.7 ROUTING ATTACK

To create routing loops, alter network traffic and source paths, and perform other malicious activities, the routing information in IoT networks can be falsified, modified, or retransmitted. The list of routing attacks is as follows:

i. The sinkhole exploit, also known as a black hole attack, allows an attacker to control all packets traveling through a base station by claiming to have a better route or path to it.
ii. In selective forwarding, an attacker can either discard a packet or send certain packets only.
iii. Wormhole attacks, in which an attacker may capture packets at a specific location in the network and tunnel them to a different location, altering the observed behavior of the network and potentially falsifying statistics, have a significant negative impact on routing functionality.
iv. The Sybil attack, in which an attacker fools other networked devices by assuming several identities.

2.4.8 PRIVILEGE ELEVATION

A low-level attacker can take advantage of further weaknesses in a device's installed authentication and authorization systems to get more privileged access to the device and its contents.

2.4.9 DENIAL OF SERVICE (DOS) ATTACK

Many IoT devices have relatively minimal memory and processing power, making them susceptible to attacks that deplete resources. Attackers can repeatedly make queries to specific devices or networks to drain their resources [17]. IoT devices are particularly at risk because attackers can hide in the backend and target low-resource devices that are part of a constrained-node network. Another method of launching a DoS attack is physically blocking the communication route. Additionally, network availability can be impeded by saturating it with a large number of packets [19]. On the other hand, attacker-compromised devices can be used to launch DDoS attacks on other networks or systems.

2.5 VULNERABILITIES IN IOT SYSTEMS

The main reason behind vulnerability in IoT systems is their lack of properly installed security mechanisms to protect against attacks. The restricted surroundings and constrained computing power of these devices are the main causes. IoT devices typically have modest power consumption, and they can only perform a limited set of tasks. Because of this, they are unable to accommodate strong security controls, processes, and data protection programs. Let's discuss a few of the vulnerabilities found in IoT systems in recent times. Figure 2.2 illustrates different vulnerabilities in IoT systems.

2.5.1 RESOURCE CONSTRAINTS

The CPU, memory, and energy capacity of IoT devices are usually limited. These factors have an immediate impact on the design of protocols for the IoT domain. For example, restrictions on the size of physical layer packets may result in limited IP-layer Maximum Transmission Units (MTU) or hop-by-hop fragmentation and reassembly. In the first scenario, excessive packet fragmentation can enable state-exhaustion attacks, which are commonly required by security systems. In the second scenario, packet loss and the need for retransmission can increase fragmentation at the IP layer, which generally degrades overall system performance.

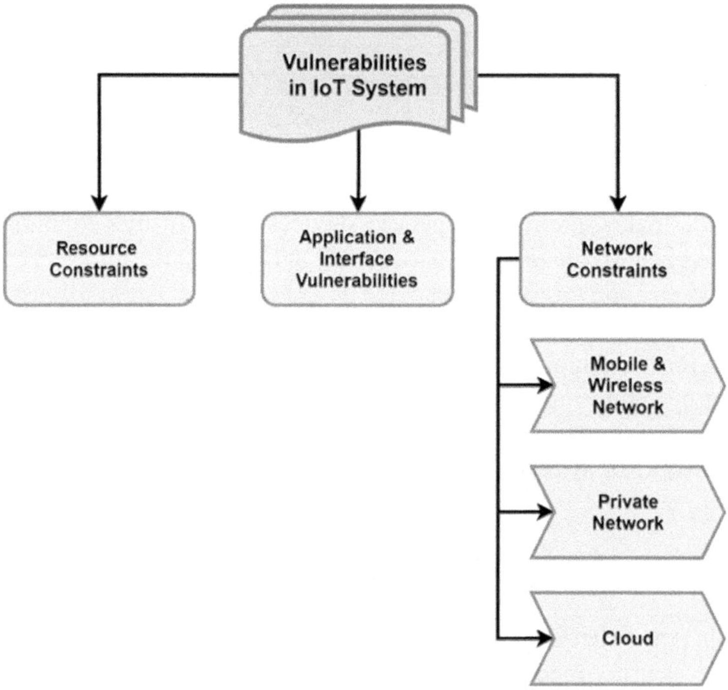

FIGURE 2.2 Vulnerabilities in IoT system

The use of resource-intensive cryptographic techniques, such as public key cryptography, which is utilized in the majority of Internet security standards, is constrained by small CPUs and limited memory. This is particularly true when low latency is required by the underlying application or when the fundamental cryptographic elements need to be utilized frequently. To minimize the need for memory and maximize bandwidth consumption, it is important to keep the size and quantity of messages to a minimum. Layered techniques, which integrate the headers of multiple protocols, may result in underperformance for devices with limited resources.

Another important consideration is the constrained energy budget of IoT nodes. To reduce energy consumption during routine operations and mitigate DoS attacks, careful protocol design and implementation are necessary.

2.5.2 VULNERABILITY IN APPLICATIONS AND INTERFACES

Mobile applications and web interfaces are frequently used by users to communicate with IoT systems. Web interfaces must function on a variety of hardware, from light switches to barcode scanners, as they integrate data and services into already-existing applications. Both web interfaces and mobile applications are susceptible to well-known attacks such as SQL injection and cross-site scripting [5]. These vulnerabilities are easily exploited by attackers, who can then gain unauthorized access to networks, steal data, and even take control of systems.

Manufacturers and software developers in the IoT domain should widely apply existing application security best practices to help mitigate some of these impacts. However, the reality remains that there are more apps and web interfaces than ever before, and each one represents a potential source of attack as the IoT industry continues to expand. Businesses must consider how to protect their data, even if their app or web interface is insecure or compromised.

2.5.3 NETWORK CONSTRAINTS

1. **Mobile and Wireless Networks:** Since large-scale device deployments would make any form of network cable impossible, the majority of IoT solutions link devices via either cellular or fixed wireless data networks. Wi-Fi and other fixed wireless networks must be adequately secured because they are frequently attacked [5].

Critical business data traveling via an IoT network can simply be exposed by a simple configuration error.

2. **The Private Network:** IoT solutions are routinely connected to business networks, which increases security vulnerabilities and challenges the traditional device-centric approach to network security. It is important to remember that vulnerabilities are not limited to external attackers; insiders, such as employees or contractors, can also pose a threat to the system. They may attempt to steal data from IoT solutions or launch cyber attacks on IoT devices from within the network, driven by malicious intent or financial motives [5]. By implementing a robust internal data encryption strategy and ensuring that all data transmitted from IoT devices to the network is constantly encrypted, this potential risk can be effectively mitigated.

   To protect data from both known and unknown vulnerabilities associated with IoT solutions, organizations should consider implementing encryption measures across their entire enterprise. The line between secure private networks and public networks becomes blurred due to the IoT. IoT data will inevitably traverse various applications, networks, and devices, each potentially having its own security flaws.
3. **The Cloud:** By utilizing the cloud, businesses can easily and affordably acquire the computing capacity required to support extensive deployments of connected devices. However, control is relinquished when certain security measures are beyond the influence of companies using cloud services. These concerns include the level of isolation between tenants on multi-tenant infrastructure, the security of open Internet connections, and the background checks conducted on employees of cloud service providers (CSPs).

The Cloud Security Alliance (CSA), an independent industry association, asserts that perimeter security measures are less effective in cloud-based IoT operations. They argue that security equipment at the network's edge is "topologically ill-suited" to defend against cloud-based attacks. In other words, when applications migrate outside the perimeter, vulnerabilities do as well.

2.6 TECHNIQUES AND BEST PRACTICES FOR IOT SECURITY

It is necessary to identify and employ appropriate risk mitigation measures, as well as adhere to certain best practices, to address the aforementioned risks and vulnerabilities. However, new threats and vulnerabilities emerge every day, and devices used in various environments are susceptible to attacks. Therefore, it is difficult to guarantee a high degree of security in an IoT system at any moment in time. IoT encompasses networks and devices, therefore security must be ensured throughout all domains. Beginning with the planning stage and continuing through the design, development, deployment, maintenance, and disposal phases, security should be seen as an ongoing process. There is not a single process that needs to be left unprotected. The strategies and best practices listed below can be used to identify and reduce any potential risks connected to IoT systems [6].

2.6.1 HIGH-LEVEL RISK MITIGATION TECHNIQUES

1. In an IoT system, a Business Impact Analysis (BIA) evaluates the effects of compromising the fundamental security properties of confidentiality, integrity, and availability. The consequences of lost data, lower sales, increased costs, regulatory fines, disgruntled consumers, etc., may all have negative repercussions for the organization. A company can determine the importance of having a robust security design by conducting a business impact study.
2. Security threats to an IoT system are analyzed using a Risk Assessment (RA), taking into account both their likelihood and severity. It also requires assigning a risk category to each of them. Risks rated as moderate or high must be reduced, which requires that the security architecture be capable of addressing certain threats. Microsoft offers an IoT Security Evaluation Framework that targets risks, use cases, and technologies seen in the real world while also offering security recommendations according to the circumstances and use cases [15].
3. A privacy impact analysis (PIA) assesses how personally identifiable information (PII) is collected, managed, or utilized in an IoT system. The objective is to fulfill any relevant legal obligations and evaluate the risks and potential consequences of data manipulation and PII loss.
4. Keep a security incident response plan up to date and practice it frequently. The approaches that enable learning about any security vulnerabilities affecting an IoT system are referred to as incident reporting and mitigation procedures. Additionally, it's important to regularly test incident response protocols. Gather security logs and use automated techniques to analyze them in real-time. Construct playbooks as a reaction to results that were unanticipated.

5. Draft and test business recovery and continuity plans. Create written instructions on how to resume regular functioning in the event of an unforeseen incident. Utilize cloud services for business continuity and backup as well. Periodically test recovery plans, and modify them based on what you've learned from testing and actual security issues.
6. Establish methods for security auditing and monitoring throughout the IoT ecosystem and pertinent IT systems in order to continually gather and report activity metrics and logs. Continuously test the security systems and controls to ensure their integrity.
7. An asset inventory should be created and maintained for all IoT assets, including a list of each item's key features. This should include information such as installed software or hardware versions, distributed certifications, etc. To effectively manage the devices, categorize them and assign appropriate tags. Focus on information that may be useful, such as the location, owner information, and the importance of the devices.

2.7 BEST PRACTICES

Recommended practices to follow at the device, application, and network levels in order to create a secure IoT configuration are listed below [3]. The different best practices followed as industry standards in IoT are shown in Figure 2.3.

2.7.1 PHYSICAL SECURITY

Before deploying devices into production, it is important to disable any unused ports and interfaces that were utilized during the development and testing phases. To ensure secure access, the administrative ports should be properly monitored. USB and Ethernet ports can be locked off, and webcam apertures can be covered with port locks and camera covers. Port locks help prevent malware attacks. Additionally, securing the device can be achieved by implementing strong boot-level passwords and allowing booting only from local storage. Endpoint hardening, which involves implementing multiple layers of security measures, can significantly delay the attack process, aid in attack detection and mitigation.

FIGURE 2.3 Best practices in IoT security.

To maintain equipment integrity, it is recommended to keep them reasonably isolated and limit physical access to a small number of authorized individuals. Endpoint hardening, as described above, can help prevent potential hackers from accessing sensitive data and protect against attacks aimed at compromising the device. For shipping, tamper-evident packaging can be used to allow the consumer to determine if the package has been opened prior to delivery.

2.7.2 FIRMWARE UPDATES/PATCHES

Due to vulnerabilities in IoT devices, the importance of device-specific firmware update procedures has increased significantly. Incorporating these update methods is crucial for addressing known vulnerabilities, enhancing device performance, and enabling additional functions such as changing configuration options and introducing new features. It is essential for devices to be patchable or upgradeable to allow for the incorporation of changes at any time, as new vulnerabilities often emerge after device deployment. To fix vulnerabilities, IoT devices must be capable of receiving over-the-air (OTA) updates. These updates enable IoT system operators, system integrators, and manufacturers to gradually add new functionality to their devices while simultaneously patching any potential security holes.

Before using a device, it is important to verify the integrity of the software and ensure it has been securely downloaded from a reputable source (signed by the manufacturer). Vendors have a responsibility to inform customers or users about any changes in vendor policy, functionality, and security by providing information on patches and updates, security threats, and privacy issues. In case of any required or failed security upgrades, appropriate notification methods should be in place to promptly alert the relevant stakeholders. Additionally, there should be controls to address any unauthorized modifications to the installed firmware or software.

2.7.3 DATA PROTECTION AT DEVICE DISPOSAL

The Internet of Things (IoT) handles a significant amount of sensitive and important data due to its connection to various types of devices. Therefore, it is crucial to ensure the security of IoT devices and the data they contain. For example, a private camera installed in a home and connected to a mobile phone stores and transmits valuable information about individuals. Proper handling of data in IoT devices when they are disposed of is a critical issue that needs to be addressed. This is a security concern because improperly discarded technology could be exploited for malicious activities. Additionally, it poses a privacy concern as it may lead to the disclosure of private information about the user or other stakeholders in the IoT system. A similar security risk arises when IoT devices change ownership.

Manufacturers should provide clear guidelines for users on how to handle their obsolete IoT equipment in terms of cleanup and disposal. Some industries prescribe Discard, Recycle, or Destroy (DRD) policies, which specify which devices should be disposed of and how to do so. Users or staff members should receive adequate training on how to handle sensitive and important data in IoT devices during disposal or ownership changes.

2.7.4 USE ROBUST AUTHENTICATION

Individual identities and credentials should be assigned to IoT systems and devices. Systems should be created to facilitate the generation, distribution, recycling, and revocation of credentials. Hardware-protected components, such as Trusted Platform Modules (TPMs), should be used as necessary to store credentials and perform authentication procedures. It is important to avoid hard coding or storing secret credentials on IoT devices. Passwords should be sufficiently complex to resist intelligent guessing and brute force techniques.

The use of two-factor authentication (2FA) is recommended, where possible. This involves using a password and an additional authentication method that does not rely on user knowledge, such as a randomly generated number sent via text messaging (SMS). For IoT applications, it is advised to implement context-aware authentication (CAA), also known as adaptive authentication. CAA utilizes contextual data and machine-learning algorithms to continuously assess the risk of malicious activity and require authentication accordingly. If there is a significant risk, the user (or hacker) will be prompted for a multi-factor token before being granted access.

2.7.5 USE ROBUST ENCRYPTION AND SECURE PROTOCOLS

Communications between devices may be vulnerable to hacking, even if device passwords are safe. Numerous protocols, including Bluetooth, Zigbee, Z-Wave, 6LoWPAN, LTE, LoRaWAN, etc., are used in IoT. The makers choose and implement the strongest encryption on devices, ideally IPsec or TLS/SSL, based on the needs and processing capacity available [3].

Precautionary measures should be taken to guarantee that fake data have no probability of harming anyone if they are transmitted unencrypted and unsigned. There are continuing efforts to use more effective underlying cryptographic primitives, including Elliptic Curve Cryptography, to lessen the resource consumption of security protocols. Additionally, most conventional security protocols have been updated recently to provide cryptographic agility, allowing the interchangeability of cryptographic primitives. These upgrades are simply the beginning of a process to minimize the computational and communication overhead of Internet protocols [1].

2.7.6 REDUCE DEVICE BANDWIDTH

The makers of the devices should restrict the quantity of network traffic that IoT devices can produce to the minimum amount required for those devices to accomplish their tasks. Vendors should utilize kernel- and hardware-level bandwidth restrictions to limit network transmission rates to values appropriate for each device's responsibilities. Even if an attacker totally compromises a device, it becomes much more difficult for him to utilize it in a DDoS assault because of these restrictions [3]. Additionally, devices must be configured to check for strange behavior on their own and reset to their default settings when alarming activity is noticed. Devices should be capable of erasing any code (by rebooting) that the attacker is executing in memory if restoring them to their factory settings is not possible.

2.7.7 SEGMENT NETWORKS

Performing network segmentation involves dividing a larger network into smaller local ones using Virtual LANs, IP address ranges, or a combination. Before handling traffic, a firewall interface needs to be allocated to a security zone. This allows businesses to assign security zones to different network segments that are managed by the firewall. For example, security administrators can place all databases containing cardholder or patient information in a single network segment designated by a security zone. The administrator can then create security rules that restrict access to the Customer Data zone to specified individuals, groups, applications, or other security zones, preventing any unauthorized internal or external access to the data within that segment.

This kind of solution is becoming more common in industrial applications and can be beneficial in various scenarios. For instance, when it comes to security systems installed in homes, a separate and independent private network might be adequate, potentially with a dedicated route. If the system needs to connect to the Internet, an implementation of a virtual private network (VPN) could be considered.

2.7.8 SET UP A PLATFORM FOR SMART CONNECTION

The job of the connection platform is to control the connectivity of each device across all networks, managing both inbound and outbound traffic.

Additionally, intelligence can be configured within this connectivity platform to ensure the security of devices and the application server, including:

1. **Identification of Unusual Behavior:** The vast majority of IoT applications exhibit highly distinct usage patterns. A device that is intended to transmit a report of 1 MB once per week can start sending more data more often, for instance, which could be a sign of malicious activity [13].
2. **Geofencing:** The only geolocation solution that is impervious to spoofing is network-based geolocation. To determine which base station each device connects to, the platform may utilize information that is 100% trustworthy.
3. **Fraudulent Use of SIM Cards:** The connection platform may recognize this and immediately disable the SIM card if it is taken out of an IoT device and put into a Smartphone for personal use.
4. **Network Firewalls, Such as IP Address and URL Filters:** This will restrict traffic to permitted apps and endpoints only.
5. **Deep Packet Inspection (DPI):** This technique may be used to find unusual data flow content.

2.7.9 PROMOTE ETHICAL HACKING AND VULNERABILITY DISCLOSURE

It is important for manufacturers and software developers to be aware of security flaws in order to address them. Researchers who find significant vulnerabilities and appropriately report them similarly serve the industry to those who identify security bugs in vehicles that are crucial for safety. Encouragement of ethical vulnerability disclosure is required. Before making a system breach public, the researcher must contact the manufacturer or relevant authorities and give them a fair amount of time to independently verify and repair the vulnerability. This is known as responsible disclosure.

The weaknesses in the devices must be found and fixed in order to increase their functioning and security, even when the producers have no financial gain from doing so. Manufacturers may find and report software problems while enhancing the quality of their products, thanks to manufacturer-paid bug bounty schemes, which are less expensive than paying penetration testers.

2.7.10 CREATE A BOARD FOR IOT SECURITY AND PRIVACY CERTIFICATION

Engineers must take ownership of their work since it is an established reality that IoT devices have security issues. For designers, builders, and suppliers of new IoT technologies that promise to adhere to accepted best practices for the development of new products, it is ideal that IEEE or comparable international organizations offer professional certification programs. The program board should have the authority to confirm that the provider complies with responsible engineering principles and would provide their endorsement to those who comply. Another aspect of this certification scheme would be a negative action, which should be limited to revoking the certification status and potentially reporting to the relevant government authority for further action.

The following components of the provider's goods, procedures, and documentation should be at least partially verified by the certification body:

i. Data is carefully managed, utilized, safeguarded, and shared.
ii. Recommended or utilized protocols do not disclose information about users beyond what is necessary to fulfill their work.
iii. Authentication is strong enough and adheres to tested methods.
iv. The devices are not under or overpowered.
v. Devices should include an identification label that is difficult to counterfeit and includes a web address so that buyers can visit it to find out the device's certification status and description (model and serial number, etc.).

These types of certification systems reduce uncertainty and provide device manufacturers, engineers, and writers with best practices to follow. These certificates are admissible as evidence of good engineering practice in court in the event of litigation.

2.8 SECURITY SERVICES FOR IOT

We already saw that security threats and vulnerabilities in the IoT system are capable of disrupting the privacy and security of client data in an IoT application. In order to defend against security incidents that can jeopardize the dependability of devices, it is crucial to address these risks by protecting IoT devices. In order to prevent customer problems or service interruptions, the cornerstone of an IoT solution is to impart security throughout the process. Setting up the proper frameworks makes it simpler to adapt to changing circumstances and verify that they do not diverge from security best practices and take action if they do. Alerts should be generated when a deviation is found so that the proper remedial action may be taken automatically. It is possible to address each component of the IoT ecosystem through implementing security services. These services also help to counter the danger emerging from the entry of connected devices into the market and the threats coming from the internet. Popular IoT solutions with end-to-end, multi-layered protection, including Microsoft® Azure® IoT Hub and AWS® IoT, successfully handle these security issues.

New security concerns have also emerged as a result of the rapid growth of the Industrial IoT (IIoT), which connects operational technologies (OT) such as industrial control systems (ICS) to the internet. For example, SKF and Volkswagen have utilized AWS IoT to enhance customer experience, reduce costs, and optimize production operations.

Customers need to evaluate the unique considerations brought by these new issues and implement the necessary security measures to address them. This section describes the security solutions available on the market to help tackle IoT security issues.

a. Securing IoT with AWS
b. Securing IoT with Azure

2.9 SECURING IOT WITH AWS

Customers can access on-demand web services from Amazon Web Services (AWS) through a global network of AWS servers. AWS makes substantial investments to ensure that security is built into every tier of its service, extending security to IoT devices. Customers in regulated sectors are among those served by AWS [8]. It offers them the resources, services, and advice they need to effectively manage their risk and understand how to comply with AWS Cloud policies. Through their adherence to well-established, widely acknowledged standards and programs, AWS offers assurance of successful risk management by assisting clients in effectively and efficiently managing risk in the IT environment.

AWS has also implemented a risk and compliance program with its IoT services throughout the entire corporation. The two main features of this program are (1) continuously enhancing and re-evaluating the organization's risk-related operations and (2) continuously managing risk across all service phases (from design to deployment). To ensure that control operations are carried out as planned, AWS routinely undergoes independent third-party attestation audits. AWS is specially evaluated in accordance with a number of international and regional security protocols, depending on the area and sector. In addition to various international, national, and sectoral accreditations, AWS participates in over 50 audit programs.

While providing a secure, scalable, and safe platform for IoT solutions, AWS places a high priority on protecting the confidentiality, integrity, and availability of client systems and data. Additionally, it offers design guidelines for using AWS to deploy IoT solutions safely. In summary, AWS adheres to the following design principles:

- **Manage the entire device security lifecycle:** Data security begins with the design phase and concludes with data and hardware erasure upon expiration. To maintain a competitive advantage and regain consumer trust, it is crucial to address the security lifecycle of the IoT solution comprehensively.
- **Ensure that all devices have the least privilege access rights:** This limits the communication channels that a device can utilize. By restricting access, the impact of an infected device on other devices is minimized.
- **Secure stored device credentials:** It is necessary to protect the stored credential information of devices using methods such as secure flash or a dedicated crypto element.
- **Implement device identity lifecycle management:** From creation to the end of their useful lives, devices should maintain their identity. An effective identity management system will monitor a device's identification, verify its authenticity, and proactively grant or revoke IoT privileges over time.
- **Approach data security comprehensively:** IoT installations involve numerous remotely deployed devices, creating a significant attack surface for data theft and privacy breaches.

2.9.1 AWS IOT PLATFORM

AWS announced the launch of its IoT platform in October 2015. AWS IoT enables safe, two-way communication between Internet-connected gadgets including smart appliances, actuators, embedded microcontrollers, and sensors. This makes it possible to securely gather, store, and analyze sensor data from various devices.

2.9.2 COMPONENTS OF AWS IOT

Here, we will discuss some of the important characteristics of AWS IoT's core components and how they are utilized in IoT applications [11].

1. **AWS IoT Core:** It offers a secure and scalable bi-directional connection to all connected devices. It can receive, route, process, and respond to messages from the devices. It also allows the creation and implementation of software for communication with offline devices. When the offline devices come back online, the cached messages are sent to them. Using SQL syntax, messages can be sent to AWS services such as Amazon Kinesis, Amazon Lambda, Amazon S3, and AWS DynamoDB. Custom logic in Lambda functions can be used to command actions to be taken by devices.
2. **AWS IoT Device Management:** This service enables mass provisioning, enrolling, and configuration of various device categories. It maintains a register of device details, including serial numbers, service tags, model numbers, and authentication information. The dynamic status of the device, including running state, firmware, temperature, RPM, and other factors, is stored in AWS IoT Device Shadow, which is a separate data structure. It allows real-time device queries based on both static and dynamic properties. It provides assistance with issue diagnosis and resolution, as well as OTA updates for all connected devices.
3. **AWS IoT Device Defender®:** This service continuously scans the fleet of connected devices for any unusual patterns that may indicate a potential security problem and notifies the authorized user. This can include excessive traffic leaving the device or going to an unapproved service, among others. After receiving a warning, the user can take corrective actions such as stopping the system or running a security patch program.
4. **AWS IoT Events:** It facilitates the fast assessment of multiple telemetry data sources to quickly determine the condition of processes, machinery, or goods, issue alarms or notifications to alert teams, trigger actions, and plan maintenance.
5. **AWS IoT Site Wise:** This service makes it possible to consistently collect data from multiple locations, organize it, and make it available and understandable—all without the need to create additional software.

6. **AWS IoT Things Graph:** It offers reusable models for displaying gadgets and online services. Low-level details, such as proprietary interfaces and communication protocols, are bridged. IoT software can be built by merging models using a graphical interface that utilizes IoT Objects Graph.
7. **IoT AWS Green grass:** This program acts as a stack that provides secure local computing and caching of data. This enables IoT devices to operate even with unreliable network connections, respond quickly to local events, and transmit only the most critical data to the cloud, resulting in cost savings. On the Green grass-enabled device, advanced features such as AWS Lambda and AWS ML can be utilized for localized intelligence gathering.
8. **FreeRTOS:** It is a free and open-source operating system for battery-powered devices that can communicate directly with the AWS Cloud, as well as via Field Gateways hosting the AWS Green grass operating system. Additionally, it provides support for transport layer encryption and the safekeeping of security credentials.
9. **AWS IoT Analytics:** This service extracts useful information from vast amounts of IoT data. AWS offers services for filtering, converting, enhancing, storing, and analyzing IoT data, much like in a conventional data pipeline. On top of this data, AWS offers advanced Big Data platforms such as Amazon ML and Amazon EMR.
10. **AWS IoT Device SDK:** This makes it simple and quick to connect devices to AWS IoT Core. Using the MQTT, HTTP, or Web Sockets protocols, devices can connect, authenticate, and send and receive messages using the SDK. Developers can utilize languages like C, C++, Java, JavaScript, and Python to create IoT modules for hardware. To assist in creating innovative IoT solutions on various hardware, it offers porting guidelines, developer guides with samples, and open-source libraries.

2.9.3 AWS IOT SECURITY AREAS

The following security aspects are addressed adequately by AWS IoT to deliver IoT security solutions:

1. Device Security;
2. Transport Security; and
3. Cloud Security.

1. **Device Security:** It discusses the protection of IoT devices while they are being used in the field. Each device requires a unique identity. AWS IoT supports the use of X.509 certificates generated by AWS IoT or issued by a Certification Authority. Each device's certificate allows for precise control, including the ability to revoke the certificate. To ensure proper functionality, devices must be capable of replacing expired certificates.

   AWS IoT offers authentication options, including AWS IoT Authentication and custom authentication processes, to verify device identities. When using AWS IoT Authentication, devices must be authenticated through the AWS IoT Message Broker. For custom authentication, a custom authorizer must be utilized. Additionally, devices can verify that they are connecting to the legitimate AWS IoT server and not a spoofed device, thanks to server-side certificates.
2. **Transport Security:** All contact with the AWS IoT Message Broker and Device Shadow service is secured using TLS 1.2, which ensures the protection of communication between IoT devices and AWS IoT servers. AWS IoT also utilizes TLS to guarantee the confidentiality of the supported application protocols and to authenticate users. TLS is also employed to encrypt the connection between the device and the broker.

   AWS IoT Device Defender aids in mitigating security threats by monitoring connected devices for any unusual behavior and evaluating the configuration of linked devices. This allows for immediate response in case of a device being compromised and ensures consistent security standards across the entire fleet of IoT devices.

   The edge analytics solution, AWS IoT Green grass, provides authorization and security for device data in both local and cloud connections, preventing any unprotected data exchange between devices and the cloud.
3. **Cloud Security:** It refers to safeguarding both data at rest in the cloud and data that is transferred between various AWS Services and AWS IoT. The AWS Config® service evaluates how well your resource settings adhere to company policies, industry standards, and legal requirements. To assist you in getting started, it offers pre-set, modifiable rules known as managed rules. You can also create your own unique custom rules. AWS Config continually monitors the configuration changes that occur across the resources to identify any breaches of the established guidelines. If a resource

violates a rule, AWS Config flags both the resource and the rule as non-compliant.

You can assess compliance with the industry's security norms and best practices using the comprehensive security state within AWS, provided by AWS Security Hub. By collecting security information from all of your AWS accounts, services, and supported third-party partner products, the Security Hub enables you to analyze your security trends and predict the most critical security issues.

2.10 SECURING IOT WITH AZURE

In order to create end-to-end IoT solutions that address diverse real-life circumstances, Azure offers a number of cloud services and SDKs [9]. Azure IoT Hub is the most fundamental and essential Azure cloud service needed to create an IoT solution. The Azure IoT Device SDKs allow programmers to create IoT device-compatible apps. Through the application, these devices can transmit telemetry data to the Azure IoT Hub. Azure IoT Hub provides bi-directional communication with IoT devices. It can control message routes to other Azure services and manage millions of device-to-cloud telemetry communications. Additionally, it can manage intermittently connected devices and transmit notifications and messages from the cloud to devices with reliability. Azure IoT Hub offers services such as per-device authentication, automated device provisioning, and registration. Figure 2.4 illustrates the various features of Azure IoT Suite.

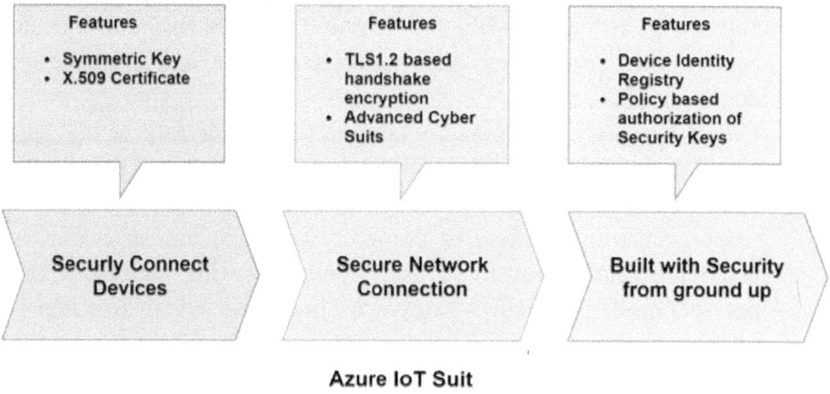

FIGURE 2.4 Azure IoT suite.

Azure provides a number of IoT security options, including the following:

- Azure Security Center for IoT is a cloud security solution based on Azure that offers complete protection against IoT device threats proliferating into the cloud. This service has the capability to unify the entire security management and also enables threat detection and analysis across different cloud workspaces and Azure IoT solutions.
- Azure Sphere is a comprehensive IoT security solution that actively protects devices, organizations, and clients. It comprises hardware, operating systems, and cloud components. By utilizing this service, top-of-the-line MCU-based IoT devices can be developed.
- Azure Sentinel is an efficient security analytics solution that utilizes AI to enhance the speed and intelligence of threat detection, mitigation steps, and investigation.
- Built on Azure IoT Hub, Azure IoT Edge is an extension and also a fully managed service. This service reduces the risk of security threats, such as physical tampering and IP hacking, by shifting the data and analytics to the intelligent edge.

2.10.1 AZURE IOT HUB PLATFORM

Due to its multi-feature portfolio and large client base, Microsoft Azure ® 's IoT Hub stands out among the many IoT platforms now available. Platform as a Service (PaaS), offered by Azure IoT, enables billions of devices to securely communicate with one another and with cloud-based applications [18]. On Azure Cloud Infrastructure, full-featured IoT applications may be created by utilizing Azure IoT Hub and a number of other Azure cloud services. With the help of Azure IoT Hub, it is possible to create systems that are extremely scalable, secure, economical, and globally replicated for various business requirements.

2.10.2 AZURE IOT SECURITY AREAS

The Microsoft Azure IoT Hub suite addresses the following security aspects in-depth to offer IoT security solutions [10]:

- Device Security;
- Connection Security; and
- Cloud Security.

2.10.3 DEVICE SECURITY

Azure IoT Hub realizes device security through the following methods:

- It provides a different identification key (SAS-based security tokens) to each IoT device or set of devices so that they can connect with Azure IoT Hub.
- It enables each device to establish a mutually authorized connection using Transport Layer Security (TLS) through:
    - X.509 certificate-based authentication (Self-signed): The Azure IoT Hub cloud service and the device use the TLS flow to verify each other's identities. The Azure IoT Device SDKs enable this functionality.
    - An X.509 certificate, created and validated by a Certification Authority (CA), is utilized to choose a device and authenticate it with Azure IoT Hub.
- It supports a wide range of Hardware Security Modules (HSMs), where each device has its private key stored in a hardware security module with an unchangeable identity that can be obtained cryptographically and verified through a secure TLS-protected communication channel.
- Countermeasures against malicious devices include:
    - IP blocking/unblocking refers to the capability of enabling or disabling device access based on the device's IP Address.
    - Using the device SDKs, the device allows connection only to services like Azure IoT Hub and blocks any other incoming connections.
    - The IoT Hub generates a device token, which can be used to customize the access control policies of the device.

2.10.4 CONNECTION SECURITY

TLS 1.2 is used to protect the internet connection between the IoT device and Azure IoT Hub.

2.10.5 CLOUD SECURITY

For each security key, Azure IoT Hub allows the creation of access control policies. Each IoT Hub endpoint is granted access using the following set of permissions. Access to an IoT Hub is restricted based on functionality.

- **Registry Read:** Read-only access to the identity register is provided. Every IoT hub includes an identity register that contains data about the devices and modules that are allowed to connect to the hub [12]. Only those devices or modules that are entered in the IoT hub's identity register are allowed to connect. The identity registry also stores the credentials of a device or module for later authentication with the IoT hub.
- **Registry Read Write:** Allows access to the identity register for both reading and writing.
- **Service Connect:** Provides access to communication and monitoring endpoints for cloud services. It enables, for instance, back-end cloud services to send and receive device-to-cloud communications, as well as transmit messages from one device to another and collect the accompanying delivery acknowledgments.

Device Connect provides access to endpoints that are directed at devices. It permits, for instance, the transmission of device-to-cloud and device-to-cloud messages.

2.11 CONCLUSION

IoT brings incredible benefits to businesses in terms of cost savings, revenue growth, and business transformation, management, and monitoring. However, protecting the enormous amounts of data and connections from scattered IoT devices is a challenge for companies. Many IoT systems claim to provide easily adaptable software components to address these problems. These pay-per-use platforms make it possible to quickly and inexpensively create IoT solutions without the need to write code or connect to a massive number of devices.

In this chapter, we have discussed the main challenges that security threats and vulnerabilities pose for IoT systems. To enhance security, we have also provided a list of strategies and best practices for IoT device makers and users. AWS IoT and Azure IoT Hub offer a variety of services to build complex IoT solutions with numerous choices for specific business requirements. Businesses must use the best methods and practices, choose the right partner, tools, and platforms that are most suitable for their customers, and consider the bigger picture and implications of privacy and security from the outset when creating new IoT solutions.

KEYWORDS

- **big data analytics**
- **distributed denial of service**
- **international data corporation**
- **internet of things**
- **open web application security project**
- **radio-frequency identification**

REFERENCE

1. Garcia-Morchon, O., Kumar, S., & Sethi, M., (2019). *Internet of Things (IoT) Security: State of the Art and Challenges*. RFC Editor, RFC8576. doi: 10.17487/RFC8576.
2. Centre for Strategic Cyberspace + Security Science, (2018). *Centre for Strategic Cyberspace + Security Science: Internet of Things (IoT) Security Whitepaper*. Retrieved from: AiSP website:https://www.aisp.sg/sig_article_IOTSecurityWhitepaper.html (accessed on 27 January 2024).
3. Corser, G., et. al., (2017). *Internet of Things (IoT) Security Best Practices Whitepaper.* Retrieved from: internetinitiative.ieee.org website: https://internetinitiative.ieee.org/images/files/resources/white_papers/internet_of_things_may_2017.pdf (accessed on 27 January 2024).
4. Noseda, M., Zimmerli, L., & Rüst, A. (2020). *Security by Design for IoT Devices Whitepaper*. Retrieved from: nxp website:https://www.nxp.com/docs/en/white-paper/ZUAS-SECURITY-BY-DESIGN-FOR-IOT-DEVICES.pdf (accessed on 27 January 2024).
5. Centri, (2016). *Securing the Internet of Things: 5 Key Areas to Protect Whitepaper*. Retrieved from: centritechnology.com website: https://www.centritechnology.com/wp-content/uploads/2019/02/CENTRI-Securing-the-IoT-White-Paper-Sept-2016.pdf (accessed on 27 January 2024).
6. CTIA–The Wireless Association, (2014). *Mobile Cyber security and the Internet of Things Whitepaper*. Retrieved from: ctia.org website: https://api.ctia.org/docs/default-source/default-document-library/ctia-iot-white-paper.pdf (accessed on 27 January 2024).
7. The OWASP IoT Security Team, (2018). *OWASP Top 10 Internet of Things*. Retrieved from: owasp.org website: https://owasp.org/www-pdf-archive/OWASP-IoT-Top-10-2018-final.pdf (accessed on 27 January 2024).
8. Amazon Web Services, (2022). *Securing Internet of Things (IoT) with AWS: AWS Whitepaper.* Retrieved from: amazon.com website: https://docs.aws.amazon.com/pdfs/whitepapers/latest/securing-iot-with-aws/securing-iot-with-aws.pdf#security-challenges-and-focus-areas (accessed on 27 January 2024).
9. Thinxtream, (2020). *Developing IoT Solutions with Azure IoT Hub Platform Whitepaper*. Retrieved from: thinxtream.com website: https://www.thinxtream.com/whitepapers/thinxtream-azure-iot-hub-solutions-wp-011.pdf (accessed on 27 January 2024).

10. Thinxtream, (2020). *Security in Internet of Things Whitepaper*. Retrieved from: thinxtream.com website: https://www.thinxtream.com/whitepapers/thinxtream-iot-security-wp-005.pdf (accessed on 27 January 2024).
11. Thinxtream, (2020). *Developing IoT Solutions with AWS IoT Platform Whitepaper*. Retrieved from: thinxtream.com website: https://www.thinxtream.com/whitepapers/thinxtream-aws-iot-solutions-wp-010.pdf (accessed on 27 January 2024).
12. Microsoft Corporation, (2022). *Understand the Identity Registry in Your IoT Hub*. Retrieved from: Microsoft.com Website: https://learn.microsoft.com/en-us/azure/iot-hub/iot-hub-devguide-identity-registry (accessed on 27 January 2024).
13. BICS, (2022). *Best Practice for End-To-End IoT Security Whitepaper*. Retrieved from: bics.com Website: https://www.bics.com/wp-content/uploads/2022/01/Bics-whitepaper-Best-practice-for-end-to-end-IoT-security.pdf (accessed on 27 January 2024).
14. Microsoft Corporation, (2017). *The Right Secure Hardware for Your IoT Deployment Whitepaper.* Retrieved from: Microsoft.com Website. https://download.microsoft.com/download/C/0/5/C05276D6-E602-4BB1-98A4-C29C88E57566/The_right_secure_hardware_for_your_IoT_deployment_EN_US.pdf (accessed on 27 January 2024).
15. Microsoft Corporation, (2017). *Evaluating Your IoT Security Whitepaper*. Retrieved from: Microsoft.com website: https://download.microsoft.com/download/D/3/9/D3948E3C-D5DC-474E-B22F-81BA8ED7A446/Evaluating_Your_IOT_Security_whitepaper_EN_US.pdf (accessed on 27 January 2024).
16. CommScope, (2020). *Ruckus IoT Security Whitepaper*. Retrieved from: commscope.com website: https://www.commscope.com/globalassets/digizuite/857777-wp-ruckus-iot-security.pdf (accessed on 27 January 2024).
17. Sundaravadivazhagan, B., & Subashini, B., (2021). A review on internet of things (IoT): Security challenges, issues, and the countermeasures approaches. *Psychology and Education, 58*(2), 6544–6560.
18. Sundaravadivazhagan, B., (2022). A novel credit grounded job scheduling algorithm for the cloud computing environment. In: *2022 International Conference on Inventive Computation Technologies (ICICT), 912*, 912–919.
19. Onyema, E. M., et.al. (2022). A security policy protocol for detection and prevention of internet control message protocol attacks in software defined networks. *Sustainability, 14*(19), 11950.

CHAPTER 3

A Fundamental Overview in Cybersecurity: Prevention and Mitigation of Cyberattacks for Securing Cyberspace

S. MAYUKHA and R. VADIVEL

Department of IT, Bharathiar University, Coimbatore, Tamil Nadu, India

ABSTRACT

The cyberspace has crossed geographical regions and borders and has spread far and wide across the entire globe. Anything that is stored or processed by any kind of computing devices comes under the cyberspace or the digital space. The surface for attacks on this cyberspace has also extended to the whole world. Today, we live in an era where one can be physically present in one region but virtually attack any point in the world. Data is considered the new currency and power in this digital era. The power play and attacks carried out by malicious hackers to gain access to the data are like an abyss. There are numerous ways a hacker can illegally gain access and manipulate data or distribute it for personal gain, such as money or fame. This chapter discusses the various possible ways cyber attacks are performed and what possible solutions can be taken to prevent or mitigate the impact of such attacks. Cyber security is a vast and interesting mix of knowledge, application, and inquisitiveness. The main aim of cyber security is CIA (confidentiality, integrity, and availability). The process of securing any digital space with these three factors intact is the holy grail of the field. Cyber security is a field where constant threats and vulnerabilities are fixed right away but still faces threats from many unknown sources and directions. This chapter also details the existing preventive measures for such cyber attacks. The evolution of technology has given rise to exponential vulnerabilities and threats.

Leveraging Artificial Intelligence (AI) Competencies for Next-Generation Cybersecurity Solutions. Pethuru Raj, B. Sundaravadivazhagan, & V. Kavitha (Eds.)

3.1 INTRODUCTION

The world is becoming digitized. Governments are going completely digital to save paper and time. There is no argument that digitizing processes eases manual work, saves time, and improves accuracy. Digitization is the right choice for the progress of our world. However, there are factors that can disrupt progress and cause losses. Digitization involves converting everything to digital data that can be stored in a computer. Once data is stored, it can be processed or manipulated as desired. Digitizing data helps people stay informed about various events worldwide. As processes scale up, constant connectivity to data becomes essential. Even metadata, which is data about data, is important. Large servers are necessary to store and process data.

Since the internet boom, there has been an exponential increase in cyberspace. Similarly, since digitization, there has been an exponential increase in the storage of digital data. The vastness of cyberspace and digital data is like a pot of gold for hackers. It is difficult for ordinary people to comprehend the intricate workings of networks and what is happening inside computers, such as how data is stored or shared. This lack of awareness becomes an advantage for malicious hackers.

This chapter discusses and raises awareness about the field of cybersecurity. It covers the stages involved in cybersecurity [1]. Additionally, this chapter sheds light on how to secure an enterprise or an individual through penetration testing.

3.2 CYBERSPACE

Cyberspace is defined as the interconnected technology of devices. Currently, internet-connected devices are not limited to computers alone. The field of the Internet of Things (IoT) has given rise to numerous components and devices that can be connected to the internet, transforming any device into a smart device. This enables the exchange of data between devices to automate certain mundane processes. In cyberspace, most things or processes described here are intangible. Due to the intangible nature of cyberspace, it becomes more challenging to protect and defend. Regardless of whether the interconnected devices are trivial or not, every single device connected to the network should be secured. Failure to do so can lead to exploitation of the device to gain access to the network, allowing traversal and hacking of the entire network. The activities that occur in cyberspace are mostly unknown to the general public. It is the responsibility of cyber security experts to safeguard the data and processes in cyberspace from malicious hackers.

Cyberspace is not only for interconnected devices. Any device that stores or processes data in digital format comes under cyberspace. For example, there is a term called an "air-gapped computer." An air-gapped computer is not connected to any network at all, making it difficult to access unless a hacker is physically touching the computer. While such computers provide a degree of security, it does not mean that air-gapped computers are invulnerable. Air-gapped computers typically use a pen drive or a floppy disk from older times to transfer data, as there is no other way to share data without the internet. However, it is contradictory when someone goes through the hassle of using an air-gapped computer for security, but then uses a pen drive or floppy disk that could potentially be exploited by an attacker with the right tools. Therefore, regardless of the type of connection or non-connection between devices, any digitized data falls under the reaslm of cyberspace.

With the advent of the metaverse, the number of devices and people connected to the internet is expected to increase exponentially, leading to a rise in cybercrimes. Cybersecurity experts will face challenges in handling this unless the right countermeasures are developed and implemented before the general population enters the metaverse. To put it simply, if cyberspace is the world, then cybersecurity is the law that governs it. Users of cyberspace are the people, cyber security experts are the police, and hackers are the criminals, as represented in Figure 3.1.

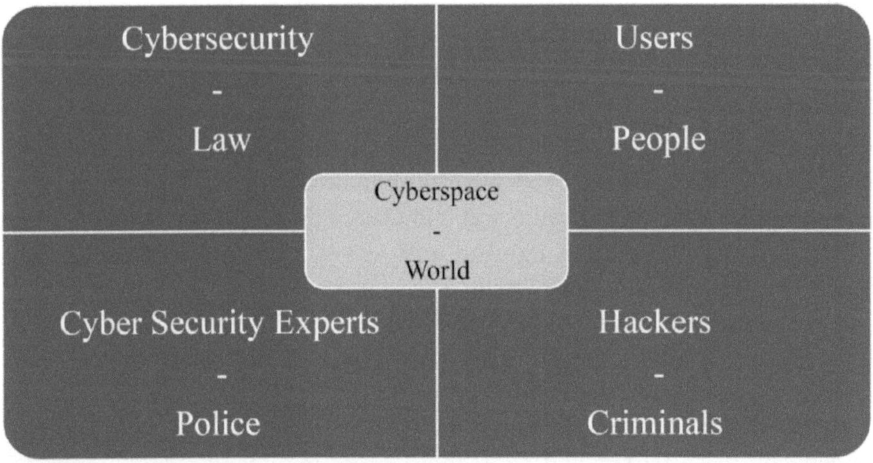

FIGURE 3.1 Cyberspace representation analogy.

3.3 CYBER SECURITY

Cybersecurity is the law governing cyberspace. There is a lot involved in cybersecurity. It is like a never-ending pit. It is not a one-size-fits-all solution in this field. Cybersecurity can be defined as securing the digital space from internal and external exploits or attacks. It mainly focuses on three things: CIA [2].

- Confidentiality;
- Integrity; and
- Accessibility.

In the field of cyber security, the object to be protected can be either a company, referred to as an enterprise, or an individual. Therefore, in order to ensure cyber security at all times, it is crucial that the Confidentiality, Integrity, and Accessibility (CIA) of both enterprises and individuals are never compromised or exploited. Cyber security enthusiasts are individuals who aspire to enter the cyber security field but may be unsure of where to begin. It is important for these enthusiasts to have a solid understanding of the fundamental principles of cyber security before embarking on a career in this field. The ultimate goal of cyber security is to uphold the principles of CIA, and enthusiasts must comprehend this in order to have a clear objective and make progress in the field.

To grasp the basics of cybersecurity, one must possess a strong knowledge of networking and computing. Becoming a cybersecurity expert requires continuous learning and staying updated in the field. The learning curve in cybersecurity is steep due to the complexity and extensive knowledge required about computer systems and networks. Despite this, there are numerous opportunities in the cybersecurity field. However, there is a shortage of cybersecurity experts compared to the increasing number of cybercrimes. Given the vastness of the cybersecurity domain, there are various levels of employment based on exposure and knowledge criteria. Figure 3.2 provides a categorization that sheds light on entry-level, mid-level, and advanced-level jobs in the cybersecurity field.

The main aim of any cybersecurity expert is to suggest and implement controls and countermeasures to prevent and mitigate cyber-attacks. This chapter discusses in detail the important factors of cybersecurity and the features of cybersecurity. It would be easy for cybersecurity enthusiasts to understand these factors and boost their career in cybersecurity. Figure 3.3 represents the foundation of cybersecurity along with what that means in security terms.

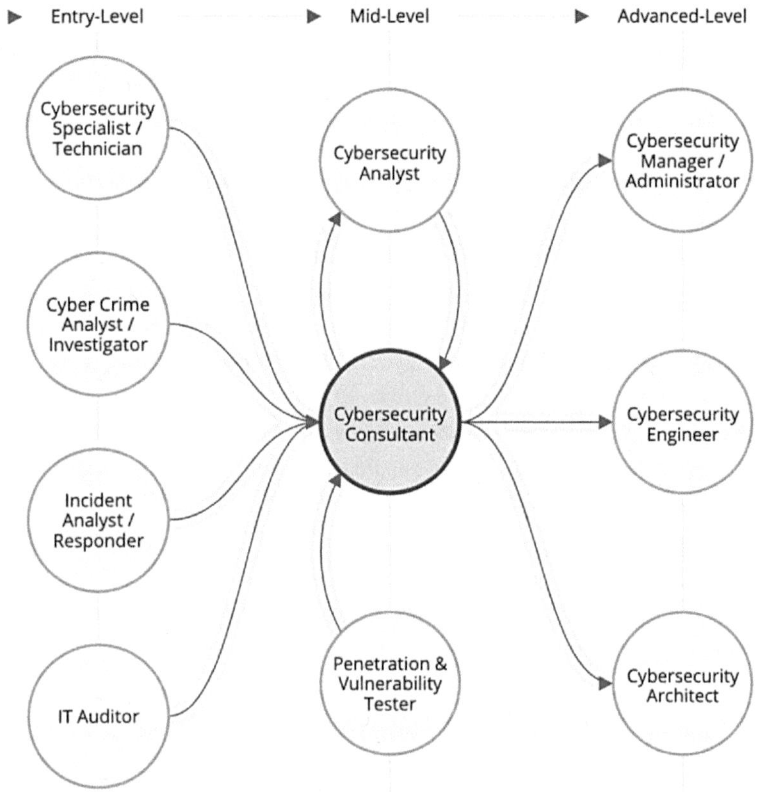

FIGURE 3.2 Cyber security job opportunities.

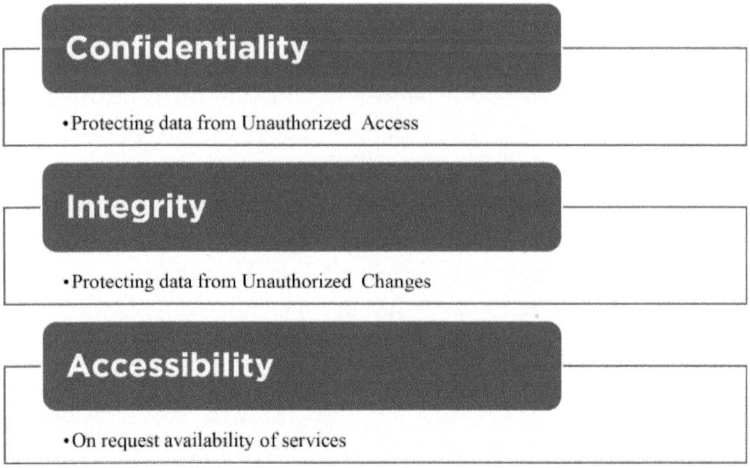

FIGURE 3.3 Overview of cybersecurity foundation.

1. **Confidentiality:** Confidentiality is defined as the secure processing and storage of data while preventing unauthorized usage. During data transfer or storage, it is crucial to maintain the privacy of the data. Depending on the type of data, access is granted to users. For example, public data can be accessed by anyone on the internet, while user-specific data should only be used by the designated user. This is the principle of confidentiality that governs and ensures cybersecurity. The software or infrastructure should strictly adhere to this principle.

   Confidentiality ensures that data is kept and processed securely, protecting it from unauthorized access by other users, the public, or malicious hackers. Regardless of the method of attack, it is essential that data remains inaccessible to anyone other than the intended user. Cybersecurity faces various attacks that can compromise the confidentiality of software or infrastructure. Some examples of these attacks are as follows:

   - Phishing;
   - Social Engineering;
   - Wiretapping;
   - Keylogger; and
   - Man in the middle.

   Steps must be taken to ensure that there is no data breach in the event of an attack. Countermeasures must be implemented in case of an attack. Confidentiality is one of the three pillars of cybersecurity. If one of the pillars is attacked and exploited, the other two can also be easily affected.

2. **Integrity:** Integrity in cybersecurity is defined as the protection of system data from unauthorized changes, whether intentional or accidental. When data is stored in software or infrastructure by an authorized person, any subsequent access to the same data should not result in any alteration of the original data by unauthorized individuals. Compromised integrity of the software or infrastructure can lead to unimaginable losses, potentially spelling the end for an enterprise or individual. Nearly all operations rely on accurate data, and any inaccuracies can result in financial and reputational losses.

   Similar to confidentiality, integrity must be maintained at all times in software or infrastructure. Users should have confidence that the software or infrastructure they are using is secure. Unauthorized usage, resulting from exploitation of bugs or access control vulnerabilities, is a real possibility. This is why software or infrastructure must undergo quality testing for integrity. Regular routine testing should also be conducted to monitor any changes that may expose the software or infrastructure.

3. **Accessibility:** Accessibility is defined as the availability of software or infrastructure whenever a user requests any resource or logs into a system. If a user attempts to access software or a network and receives an error instead, there is an accessibility issue. The accessibility of any software or network can be easily compromised by launching a Denial of Service (DoS) attack. Accessibility issues can result in losses for enterprises or individuals in various ways, such as redirecting users to fake software or infrastructure when the original one is down. It can also cause disruptions in global processes, for example, when social media giants like Facebook, Instagram, WhatsApp, etc., experience downtime for almost a day, resulting in millions of dollars in losses for the company.

   Accessibility can easily be disrupted. The main aim of attacks based on accessibility is the disruption of service rather than gaining access or manipulating data. For this, hackers can simply release a virus or trojan to overload the server by using its resources. This would be an easy target unless the necessary prevention and mitigation steps have been carried out.

3.4 CYBER SECURITY FEATURES

There are no defined sets of scope in cybersecurity. Attackers or hackers can gain access to any kind of network, device, or software using various techniques and tactics [4]. Although the scope is not defined, the stages or processes that attackers follow to exploit or gain access to a network can be mapped. Cybersecurity features refer to the factors that affect security in cyberspace. In the field of cybersecurity, the main concern is the BES. In addition to these concerns, factors such as time, value, and frequency should also be considered. There is a myth that attacks in cybersecurity cannot be measured. The book "How to Measure Anything in Cybersecurity" clearly proves that cybersecurity attacks can be measured. Therefore, the factors that need to be taken into account are provided in Table 3.1:

The features mentioned here should be accompanied by the Bayesian model. This model takes into consideration both the hypothesis and the data when calculating probabilities. The Bayesian model is commonly used in statistics to provide approximate estimates for these features.

TABLE 3.1 Features of Cyber Security

Parameter	Description
Confidentiality	Risk of unauthorized access
Integrity	Risk of unauthorized changes
Accessibility	Risk of services being down
Frequency	Number of times an attack can be performed
Time	Time taken to attack and time taken to recover
Value	Approx. loss of monetary and reputation value

In the field of cyber security, the predominant method used is the risk indexing matrix. This matrix plots the risk of any attack on a heat map or matrix, indicating whether the risk is Severe, High, Medium, or Low. The main objective of this method is to minimize the risk of any given attack by taking necessary precautions. However, one of the main drawbacks of this method is that the risk indexing matrix is not quantitative. It cannot specifically convey how much the risk has been reduced if a system has been secured through penetration testing. Another significant challenge in cyber security is effectively communicating the risks and associated costs to the management board of an enterprise. It is crucial to accurately convey the results of any security implementation to ensure transparency.

3.5 VULNERABILITY

Vulnerability in cybersecurity is the main factor to consider. Vulnerabilities arise with complexity, as the level of vulnerability is directly proportional to complexity. It is easier to control or reduce vulnerabilities if the system is simple. Vulnerabilities are only detected during testing, which can be done through system testing or penetration testing. Testing the system is a process that ensures the quality assurance of the software or system. Deploying software or a system without testing would be disastrous, as once it is released to the general public, the vulnerabilities become the attack surface for hackers, as shown in Figure 3.4. The attack surface is the presentation layer of the software or system where vulnerabilities are exposed to the general public. Any hacker can enumerate basic attacks and exploit the vulnerabilities on the attack surface.

Vulnerabilities are an integral part of any software or system. Nobody can assure that the software or system that has been developed is free of vulnerabilities. Unless the software or system is repeatedly tested for bugs and vulnerabilities, there will always be hidden vulnerabilities that can compromise the system.

FIGURE 3.4 Vulnerabilities are proportional to attack surface.

Sometimes, enterprises or individuals may engage penetration testers to audit the developed software or system. The outcome of the pentester's audit would be a report containing the severity of the vulnerabilities and the pentester's suggestions on how to mitigate the discovered vulnerabilities. The final decision to implement the suggestions lies with the enterprise or the individuals. The severity of the vulnerabilities is ranked in the following order:

- Fatal;
- Severe;
- High;
- Medium; and
- Low

The current measurement of vulnerability is in the form of a risk rating matrix. This may evolve into quantitative analysis in the future. The quantitative method of measurement is necessary to make informed decisions regarding the security of the software or system. There is an entire field dedicated to the identification of vulnerabilities. Vulnerabilities can turn into threats if hackers are able to exploit them. Through these vulnerabilities, hackers may gain unauthorized access to the software or system. Therefore, the pentester must enumerate the types of threats that the software or system may face in the current technological scenario. This varies based on the platforms or programming languages and frameworks chosen for the software or system. Hence, the pentester should be capable of creating threat models for the software or system.

3.6 THREAT MODELING

Threat modeling [5] is the process of identifying potential threats in a system, determining their presence, assessing the potential losses if the threats are exploited, and defining the necessary mitigation steps. This process follows a step-by-step approach, similar to a lifecycle, which pentesters adhere to. The fundamental phases of any threat modeling are as follows:

- Identify;
- Attack;
- Analysis;
- Prioritization; and
- Mitigation

Depending on the framework of threat modeling used, the phases can vary, as each threat modeling approach is unique. The phases should be implemented continuously, as depicted in Figure 3.5 of the threat modeling process. Following the completion of a full cycle, there should be ongoing monitoring to ensure that the implemented mitigation measures are effective and do not introduce any new threats or vulnerabilities to the system.

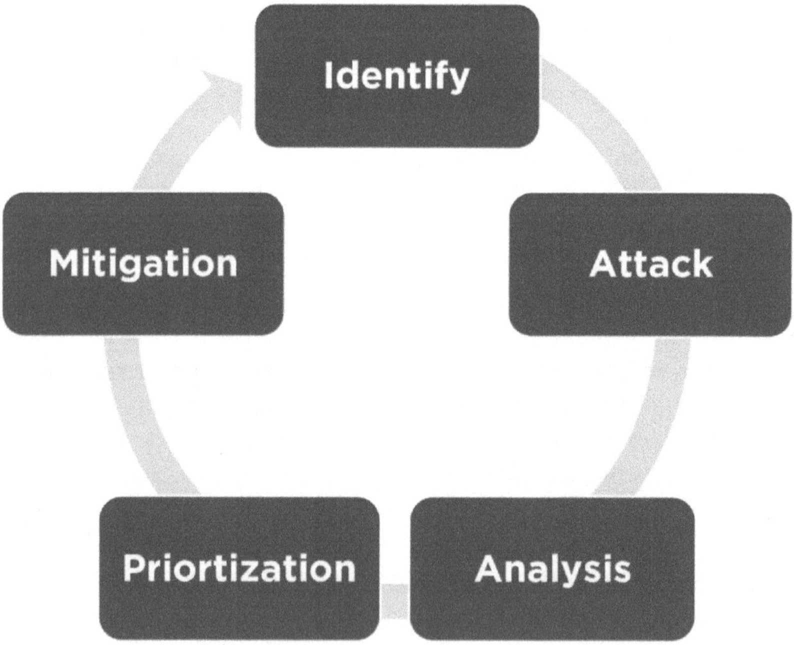

FIGURE 3.5 Threat modeling generic process.

There are associated tasks that need to be carried out in each phase. The tasks that need to be executed in each phase are described in Table 3.2. Each task of the phases should be carried out, documented, and presented as a report. This report will serve as a basis for future threat modeling.

TABLE 3.2 Threat Modeling Tasks Associated with Phases

Phase	Task
Identify	Identification of possible threats and threat actors if any in the system
Attack	Enumerate attacks on the threats to see if the exploitation of the threat is positive
Analysis	Analyzing the success of the threat exploitation and associated depth of loss
Prioritization	Prioritizing all the threats in rank of severity to address the highest first
Mitigation	Steps taken to neutralize the threat that has been successfully exploited.

According to the phases and the tasks associated with them, there are 12 types of threat modeling frameworks in threat modeling.

- STRIDE
- OCTAVE
- PASTA
- VAST Modeling
- LINDDUN
- Quantitative Threat Modeling Method
- CVSS
- Persona non grata
- hTMM
- Attack Trees
- Security Cards
- Trike

There are some terms associated with threat modeling. A pentester should be aware of these terms to better understand threat modeling. These terms are explained in Figure 3.6 and listed in Table 3.3 for easy reference. Each term in the threat modeling landscape has defined interactions with other terms, with varying levels of impact or purpose. This helps in identifying where threat modeling becomes critical. For instance, a successful attack [6] is a cause for concern for the pentester, who approaches threat modeling from a red team perspective [7].

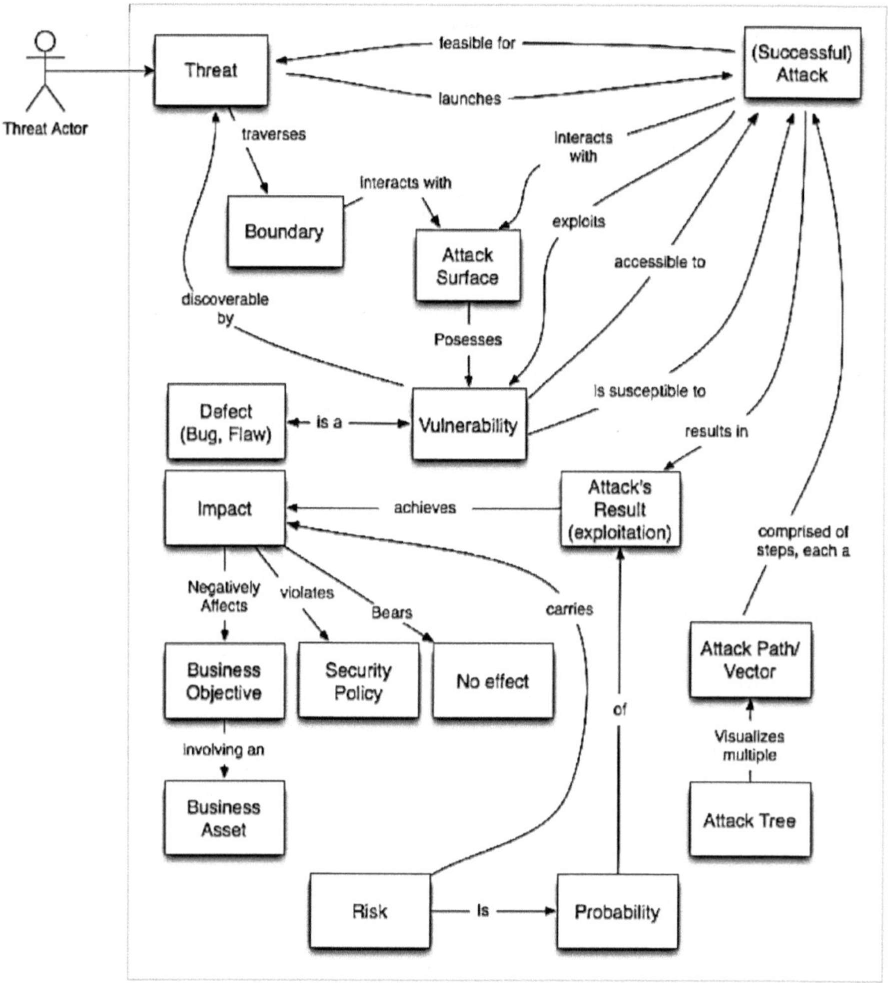

FIGURE 3.6 Terms in threat modeling.

Red teams in penetration testing are offensive testers of the software or system. Offensive testers test the software or system by attacking it, similar to a hacker. Red teaming is the best way to ensure the strongest defense for a software or system.

With the right terms and the right framework, threat modeling can be simple to understand and execute. There are many tools [8] available for the threat modeling process. These tools can assist pentesters in efficiently designing threat models for software or systems.

TABLE 3.3 Glossary of Threat Modeling

Sl. No	Terms	Description
1	Threat Actor	The person due to which the threat arises
2	Threat	Vulnerability or exploit
3	Boundary	Scope of the testing ground
4	Attack Surface	Interaction layer of the software or system
5	Vulnerability	Weakness in the software or system
6	Defect	Error in the software or system
7	Attack	Enumeration attack of the pentester
8	Impact	degree of the attack
9	Attack's Result	Result of the performed attack
10	Business Objective	In terms of loss to business
11	Security Policy	Possible violation of policy
12	No Effect	If the threat has no cause for concern
13	Business Asset	Preceded by business objective
14	Attack Path / Vector	Steps comprising of the attack
15	Attack Tree	Mapping the steps of the attack
16	Probability	Possibility of the attack happening real time
17	Risk	Possible loss

3.7 RISK

Risk is the possibility of potential loss in the event of an attack on the software or system. Risk assessment [9] is crucial in cyber security, as it determines the subsequent procedures. Based on the severity of the potential loss and the probability of impact, the risk is classified as either severe or not. Once the risk assessment is completed, steps are taken to mitigate the risk. The process for this is described in Figure 3.7.

Risk assessment should be conducted periodically to ensure that no vulnerabilities are overlooked. Even if a system is functioning well without any issues, it is important to continuously evaluate the risk assessment. This will help identify any hidden vulnerabilities or exposures. The formulation of risk is as follows:

$$Risk = Vulnerability * Threat$$

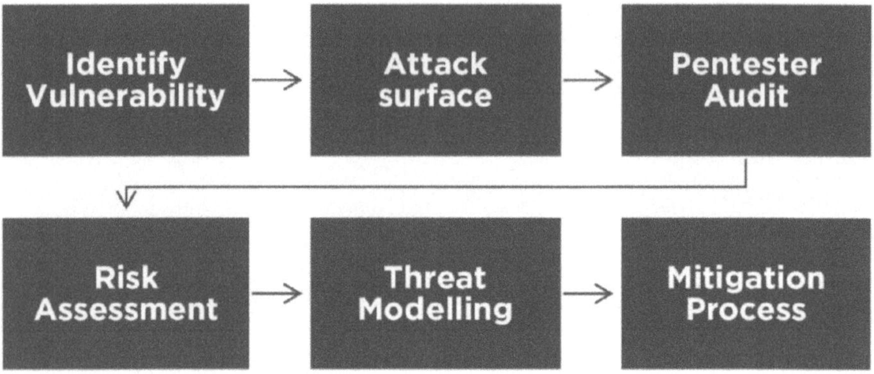

FIGURE 3.7 Process of risk assessment with mitigation.

As vulnerabilities are exposed through the attack surface, which can be multiplied by threats carried out by various threat actors, the risk increases as a result of the combination of vulnerability and threat. This is undoubtedly a major factor that needs to be addressed. Everything begins with vulnerability in the application, and hackers attempt to exploit it through attacks and exploits, while pentesters aim to mitigate it.

3.8 DETECTION

In the field of cybersecurity, it is crucial for enterprises and individuals to be aware of potential attacks on their software or systems. Cyber-attacks are intangible in nature, making it difficult to identify them. Therefore, detecting whether a software or system is under attack, or if any malicious entities have gained unauthorized access, can be challenging. Typically, hackers gain initial illegal access to the software or system and then proceed to gather as much information as possible to maximize the impact of their attack.

To effectively safeguard against attacks, constant monitoring of logs, access, and controls is necessary to capture any intrusion before an attack occurs. Additionally, hosting an Intrusion Detection Service (IDS) tool can help identify any intrusions in the software or systems. Regular detection operations are essential for the Chief Information Security Officer (CISO) to mitigate attacks well in advance. It is important to note that malicious hackers require time to prepare their attacks, so immediate attacks upon initial access are unlikely. If the pentester or CISO can detect the initial illegal intrusion into the software or system, immediate mitigation steps can be deployed to neutralize the threat.

Ultimately, the timely and effective response from the owner, whether an enterprise or an individual, is crucial in mitigating these threats.

It is also not enough if only detection is in place. After the identification of the intrusion, the system should be loaded with scripts or tools to process the intrusion immediately upon discovery. Next comes prevention, followed by mitigation. This is the best order to countermeasure a cyber-attack before the threat becomes severe.

3.9 PREVENTION

Prevention is the next phase of securing software or a system once an intrusion is detected. Prevention involves executing scripts or commands to prevent any attack [11] by malicious hackers. By implementing prevention measures, the initial access of hackers to the software or system is deflected, thereby preventing the attack from occurring. This type of security is crucial in the current trend, emphasizing the importance of prevention before an attack occurs. There are tools available now for continuous monitoring of spaces, which can be upgraded to incorporate AI or other technologies to automatically prevent unauthorized access by hackers.

Prevention of attacks or threats can be implemented, but automating the prevention would have limited effectiveness. With technology advancing rapidly, hackers constantly discover new and innovative methods to exploit vulnerabilities. Therefore, the prevention system should automatically update to address the latest vulnerabilities. There is potential to incorporate AI/ML into cybersecurity threat prevention.

3.10 MITIGATION

Mitigation is the process of neutralizing or reducing the impact of a threat or attack [12]. Mitigation can also be implemented prior to an attack. For instance, if a vulnerability is discovered in the underlying software, one mitigation approach is to deploy monitoring systems to detect initial access attempts. This is the chosen mitigation strategy by the pentester. Mitigation can involve fixing the vulnerability permanently or preventing unauthorized access to the software or system through the known vulnerability. However, if the mitigation setup does not cover all possible breaches in the software or system resulting from the vulnerability, hackers may find alternative attack methods, rendering the mitigation ineffective.

Another misconception is the belief that once mitigation measures are in place, the software or system is completely secure. While mitigation may deter malicious hackers initially, persistent hackers [13] will continue to explore new and innovative ways to gain unauthorized access. In such cases, the pentester or the CISO should be prepared with appropriate mitigation steps to address these unwarranted actions.

3.11 PROTECTION

The path to achieving protection begins with detection. Detection is then followed by prevention, mitigation, and ultimately, protection. The protection phase represents the highest level of security that a software or system can attain against vulnerabilities. If a tool is implemented to automatically monitor and identify vulnerabilities in the software or system, it becomes a true form of protection. If the Chief Information Security Officer (CISO) or pentester has to repeatedly implement controls to defend against the same attack, it indicates that the current solution may not be the most effective. The ideal tool should incorporate the detection, prevention, and mitigation measures for a specific attack and automatically apply them to protect the software or system from future instances of the same attack. By following this approach, the protection system will evolve into the most robust tool for safeguarding the software or system.

Another concept to consider is in-depth defense, which involves implementing protection measures at each layer of the OSI Model. The OSI Model, which stands for "Open Systems Interconnection," is a framework used by computers to exchange data between client and server. It consists of seven layers, as shown in Figure 3.8.

i. Physical Layer
ii. Data Link Layer
iii. Network Layer
iv. Transport Layer
v. Session Layer
vi. Presentation Layer
vii. Application Layer

The protection [14–17] of any software or system should be implemented in all seven layers of the OSI Model. This serves as a reliable benchmark for cyber security. With the advancements in technology, processes can now be automated, and with the introduction of ML/AI, there is an opportunity to make software and system protection fail proof at each layer of the OSI model by automating the process and allowing the tools to determine the necessary countermeasures. The OSI model layers and their respective operations are provided in Figure 3.9.

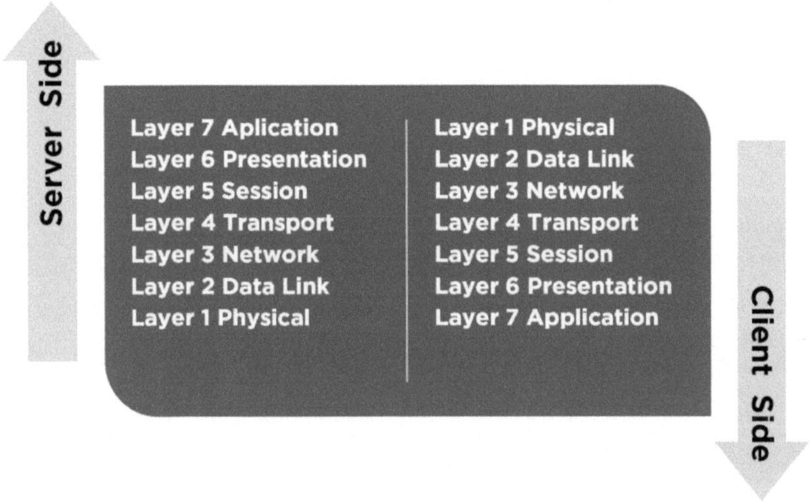

FIGURE 3.8 Order of the layers in the OSI model.

Each and every layer of the OSI Model is vital and has its own responsibilities and roles in transferring data. This is what happens at the backend of the communication between two computers. Hackers could inject into any one of these layers to gain illegal access to unauthorized data. That is why in-depth defense is required. Protection in cybersecurity should be automated so that redundant processes are taken care of by the tools. As days pass, there are many new tools and attacking methods that are gaining momentum. Most of the hacking tools are open source, and hence even beginner hackers would be able to perform some basic attacks against the software or system. Therefore, care should be taken to reject the usual kinds of attacks launched on the software or system.

3.12 CYBER SECURITY FRAMEWORKS

Since cyberspace has become an integral part of people's day-to-day lives, there are companies that have developed a variety of processes and tasks to safeguard data in cyberspace. These processes ensure that the enterprise adheres to a set of standards that can protect data from hackers. Various companies around the world propose cyber security frameworks. Some of the cyber security frameworks are listed below:

- NIST
- ISO 27001
- FISMA
- SOC2
- GDPR
- NERC-CIP
- HIPAA

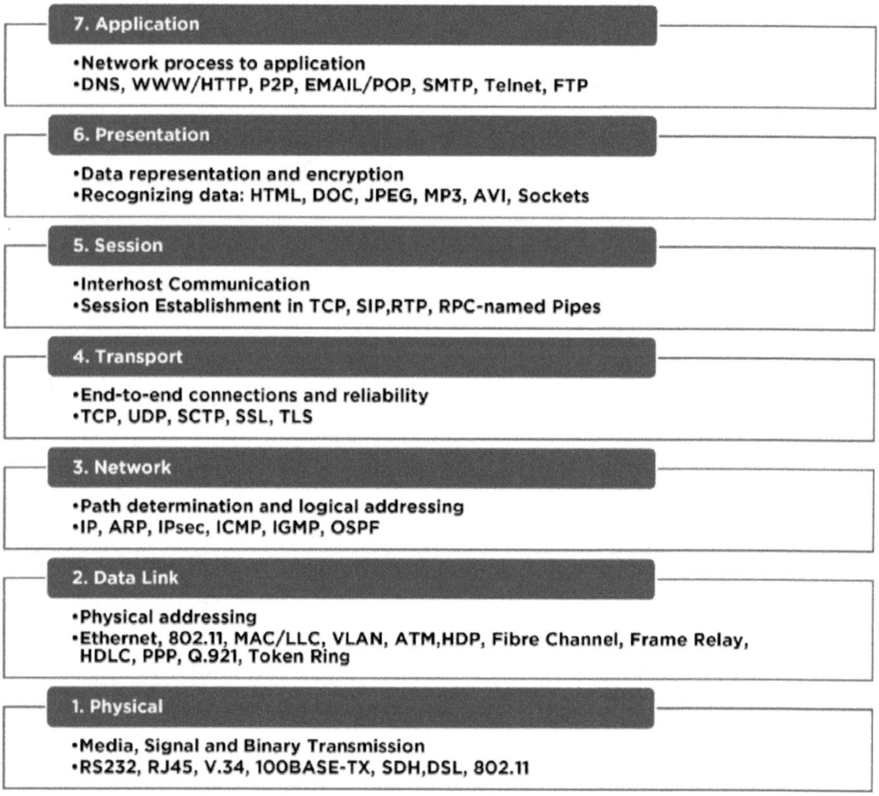

FIGURE 3.9 OSI model layer and its operations.

These frameworks are popular in the world of cybersecurity. If an enterprise is audited and certified that the processes in the enterprise are carried out according to a certain framework, it would increase trust in the enterprise. These frameworks primarily cover data security. The ways in which the frameworks achieve data security differ. The brief description of each framework is as follows:

i. **NIST:** NIST is the acronym for the National Institute of Standards and Technology. NIST is a globally recognized and widely used framework. It serves as a valuable benchmark for data security in any enterprise. This framework encompasses a total of five functions:

   - Identify
   - Protect
   - Detect
   - Respond
   - Recover

   Each of these functions will cover a range of operations to be performed as described in Figure 3.10. The main aim of implementing this framework is to reduce the cyber risk of an enterprise. Depending on the needs of the enterprise, the functions can be chosen from the NIST framework.

ii. **ISO 27001:** ISO is a globally recognized organization for standards. ISO stands for International Organization for Standardization. ISO 27001 focuses on securing information assets and managing information security. This is crucial for enterprises operating in the IT sector or those dealing with digital data processing. The ISO 27001 certification is obtained through an audit process, following the compliance steps outlined in Figure 3.11. Once the enterprise aligns its processes with the compliance steps, the auditor issues the ISO 27001 certificate. This certification is important to establish credibility and trust with the enterprise's customers. Additionally, it ensures the implementation of correct procedures in information security management, minimizing cyber risks in the event of an attack. ISO 27001 certificates can be issued by various governing bodies worldwide, such as IAF, Dakks, UKCS, and others.

   Any enterprise that follows this compliance cycle has minimized risk. This is not a guarantee that there won't be any attacks at all, but it would make the enterprise a little more secure.

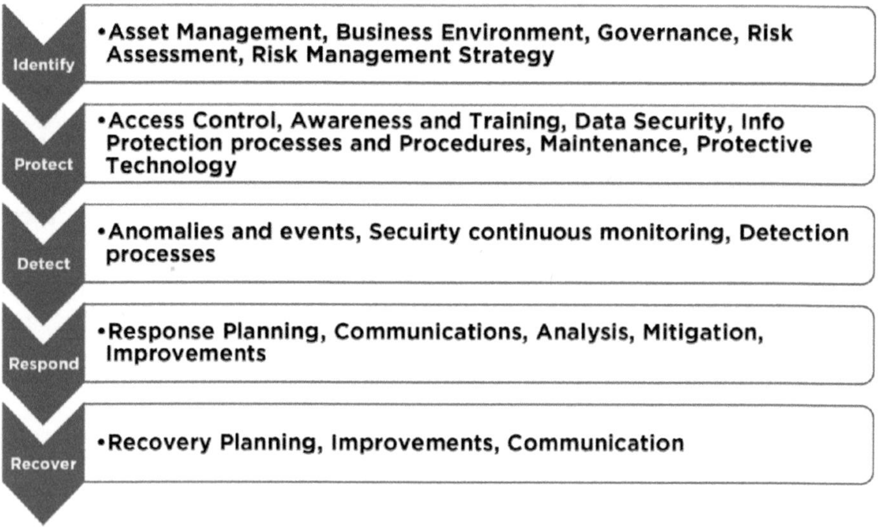

FIGURE 3.10 NIST cyber security framework.

FIGURE 3.11 ISO 27001 compliance cycle.

iii. **FISMA:** FISMA is the acronym for the Federal Information Security Management Act of 2002 by the United States of America. The requirements of FISMA are as follows.

- Inventory information systems
- Prepare a system security plan
- Get Certified and accredited
- Categorize risks
- Evaluate security controls
- Assess risks
- Ensure continuous monitoring

FISMA is a federal law in the United States of America and must be adhered to by enterprises based in America.

iv. **SOC2:** SOC2 was formerly known as Service Organization Control 2. It is similar to other cybersecurity frameworks in terms of protecting information. The processes involved in this cybersecurity framework are described in Figure 3.12.

FIGURE 3.12 SOC2 compliance stages.

v. **GDPR:** GDPR is the acronym for General Data Protection Regulation. This is slightly different from other frameworks. The European Union introduced GDPR to safeguard the general public from data breaches and the misuse of data on enterprise platforms (Figure 3.13).

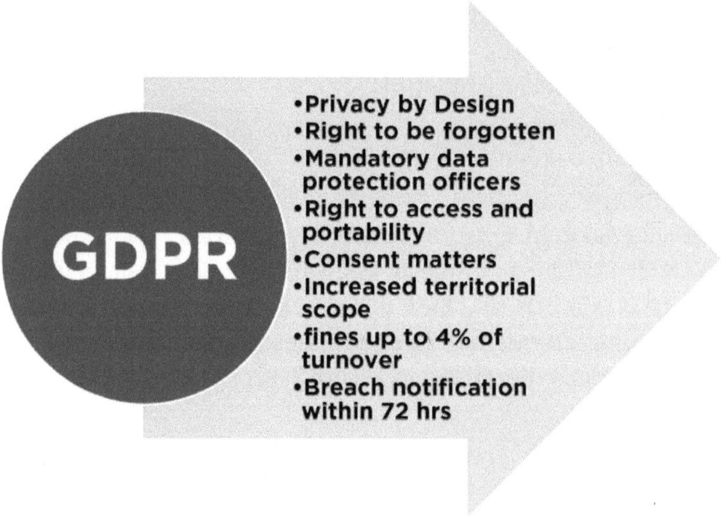

FIGURE 3.13 GDPR compliance steps.

vi. **NERC-CIP:** NERC-CIP is the acronym for North American Electric Reliability Corporation Critical Infrastructure Protection. This framework is used to regulate and manage the bulk electric system (BES) in North America. There are approximately 11 standards that are part of the CIP.

vii. **HIPAA:** HIPAA is the acronym for the Health Insurance Portability and Accountability Act of 1996. This act was mandated by the United States Department of Health and Human Services (HHS) to protect the health information of patients.

3.13 FUTURE OF CYBER SECURITY–AI POWERED

Cybersecurity is one of the most demanded jobs in the world. The world is currently undergoing a technological revolution. There are hackers who exploit this for malicious purposes and profit from it, causing genuine users and enterprises to suffer the consequences of these attacks. The future of cybersecurity relies on experts and enthusiasts in the field. Only through awareness and continuous evolution can cybersecurity be established worldwide, including in space. Unfortunately, there is a lack of awareness about cybersecurity among users, but this is slowly changing as governments and countries implement various cybersecurity initiatives. This momentum should be sustained in the future. AI-based tools for cybersecurity, such as Real Intelligence Threat Analytics (RITA) by Active Countermeasure, utilize constant monitoring and traffic analysis to detect threats. Such AI tools should be developed to actively monitor and protect systems, rather than relying on passive methods. Experts with intricate knowledge of security systems can develop these AI-powered tools.

3.14 CONCLUSION

Cybersecurity is a vast area that encompasses various areas of expertise. Persistence and resilience are essential qualities for survival in this field. Pentesters and CISOs should continuously learn about the latest technologies, vulnerabilities, and mitigation processes to better serve the cybersecurity community. There are numerous resources available for beginners to enhance their skills, such as the MITRE ATT&CK Framework, which provides a comprehensive overview of different attack stages. It is considered essential knowledge for cybersecurity experts.

Cybersecurity covers all fields of computer science. A cybersecurity expert should have a deep understanding of hardware, software, and their internal workings. Capture the Flag (CTF) challenges can help beginners gain practical experience and understand different types of attacks and hacking processes. "Hack the Box" is one such CTF platform.

This chapter serves as a basic guide to cybersecurity, introducing the most common and fundamental knowledge required by cybersecurity experts and enthusiasts. It should be followed by practical approaches in the field of cybersecurity.

3.15 CYBER SECURITY DICTIONARY

A glossary of terms in cybersecurity should be known by any beginner-level cybersecurity enthusiasts. Here are some of the most commonly used keywords [20–46].

Sl. No.	Keyword	Description
1.	Cyber space	Digital World
2.	Cyber security	Security for the digital world
3.	CIO	Chief Information Officer
4.	CISO	Chief Information Security Officer
5.	Pentester	Penetration Tester
6.	Hacker	Malicious Criminal
7.	Ethical Hacker	Hacker with Morals
8.	Red Team	Offensive Techniques
9.	Blue Team	Defensive Techniques
10.	Purple Team	Both Offensive and Defensive Techniques
11.	Vulnerability	Weakness in system
12.	Bug	Unintended features or misbehaviors of system
13.	Exploit	Attacking the vulnerability
14.	Threat	Way to intrude the system
15.	Cyber Attack	Attack on the digital world
16.	Mitigation	Countermeasures
17.	Risk	Possible loss
18.	OSI Model	Open systems interconnect model
19.	Layer 1	Physical layer
20.	Layer 2	Data link layer
21	Layer 3	Network layer
22	Layer 4	Transport layer
23	Layer 5	Session layer
24	Layer 6	Presentation layer
25	Layer 7	Application layer
26	ML/AI	Machine Learning / Artificial Intelligence
27	MAC Address	**Media access control address**
28	IP Address	Internet Protocol Address
29	VPN	Virtual Private Network
30	Router	connects networks
31	Network Switch	connects devices
32	Network Packets	Small segment of large message
33	TCP	Transfer Control Protocol
34	UDP	User Datagram Protocol
35	FTP	File Transfer Protocol
36	Ports	Software defined number to network protocol
37	Proxy server	Intermediary between client and server

(Continued Table)

Sl. No.	Keyword	Description
38	TOR Browser	The onion ring browser
39	Dark Web / Deep Web	Hidden collective of website accessible by TOR
40	Phishing	Fake interface or communication
41	Vishing	Voice phishing
42	Spam	Unsolicited digital communication
43	Risk Rating matrix	To measure risk
44	Kali Linux	Penetration Testing OS
45	Protocols	Set of rules and regulations
46	GDPR	General Data Protection Regulation
47	Server	Provides service
48	Client	Requests service
49	Open Source	Free and source code available
50	Terminal	Command line interface
51	GUI	Graphical user interface
52	GNU	Unix like system
53	NIST	National Institute of standards and Technology
54	GRC	Governance Risk Compliance
55	ISO 27001	Cyber security framework of ISO
56	SOC2	Cyber security framework
57	NERC-CIP	Cyber security framework
58	HIPAA	Cyber security framework
59	FISMA	Cyber security framework
60	Cybrary	Ethical Hacking Platform
61	CTF	Capture the Flag
62	Hack the Box	CTF Platform
63	ctf.live	CTF Platform
64	Bug Bounty	Hunting
65	Kevin Mitnick	Greatest Hacker in History
66	Encryption	Plain text to cipher text
67	Decryption	Cipher text to plain text
68	End to End encryption	Encrypted communication
69	Security Audit	Checking security Compliance
70	Confidentiality	Authorized access
71	Integrity	Authorized changes
72	Accessibility	Availability

(Continued Table)

Sl. No.	Keyword	Description
73	MITRE ATT&CK	Framework of attacks
74	CAPEC	Common attack pattern enumeration and classification
75	CVE	Common vulnerability and Exposures
76	CWE	Common weakness Enumeration

KEYWORDS

- **chief information security officer**
- **confidentiality, integrity, and**
- **cyber security & attacks**
- **cyber security frameworks**
- **internet of things**
- **threat modeling**

REFERENCES

1. Jangirala, S., Ashok, K. D., & Neeraj, K. (2019). *Future Generation Computer Systems, 92*, 178–188, https://doi.org/10.1016/j.future.2018.09.063.
2. Popescul, D., (2011). The confidentiality–integrity–accessibility triad into the knowledge security. *A Reassessment from the Point of View of the Knowledge Contribution to Innovation, 4,* 1338–1345.
3. Munaiah, N., Rahman, A., Pelletier, J., Williams, L., & Meneely, A., (2019). Characterizing attacker behavior in a cybersecurity penetration testing competition. In: *ACM/IEEE International Symposium on Empirical Software Engineering and Measurement (ESEM).*
4. Mancini, S., Iacono, L., Hartle, F., Garfinkel, M., Horn, D., & Sullivan, A., (2021). Introducing the common attack process framework for incident mapping. *Int. J. Cyber. Res. Edu. (IJCRE).*
5. Kolokotronis, N., & Shiaeles, S., (2021). *Cyber-Security Threats, Actors, and Dynamic Mitigation* (1st edn.). CRC Press. https://doi.org/10.1201/9781003006145.
6. Asaad, R. R., (2021). Penetration testing: Wireless network attacks method on kali Linux OS. *Academic Journal of Nawroz University, 10*(1), 7–12. https://doi.org/10.25007/ajnu.v10n1a998.
7. Xiong, W., Legrand, E., Åberg, O., & Lagerström, R., (2021). Cyber security threat modeling based on the MITRE enterprise ATT&CK Matrix. *Software Syst. Model.*
8. Upadhya, A., & Srinivas, B. K., (2020). *A Survey on Different Port Scanning Methods and the Tools Used to Perform Them*. academia.edu.

9. McNabC, (2007). *Network Security Assessment* (2nd edn.). O'Reilly, USA.
10. Velu, V. K., (2017). *Mastering Kali Linux for Advanced Penetration Testing.* Packt Publishing Ltd.
11. Najera-Gutierrez, G., & Ansari, J. A., (2018). *Web Penetration Testing with Kali Linux: Explore the Methods and Tools of Ethical Hacking with Kali Linux* (3rd edn.). Packt Publishing Ltd.
12. Jon, E., (2008). *Hacking: The Art of Exploitation* (2nd edn.).
13. Kevin, M., & William, S. L., (2012). Ghost in the Wires: My Adventures as the World's Most Wanted Hacker Paperback–Illustrated. ISBN-10–0316037729 ISBN-13, 978-0316037723.
14. Yuri, D., & Erdal, O., (2018). *Cybersecurity? Attack and Defense Strategies: Infrastructure Security with Red Team and Blue Team Tactics.* Packt Publishing Ltd.
15. Ivan, K., & Stjepan, G. (2020). Red Teams–Pentesters, APTs, or Neither. pp. 1242–1249.
16. John, Y., (2013). *Using Penetration Testing to Enhance Your Company's Security.* Trust wave.
17. https://www.synopsys.com/blogs/software-security/threat-modeling-guide.html (accessed on 27 January 2024).
18. FS (2021). TCP/IP vs. OSI: What's the Difference Between the Two Models? https://community.fs.com/article/tcpip-vs-osi-whats-the-difference-between-the-two-models.html (Accessed on 27 January 2024)
19. https://bytesofgigabytes.com/networking/osi-model/(accessed on 27 January 2024).
20. MITRE Homepage. https://www.mitre.org/(accessed on 27 January 2024).
21. ATT&CK Homepage. https://attack.mitre.org/(accessed on 27 January 2024).
22. CAPEC Homepage. https://capec.mitre.org/(accessed on 27 January 2024).
23. CVE Homepage. https://cve.mitre.org/(accessed on 27 January 2024).
24. CWE Homepage. https://cwe.mitre.org/(accessed on 27 January 2024).
25. Kali Linux Homepage. https://www.kali.org/ (accessed on 27 January 2024).
26. Nmap Homepage. https://nmap.org/(accessed on 27 January 2024).
27. Metasploit Homepage. https://www.metasploit.com/(accessed on 27 January 2024).
28. Vscan Github. https://github.com/xen0vas/vscan (accessed on 27 January 2024).
29. https://blog.convisoappsec.com/en/code-review-and-pentest-what-they-are-and-when-to-use-them/ (accessed on 27 January 2024).
30. https://flatironschool.com/blog/how-to-get-into-cyber-security/ (accessed on 27 January 2024).
31. https://www.zdnet.com/article/google-paid-6-7-million-to-bug-bounty-hunters-in-2020/(accessed on 27 January 2024).
32. https://www.ptsecurity.com/ww-en/analytics/external-pentests-results-2020/(accessed on 27 January 2024).
33. https://www.hackthebox.eu/ (accessed on 27 January 2024).
34. https://www.ctf.live/ (accessed on 27 January 2024).
35. https://tryhackme.com/ (accessed on 27 January 2024).
36. https://www.hackerone.com/ (accessed on 27 January 2024).
37. https://www.bugcrowd.com/bug-bounty-list/ (accessed on 27 January 2024).
38. https://medium.com/omise/an-analysis-of-omises-bug-bounty-program-b242c069a531 (accessed on 27 January 2024).
39. https://www.exabeam.com/security-operations-center/2020-red-and-blue-team-survey/ (accessed on 27 January 2024).

40. https://www.compact.nl/en/articles/purple-team-drive-defense-with-offense/ (accessed on 27 January 2024).
41. https://pentestmag.com/red-teaming-10000-feet/ (accessed on 27 January 2024).
42. https://gomindsight.com/insights/blog/red-team-vs-blue-team/ (accessed on 27 January 2024).
43. https://sectigostore.com/blog/42-cyber-attack-statistics-by-year-a-look-at-the-last-decade/ (accessed on 27 January 2024).

CHAPTER 4

A Review of Big Data Analytics Techniques to Collar Cybersecurity Concerns

S. GEETHANJALI and S. SASIKALA

Department of Computer Science, IDE, University of Madras, Chennai, Tamil Nadu, India

ABSTRACT

Cyber security has been a growing concern since the inception of the internet. With the increasing number of activities taking place online, the exponential growth of data has led to a surge in potential risks in the form of cyber attacks. Businesses and individuals worldwide are experiencing significant losses due to various cyber attacks targeting their assets. In light of this situation, it is evident why cyber security experts are turning to big data analytics (BDA) to detect, recover from, and prevent various forms of cyber attacks. This chapter aims to educate readers on recent promising applications of BDA in addressing different issues in the cyber threat landscape, such as malware detection, intrusion detection, phishing, ransom ware, and denial of service (DoS) attacks. Understanding how Artificial Intelligence (AI), Machine Learning (ML), and Deep Learning (DL) technologies impact the cyber security process is crucial for developing big data models that can enhance threat detection and mitigation. Additionally, this chapter provides a brief discussion of notable research works in the areas of intrusion detection, malware analysis, fraud detection, and cloud security. Furthermore, it introduces the significant role played by DL in today's cyber security and how it paves the way for future industrialists and researchers to develop efficient solutions for the challenges faced in the digital world. In conclusion, the chapter discusses the challenges involved in the analysis process and

Leveraging Artificial Intelligence (AI) Competencies for Next-Generation Cybersecurity Solutions.
Pethuru Raj, B. Sundaravadivazhagan, & V. Kavitha (Eds.)

outlines the future directions that the field of BDA needs to pursue in order to create more accurate and user-independent models.

4.1 INTRODUCTION

4.1.1 WHAT IS BIG DATA ANALYTICS?

According to Wise from Earthweb [1], "2.5 quintillion bytes of data are being generated every day in 2022." Such a vast amount of data has the ability to transform society and the way it functions. Among the different variety, size, and complexity of data collected and processed lies fruitful information that can be beneficial to millions if it can be properly and correctly harnessed. This is the basic idea behind Big Data Analytics (BDA). So, what exactly is this big data?

The TechAmerica Foundation defines big data as follows: "It is a term that describes large volumes of high velocity, complex, and variable data that require advanced techniques and technologies to enable the capture, storage, distribution, management, and analysis of the information [2]."

With the advancement in computing devices in terms of storage and processing capabilities, the definition of big data has also changed. Though originally characterized by the 3Vs – volume, velocity, and variety, data can now be categorized as big data if it satisfies the six important characteristics. These are Volume, Velocity, Variety, Veracity, Value, and Visualization as depicted in Figure 4.1.

The primary characteristic of big data is its volume, measured in Terabytes (TBs)/Petabytes (PBs)/Zettabytes (ZBs), or by the number of instances, transactions, or files. Furthermore, one of the things that makes data big data is that it comes from multiple sources. This can include anything ranging from values generated by single system software to multiple edge devices. The variety of data used for analytics includes common structured data, unstructured data like text and human language, as well as semi-structured data such as Rich Site Summary (RSS) feeds or eXtensible Markup Language (XML). Data that is difficult to classify, such as audio, video, and medical data, are some of the different types of data that big data is capable of handling. Furthermore, the velocity or speed of massive data acts as an important factor. This indicates the frequency of data creation or data transmission. Streaming data, which is gathered in real-time from websites, is at the cutting edge of big data [4]. Adding on to what was defined by the authors of [2], Table 4.1 gives a brief description of the seven characteristics.

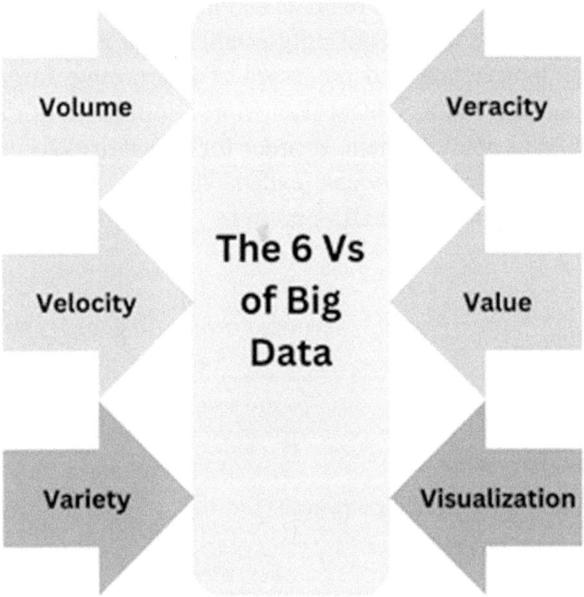

FIGURE 4.1 The six essential Vs of big data.

TABLE 4.1 Characteristics of Big Data

Characteristic	Description	Metrics
Volume	The entire amount of data that is being generated by one/more device (s) to be taken up and processed by a computer for analysis and decision-making	Petabytes (PB), Exabytes (EB) or Zettabytes (ZB)
Velocity	The rate at which data is being produced, accumulated, and consumed to become accessible	Thousands of transactions per hour (TTPH) or Millions of transactions per hour (MTPH)
Variety	The structural heterogeneity of the data being produced from various sources and sent to a device for processing	Structured (tabular), Unstructured (raw text, image, etc.), or Semi-structured (XML)
Veracity	The assurance of credibility of the data being produced from various sources	Good, Bad, Incomplete, Inconsistent, Ambiguous, Latent, or Deceptive
Value	The efficacy of the data collected for processing and converting into information to aid in decision-making	Valuable, Invaluable, Expired (time sensitive)
Visualization	The ability to process the data and depict the information gained using visual aids such as charts and graphs	Correct or Misleading

Understanding the nature of big data and interpreting the analytics results is an art in itself. The six Vs not only establish the nature of the data but also the tools, approaches, and processes of interpreting information. The ultimate goal of BDA is to collect data from a multitude of devices, filter/clean, process, and analyze them in order for enterprises to use the results to provide better products/services, expand themselves, and develop their clientele. Figure 4.2 shows the BDA process.

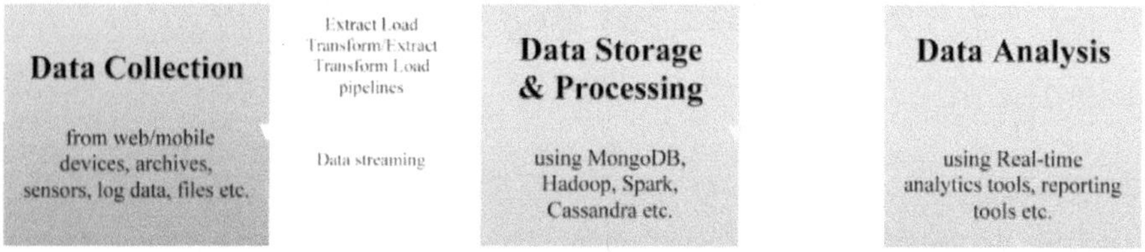

FIGURE 4.2 Simplified view of big data analytics.

4.1.2 OVERVIEW OF CYBER SECURITY

Threats have always loomed over mankind since the dawn of time. They manifest in different ways depending on human activities at that particular point in time. The invention and evolution of programmable computers into modern computers have witnessed various forms of threats attempting to disrupt the ongoing work. As technology advanced, its role in processing information grew exponentially. What initially began as a harmless joke now poses a significant risk to the cyber landscape. It all started with the creeper-reaper, followed by the Morris worm, viruses, trojans, and eventually culminated in ransom ware [20].

A team of researchers and engineers from the Advanced Research Projects Agency (ARPA), the U.S. Air Force, and other organizations collaborated to develop a security solution for detecting software vulnerabilities [20]. This paved the way for the creation of the first Anti-virus software in 1987. As cybercriminals felt the impact of these measures, they began to outsmart these tools with newer, stealthier, and more dangerous threats.

With the advent of connected devices, such as the Internet of Things (IoT), Edge Computing, and Cloud Computing, the security and privacy risks associated with their usage have grown dramatically due to the escalation in the amount of data generated and utilized. News about various cyber-attacks, such as Ransomware attacks, viruses, network intrusions, and data theft, has become quite common. Tackling these cyber-attacks is a challenging task that involves the use of cybersecurity systems.

Cyber security refers to "the act of protecting computing systems – hardware, software, networks, data, and the humans using these systems – from malicious agents by ensuring all security vulnerabilities are properly dealt with."

Microsoft defines a security vulnerability as: "A weakness in a product that could allow an attacker to compromise the integrity, availability, or confidentiality of that product [16]."

Below are a few other definitions of cybersecurity from various sources.

i. According to [17], cybersecurity in the context of big data can be defined as "the component that manages the security of cyberspace, including the data it contains."
ii. "Cybersecurity is the collection of tools, policies, security concepts, security safeguards, guidelines, risk management approaches, actions, training, best practices, assurance, and technologies that can be used to protect the cyber environment, organization, and user's assets [18]."
iii. "Cybersecurity is the processes employed to safeguard and secure assets used to carry information of an organization from being stolen or attacked [19]."

The evolution of threats and the resulting novel threat vectors pose a challenge for security and forensic analysts in identifying and mitigating the havoc they cause. Security analysts and analysis tools need a solid understanding of these threat vectors to uncover and prevent any harm. As Eric Schmidt, former CEO of Google, once said, "*In our digital age, the issues of cybersecurity are no longer for the technology crowd; they matter to us all [13].*"

The impact of COVID-19 across the globe and the resultant lockdowns drove all businesses into a fully online operational mode. Specifically, as employees shifted their workforce to their homes, the usage of public internet services for all activities rose. Countermeasures need to be developed keeping in mind the changing security landscape to integrate work-from-home [14]. Ransom ware was ranked as the number one threat in 2020. Supply chain attacks grew and have impacted more than 18,000 companies, including Fortune 500 companies and government agencies [14]. Banks in the EU and the US were under severe censure after a massive online fraudulent operation was unearthed. At least 16,000 accounts were compromised, and the offenders were able to get away with millions of dollars [15].

4.1.3 A QUICK LOOK AT CYBERSECURITY THREATS

As cybercriminals began to develop new technologies to outsmart and bypass various security measures, such as firewalls and weak antivirus software, companies started to hire more experts in this field and formed internal teams to detect, address, and report threats. With an increasing number of companies collecting Personally Identifiable Information (PII) from their customers, the risk of this data being exposed is significant.

PII is defined as: *"any information about an individual maintained by an agency, including, but not limited to, education, financial transactions, medical history, and criminal or employment history, as well as information that can be used to distinguish or trace an individual's identity, such as their name, social security number, date and place of birth, mother's maiden name, biometric records, etc., including any other personal information that is linked or linkable to an individual [21]."*

Apart from PIIs, other data that are at risk include medical research, defense research, military data, financial records, etc. Some of the most confronted threats are listed below in this section.

4.1.3.1 MALWARE THREATS

Malicious software, better known as malware, includes a range of attack vectors that infiltrate a system through infected links, untrusted downloads, and malicious email attachments [22]. Once inside a system, these programs tend to infect the host machine by replicating themselves, collecting sensitive information, manipulating files, creating backdoors for attackers to access other system contents, blocking access to files, corrupting the system by deleting kernel files, and rendering the system unusable. Some common types of malware are:

- **Virus:** A type of malware that, upon opening or triggering, replicates itself to modify or destroy other files in the system. The ILOVEYOU virus was created to steal passwords of online services, while the Melissa virus was used to spread and overload servers. These are historically significant examples.
- **Worms:** These types of malware replicate themselves in the host machine and create backdoors by exploiting vulnerabilities. They differ from viruses in that worms do not need to be triggered by the host to wreak havoc. Mydoom and Sobig, which led to denial of

several services, are some of the most expensive worms the digital world has faced.

- **Trojans:** A type of malicious code that hides itself in unsuspecting programs and, upon downloading, causes extensive damage to the host. Emotet and Zeus are two notorious trojans that have caused the loss of millions of dollars across financial institutions.
- **Botnets:** This includes a series of connected devices that have already been compromised and contain bots that perform various malicious tasks. The presence of these bots is largely unknown to users, as they are well hidden and often go undetected by threat-detecting tools.
- **Spyware:** These malwares allow access to sensitive information stored in a device. They are nefarious as they mainly target personal information such as contacts, passwords, and payment details using keyloggers. CoolWebSearch is one such spyware that hacks Internet Explorer and sends data to the attacker.
- **Adware:** A type of spyware that records a user's web activities to display more targeted advertisements. While mostly harmless, these operate without the user's consent and often involve selling the user's information. Fireball serves as a notable example of adware, although it posed significant risks due to its ability to function remotely and facilitate the download of malicious code.
- **Ransomware:** This is the most serious type of threat in the list of malware. Upon entering a host system, it encrypts all the files and blocks users from accessing their data or system. The key required to decrypt and retrieve the data is held by the attackers, who demand a ransom for the key. However, whether the attackers will share the correct key or not is a question of concern. The infamous WannaCry and Crypto Locker ransomware were able to extract millions of dollars' worth of bitcoins from users since their release.
- **Cryptojacking:** This type of malware became common after the growth of cryptocurrency mining. Attackers attempt to gain access to a host computer and utilize its resources for mining cryptocurrencies.

4.1.3.2 MAN-IN-THE-MIDDLE (MITM) ATTACKS

Attacks that involve compromising the communication between two authenticated parties are termed Man-in-the-Middle (MitM) attacks [22]. Some of the common types of MitM attacks include:

- **Wi-Fi Eavesdropping:** The attacker poses as a genuine Wi-Fi connection agent and allows users to join and use the Wi-Fi to track their movements across the web, targeting monetary transactions.
- **Email Hijacking:** The assailant poses as a legitimate company and creates fake emails to gather sensitive data from users.
- **DNS Spoofing:** A spoofed Domain Name Server (DNS) redirects the users of a website to a malicious site.
- **IP Spoofing:** Similar to DNS spoofing, the only difference is the IP address that is spoofed.

4.1.3.3 DENIAL OF SERVICE (DOS) / DISTRIBUTED DENIAL OF SERVICE (DDOS) ATTACKS

DoS bots are trained to overload servers to their maximum capacity. This results in a server crash and renders it temporarily unavailable. The attackers make use of this temporary window to conduct any malicious activities. When such an attack happens with multiple devices, it is termed a DDoS.

4.1.3.4 SOCIAL ENGINEERING

This class of attacks involves misleading or cheating users into revealing sensitive information to gain access to systems or data [22]. Some common attacks include:

- **Phishing:** The attacker sends emails, text messages, or voice messages, but conceals them to appear as if they are from the original authentic source.
- **Blackmail:** The attacker gains an advantage over authenticated users and uses it to blackmail them into providing access to sensitive information.

4.1.3.5 INJECTION ATTACKS

This class of attacks attempts to introduce malicious content into a program by exploiting known vulnerabilities. The primary objective of these attacks is to gain access to sensitive data or execute other types of attacks on a system or network. The following are some significant injection attacks.

- **SQL Injection:** If user inputs into forms are not properly validated, malicious agents may write critical Structured Query Language (SQL) code. When this code is processed, it executes the query and can cause unintended reactions on the stored data. As No-SQL databases are becoming more popular, a similar technique is also employed on those as well.
- **Code Injection:** Attackers attempt to insert malicious code into programs in an effort to disrupt the natural flow of execution.
- **Cross-site scripting (XSS):** A type of attack where JavaScript code is inserted into web forms, allowing attackers to gain access to the system or steal information.

4.1.3.6 ZERO-DAY EXPLOITS

This is an attack on systems whose software or hardware vulnerabilities are unknown to the users or developers. These attacks are difficult to identify, as traditional Anti-Malware Tools (AMT) detect malware or other similar attacks using existing signatures [23].

4.1.3.7 ADVANCED PERSISTENT THREATS (APT)

As the name suggests, these types of attacks are persistent. All highly secure organizations fall victim to this attack [24]. Though this requires a great deal of patience and stealth from the hackers, the bounty is proportionally worth it.

4.1.4 CYBERATTACK SOLUTION STRATEGIES

Counteracting the aforementioned threats are various solution strategies, some of which are listed below. It should be noted that no single solution may be sufficient to provide complete protection against threats, given the vastness of the threat landscape. Therefore, individuals and organizations should employ a combination of multiple defense solutions to safeguard themselves.

4.1.4.1 INTRUSION DETECTION SYSTEM (IDS)

IDSs are essential for detecting if an attacker has compromised systems and gained access to private data. Knowledge-based IDSs [29] employ a database of signatures or footprints of known attacks and use them to compare with new signatures to determine if an activity is legitimate or not. Their downside is that these systems can only identify known threats and also require endless database updates to include the signatures of newly emerged threats. In contrast, anomaly-based IDSs [29] are much simpler to work with as they store the profile of the network under normal usage and compare this signature with others to detect a deviation from the normal baseline behavior and hence recognize old and emerging threats efficiently.

4.1.4.2 FIREWALLS

Firewalls are employed to protect private networks from unwanted intrusions. They employ a screening router that monitors the network traffic and filters the data packets it receives, granting or revoking access to the network [57]. All whitelisted IPs are granted access based on the firewall configuration. Anything that looks suspicious is blocked from accessing the network. Web Application Firewalls (WAF) are a type of firewall that is tailored to protect web applications hosted on a server from attacks over the internet, such as XSS and SQL Injection, among others. They have become common in many web applications over the past decade.

4.1.4.3 ANTI-VIRUS

An antivirus is software specially designed to detect, isolate, terminate, and remove any malware detected on a computer. It can be used to run periodic scans to identify security loopholes and rectify them immediately. It also scans web pages in advance and blocks access to suspicious sites. These tools detect malware using available malware signatures stored in their database. Antivirus software needs to be regularly updated to include the signatures of the latest threats in order to provide protection against upcoming dangers.

4.1.4.4 HONEYPOTS

These tools serve as decoys manufactured to lure criminals towards them and therefore away from legitimate targets [58]. When deployed in a secure environment, honeypots are also capable of capturing information about the attackers, their methods of attack, and other details that can assist cybersecurity specialists in reconnaissance of potential threats that the network or server may face. They are commonly filled with enticing fake information that hackers attempt to obtain, and generally have some security vulnerabilities to facilitate easy access for the attackers, although this may not always be the case [58].

4.1.4.5 *SOFTWARE PATCH UPDATES*

Software patches and updates are critical components of the digital safety and security landscape. They are designed to fix vulnerabilities and backdoors in software while also introducing new features [59]. It is important to address software vulnerabilities identified by vendors or customers promptly, in order to prevent attackers from exploiting them. These security tools are generally simple and often available for free.

4.1.4.6 *VIRTUAL PRIVATE NETWORKS (VPN)*

Initially established to protect sensitive information shared with remote workers, VPNs are a means of establishing secure and private connections to the internet [60]. Data is passed around through an encrypted tunnel, hiding the originating IP address and physical location. They are a great means of security as they protect the identity of individuals, especially over public Wi-Fi networks, which are desired by attackers to steal sensitive credentials via MitM attacks [60].

4.1.4.7 *MULTI-FACTOR AUTHENTICATION (MFA)*

Authentication is the practice of verifying the identity of a user/entity to gain access to data/services. Passwords are the simplest means of digital safeguarding against illegitimate access. However, passwords alone as a single-factor authenticator may not provide the necessary protection for highly sensitive data [61]. Hence, MFAs are a great improvement in ensuring security as they require two or more forms of authentication by legitimate users to gain access. Typically, a knowledge factor (passwords), an ownership factor (tokens, identity cards), and a biometric factor (fingerprint, face recognition) are employed in an MFA system [61]. An attacker who wants to compromise a system needs to bypass multiple layers of security, some of which are frankly difficult to obtain.

4.1.5 THE IMPORTANCE OF BIG DATA ANALYTICS IN CYBERSECURITY

Cyber-attacks are a growing concern in the modern world as different threat agents seek to disrupt the smooth flow of data within organizations. As security measures continue to evolve, so do the threats, and at an incredibly rapid pace. Since the advent of data analytics, researchers have been exploring the various fields where this new technology can be applied. Big Data Analytics (BDA) is one of the most promising areas of research, enabling easy monitoring, identification, and prevention of cyber-attacks using tools such as Machine Learning (ML), Artificial Intelligence (AI), and Deep Learning (DL). It is clear that large volumes of historical data on various threat vectors and their signatures play a crucial role in identifying both existing and emerging threats. Manual analysis of this data is nearly impossible due to its sheer size and complexity.

Conventional cybersecurity systems tend to focus on the current state of a system and how to maintain it. The vast amount of data available for processing, although complex, presents a significant opportunity to detect and prevent malicious activity. By utilizing an intelligent system powered by Big Data Analytics (BDA), anomalies can be detected and systems can be protected from unintended usage by malicious agents [12]. Therefore, automating the analysis process became essential. This is how data analytics was applied to the field of cybersecurity, and as the database grew to exabytes and petabytes, BDA emerged as the solution. Analysts can uncover valuable insights by leveraging data from networks and computers.

The relevance of big data lies in its potential to provide valuable information and expertise for decision-making. Big data is a valuable asset with a wide range of applications in the field of cybersecurity, particularly in managerial decision-making activities. These decisions are driven by factors such as monetization, growth, and research. The vast amount of data from different sources such as scanners, sensors, social media platforms, and the web has the potential to assist in decision-making. However, this can only be achieved if the data is properly processed and within any time constraints that may exist. This data enables organizations to leverage historical and real-time data to make informed decisions and counter any adversaries. Businesses that can effectively harness the implications of big data may outperform their competitors [5]. The following is a list of recent applications of BDA (Figure 4.3) in combating various cyber threats.

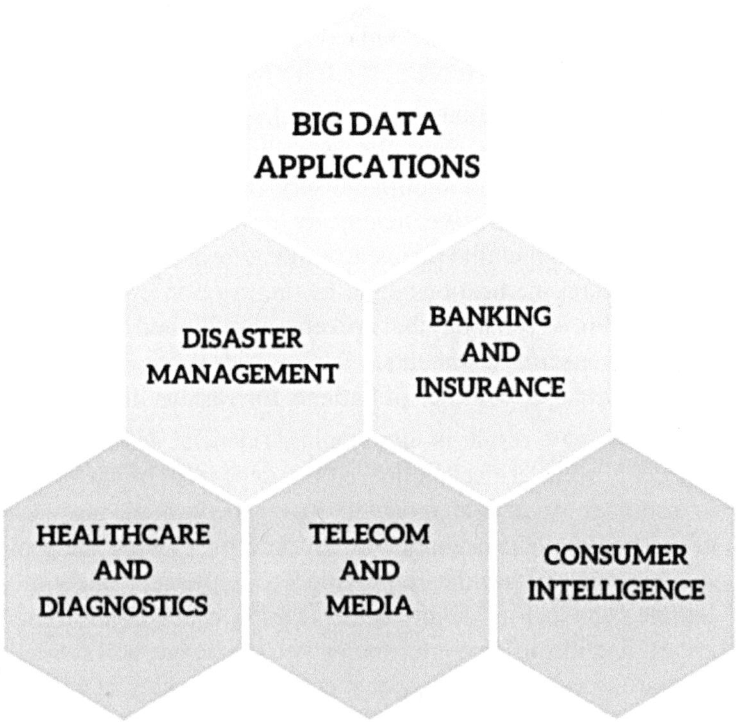

FIGURE 4.3 Big data applications.

1. **Big Healthcare and Diagnostics Data:** The digitization of all clinical examinations and medical records in healthcare systems has become a widely adopted practice since the emergence of computer systems. These Electronic Health Records (EHR) enable different doctors to access a patient's complete medical history. Big healthcare data includes medical diagnoses, prescriptions, information about known allergies, demographics, clinical narratives, and results from various laboratory tests [6]. These data are utilized to enhance patient health, disease surveillance, drug research, and managerial-level policy-making [25]. However, these data are vulnerable to cyber attacks due to their valuable information, which can be exploited by unauthorized access to a healthcare provider's database. Additionally, these data

are also susceptible to attacks like ransomware, as discussed in Ref. [26], which can result in data or financial loss for hospitals, or even disrupt their operations. Individual patient records may be sold on the dark web for significant sums of money, leading to the growing threat of medical identity fraud. Medical identity fraud involves the illegal acquisition of one or more of the following [63]:

i. To access complimentary medical services such as costly laboratory tests, treatment for incurable illnesses that may cause financial strain, or significant surgical procedures.
ii. To access expensive healthcare devices such as pacemakers, stents, and metallic plates, to name a few..
iii. To obtain medications such as prescription opioids (morphine, Vicodin, oxycodone) that are otherwise limited to patients undergoing specific treatments.
iv. to steal false identities of patients for various illegal uses.

These may result in devastating financial losses for patients, especially when critical clinical care is dependent on their life savings or insurance. Apart from monetary loss, patients might also face denial of desperate medical treatments. In order to prevent these mishaps, it is essential for healthcare providers to implement safety measures. Simple steps such as identity authentication, access control, encrypting critical and identifiable information before storage, and storing data in separate servers go a long way in ensuring the safety of EHRs.

2. **Disaster Management:** The damage to life and property caused by various disasters can be managed or, at times, prevented by detecting them as early as possible. In this event, the BDA (Big Data Analytics) plays a major role in the early detection of disasters from meteorological or industrial data, thus helping the community prepare to face the disaster while preventing or minimizing losses [7]. Therefore, a reliable real-time analytics platform is necessary to monitor and detect disasters [8].

   Social media data, such as tweets, posts, statuses, etc., serve as the primary source of data for natural disasters. Man-made disasters can be effectively averted by implementing proper access controls in the systems. In industries, multi-agent systems like Supervisory Control and Data Acquisition (SCADA) and Digital Fault Recorders (DFR) are utilized for controlling, monitoring, and automating power systems, grids, and plants to prevent unauthorized usage and rely on big data [27]. By employing appropriate analytics on this data, engineers can identify any faults and trace their origin to rectify them before any catastrophe, such as power outages, radiation leakage from power plants, or even nuclear disasters, could occur when sensitive information is compromised.

3. **Consumer Intelligence:** BDA has a ton of potential for consumer intelligence and can help businesses like retail, finance, and telecoms. Big data has the potential to increase transparency and make important data more easily accessible to stakeholders [9]. Customer-generated data can help organizations profile the mindset and choices of consumers. This results in an increase in consumer satisfaction and therefore consumer retention. The likes or dislikes of consumers can also help companies in product quality control and take action to save their business [4]. Though such business intelligence is great from a business perspective, from an end user's perspective, the algorithms utilized in making predictions or recommendations for the user tend to deprive the consumers of their autonomy in choosing what they want [67]. The very welfare of users is at stake when technology overtakes humans and algorithms start to dictate what we need to do. It is crucial to understand that such algorithms give suggestions to users with the aim of engaging them for a longer period [67]. But a great deal of data is required at the back of it all. This could be extended to a variety of areas such as users' online footprints – such as watch history, and browsing history in the case of web applications. Smart IoT home devices that are targeted to ease a person's life are also tracking every movement and decision taken by the users. Sensors track temperature preferences, voice, pulse, and physical movement, to name a few. Social media platforms collect millions of gigabytes per day, and most of it consists of personally identifiable information (PII). With the potential invasion of user privacy, any improper use of this data calls for action from governments. Although laws are established and incorporated into the constitution, it is ultimately the responsibility of companies to protect this data. Typically, all this data is stored on a server and then utilized by algorithms to provide recommendations. The primary threat in this area is the exposure of customer data, such as personal and banking information. This can occur due to voluntary data release by companies or compromising cyberattacks on company servers. Another concern is how companies tend to share their data with third-party service providers, which violates user privacy. Each piece of data out of the 736 data pieces generated by a user in just a

day can contribute to building a comprehensive profile of who they are [68]. Data theft of consumers leads to identity theft, fake profile creation, impersonation, social engineering attacks, and other societal issues.

4. **Banking and Investment:** One of the crucial sectors that has always been under the eye of attackers is the banking, insurance, and investment industries. As banks rely on technology to conduct their day-to-day transactions, financial organizations across the globe face one problem or another owing to cyber attacks. Using multiple devices such as computers and mobile devices to perform banking transactions increases the risk [28]. Typically, these organizations under-report attacks due to fear of loss of reputation. Fraud and theft are common issues, with phishing being the most prevalent. Big data can be used to detect and monitor money laundering, documentation fraud, fraudulent transactions, and thereby suspend these activities by blocking the transaction and the associated accounts [28].
5. **Telecommunications and Media:** The telecom and media sectors have greatly benefited from the growth of the internet. However, they are also vulnerable to cyber threats. Scammers attempt to gain unauthorized access and manipulate or hold hostage sensitive telecom data, including customer records, encryption keys, access credentials, personally identifiable information (PII), resource usage logs, and the resources themselves. The media segment also faces these challenges, in addition to being targeted for data theft due to the valuable nature of the information they handle. The consequences of such attacks are similar to those discussed in the customer intelligence section.

4.1.6 A REVIEW OF BIG DATA TECHNIQUES FOR CRACKING CYBERSECURITY CONCERNS

In this section, some of the latest analytics techniques and models used to address cybersecurity concerns are discussed. These techniques are broadly classified into four categories (Figure 4.4) based on their types:

i. Intrusion detection strategies;
ii. Malware detection strategies;
iii. Fraud detection strategies; and
iv. Cloud/IoT security strategies.

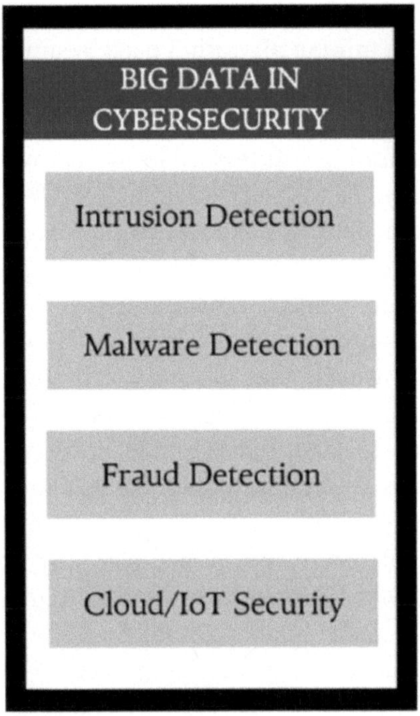

FIGURE 4.4 Applications of big data in cybersecurity.

4.1.6.1 USAGE OF BIG DATA ANALYTICS FOR INTRUSION DETECTION

The rest of this section provides a concise overview of recent advancements in the utilization of IDSs for anomaly detection.

A basic IDS can be enhanced by using BDA that could utilize a simple pattern-matching scheme similar to rule-based expert systems, where the knowledge of humans is replicated in the form of rules. Siadati & Memon [31] proposed a similar approach that can effortlessly detect malicious logins from within an enterprise network. This is achieved using a Network Login Structure, which models authentic user login characterization. When attackers, who gain access to networks via Credential-based Lateral Movement (CLM), try to log in to a different server, these login attempts do not fall under the normal access characterization. This could be due to the fact that the attacker can only access compromised systems which may be part of a different cluster or connected with a different sub-network. Hence, they are flagged as anomalous logins. The system architecture uses two main components – a pattern miner and a semi-supervised classifier. The pattern miner is capable of taking login data and devising a pattern for all normal logins. The fast pattern mining algorithm used is similar to association rule mining. The output of the pattern miner is passed to the classifier, which makes use of an algorithm to detect any deviation from the said pattern and flags the login transactions accordingly. The system was subject to synthetic attacks and was found to be 82% effective in detecting malicious logins.

Traditional intrusion detection models employ Support Vector Machines (SVM) as the base model. This can be attributed to their speed, simple generalization nature, and ability to handle higher dimensions of data with ease. They are much more capable of learning large sets of patterns than rule-based models. However, they do pose a problem since they belong to the category of supervised learning algorithms. This means that human intervention is needed to label existing data to ensure smooth operation. SVMs also require large training sets and consume more time, making them computationally expensive.

In Ref. [32], Othman et al. developed a system that can detect intrusions using the Spark-chi-SVM model. The first part of the model, Spark, involves the usage of Apache Spark to build Resilient Distributed Databases (RDD) from the given data. Preprocessing procedures are applied, followed by feature selection approaches achieved by engaging the ChiSq Selector, which is then trailed by an SVM classifier. The model was evaluated using the area under the receiver operating characteristic curve (AU-ROC) and was found to be better than its peers reviewed in the literature.

Du et al. [30] proposed the architecture of Deep Log, a deep neural network (DNN) that utilizes a stream of system log information such as timestamps, log keys, and other parameters to detect anomalies. It achieves this by modeling the sequence of log entries using Long Short-Term Memory (LSTM), a type of Recurrent Neural Network (RNN) that can remember the sequence of decisions taken by the model. Density clustering is also employed to cluster similar log entries.

This model is capable of adapting to new changes in the environment and producing accurate results. It achieves this through a feedback system that informs the model which parameters need to be adjusted to improve the network weights. As the system behavior evolves over time, the model ensures that the behavioral patterns are stored online, preferably in a cloud environment, as a database of signatures.

Deep Log, implemented using Keras and TensorFlow, can be trained using less than 1% of the total instances. Traditional log analytics tools such as Principal Component Analysis (PCA), Invariant Mining, and N-gram are compared with the performance of Deep Log. It was found that Deep Log outperformed the others by generating the fewest number of false positives and false negatives.

This model can also be utilized in network Intrusion Detection Systems (IDSs) to detect intrusions and identify Blind Return Oriented Programming (BROP) attacks on servers.

An even better intrusion detection rate can be achieved using a hybrid deep learning (HDL) approach. As implemented in their paper, Al and Dener [34] apply a combination of Convolutional Neural Network (CNN) and LSTM as an HDL network. It is also worth mentioning that the authors have also applied a balancing algorithm as most of the data concerning IDS logs tend to be highly imbalanced as the scales weigh more towards the normal transactions whereas they seldom weigh down on the other side. For this purpose, again a combination of algorithms was used, those being the Synthetic Minority Oversampling Technique (SMOTE) and Tomek-Links, collectively referred to as STL. The authors were also able to show good performance of the HDL-STL model in terms of accuracy and F-Score.

Sriram et al. [35] conducted an experiment on implementing Deep CNNs (DCNN) to detect malignant transactions in an IDS. The CNN component of the model handles feature extraction and utilizes a 1D filter for dimensionality reduction through max pooling. The model's performance is further improved by applying batch normalization to enhance training speed. Additionally, a dropout layer is implemented to deactivate unnecessary neurons in the model, and all layers use a ReLU activation function. The results demonstrated that DCNN outperformed IDSs that employed regular CNN, DNN, and even LSTM models in detecting attacks.

In one of the latest techniques proposed, Ponmalar and Dhanakoti [36] devised an ensemble model that combines Chaos Game Optimization (CGO) with Support Vector Machines (SVM) to create a successful decision-making strategy. This strategy reduces processing time by almost half, while improving attack detection accuracy and speed. The use of CGO aims to avoid caching irrelevant instances and only utilize those near the SVM boundary that contribute significantly to instance classification. As a result, the dimensionality of the data is also significantly reduced. CGO also plays a crucial role in tuning the hyperparameters of SVM classifiers.

A more cost-sensitive approach was proposed by Gupta et al. [37], which utilizes a 3-layer Cost Sensitive Deep Learning and Ensemble algorithms in IDS (CSE-IDS). Layer 1 is responsible for segregating normal and suspicious network traffic using Cost Sensitive DNN. The suspicious traffic is then processed by Layer 2, where it is categorized into major attack classes or minor classes of attacks that are less common or not evident to the community.

This is achieved with the assistance of a Boosting ensemble consisting of eXtreme Gradient Boosting (XGBoost). The set of attacks classified as minor undergo further classification in Layer 3, which determines the exact type of threat. This step is accomplished by a Bagging ensemble, typically using Random Forest (RF). Data imbalances are addressed by employing Random Oversampling in Layer 2, while SVM-SMOTE is used in Layer 3.

Figure 4.5 represents an environment that utilizes IDS for detecting unwanted or harmful requests and blocking access to them. The IDS is connected to the Local Area Network (LAN). The firewall between the internal and external organizations filters and forwards only whitelisted IP addresses. When uncertain, the IDS relies on an ML or DL model trained using historical big data.

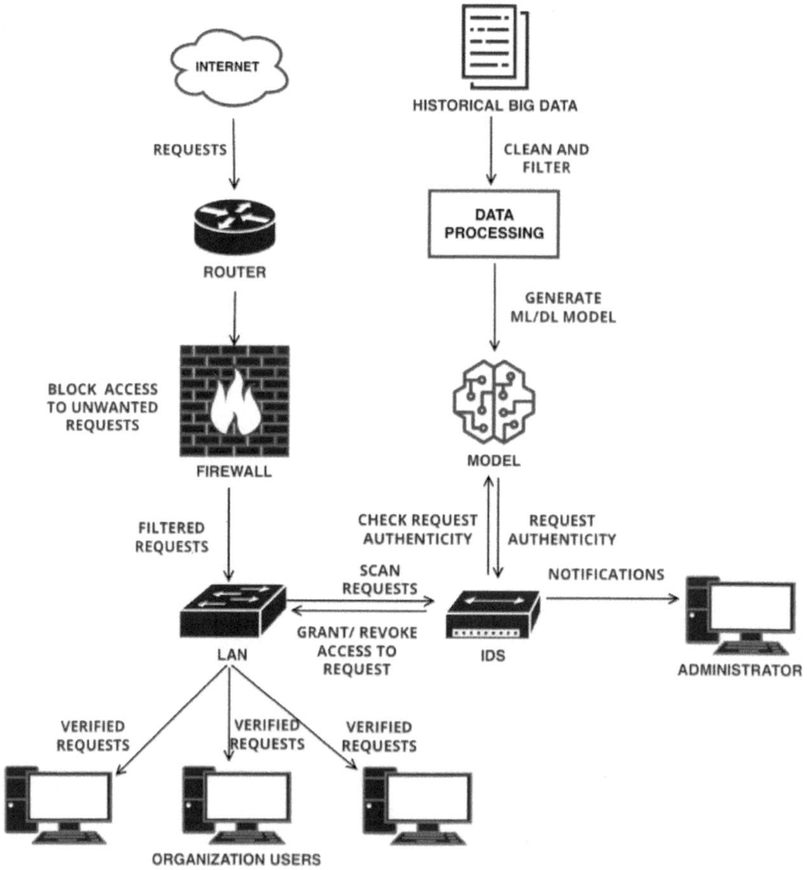

FIGURE 4.5 Working on an intrusion detection system using big data.

4.1.6.2 USAGE OF BIG DATA ANALYTICS FOR MALWARE DETECTION

The presence of malware can have catastrophic consequences on systems, networks, or an organization if left unnoticed and unmonitored. While it is nearly impossible to prevent all devices from encountering malware, there is a high chance that these malware can be prevented from causing harm if they are detected well in advance of their execution. The following sections present a few carefully selected techniques that utilize BDA for this purpose. A wide range of strategies are discussed, along with their benefits and drawbacks.

One well-known approach that has been in use for a long time, and has also been referenced in Ref. [41], is the use of a cloud-based hybrid of static and dynamic methods for detecting malware. The static part involves a DNN that utilizes an unsupervised stacked denoising autoencoder to estimate the model parameters. The obtained model is then trained using a supervised back propagation approach, implementing Adam stochastic optimization. Dynamic analysis is employed if the static analysis does not provide accurate classification results. The uncertain samples undergo dynamic analysis using Cuckoo sandbox with API-n-grams for call sequencing, along with an RF classifier to detect malware samples. As new samples are analyzed, their signatures become more familiar to the model and are added to the list of existing signature samples. Thus, a continuous learning process is employed here, which the authors suggest is the most effective approach.

Azmoodeh et al. [45] proposed a method to detect botnets using Eigen-value weighting. In this work, graph clustering, a common approach to identifying botnets, is implemented through spectral clustering. Two popular algorithms, S&T and IU-RED, are deployed. To ensure clarity, the proposed method expands its active eigenspace to k eigenvectors. Their work has demonstrated good results when tested on benchmark datasets. The proposed methodology incorporates three variations of the optimization functions. The authors introduce two measuring metrics, namely Purity and Proportionate Eigenvector Direction Keeping (PEDK), which effectively estimate the functionality of the distinct optimization functions.

Usman et al. [38] designed a novel hybrid method that combines dynamic malware analysis, cyber threat intelligence, and data forensics with data analytics to identify malicious IP addresses. This method was conceived to combat various drawbacks faced by the existing systems, such as identifying and classifying zero-day attacks and reducing the rate of false positives (FPR). To this extent, the authors proposed an Intelligent Dynamic Malware

Detection using Machine Learning (ML) in IP Reputation for Forensics Data Analytics (IP Rep–FDA).

Any suspicious behavior and the pings made by IPs in the network are chronicled by honeypots deployed in Docker and analyzed in a Cuckoo Sandbox. They are then compared with existing lists of malware to extract zero-day malware. The zero-day malware is subject to a decision tree algorithm, which is able to label the records. The severity scores of these labeled zero-day malware are determined by employing a kill chain methodology.

To determine the trustworthiness of an IP address, a few factors such as the risk associated, confidence, lifespan, and general behavior are also analyzed. This analysis helps in reducing false alarms to a great extent. The proposed framework is much more efficient in detecting malware than the existing approaches by the authors.

In reference [39], the authors proposed a novel approach to detect zero-day malware in an Android environment with minimal human interaction during the detection process. They developed a multi-view deep CNN that utilizes three sets of raw inputs: app permissions, low-level opcodes, and Android API packages, all extracted from individual apps. Four benchmark datasets were used, with Intel Security and Malgenome datasets used for training and hyper parameter tuning, and AMD and Drebin datasets used for testing the feature sets. The first CNN learns to distinguish malicious opcodes from benign ones using opcodes as input, while the second CNN uses API sequences. The third CNN focuses on learning app permissions from the corresponding manifest file. The performance of single-view and multi-view models was compared based on their ability to classify apps as malicious or not, with the multi-view models outperforming the former in terms of accuracy, F1-score, and FPR. The model was also used to demonstrate the effectiveness of zero-day experiments and showed excellent performance.

Bozkir et al. [40] conducted a study on identifying and detecting malware, employing a two-phased approach. In the first phase, they utilized GIST and Histogram of Gradients (HOG) image descriptors on resized and squared RGB binary memory dump images to extract labeled features. These features were then fed to different classifiers to ensure accurate classification. GIST outperformed HOG, but the combination of GIST+HOG yielded even better results. The second phase of their work focused on identifying zero-day attacks. To address the inherent imbalance in the dataset and prevent overfitting, the researchers split the dataset into three folds and applied RF, XGBoost, and SVM classifiers to each fold. As a novel solution, Bozkir et al. employed Uniform Manifold Approximation and Projection (UMAP) for dimension reduction before performing classification. The results demonstrated a consistent improvement in accuracy for each fold.

Entropy analysis for classifying malware is a sophisticated approach that is rarely used in research, as seen in the case of [44]. In this study, Android malware samples are extracted from APK files and analyzed in a secure sandbox to obtain the most effective features. These features are then used as input for the proposed EntropLyzer – a dynamic behavioral analysis system based on the entropy of the extracted features. The EntropLyzer utilizes Shannon entropy for execution. To classify the detected malware, conventional algorithms such as Naïve Bayes (NB), SVM, RF, and DT were employed, with DT emerging as the best performer, surpassing the others by a significant margin.

Figure 4.6 provides a clear depiction of how a malware detection system can work in conjunction with big data. Malware samples that have been collected are extracted along with their respective malware signatures. The extracted data is then pre-processed, filtered, and used to train a machine learning model. Whenever new data emerges within the network, it is sent to the machine learning model to predict its authenticity.

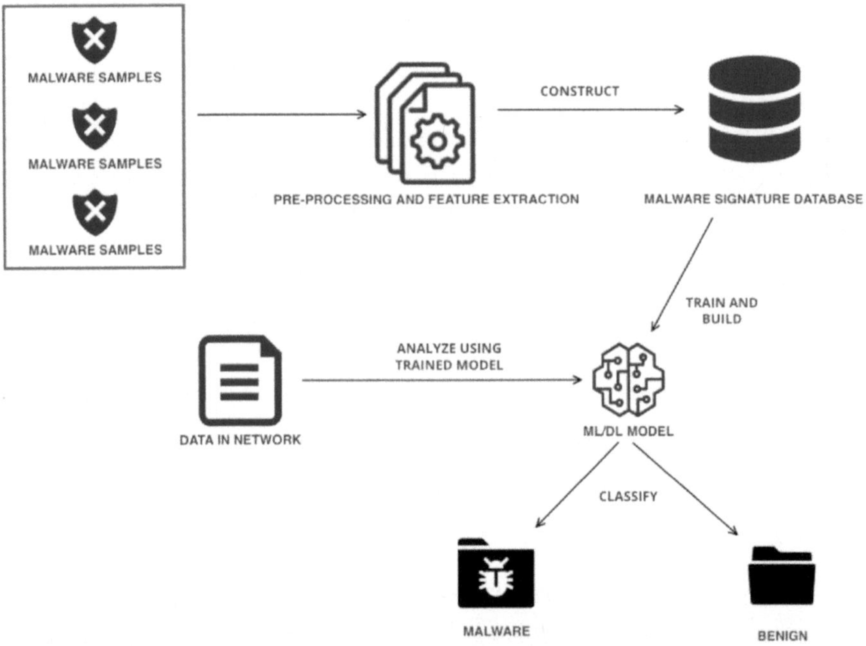

FIGURE 4.6 General architecture of a big data-based malware detection system.

4.1.6.3 USAGE OF BIG DATA ANALYTICS FOR FRAUD DETECTION

Fraud has become a common term due to its widespread nature. It has gained popularity, particularly in the financial sector, including banks, stock exchanges, and credit unions, among others. Despite the existence of various countermeasures such as Personal Identification Numbers (PINs), passwords, and one-time passwords (OTPs) [46], designed to ensure user safety and protect their assets, they are gradually proving inadequate in the face of emerging threats. Fraud can occur at two levels - internal and external. Internal fraud refers to acts committed by insiders such as employees or business partners, while external fraud is perpetrated by clients, users, hackers, or any other individuals external to the institutions [46]. By leveraging technologies like big data and ML, fraudulent transactions can be identified and prevented before they are completed, thus ensuring the security of assets. This section discusses some notable research works on fraud detection using BDA.

Yang et al. [50] proposed a framework that can address the challenge of recognizing credit card fraud. The proposed framework, called FFD: federated learning for fraud detection, utilizes a federated approach where different distributed devices participate in a network and learn from each other. Each participating node performs computations and attempts to update the globally shared data. This shared data is immediately accessible to the remaining nodes in the network. To detect credit card fraud, diverse financial institutions can share their datasets without revealing customers' private details. A Deep CNN model is maintained in a central server and trained using the datasets made visible by the participating nodes. Each node applies gradient descent to correct classification errors, thereby minimizing the loss rate. After undergoing a series of training and correction, the DCNN model becomes stable and can be used by banks to identify fraudulent transactions. SMOTE is applied to balance the raw data at each node. The authors have demonstrated that the proposed framework outperforms traditional federated learning approaches.

In a groundbreaking study, Johnson and Khoshgoftaar [48] conducted research on detecting fraudulent activities in Medicare, the affordable health insurance provided by the Federal Government of the United States. Medicare fraud has resulted in the loss of billions of taxpayers' money, leading to increased healthcare costs nationwide. To identify and combat abusive and fraudulent events, utilizing Big Data Analytics (BDA) is crucial due to the vast amount of available healthcare data. Employing an appropriate analysis technique is essential to leverage this data, gain insights, and effectively prevent fraud.

The primary challenge in the dataset is its severe imbalance, with only 0.03% of the data representing the positive class. The authors explored various techniques to address this imbalance without compromising the model's accuracy. At the data level, they experimented with Random Over-sampling (ROS), Random Under-sampling (RUS), and a combination of both (ROS-RUS). The combination approach yielded better results in terms of the Geometric Mean (G-Mean) compared to the previous methods. To tackle the class imbalance issue at the algorithm level, the authors also investigated cost-sensitive learning using weighted Cross-Entropy (CE) loss function, Mean False Error (MFE), and Focal Loss (FL). All three methods demonstrated similar performance when measured using G-Mean. However, it was evident that ROS-RUS is the preferred method for addressing such severe imbalance in the dataset when employing a Deep Neural Network (DNN), considering both G-Mean and AUC-ROC metrics.

The authors of [49] proposed to address the growing risk of credit card fraud by utilizing a Deep Learning (DL) model that incorporates LSTM. Various optimizers, namely RMSprop, Adam, and Adagrad, were tested to determine the optimal choice. Adam outperformed the others by a small margin in terms of accuracy, but significantly reduced the associated loss rate by more than half. The effectiveness of the Adam optimizer was evaluated with different numbers of layers, and a model with just three layers appeared to be slightly better. Additionally, the performance of the optimizers was influenced by the number of iterations. It was found that increasing the number of iterations did not have a severe impact on the model's accuracy, but rather resulted in minimal changes. The proposed model was compared to state-of-the-art auto encoder models and various other traditional ML algorithms to demonstrate its enhanced efficiency.

Javaid et al. [51] proposed a framework to identify electricity theft in smart grids using a Deep Siamese Network (DSN). The collected data is massively imbalanced, as expected, and is therefore subjected to multiple methods to overcome this. Adaptive synthesis (ADASYN) was found to provide the required amount of balancing. The DSN is built by combining a CNN and a LSTM model. The dataset undergoes other preprocessing steps, such as feature extraction, before being sent to the model to deliver accurate results. Additionally, other traditional supervised learning algorithms are employed and the results are compared. As expected, the proposed framework was able to deliver quality results.

Mururunkwere et al. [47] demonstrated the usage of Artificial Neural Networks (ANN) in detecting income tax fraud. The study was primarily conducted using data from Rwanda and aimed to identify individuals who attempt to evade proper tax payment. The authors found that ANNs, which simulate the functioning of the human brain, outperformed other traditional supervised and unsupervised machine learning algorithms. Hyperparameter tuning was performed to determine the optimal activation function, number of epochs, and number of layers for the ANN. The Adamax optimizer was employed to minimize prediction errors. The study achieved high accuracy and F1-score in terms of performance. Additionally, the authors focused on identifying the important features from the dataset, which influenced the final outcome. For this purpose, they utilized a Variance-based feature Importance of ANNs (VIANN).

Figure 4.7 depicts the architecture of a BDA-based Fraud Detection System. It works similarly to the previously discussed architecture.

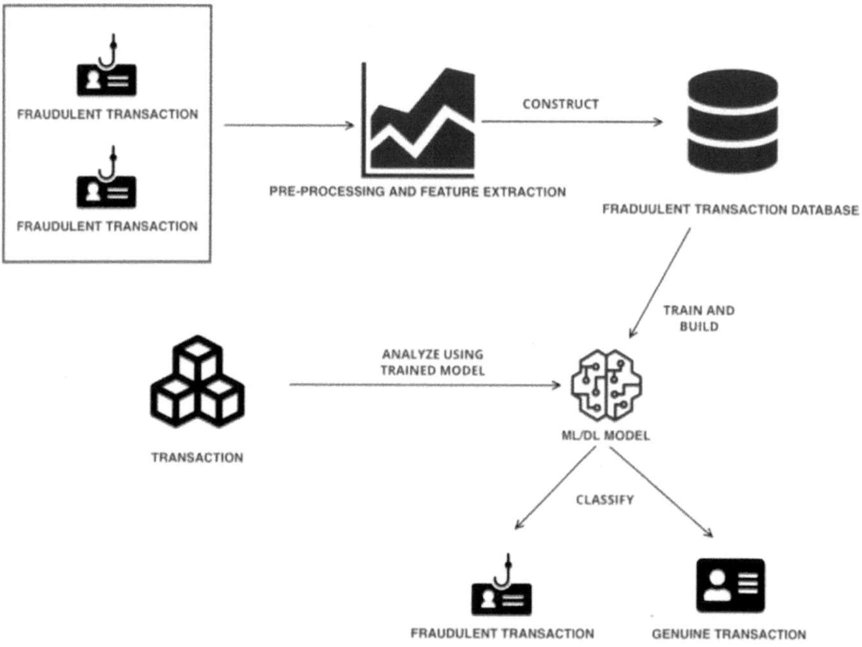

FIGURE 4.7 Generalized architecture of a fraud detection system using big data analytics.

4.1.6.4 USAGE OF BIG DATA ANALYTICS FOR CLOUD/IOT SECURITY

As businesses, small and large, leverage the advantages of cloud computing, it has become essential for every online user to be a part of the cloud environment. Regardless of their types, cloud networks are an easy target for attackers, as they provide a 24-hour service that serves millions of customers simultaneously. The massive amount of data stored in these servers is highly vulnerable, especially in the public cloud. Attackers who target cloud environments often aim to gain access to sensitive user information or cloud resources.

Balamurugan and Saravanan [33] attempt to address security concerns in cloud computing, specifically focusing on intrusion detection and prevention. Their proposed approach involves the design of multiple cloudlets, which assist in collecting data packets from users through routers. These packets are then monitored by Cloudlet Controllers (CC) using a Packet Scrutinization algorithm. To detect attacks on the servers, a Virtual Machine Manager (VMM) is employed, which incorporates a hybrid Normalized K-means clustering algorithm and an RNN (NK-RNN) classifier. Additionally, a Time Signature (OTS) is generated for each user to authenticate any transactions initiated by them, ensuring an extra layer of security. All verified transactions are then allocated the requested resources.

An interesting approach was taken by Dovom et al. [42], who used a Fuzzy Pattern Tree (FPT) for malware detection and classification in edge devices in IoT. An FPT is structurally similar to a binary decision tree, with the crucial difference being the use of fuzzy logical operators and fuzzy predicates at the inner and leaf nodes, respectively. A fast FPT is an alternative to FPTs when the complexity or dimension of the data increases. Four datasets were used in this study, with their opcodes extracted and analyzed. Both FPT and fast FPT were compared with state-of-the-art classification algorithms such as SVM, decision trees, RF, and k-nearest neighbors in Python. The results were remarkably similar to the traditional classifiers, demonstrating the effectiveness of FPTs as reliable classifiers. The identified malware was also categorized using fast FPT, yielding accurate results.

Visualization of threats is a hot topic that is being implemented in various research, including [43], where the authors aim to uncover malware in the Industrial Internet of Things (IIoT). The proposed architecture utilizes a sniffer to capture network traffic and pass it to the analyzer. The analyzer unit consists of two modules. The first module converts the associated malware binary into a 2D matrix, which is then transformed into a color image. The image is then fed to the second module, which comprises a Deep CNN (DCNN) that classifies the data. If a malicious sample is recognized, a response unit is alerted, and the administrator is notified.

When it comes to cloud computing, the significance of the various services offered by cloud providers to their clients cannot be ignored. As more and more organizations transition to a cloud-based pay-as-you-go service or resource, attackers seize the opportunity to steal these resources and make legitimate users bear the cost. This is commonly known as Fraudulent Resource Consumption (FRC). Agarwal et al. [52] have proposed a detection scheme called P-estimation in their work, which examines the HTTPS web traffic logs entering a network for any suspicious behavior induced by botnets, as botnets remain the preferred tool for such attacks. The server logs are carefully filtered, and a 1-D Discrete Wavelet Transformation technique is applied to remove any noise. The processed data is then fed into a DNN consisting of an LSTM with three layers. The attack percentage for a week is estimated to determine if an FRC attack has occurred. If an attack is detected, it is crucial to be able to mitigate it. The authors have also devised a simple yet effective solution to tackle this issue. By identifying IP addresses that generate abnormal requests, these IPs are added to a list. To gain access, an IP must be included in the whitelist. Otherwise, the IP is checked against the graylist. IPs in the graylist are granted temporary access but are subjected to frequent challenges that require completion by the human user in order to move the IP to the whitelist. If the user fails, their IP is blacklisted. Similarly, IPs in the blacklist have the opportunity to be moved back to the graylist if they are legitimate or denied access altogether. The proposed scheme was tested and compared to existing FRC models, outperforming them all. Another notable feature of this model is its ability to be regularly updated based on data flow and self-learn to classify.

4.1.7 DEEP LEARNING–THE GAME CHANGER

An analysis of the applications of BDA in intrusion detection, malware detection, fraud detection, and cloud/IoT security is in Section 4.1.6. It has been revealed that DL techniques are fast becoming the go-to method for building viable models that can easily be trained and deployed without the hassle of re-training them with new instances.

The simplest definition states that: "Deep Learning is a subfield of AI that predominantly centers around producing large neural networks that are capable of making accurate data-driven decisions [54]."

DL is not a single approach but rather represents a class of algorithms and topologies that can be applied to solve a broad spectrum of real-world problems. Like any other form of neural network, DLs mimic the workings of a human brain and relay the data through a network of neurons. The primary difference between DL and neural networks lies in the fact that the former is highly scalable with respect to the data and system performance. It is best suited when dealing with fairly complex or enormous datasets and therefore consists of a greater number of hidden layers that enable the model to learn minor details contributing to an output [54]. DL also relies more on high-performance machines with Graphical Processing Units (GPU).

A matter of contention regarding DL is determining the problem category it falls into. It can be argued that DL can be applied to any category–supervised, unsupervised, semi-supervised, or reinforcement learning–depending on the context of its use. A prime example of utilizing supervised DL is in image classification or object detection using CNN, where both the input and expected output are provided to the model. Unsupervised DL models, such as auto encoders, are employed in tasks like image clustering and compression, where they learn to identify patterns in the data. In the military and gaming industries, reinforced DL is commonly used, as demonstrated by Google's Deep Mind defeating human expert players in games like Go and chess [56]. A hierarchy of the DL architecture with a few algorithms is shown in Figure 4.8.

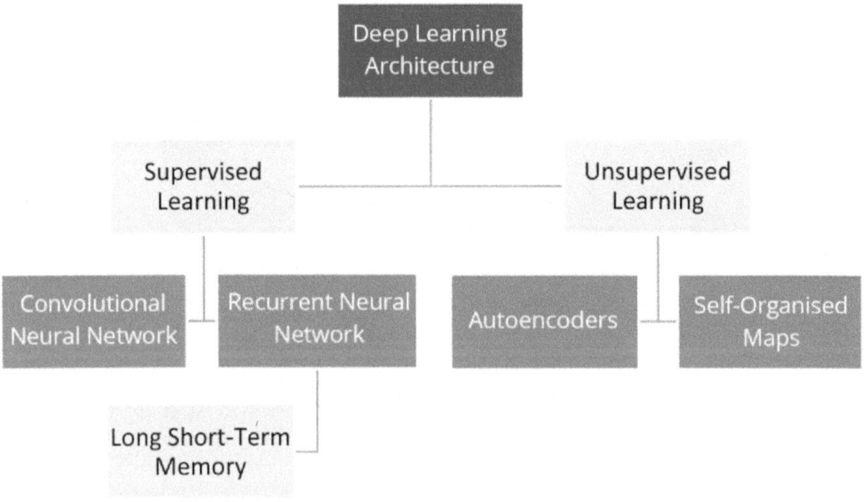

FIGURE 4.8 Hierarchy of deep learning categories.

One of the reasons for the success and the resultant popularity of DL models is their ability to know what to measure from the given data and how to measure it with the help of Representation Learning (RL). RL is a set of procedures that allow a machine to be fed with raw data and to automatically discover the representations needed for detection or classification [55]. A DL model comprises multiple hidden layers of representation that diverge across the levels with the aim of breaking down the raw data into abstract and complex representations. The amplification of the representation at each level enables the model to discover minor details of input that govern the outcome. This allows the models to learn and make accurate output predictions. A model architecture of DL is depicted in Figure 4.8.

The number of layers in a model is determined either by the type of model being used or by the input features. The nodes at each layer represent information that is used to estimate the nodes in the subsequent layer with the help of an activation function and a bias. Activation functions help convert weighted inputs of the neurons into their output [70]. Some common examples of activation functions are SoftMax, Linear, Identity, Rectified Linear Unit (ReLU), and Exponential Linear Unit (ELU). Weights are assigned to each layer, initially randomly but then weighted and corrected in every subsequent iteration [62]. Loss functions are also used to rectify the values of weights or hyperparameters used in the neural network when back propagation is applied [70]. The architecture of the DL model is shown in Figure 4.9.

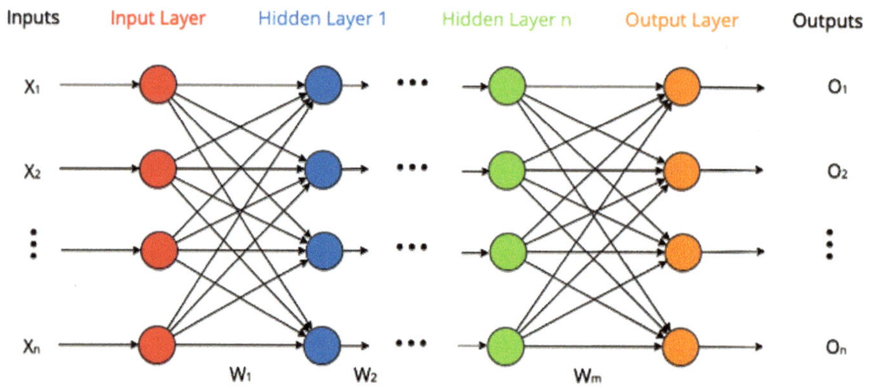

FIGURE 4.9 Deep learning model architecture.

The surge in using DL as the primary go-to technique in analytics and AI can be attributed to a number of factors, such as the sheer size and availability of labeled datasets, advanced processing capabilities like parallel and multi-core processors, wide availability of GPUs, better optimization algorithms like Adam, AdaGrad, Swarm Intelligence, etc., and improved open-source platforms like TensorFlow, Keras, PyTorch, BigDL, etc. [62]. It is expected to see many further variations of DL models that may outperform the existing models.

In the following sections, famous DL models such as CNN, RNN, Auto-encoders, and Deep Reinforcement Learning are discussed.

4.1.7.1 CONVOLUTIONAL NEURAL NETWORK (CNN)

A CNN is a method of DL that works on images. Most commonly used in the field of Image Processing and Computer Vision, CNNs take an input image and train a model to distinguishable features by assigning importance (bias and weights) to objects in the image [64]. These require comparatively less preprocessing and are therefore claimed to be the best-suited algorithm for image-based learning. Some of its common areas of application include image classification and segmentation, object detection, and video processing, and have also shown excellent performance in speech recognition and Natural Language Processing (NLP) [64]. Being a hierarchical feed forward multilayer network, CNNs use convolutional kernels as different layers to execute several conversions on the input data [65]. These conversions help in learning different features of the image and thereby generate a feature map. Pooling layers are used to reduce the size of the convolved feature maps. These operations are repeated as required and finally, it is fed to an activation function as a flattened layer, which aids in learning the components that make up an image. Lastly, sub sampling is applied to generate the classification result [65]. The architecture of a CNN is depicted in Figure 4.10. CNNs have wide usage in cybersecurity threat detection [34, 35, 39]. Binary images, signature images, opcodes, and text are used to train and deploy CNN models. CNNs may also be used to analyze video feeds frame by frame to detect anomalies. As in the case of [77], where video frames are analyzed using CNN for face forensic detection.

4.1.7.2 RECURRENT NEURAL NETWORK (RNN)

RNNs are a special form of neural networks used exclusively for processing sequential data of variable length [66]. These networks are analogous to other neural networks in that they work with multiple hidden layers. Their primary uniqueness comes from the fact that the output of the previous layers is fed as the input to the current layer. A "memory" is used to store the previous output. RNNs standardize various activation functions, weights, and biases, resulting in identical values for each hidden layer [66]. Then, rather than constructing numerous hidden layers, it will generate one and

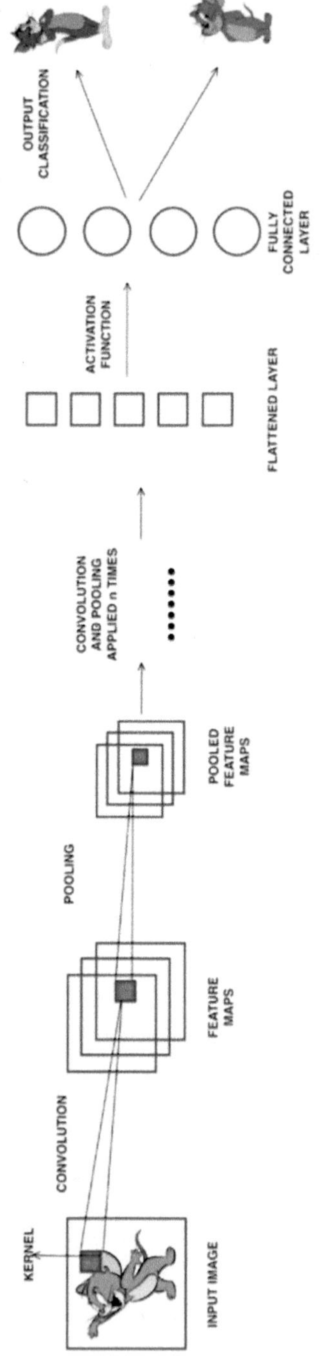

FIGURE 4.10 CNN architecture.

loop over it as many times as necessary. They find their use in language translation, voice classification, and sentiment classification, to name a few. The architecture of an RNN is given in Figure 4.9. X_t represents the input at time step t, which, when supplied to an activation function, yields Y_t, its output. A loop over the activation function indicates its recurrent nature. Upon unfolding the representation, it can be seen that at each time step t, the input of the current time step is the output of the previous time step t-1. Upon its completion, the current output is sent as the input to the successive step. RNNs were implemented in Refs. [30, 33] for intrusion detection and IoT security (Figure 4.11).

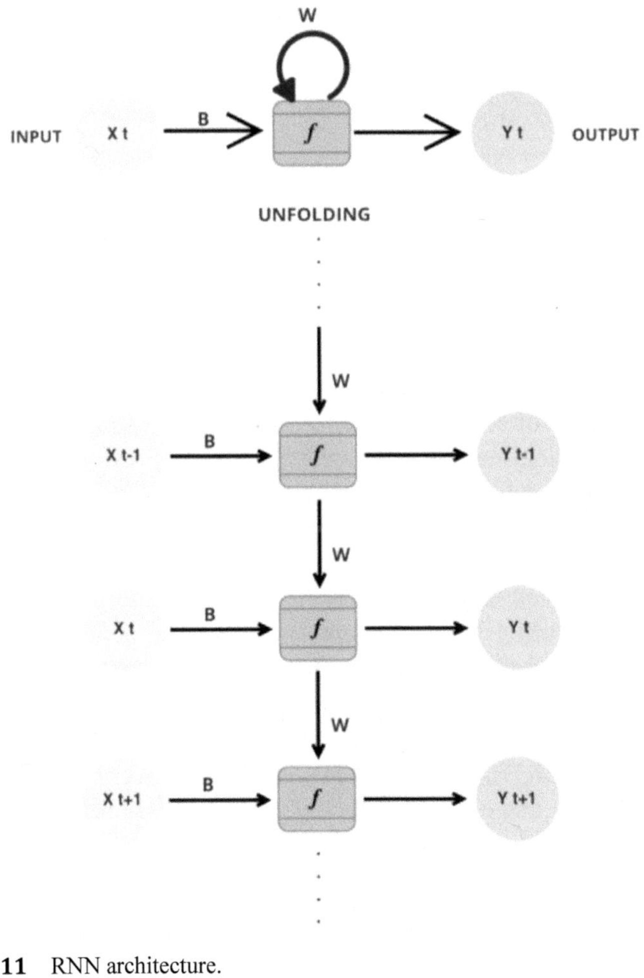

FIGURE 4.11 RNN architecture.

4.1.7.3 AUTOENCODERS

These are a class of tools intended for non-linear dimensionality reduction and extend from the work of the standard linear Principal Component Analysis (PCA) [75]. They consist of two neural networks: an encoder and a decoder. An input fed to the encoder will be mapped to a latent encoding space and result in the creation of the latent code. This is then passed on to the decoder, whose job is to decode the latent code and reconstruct the original input accurately [76]. A loss function is utilized to minimize the encountered loss. Figure 4.12 illustrates the working of an autoencoder. Papers [41, 49] demonstrate the advantage of using auto encoders.

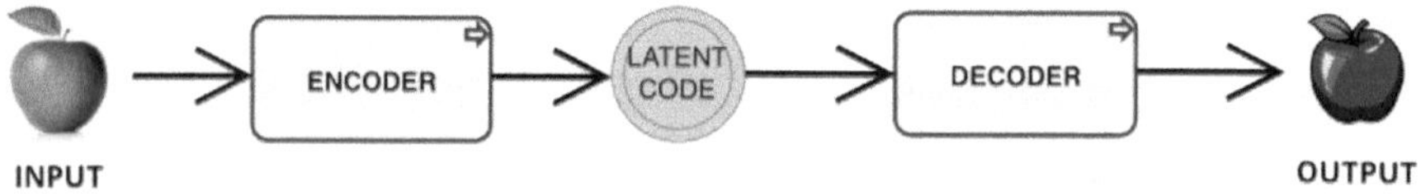

FIGURE 4.12 Working as an autoencoder.

4.1.7.4 DEEP REINFORCEMENT LEARNING

Deep reinforcement learning combines reinforcement learning with deep learning. This area of study has recently been able to address a wide range of complex decision-making problems that were previously beyond the capabilities of computers. Introduced in 2015, this class of models allows for a cumulative reward mechanism that needs to be maximized through the model's performance in learning actions on its own. Higher dimensional data is reduced using DNNs [71]. The architecture of such a model is depicted in Figure 4.13. The agent learns desired actions and interacts with the environment through a rewarding mechanism that employs a greedy approach. Deep Reinforcement Learning has found numerous applications, including cybersecurity. It is well-suited for network intrusion detection, as demonstrated in Ref. [72]. The model used a Multi-Layer Perceptron (MLP) for data reduction. The developed model yielded good results. In Ref. [73], the authors took a similar approach to [72], and the classifier was tested on real-time network traffic data, showing promising results. Similarly, in Ref. [74], the proposed IDS demonstrated competitive performance with other state-of-the-art models.

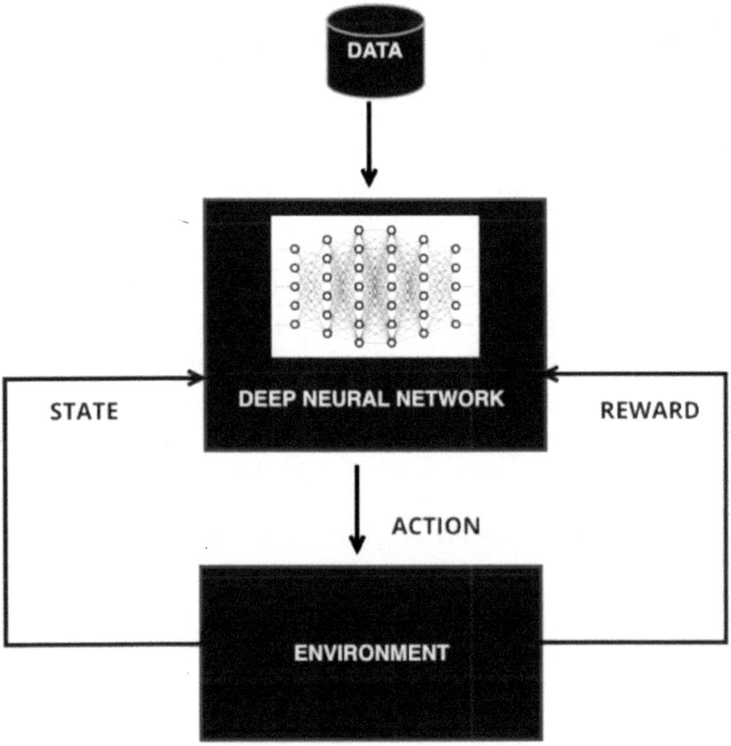

FIGURE 4.13 Deep reinforcement learning model.

4.1.8 CYBERSECURITY DATA SCIENCE

Data Science refers to the broad area of information extraction by utilizing mathematics, statistics, programming, analytics, ML, AI, and deep learning [69]. As the entire security industry is moving towards data science, data-driven decisions control all aspects of a business. The combination of data science and behavioral analytics has given rise to the term "cybersecurity data science," which refers to the collection and analysis of large amounts of security event data from various sources to detect security risks or attacks by discovering useful insights [69].

A more formal definition of cybersecurity data science is stated in Ref. [69]: "Cybersecurity data science is a research or working area that exists at the intersection of cybersecurity, data science, and ML or AI. It is mainly focused on security data, applies ML methods, attempts to quantify cyber-risks or incidents, and promotes inferential techniques to analyze behavioral patterns in security data. It also focuses on generating security response alerts and ultimately aims to optimize cybersecurity solutions to build automated and intelligent cybersecurity systems."

All the works discussed in Section 4.1.6 can be categorized under the domain of Cybersecurity Big Data Analytics (CBDA). CBDA, in turn, is a sub-field of Cybersecurity Data Science. The current state of the world mandates that every cybersecurity professional be well-versed in data science techniques. In addition to providing security to applications or systems during the development phase, they must also be vigilant in identifying security vulnerabilities and risks, handling and mitigating attacks, recognizing potential breaches, designing security protocols, and issuing reports and documents. They must be capable of performing these tasks using both manual tools and advanced tools such as BDA.

4.1.9 CHALLENGES IN CBDA

Like any other field, CBDA also faces many obstacles that may limit its functionality. In the following, we summarize the challenges encountered while engaged in CBDA.

- **Availability of Datasets:** Primary datasets are essential for enabling model generation and training. However, the availability of proper datasets, especially those related to different types of attack vectors, is often limited. Many threats or security-related datasets are private to the collecting organization and therefore not accessible for public use. Another challenge is the availability of recent datasets that accurately reflect the behaviors of prevalent threats.
- **Data Quality Problems:** The obtained datasets may be noisy, imbalanced, incomplete, or inconsistent. These factors can skew and compromise the quality of the generated model and degrade the learning process. The quality issues can be addressed by applying appropriate pre-processing methods.
- **Sensitivity of Information:** Cybersecurity breaches could result in the leakage of valuable and sensitive data, which is highly undesirable. When it comes to protecting or assessing sensitive information, the ML models need to be fed with this data. Using sensitive data, even for security reasons, defeats the ultimate purpose of a safe and secure storage environment. Unless proper encryption techniques are applied, this data must never be used in building datasets.

- **Timeliness of the Data:** All data contributes to decision-making at some level in an organization. Timely data can have a great positive impact on the decision being made. The timeliness of data is one of the central aspects contributing to the quality of data. Data that isn't timely can lead to negative impacts through wrong decision-making.
- **Validity of the Models:** ML or DL models built and trained on traditional datasets cannot guarantee model accuracy, especially since new and evolving threats have different patterns from older ones. Hence, a model built once is not suitable for long-term use. Models must be able to adapt and train with newer data to recognize new threats.
- **Priority-Based Alert Generation:** A security system that is not well-defined may generate a significant number of false alarms. These false alarms can gradually degrade the overall performance of the system as the trained model becomes outdated. For instance, a typical IDS might generate thousands of alerts, with the majority being false positives. Manual intervention is required in most cases to verify and check these false positives. However, this process is highly time and resource-intensive, resulting in resource wastage. A good model should be capable of prioritizing threats and alerting administrators accordingly.

4.10 CONCLUSION

This chapter presents an overview of BDA and cybersecurity and how these two concepts came to work together. It addresses some of the applications of big data in the field of cybersecurity, highlighting key threat vectors and commonly employed digital security measures. To assess the performance of cybersecurity mechanisms aided by big data, an extensive review of notable research works is provided. These works are categorized as BDA for intrusion detection, malware detection, fraud detection, and cloud/IoT security measures. Each category offers a glimpse into prominent work and provides a general architecture for better understanding. Among various techniques, it becomes apparent that Deep Learning techniques, such as CNN, RNN, Autoencoders, and Deep Reinforcement Learning, are estimated to outperform traditional ML algorithms. The chapter concludes with a brief overview of the emerging field of Cybersecurity Data Science and the challenges it faces.

KEYWORDS

- **artificial intelligence**
- **big data analytics**
- **cybersecurity**
- **deep learning**
- **denial of service**
- **machine learning**
- **structured query language**
- **web application firewalls**

REFERENCES

1. Wise, J., (2022). *How Much Data is Created Every Day in 2022?* Retrieved from: https://earthweb.com/how-much-data-is-created-every-day/#:~:text=2.5%20quintillion%20bytes%20of%20data%20is%20created%20every%20day (accessed on 27 January 2024).
2. TechAmerica Foundation's Federal Big Data Commission, (2012). *Demystifying Big Data: A Practical Guide to Transforming the Business of Government.* Retrieved from: http://www.techamerica.org/Docs/fileManager.cfm?f=techamerica-bigdatareport-final.pdf (accessed on 27 January 2024).
3. Gandomi, A., & Haider, M., (2015). Beyond the hype: Big data concepts, methods, and analytics. *International Journal of Information Management, 35*(2), 137–144.
4. Elgendy, N., & Elragal, A., (2014). Big data analytics: A literature review paper. In: *Industrial Conference on Data Mining* (pp. 214–227). Springer, cham.
5. SAS Reports, (2012). *CEBR: Data Equity, Unlocking the Value of Big Data.* Retrieved from: https://www.sas.com/content/dam/SAS/en_gb/doc/analystreport/cebr-value-of-big-data.pdf (accessed on 27 January 2024).
6. Dash, S., Shakyawar, S. K., Sharma, M., & Kaushik, S., (2019). Big data in healthcare: Management, analysis, and future prospects. *Journal of Big Data, 6*(1), 1–25.
7. Ghosh, S., & Gosavi, A., (2017). A semi-Markov model for post-earthquake emergency response in a smart city. *Control Theory and Technology, 15*(1), 13–25.
8. Akter, S., & Wamba, S. F., (2019). Big data and disaster management: A systematic review and agenda for future research. *Annals of Operations Research, 283*(1), 939–959.
9. Manyika, J., Chui, M., Brown, B., Bughin, J., Dobbs, R., Roxburgh, C., & Hung, B. A., (2011). *Big Data: The Next Frontier for Innovation, Competition, and Productivity.* McKinsey Global Institute.
10. Gunasekaran, A., Papadopoulos, T., Dubey, R., Wamba, S. F., Childe, S. J., Hazen, B., & Akter, S., (2017). Big data and predictive analytics for supply chain and organizational performance. *Journal of Business Research, 70*, 308–317.

11. Talwar, S., Kaur, P., Fosso, W. S., & Dhir, A., (2021). Big data in operations and supply chain management: A systematic literature review and future research agenda. *International Journal of Production Research, 59*(11), 3509–3534.
12. Alani, M. M., (2021). Big data in cybersecurity: A survey of applications and future trends. *Journal of Reliable Intelligent Environments, 7*(2), 85–114.
13. Singer, P. W., & Friedman, A., (2014). *Cybersecurity: What Everyone Needs to Know.* OUP USA.
14. Purplesec, (2022). *Cyber Security Statistics the Ultimate List of Stats Data, & Trends For 2022.* Retrieved from: https://purplesec.us/resources/cyber-security-statistics/ (accessed on 27 January 2024).
15. Goodin, D., (2020). *"Evil Mobile Emulator Farms" Used to Steal Millions from US and EU Banks.* ARS Technica. Retrieved from: https://arstechnica.com/information-technology/2020/12/evil-mobile-emulator-farms-used-to-steal-millions-from-us-and-eu-banks/ (accessed on 27 January 2024).
16. Microsoft. (2014). *Definition of a Security Vulnerability.* Retrieved from: https://msdn.microsoft.com/enus/lib rary/cc751383.aspx (accessed on 27 January 2024).
17. El Hamzaoui, M., & Bensalah, F., (2021). Toward a knowledge-based model to fight against cybercrime within big data environments: A set of key questions to introduce the topic. In: *Machine Intelligence and Big Data Analytics for Cybersecurity Applications* (pp. 75–101). Springer, Cham.
18. ITU, (2008). *Overview of Cybersecurity, in Series X: Data Networks, Open System Communications and Security.* International Telecommunication Union. Retrieved from: https://www.itu.int/rec/T-REC-X.1205-200804-I (accessed on 27 January 2024).
19. Cybrary. *Cybersecurity Glossary.* Retrieved from: https://www.cybrary.it/resources/glossary/#c(accessed on 27 January 2024).
20. Daily, S. (2021). The Evolution of Cybersecurity Solutions and Threats. SecurityScorecard. Retrieved 24-July-2022 from https://securityscorecard.com/blog/the-evolution-of-cybersecurity-solutions-and-threats/ (accessed on 27 January 2024).
21. GAO, (2008). *Privacy: Alternatives Exist for Enhancing Protection of Personally Identifiable Information.* GAO Report 08-536. Retrieved from: https://www.gao.gov/assets/gao-08-536.pdf (accessed on 27 January 2024).
22. Narwal, B., Mohapatra, A. K., & Usmani, K. A., (2019). Towards a taxonomy of cyber threats against target applications. *Journal of Statistics and Management Systems, 22*(2), 301–325.
23. Ciancioso, R., Budhwa, D., & Hayajneh, T., (2017). A framework for zero day exploit detection and containment. In: *2017 IEEE 15th Intl. Conf. on Dependable, Autonomic and Secure Computing, 15th Intl. Conf. on Pervasive Intelligence and Computing, 3rd Intl. Conf. on Big Data Intelligence and Computing and Cyber Science and Technology Congress (DASC/PiCom/DataCom/Cyber SciTech)* (pp. 663–668). IEEE.
24. Liu, J., Liu, Y., Li, J., Sun, W., Cheng, J., Zhang, R., & Pang, J., (2022). Two statistical traffic features for certain APT group identification. *Journal of Information Security and Applications, 67*, 103207.
25. Abouelmehdi, K., Beni-Hessane, A., & Khaloufi, H., (2018). Big healthcare data: Preserving security and privacy. *Journal of Big Data, 5*(1), 1–18.
26. Landi, H., (2019). *Ransomware Attack at Brooklyn Hospital Center Results in Permanent Loss of Some Patient Data.* Fierce Healthcare. Retrieved from: https://www.fiercehealthcare.com/tech/ransomware-attack-at-brooklyn-hospital-center-results-permanent-loss-some-patient-data(accessed on 27 January 2024).

27. Davidson, E. M., McArthur, S. D., McDonald, J. R., Cumming, T., & Watt, I., (2006). Applying multi-agent system technology in practice: Automated management and analysis of SCADA and digital fault recorder data. *IEEE Transactions on Power Systems, 21*(2), 559–567.
28. Soni, V. D., (2019). Role of artificial intelligence in combating cyber threats in banking. *International Engineering Journal for Research & Development, 4*(1), 7–7.
29. Sun, C. C., Hahn, A., & Liu, C. C., (2018). Cyber security of a power grid: State-of-the-art. *International Journal of Electrical Power & Energy Systems, 99*, 45–56.
30. Du, M., Li, F., Zheng, G., & Srikumar, V., (2017). Deep Log: Anomaly detection and diagnosis from system logs through deep learning. In: *Proceedings of the 2017 ACM SIGSAC Conference on Computer and Communications Security* (pp. 1285–1298).
31. Siadati, H., & Memon, N., (2017). Detecting structurally anomalous logins within enterprise networks. In: *Proceedings of the 2017 ACM SIGSAC Conference on Computer and Communications Security* (pp. 1273–1284). ACM, New York.
32. Othman, S. M., Ba-Alwi, F. M., Alsohybe, N. T., & Al-Hashida, A. Y., (2018). Intrusion detection model using machine learning algorithm on big data environment. *Journal of Big Data, 5*(1), 1–12.
33. Balamurugan, V., & Saravanan, R., (2019). Enhanced intrusion detection and prevention system on cloud environment using hybrid classification and OTS generation. *Cluster Computing, 22*(6), 13027–13039.
34. Al, S., & Dener, M., (2021). STL-HDL: A new hybrid network intrusion detection system for imbalanced dataset on big data environment. *Computers & Security, 110*, 102435.
35. Sriram, S., Shashank, A., Vinayakumar, R., & Soman, K. P., (2019). DCNN-IDS: Deep convolutional neural network-based intrusion detection system. In: *International Conference on Computational Intelligence, Cyber Security, and Computational Models* (pp. 85–92). Springer, Singapore.
36. Ponmalar, A., & Dhanakoti, V., (2022). An intrusion detection approach using ensemble support vector machine-based chaos game optimization algorithm in big data platform. *Applied Soft Computing, 116*, 108295.
37. Gupta, N., Jindal, V., & Bedi, P., (2022). CSE-IDS: Using cost-sensitive deep learning and ensemble algorithms to handle class imbalance in network-based intrusion detection systems. *Computers & Security, 112*, 102499.
38. Usman, N., Usman, S., Khan, F., Jan, M. A., Sajid, A., Alazab, M., & Watters, P., (2021). Intelligent dynamic malware detection using machine learning in IP reputation for forensics data analytics. *Future Generation Computer Systems, 118*, 124–141.
39. Millar, S., McLaughlin, N., Del Rincon, J. M., & Miller, P., (2021). Multi-view deep learning for zero-day Android malware detection. *Journal of Information Security and Applications, 58*, 102718.
40. Bozkir, A. S., Tahillioglu, E., Aydos, M., & Kara, I., (2021). Catch them alive: A malware detection approach through memory forensics, manifold learning, and computer vision. *Computers & Security, 103*, 102166.
41. De Paola, A., Gaglio, S., Re, G. L., & Morana, M., (2018). A hybrid system for malware detection on big data. In: *IEEE INFOCOM 2018-IEEE Conference on Computer Communications Workshops (INFOCOM WKSHPS)* (pp. 45–50). IEEE.
42. Dovom, E. M., Azmoodeh, A., Dehghantanha, A., Newton, D. E., Parizi, R. M., & Karimipour, H., (2019). Fuzzy pattern tree for edge malware detection and categorization in IoT. *Journal of Systems Architecture, 97*, 1–7.

43. Naeem, H., Ullah, F., Naeem, M. R., Khalid, S., Vasan, D., Jabbar, S., & Saeed, S., (2020). Malware detection in industrial internet of things based on hybrid image visualization and deep learning model. *Ad Hoc. Networks*, *105*, 102154.
44. Keyes, D. S., Li, B., Kaur, G., Lashkari, A. H., Gagnon, F., & Massicotte, F., (2021). EntropLyzer: Android malware classification and characterization using entropy analysis of dynamic characteristics. In: *2021 Reconciling Data Analytics, Automation, Privacy, and Security: A Big Data Challenge (RDAAPS)* (pp. 1–12). IEEE.
45. Azmoodeh, A., Dehghantanha, A., Parizi, R. M., Hashemi, S., Gharabaghi, B., & Srivastava, G., (2020). Active spectral botnet detection based on eigenvalues weighting. In: *Handbook of Big Data Privacy* (pp. 385–397). Springer, Cham.
46. Gyamfi, N. K., & Abdulai, J. D., (2018). Bank fraud detection using support vector machine. In: *2018 IEEE 9th Annual Information Technology, Electronics and Mobile Communication Conference (IEMCON)* (pp. 37–41). IEEE.
47. Murorunkwere, B. F., Tuyishimire, O., Haughton, D., & Nzabanita, J., (2022). Fraud detection using neural networks: A case study of income tax. *Future Internet*, *14*(6), 168.
48. Johnson, J. M., & Khoshgoftaar, T. M., (2019). Medicare fraud detection using neural networks. *Journal of Big Data*, *6*(1), 1–35.
49. Alghofaili, Y., Albattah, A., & Rassam, M. A., (2020). A financial fraud detection model based on LSTM deep learning technique. *Journal of Applied Security Research*, *15*(4), 498–516.
50. Yang, W., Zhang, Y., Ye, K., Li, L., & Xu, C. Z., (2019). FFD: A federated learning-based method for credit card fraud detection. In: *International Conference on Big Data* (pp. 18–32). Springer, Cham.
51. Javaid, N., Jan, N., & Javed, M. U., (2021). An adaptive synthesis to handle imbalanced big data with deep Siamese network for electricity theft detection in smart grids. *Journal of Parallel and Distributed Computing*, *153*, 44–52.
52. Agarwal, A., Prasad, A., Rustogi, R., & Mishra, S., (2021). Detection and mitigation of fraudulent resource consumption attacks in cloud using deep learning approach. *Journal of Information Security and Applications*, *56*, 102672.
53. Sahu, A. K., Sharma, S., Tanveer, M., & Raja, R., (2021). Internet of things attack detection using hybrid deep learning model. *Computer Communications*, *176*, 146–154.
54. Kelleher, J. D., (2019). *Deep Learning*. MIT press.
55. LeCun, Y., Bengio, Y., & Hinton, G., (2015). Deep learning. *Nature*, *521*(7553), 436–444.
56. Gibney, E., (2015). Deep Mind algorithm beats people at classic video games. *Nature*, *518*(7540), 465, 466.
57. Ranum, M. J., (1993). Thinking about firewalls. In: *Proceedings of Second International Conference on Systems and Network Security and Management (SANS-II)* (Vol. 8).
58. Crowd strike. (2022). *Honeypots in Cybersecurity Explained.* Retrieved from: https://www.crowdstrike.com/cybersecurity-101/honeypots-in-cybersecurity-explained/ (accessed on 27 January 2024).
59. Symanovich, S., (2021). *5 Reasons Why General Software Updates and Patches Are Important.* Norton. Retrieved from: https://us.norton.com/internetsecurity-how-to-the-importance-of-general-software-updates-and-patches.html(accessed on 27 January 2024).
60. Belcic, I., & Edwards, C., (2022). *What Is a VPN? and How Does It Work?* AVG. Retrieved from: https://www.avg.com/en/signal/what-is-a-vpn-and-why-should-you-use-one(accessed on 27 January 2024).

61. Ometov, A., Bezzateev, S., Mäkitalo, N., Andreev, S., Mikkonen, T., & Koucheryavy, Y., (2018). Multi-factor authentication: A survey. *Cryptography*, *2*(1), 1.
62. Sengupta, S., Basak, S., Saikia, P., Paul, S., Tsalavoutis, V., Atiah, F., & Peters, A., (2020). A review of deep learning with special emphasis on architectures, applications and recent trends. *Knowledge-Based Systems, 194*, 105596.
63. Gordon, G. R. (2013). The growing threat of medical identity fraud: a call to action. Medical Identity Fraud Alliance (MIFA). Retrieved on 21-Aug-2022 from https://www.medidfraud.org/wp-content/uploads/2013/07/MIFA-Growing-Threat-07232013.pdf.
64. Khan, A., Sohail, A., Zahoora, U., & Qureshi, A. S., (2020). A survey of the recent architectures of deep convolutional neural networks. *Artificial Intelligence Review, 53*(8), 5455–5516.
65. LeCun, Y., Kavukcuoglu, K., & Farabet, C., (2010). Convolutional networks and applications in vision. In: *Proceedings of 2010 IEEE International Symposium on Circuits and Systems* (pp. 253–256). IEEE.
66. Goodfellow, I., Bengio, Y., & Courville, A., (2016). *Deep Learning*. MIT Press.
67. André, Q., Carmon, Z., Wertenbroch, K., Crum, A., Frank, D., Goldstein, W., & Yang, H., (2018). Consumer choice and autonomy in the age of artificial intelligence and big data. *Customer Needs and Solutions, 5*(1), 28–37.
68. Kurlwich, R., (2012). *How Much Do They Know About Me in the 'Cloud'?* NPR. Retrieved from: https://www.npr.org/sections/krulwich/2012/02/27/147497042/how-much-do-they-know-about-me-in-the-cloud (accessed on 27 January 2024).
69. Sarker, I. H., Kayes, A. S. M., Badsha, S., Alqahtani, H., Watters, P., & Ng, A., (2020). Cybersecurity data science: An overview from machine learning perspective. *Journal of Big Data, 7*(1), 1–29.
70. Sewak, M., (2019). *Deep Reinforcement Learning*. Singapore: Springer Singapore.
71. Dixit, P., & Silakari, S., (2021). Deep learning algorithms for cybersecurity applications: A technological and status review. *Computer Science Review, 39*, 100317.
72. Gülmez, H. G., & Angın, P. (2021). A study on the efficacy of deep reinforcement learning for intrusion detection. *Sakarya University Journal of Computer and Information Sciences, 4*(1), 11–25.
73. Hsu, Y. F., & Matsuoka, M., (2020). A deep reinforcement learning approach for anomaly network intrusion detection system. In: *2020 IEEE 9th International Conference on Cloud Networking (Cloud Net)* (pp. 1–6). IEEE.
74. Sethi, K., Sai, R. E., Kumar, R., Bera, P., & Venu, M. Y., (2020). A context-aware robust intrusion detection system: A reinforcement learning-based approach. *International Journal of Information Security, 19*(6), 657–678.
75. Kramer, M. A., (1991). Nonlinear principal component analysis using auto associative neural networks. *AICHE Journal, 37*(2), 233e243.
76. Pinaya, W. H. L., Vieira, S., Garcia-Dias, R., & Mechelli, A., (2020). Autoencoders. In: *Machine Learning* (pp. 193–208). Academic Press.
77. Sedik, A., Faragallah, O. S., El-sayed, H. S., El-Banby, G. M., El-Samie, F. E. A., Khalaf, A. A., & El-Shafai, W., (2022). An efficient cybersecurity framework for facial video forensics detection based on multimodal deep learning. *Neural Computing and Applications, 34*(2), 1251–1268.

CHAPTER 5

Deep Neural Networks for Automated Cyber Security: Identification of Dynamic Malware in Cloud Computer Networks

P. S. VIJAYALAKSHMI[1] and M. JAYAKUMAR[2]

[1]*Assistant Professor, Department of Computer Applications, Dr. N.G.P Arts and Science College, Coimbatore-48, India*

[2]*Assistant Professor, Department of Information Technology SNMV College of Arts and Science, Coimbatore-50, Tamil Nadu, India*

ABSTRACT

As more organizations complete their digital transformations and shift their systems to the cloud, cybersecurity has become a major challenge. Because information security is continuously changing, it is critical to mitigate the risks of cloud computing and ensure that systems and networks are secure at rest, in use, and in transit. The goal of cloud security is to protect data in the cloud, operations, and infrastructure against cyberattacks. It consists of a number of technologies, services, rules, and best practices. Automated cybersecurity is essential for the company to adhere to data privacy rules and prevent data loss. In most situations, the Cloud Service Provider (CSP) provides customers with the most up-to-date cloud cybersecurity technologies, ensuring they are protected against both known and unknown threats. The main threats we will encounter are data and privacy compliance, data theft, unauthorized access, malware attacks, and hacking attacks. In this paper, we present deep neural network (DNN) technologies such as CNN, RNN, and Autoencoders algorithms for automated cybersecurity in cloud

Leveraging Artificial Intelligence (AI) Competencies for Next-Generation Cybersecurity Solutions.
Pethuru Raj, B. Sundaravadivazhagan, & V. Kavitha (Eds.)

computing networks. Here, we include malware classification using CNN and AE, which provide better performance results.

5.1 INTRODUCTION

The deployment of diverse technologies and procedures to safeguard digital systems, programs, information, networking, and all the devices contained inside them is known as cybersecurity. For each of these areas, cybersecurity seeks to prevent or, at the very least, minimize digital threats. It safeguards businesses from online fraudsters and the hazards they pose. Automation in cybersecurity is more sophisticated than these common instances. By utilizing artificial intelligence (AI) technologies, security systems can be created that automatically identify, investigate, and resist cybersecurity threats.

A computerized cyber security technology could identify a perceived threat and be able to eliminate it rather than just sounding an alarm to summon a human security expert to take appropriate action. Automated cyber security systems can choose the optimal course of action to neutralize the attack due to AI and machine learning (ML). Key infrastructure information systems, network monitoring, cloud security, and internet of things security are a few areas where cyber security should always be applied. The cloud should be utilized to implement cyber security. For data storage, businesses often employ a variety of methods, including an internal on-site data center, networking equipment from a data center provider, colocation data centers, and the cloud. Since the cloud often manages storage and keeps track of data, cyber security is crucial. The management of cloud systems requires the implementation of cyber security (Figure 5.1).

By integrating AI and ML, automation cybersecurity systems can anticipate and protect against attacks. The gathered information is used to predict and prevent future attacks. Once the problem is identified and resolved, the system may experience long-term effects. As a final step in cybersecurity automation, the system is scanned to ensure there are no remaining threats. Re-evaluating the system confirms its security against the initial threat discovered in the initial analysis.

To keep up with the rapid growth of malware, it is crucial to develop reliable approaches for classifying malware that can accommodate variations within the same family. As a first step in this direction, we propose a deep learning framework for malware categorization. In contrast to conventional methods, our study adopts a different strategy to evaluate and categorize malware. We utilize Convolutional Neural Networks (CNN) and Autoencoders, which are deep learning architectures. Deep learning has shown exceptional performance in various domains such as bioinformatics, computer vision, speech synthesis, and natural language processing (NLP). However, the potential of implementing CNNs in other domains, including cybersecurity, has not been thoroughly explored. CNNs have demonstrated remarkable success, particularly in tasks involving image analysis.

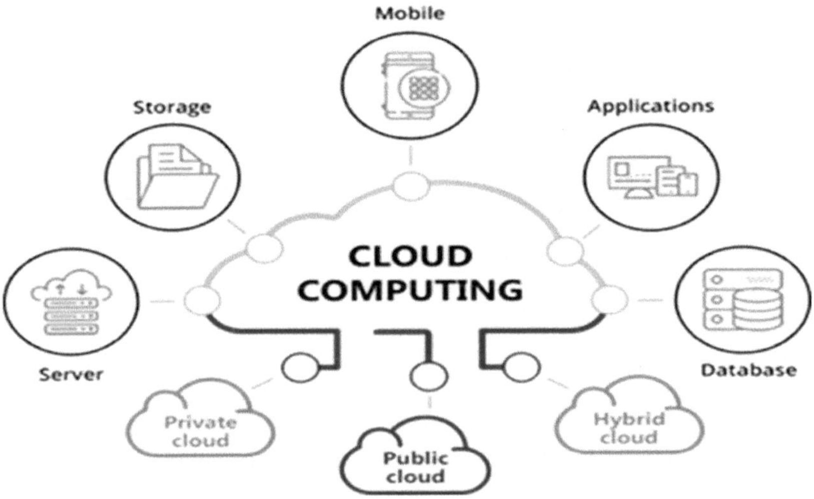

FIGURE 5.1 Cloud computing.

5.2 LITERATURE REVIEW

Network security protocols and computer security systems are composed of various components. At a minimum, these include a firewall, antivirus application, and intrusion detection system (IDS). IDSs are utilized to detect, determine, and identify instances of information management misuse, duplication, alteration, and destruction [1].

Data protection of cloud storage is an important consideration because it is provided via the Internet. According to a recent white paper on cloud computing security, intrusion identification and monitoring in cloud infrastructures are the second-biggest security problems after data security [2]. Virtualization techniques, integrated technologies, and common Internet protocols are all used in cloud infrastructure. These have a lot of vulnerabilities, which makes them attractive to attackers.

Various conventional threats, including Internet Protocol spoofing, Address Resolution Protocol (ARP) spoofing, Routing Protocol spoofing, flooding, Refusal of Service (DoS), DNS poisoning, Distributed Rejection of Service (DDoS), etc., also affect cloud computing. For instance, a DoS assault on the underlying infrastructure of Amazon Cloud caused BitBucket.org, a website hosted on AWS, to be offline for a short period [3]. When it comes to Cloud computing, the cost of implementing modern encryption algorithms cannot be overlooked [4]. While it may not be effective against cyber-attacks, a firewall can serve as a valuable alternative for preventing external attacks. To mitigate these threats, it is crucial to incorporate effective security mechanisms such as IDS (Intrusion Detection System) and IPS (Intrusion Prevention System) technologies into the cloud infrastructure.

Revathi and Malathi [5] focus on machine-learning intrusion strategies. They also describe a comprehensive collection of machine-learning techniques using the NSL-KDD vulnerability scanning dataset. However, their analysis only considers a misuse-based context, whereas this study discusses both abuse detection and anomaly detection. Sahoo et al. [6] identify and review various works that address different aspects of this subject, providing a formal definition of Malware Detection as a machine-learning task. However, they do not provide technical details of the algorithm, unlike this study.

ML techniques and their implementations for intrusion detection are the main topics covered by Buczak and Guven [7]. In-depth descriptions are provided for algorithms including Decision Trees, Fuzzy Logic, Bayesian Networks, Support Vector Machines (SVM), Genetic Algorithms, and Fuzzy Networks. Major ML and/or DL techniques like clustering, Intelligent machines, and swarm optimization are not mentioned, though. They concentrate on IDSs in their article. Attackers trying to enter a wired network must get past several operating systems and firewall defenses or physically access the network. However, a wireless network is more susceptible to malicious assaults and is more challenging to defend than a wired network because every node might be a target in one.

The development of virtual servers, service-oriented architectures, autonomous computing, and utility computing led to the creation of what is now commonly known as cloud technology. Malware monitoring was specially built into a platform for cloud computing. Information like a component's or piece of infrastructure's location [13].

In their thesis titled "N-Version Antivirus in the Networking Cloud," Oberheide [14] proposed a new strategy for antivirus deployment by offering antivirus capability as a network service while utilizing N-version security. This unique paradigm offers several benefits compared to standard host-based antivirus, including increased malware detection, greater forensic capabilities, and improved deployable and managed retroactive detection. The Cloud AV platform is used in a real-world deployment and production setting. Additionally, Schmidt et al. [15] presented a method for combining detection techniques and kernel rootkit avoidance in virtualization cloud computing environments, where all executing binaries in virtual instances are intercepted and transmitted with one or more analysis engines.

Since the late 1980s, malware identification has been a significant problem in computing. Scanning a computer network for infection by comparing malware identifiers to documents on the computer has been the main method of malware detection. Despite the high degree of reliability in the detection of known samples, signature-based detection is only effective for malware that has been obtained, examined, and for which an appropriate signature has been found. Murad et al. [16] demonstrated that it is possible to avoid signature-based detection by examining the malware's instructions and figuring out which instructions make up the signature.

Nataraj et al. [17] first investigated the use of byte plotting projection as binary images for automatic classification. They utilized a library of malware images consisting of approximately 9,342 malware samples divided into 25 different categories. GIST [18] features were extracted from the grayscale images and classified using K-nearest neighbor with the Distance function as the metric. Their approach requires a significant computational investment. Mirza et al. incorporated decision trees, SVM, and boosting techniques using features extracted from malware files [19] to detect malware.

For the classification of malware families, Vinayakumar et al. proposed a deep-learning method utilizing CNN and LSTM. Experiments on the Malimg dataset demonstrated an effectiveness of 96.3% [20]. Su et al. conducted research where they developed grayscale images from accessible binaries in two families and used a lightweight CNN to classify the images into their respective families. They achieved accuracy rates of 94.0% and 81.8% for malware and benign samples, respectively [21].

The vulnerabilities described by Modi et al. [22] pertain to cloud services and include issues with virtualized access technologies and unauthorized administration interfaces. They also discuss the impact of exploiting these vulnerabilities. Additionally, the paper explores the dangers posed to cloud services and potential fixes. It provides a list of relevant attack types, their impacts, and mitigation techniques that put the confidentiality, integrity, and availability of cloud resources at risk. Finally, the security vulnerabilities are categorized according to the various cloud infrastructure tiers they affect.

More security mechanisms that arise in cloud computing settings are listed by Khalil et al. [23] and are divided into five categories: specific reasons, networking, security systems, cloud hosting, and data. Accordingly, nine recognized attack groups have been identified, including cross-VM side channels, cloud malware insertion, and denial-of-service (DoS). The defenses against the attacks that were described have been explored and assessed. The authors also provide an outline of earlier work in the area of cloud computing security.

Using three different levels of classification, Ardagna et al. [24] present a comprehensive overview and classification of 306 scientific publications addressing vulnerabilities, risks, intrusions, and corresponding solutions in clouds. The security features of confidentiality, integrity, availability, authenticity, and privacy were taken into account when classifying the system. The authors describe security assurance as a more general phrase that encompasses strategies for gathering and assessing evidence that supports security attributes, in addition to discussing security-related challenges. Testing, monitoring, certification, audit/compliance, and service level agreements are the categories into which the security assurance methodologies considered in the publications surveyed have been divided.

5.3 CLOUD COMPUTING NETWORKS

Cloud computing is a technique for providing computer services such as hosting, storage, networking, applications, and data analysis over the internet. Businesses utilize the cloud to reduce costs, increase agility, and enhance cloud security. The scalability of cloud storage, including cloud security, allows for uninterrupted operations even during periods of rapid expansion.

Depending on the services offered to clients, cloud services can be categorized into three types: Software as a Service (SaaS), Platform as a Service (PaaS), and Infrastructure as a Service (IaaS).

SaaS is commonly used by businesses to distribute their applications to end users, with a private cloud being managed by a third party to maintain the software.

PaaS, a public cloud service, allows developers to work on their applications remotely. This simplifies the development management process, allowing them to focus solely on development while the company or a third party takes care of servers, storage, and networking.

Organizations can obtain a complete set of operational infrastructure, including security, communications, surveillance, and other services, through

Cloud services (IaaS), all on a cloud infrastructure. This simplifies organizational management procedures and frees up additional resources that would have been required with outdated infrastructure (Figure 5.2).

The most significant dangers we'll encounter are:

- Data privacy compliance
- Data breaches
- Unapproved entry
- Virus infections
- Cyberattacks

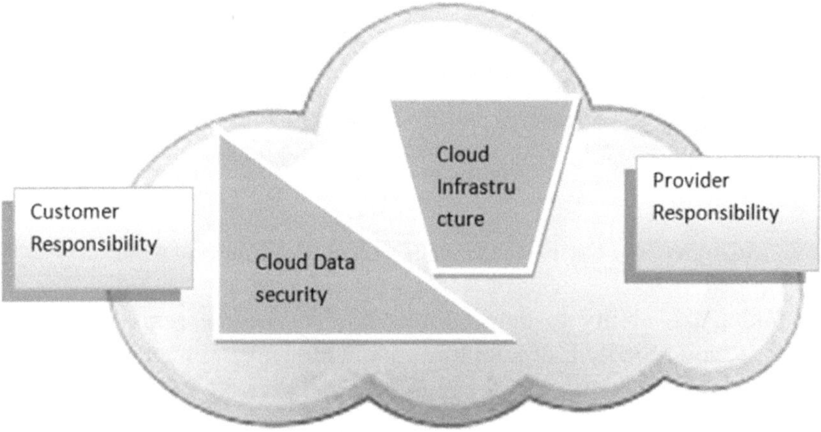

FIGURE 5.2 Cloud data security.

To protect data in the cloud environment from cyber threats and attackers, a range of technologies, services, policies, and best practices are consolidated under the umbrella of cloud security. Cloud-based solutions for cybersecurity are essential to prevent data loss and assist the organization in maintaining compliance with data privacy regulations.

5.4 CYBER SECURITY IN CLOUD

Tools, information, and infrastructure are needed for cloud cybersecurity to safeguard cloud applications and services from online threats. To prevent attacks, manage your identity, and protect the company's applications and infrastructure, it is necessary to integrate these cybersecurity tools.

5.4.1 AUTHENTICATION WITH MANY FACTORS

Investing in inter-authentication is crucial. To obtain service for an account, application, or network access, authentication requires users to provide two or even more verification methods. A solid identity and access (IAM) policy must include multi-factor authentication (MFA).

5.4.2 MONITORING OF TRAFFIC IN FULL AUTOMATION

By investing in monitored detection and response solutions, organizations can monitor internal traffic to the solution and receive real-time alerts about their source and shared data. These solutions use ML and AI to evaluate vast volumes of data from numerous sources and discover and notify you of issues in the surroundings. To shorten our mean time to find problems, this rate of deployment requires a level of current traffic and vulnerability capabilities.

5.4.3 SCANNING FOR VULNERABILITIES AUTOMATICALLY

Automating the ability to detect faults and implement remedies quickly, efficiently, and reliably is crucial for reducing the mean time to remediation. IT teams can manage most of the reporting and security patch management with the assistance of solutions.

5.4.4 UTILIZE AI SOLUTIONS TO REACT QUICKLY

AI is a vital tool for detecting breaches of cloud best practices, preventing attacks, and enhancing the effectiveness and precision of security analysts so they can react more quickly to situations. Implementing solutions with lower complexity and higher return on investment is crucial. The AI must be able to access its applications, technology, and data storage completely in order to succeed. To improve the output of the AI, it is important to keep such accessibility up to date.

5.4.5 MONITORING USER ACCESS

Essential to cloud cyber security is continuous monitoring of your network endpoints, internal and external users, and activities across them. Utilize cloud computing consultancy to assist you in recognizing and keeping an eye on all the ways that online predators can attack you, as well as to notify you of any unusual behavior (Figure 5.3).

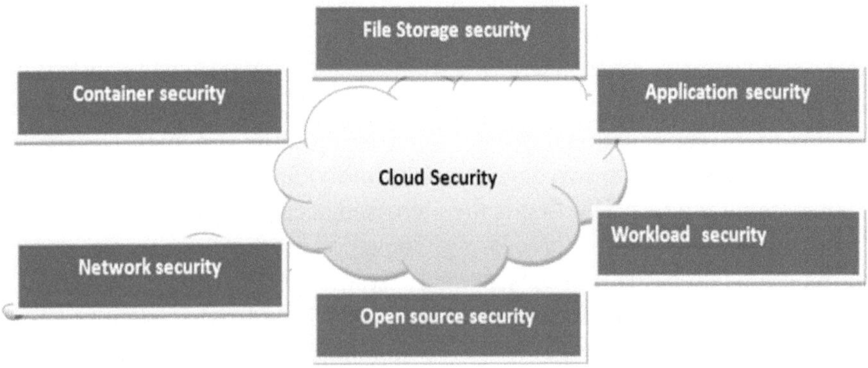

FIGURE 5.3 Cloud computing security.

5.4.6 *SAFEGUARD VALUABLE DATABASES*

Upgrading your firewall, implementing email server gateways, and keeping an eye on all of the access risk categories are essential elements of database protection. There are machine learning tools that can filter and evaluate network threats and warn you of suspicious activities.

5.4.7 *DATA ENCRYPTION BOTH IN TRANSIT AND AT REST*

Utilizing secure protocols to protect data originating from the internet and traveling through the infrastructure adds another layer of security. With this security, hackers find it more challenging to intercept your secure data transmission. Encrypting data on a disk or even in the backup media is also essential.

5.5 DEEP LEARNING

A basic explanation of deep learning is a collection of all machine learning techniques that aim to learn at multiple levels, corresponding to various degrees of abstraction. Higher-level categories are discussed based on lower-level ones, and numerous higher-level concepts can be defined using the same lower-level notions [8]. The initial levels of a deep network are responsible for feature extraction. Deep learning architectures come in unsupervised, supervised, and mixed varieties. Shallow neural networks, which only have one hidden layer, are unable to perform complex feature extraction and cannot capture the more intricate ideas that neural networks can. This holds true for many other machine learning techniques as well. To train deep learning models within a reasonable amount of time, deep learning approaches require more computational power, such as multiple graphics processing units (GPUs). Thanks to two developments, individuals can now easily create deep learning models. The first is the increased accessibility of GPUs, which enables significantly faster processing. The second is that the layers of a deep learning model can be trained separately from each other [9]. As a result, a complex model with a set of parameters can be optimized in smaller, more manageable portions, using much less computing power (Figure 5.4).

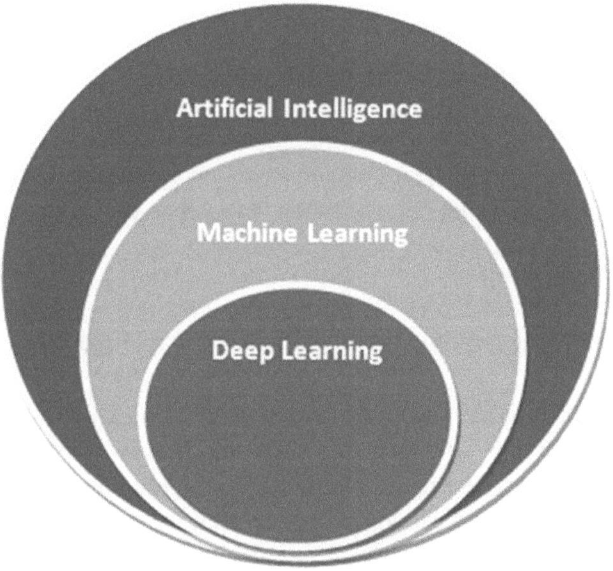

FIGURE 5.4 AI, ML, and DL.

5.6 DEEP NEURAL NETWORK ALGORITHMS

The following methods are used for a variety of cybersecurity activities, such as phishing, malware identification, intrusion detection, and cyber-attack prediction, including denial of service (DoS) assaults, fraud detection, and cyber anomalies, among others. In the context of cybersecurity, neural network and deep learning methods encompass supervised, semi-supervised, unsupervised, and reinforcement learning. Convolutional neural networks are referred to as ConvNets (CNN), while recurrent neural networks are known as RNN or long short-term memories (LSTM). Other examples of artificial intelligence techniques include self-organizing maps (SOM), restricted Boltzmann machines (RBM), auto-encoders (AE), deep belief networks (DBN), and generative adversarial networks (GANs). These neural network-based learning methods, along with their ensembles and optimization techniques, can be used to intelligently address various security problems, including intrusion detection, malware recognition, phishing, cyber-attack anticipation (including denial-of-service assaults), fraud detection, and identification of cyber abnormalities (Figure 5.5).

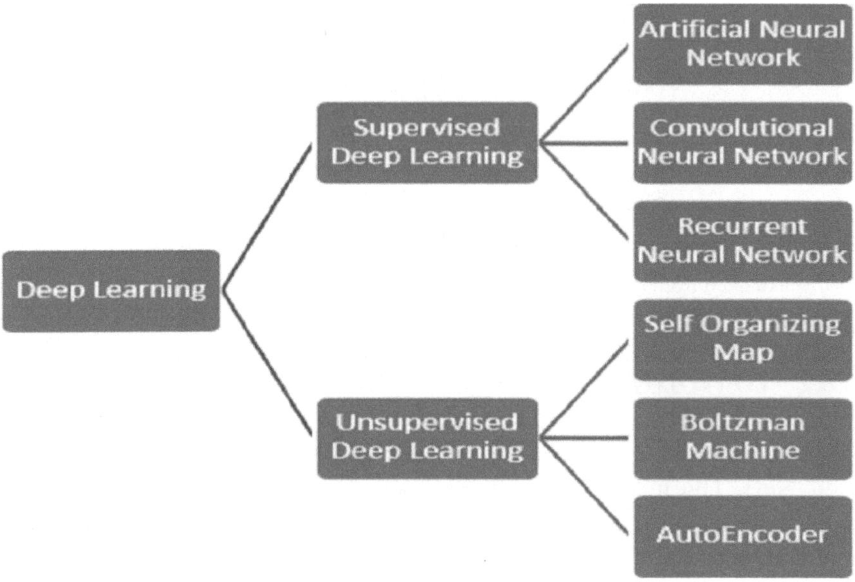

FIGURE 5.5 Deep Learning algorithms

5.6.1 MULTI-LAYER PERCEPTRON (MLP)

A class of feed-forward artificial neural systems(ANN) is based on the multilayer perceptron (MLP)—the most fundamental deep learning model, which consists of several fully connected layers, is an MLP model. Traditional deep learning architectures require significant computational resources, but MLP ML techniques can circumvent this limitation. Each additional layer is composed of a group of nonlinear functions that represent the weighted sum of all the outputs from the preceding layer (Figure 5.6).

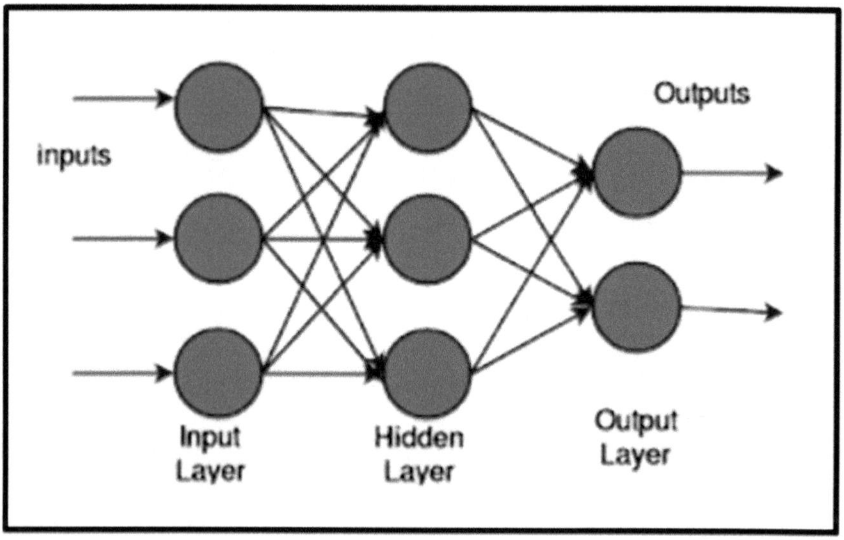

FIGURE 5.6 Multi-layer perceptron.

Source: Adapted from Ref. [36]

5.6.2 CONVOLUTIONAL NEURAL NETWORK

Another type of DL model is indeed a convolutional layer (CNN, or ConvNet). The most typical application of CNNs is in computer vision. The AI algorithm learns and automatically generates the properties of these inputs to achieve a specified goal, such as picture classification, face identification, or image semantic segmentation, given a sequence of real-world images or videos and using CNN. In contrast to fully connected layers in MLPs, one or more convolution layers in CNN models use convolution operations to extract simple characteristics from the input. Each layer consists of a collection of nonlinear functions that compute weighted sums of spatially nearby subsets of the outputs from the preceding layer at various positions (Figure 5.7).

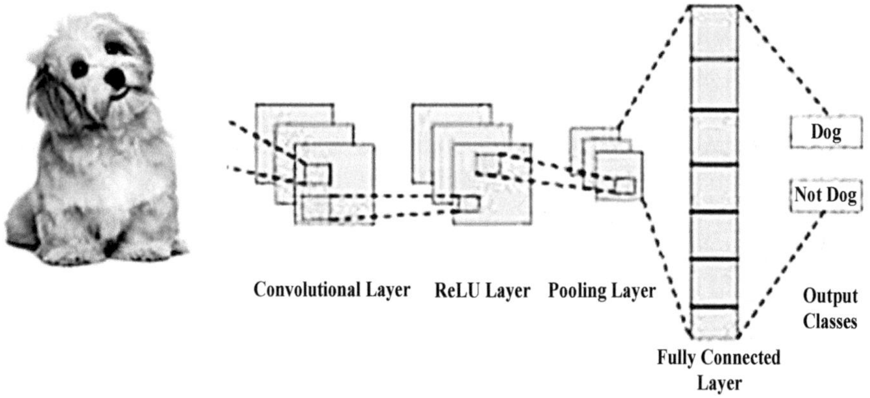

FIGURE 5.7 Convolutional Neural Network

5.6.3 RECURRENT NEURAL NETWORK (RNN) AND LONG SHORT-TERM MEMORY (LSTM)

An artificial neural network (ANN) that utilizes sequential data feeding is known as a Recurrent Neural Network (RNN). RNNs have been developed to handle the time-dependent nature of sequential input data. The input of an RNN consists of the current sample as well as the prior samples, forming a directed graph that represents the connections between the nodes in a time order. Each neuron in an RNN has its internal memory, where it stores computation-related data from previous samples (Figure 5.8).

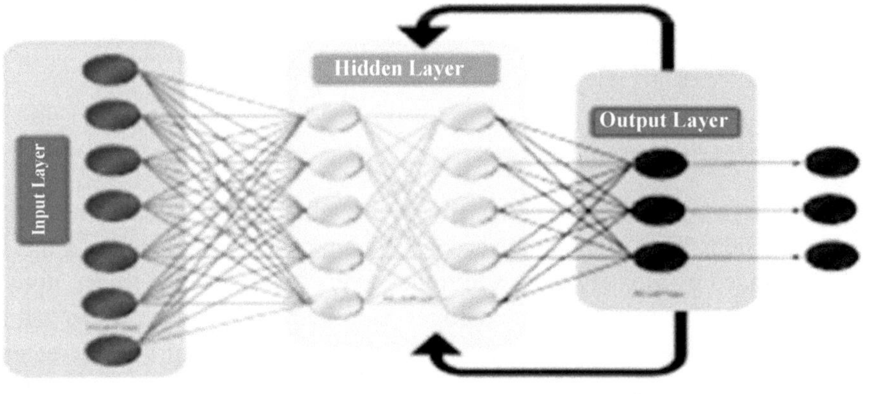

FIGURE 5.8 RNN.

RNNs of the LSTM variety are capable of learning order dependency. The current RNN step uses the output from the previous step as its input. Hochreiter & Schmidhuber created the LSTM. It addressed the issue of RNN long-term dependency, where the RNN can forecast words based on current data but is unable to forecast words in the future. As the gap length increases, RNN's performance becomes inefficient. By default, the LSTM can retain data for a very long time. It is used in the analysis, prediction, and classification of time-series data.

The LSTM is composed of multiple neural networks and several memory cells connected in a chain arrangement. A neuron, an input gate, an output unit, and a memory gate make up a typical LSTM unit. Three gates regulate the information flow into or out of the cell, and the cell retains values for arbitrary periods. The LSTM method is ideal for classifying, examining, and forecasting time series with various durations.

5.6.4 SELF-ORGANIZING MAPS (SOM)

An artificial neural network (ANN) called the Self-Organizing Map (SOM) creates a two-dimensional representation of a problem area through unsupervised learning. Unlike error-correction learning methods like back propagation with gradient descent, the self-organizing method involves competition for training. This is the main distinction between a self-organizing map and other problem-solving approaches. A self-organizing map can generate visual presentations on both hexagonal and rectangular grids. Its applications include oil and gas exploration, project prioritization, oceanography, meteorology, and more. The self-organizing map is also known as a Kohonen map or a Self-Organizing Feature Map (SOFM) (Figure 5.9).

5.6.5 AUTO ENCODER (AE)

An encoder, a code, and a decoder are the three fundamental components of an autoencoder. The original data is converted into a codified result, which is then expanded by the network's additional layers to produce the final output. Examining a "denoising" autoencoder can help you better understand autoencoders. The denoising autoencoder refines the output and recreates something that accurately uses a single set of inputs, incorporating both the original inputs and the noisy input. Autoencoders are useful in image processing, classification, and other areas of machine learning (Figure 5.10).

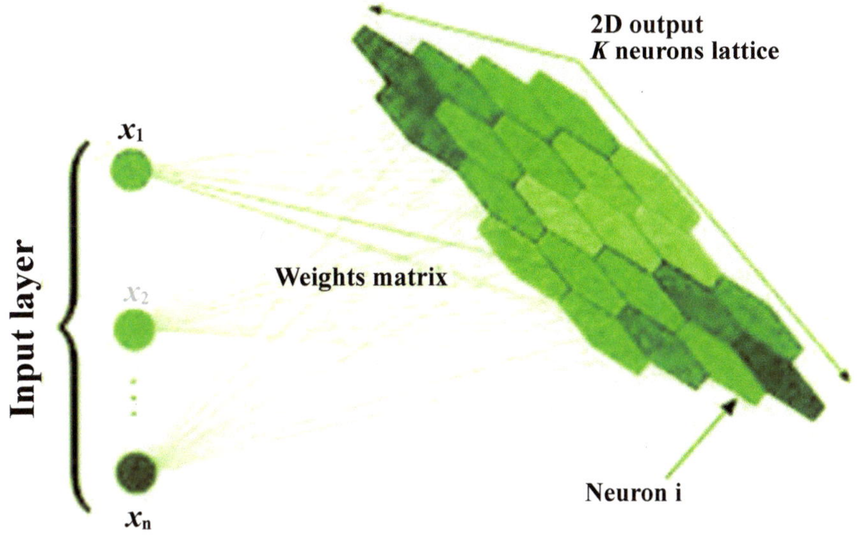

FIGURE 5.9 SOM.

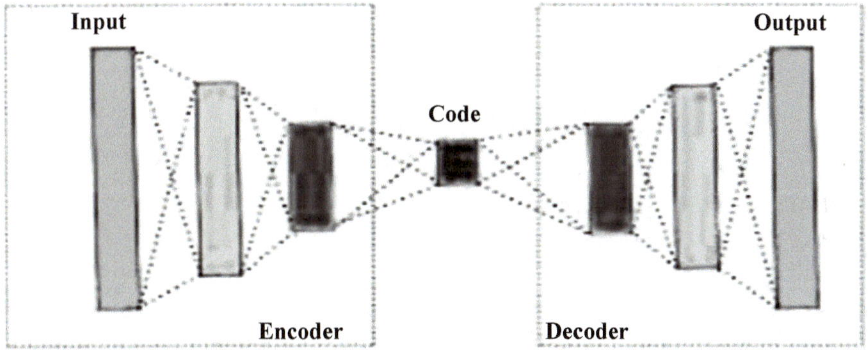

FIGURE 5.10 Auto encoder.

5.6.6 RESTRICTED BOLTZMANN MACHINES (RBM)

RBM is also a stochastic two-layered neural network that can automatically identify underlying patterns in data by recreating input. It is a subset of energy-based models. RBMs have two layers, one of which is hidden. At the hidden layer, a weighted total of the input layers is produced. The hidden layer is constructed by nodes that collect feature knowledge from big data and are buried beneath the visible layer. RBMs do not have any output nodes, which might appear unusual, and they do not have the typical binary output that allows for the learning of patterns. They vary because learning cannot take place without that capacity. We are not concerned about hidden nodes; we only care about input nodes. RBMs automatically track all the patterns, parameters, and relationships among the data once input is given (Figure 5.11).

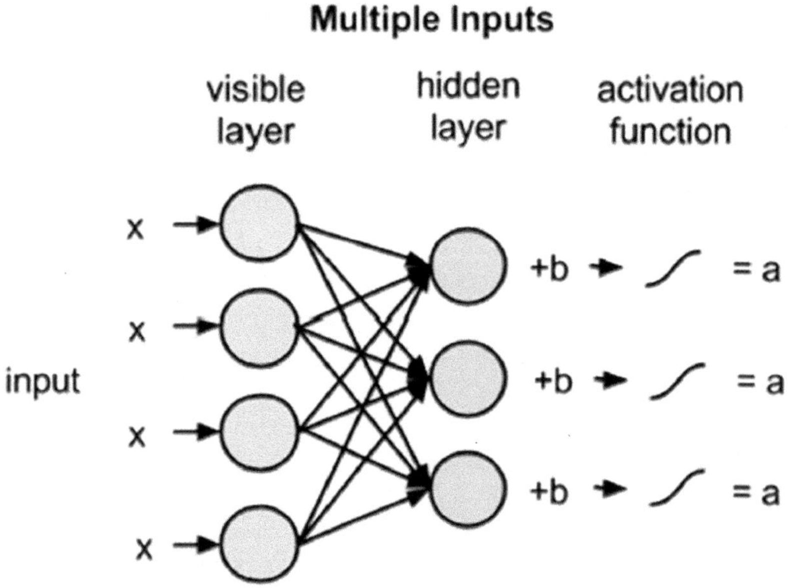

FIGURE 5.11 Restricted Boltzmann machines.

Source: Adapted from Ref. [37]

5.6.7 GENERATIVE ADVERSARIAL NETWORKS (GAN)

In an ML model called the GAN, two neural networks compete with each other to improve their prediction accuracy. GANs typically operate in an unsupervised manner and learn through cooperative zero-sum games. The producer and the discriminator are the two ANNs that make up a GAN. The discriminator is a coevolutionary neural network, while the generator consists of a convolutional layer. The objective of the generator is to generate outputs that closely resemble real data. On the other hand, the discriminator's objective is to determine whether the received outputs are real or generated intentionally. The first steps in creating a GAN involve selecting the desired output and creating a basic training dataset based on specific parameters. The generator is then trained using this randomized data until it achieves a certain level of output accuracy. Subsequently, the discriminator is trained using real data points from the original concept as well as the generated images. The discriminator evaluates the data and assigns a likelihood value between 0 and 1 to indicate the authenticity of each image. These likelihood values are manually verified for accuracy, and the process is repeated until the desired result is achieved (Figure 5.12).

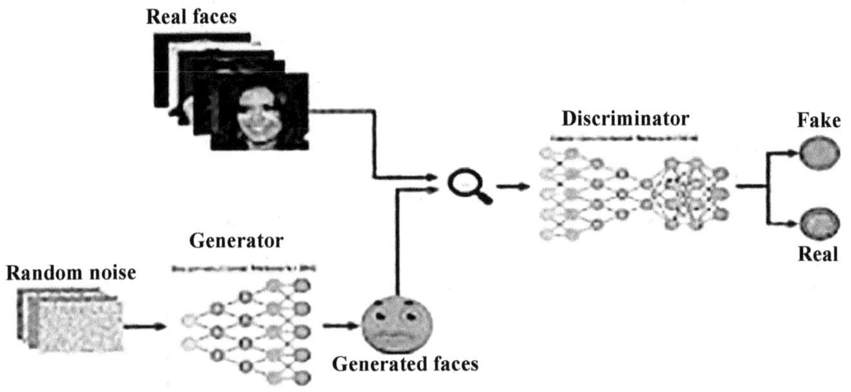

FIGURE 5.12 GAN.

5.7 DEEP LEARNING APPLICATIONS IN CYBER SECURITY

Deep neural networks (DNN) can learn using unsupervised data and tackle complex problems thanks to ML techniques. They have a wide range of applications in cybersecurity, defending businesses from dangers including phishing, spear-phishing, drive-by attacks, password attacks, DoS, etc. Explore the top 10 uses of deep learning within cybersecurity.

5.7.1 FINDING EVIDENCE OF AN INVADER

To enhance the intelligence of ID/IP systems, learning techniques such as CNNs, including RNNs, can be utilized to analyze network traffic with greater accuracy. This can lead to a reduction in false alarms and aid security teams in distinguishing between malicious and legitimate network activity. Notable solutions in this regard include Next-Generation Firewalls (NGFW), Web Application Firewalls (WAF), and User Identity and Behavior Analytics (UIBA).

5.7.2 THE FIGHT AGAINST MALWARE

A signature-based detection technique is used by conventional malware solutions, such as common firewalls, to identify malware. The company maintains a set of known risks, which is regularly updated to include new threats that have recently emerged. While this approach is effective against basic threats, it struggles with more advanced attacks. Deep learning algorithms can detect more complex threats because they are not reliant on the memory of well-known signatures and typical attack patterns. Instead, they learn about the system and identify unusual behavior that may indicate malware or malicious actors.

5.7.3 SOCIAL ENGINEERING AND SPAM DETECTION

A deep learning method called NLP can assist you in quickly identifying and dealing with spam and other types of social engineering. NLP uses numerous statistical models, natural methods of communication, and linguistic patterns to identify and block spam.

5.7.4 ANALYSIS OF NETWORK TRAFFIC

To detect suspicious activity, deep learning ANNs are analyzing HTTPS network traffic and yielding promising results. This technology greatly simplifies the management of various online risks, such as SQL injection attacks and DOS attacks.

5.7.5 ANALYTICS FOR USER BEHAVIOR

An essential deep learning-based security strategy for any firm is to track and examine user activities and habits. Since it goes beyond security mechanisms and often does not trigger any flags or alerts, it is significantly harder to detect than regular harmful operations against networks. One effective defense against these attacks is User and Entity Behavior Analytics (UEBA). After a period of adjustment, it can learn the typical patterns of employee behavior and identify suspicious activity that may indicate an insider attack, such as accessing the system at unusual hours and raising alarms.

5.7.6 EXAMINING E-MAILS

Monitoring individuals' official email accounts is crucial in preventing any type of cybercrime. For instance, phishing attacks often involve sending emails to staff members and requesting sensitive information from them. Deep learning and cybersecurity software can be utilized to prevent certain types of attacks. By using NLP, emails can be examined for any suspicious activity.

5.7.7 MOBILE ENDPOINT ANALYSIS

Deep learning is already commonplace on mobile platforms and powers voice-based interactions with mobile assistants. Therefore, when the company wants to mitigate the increasing amount of malware on mobile devices, deep learning can be used to discover and analyze threats against mobile endpoints.

5.7.8 IMPROVED HUMAN ANALYSIS

Deep learning and cybersecurity can aid in conducting vulnerability assessments, detecting malicious attacks, and providing endpoint protection. By identifying alternative approaches to solving challenges, individuals can make more informed decisions as a result.

5.7.9 AUTOMATING WORK

The key advantage of deep learning is its ability to automate repetitive operations, allowing staff to focus on more crucial work. Some cyber security jobs can be automated with the help of machine learning. Deep learning can help enterprises perform activities more quickly and effectively, and prevent the increasing number of viruses on mobile devices.

5.7.10 WEBSHELL

A maliciously added piece of code called WebShell provides unauthorized access to the system's websites, allowing changes to be made. Consequently, intruders can gain access to the database. Deep learning can aid in detecting typical shopping cart behavior, and the models can be trained to differentiate between lawful and unlawful conduct.

5.7.11 SCORE NETWORK RISK

Deep learning can also be used to evaluate datasets from prior cyber-attacks and identify the parts of the network that were targeted by a specific attack. This can aid in thwarting an attack concerning a particular network area.

5.8 MALWARE ATTACKS IN CLOUD COMPUTER NETWORKS

Any malicious program or code that has the potential to damage systems is referred to as malware. Malware is harmful software that disrupts or interferes with computer systems, such as viruses, worms, trojans, and spyware [10]. Software designed to carry out any form of unpleasant or destructive behavior on computer infrastructure is known as malware. These activities aim to damage or modify the compromised system, steal sensitive information, or utilize spyware. Malware comes in various forms, including viruses, worms, trojan horses, rootkits, backdoors, spyware, adware, ransomware, and more [11, 12].

5.8.1 CLOUD MALWARE

Security in the cloud is challenging. While cloud service providers (CSPs) are responsible for the security of the technology they manage, cloud customers are responsible for properly configuring cloud security and protecting their workloads and applications. Malware infection in a cloud computing environment is one of the most significant security risks that can arise from incorrect configuration and a lack of safeguards at the application level.

Although cloud malware is still a relatively new problem, cybercriminals have already recognized the suitability of cloud platforms for distributing viruses. Systems hosted in the cloud are

- Usually accessible over the Internet.
- Standardized and simple for an attacker to learn.
- Consisting of various components, such as virtual machines (VMs), containers, or storage buckets, each of these elements can potentially serve as a vulnerability that attackers can exploit.

5.8.2 INCREASING CLOUD MALWARE

According to studies, as the use of the cloud expands, data breaches are now more likely to occur in nearly 90% of firms. The use of malware is common in many of these breaches, much like in conventional data centers. Cloud use is on the rise, and with it comes greater threats. Therefore, cloud security is essential for any enterprise.

Businesses utilize an average of 1,181 cloud services, but according to a Netskope survey, 92.7% of them are neither secure nor prepared for enterprise demands. Malware in cloud systems can propagate to collaborators, whether they are clients, partners, or contractors, survive system cleanups, and pose a risk to sensitive data repositories linked to the infected system.

5.8.3 ATTACKS VIA CLOUD MALWARE

Here are a few frequent cyberattacks that utilize cloud viruses.

5.8.3.1 DDOS THREATS

Millions of hacked devices are combined to form massive botnets, which are increasingly available to attackers. Threat actors are offering botnets as a service at low prices, making it easier for anyone to launch a DDoS attack.

A DDoS assault on your business or any of the neighbors in the cloud platform can have an impact on the entire "neighborhood" as well as the underlying cloud architecture. Additionally, there is a persistent risk that uncontrolled VMs or containers will be infiltrated by hackers, exposing the cloud computing resources to the risk of misuse.

5.8.3.2 ATTACKS THROUGH HYPERCALL

During a hypercall assault, an attacker compromises an organization's VMs using the hypercall handler. This is a part of virtual machine management that is set up on each cloud computing system used by solutions like Amazon EC2. In some circumstances, the technique also enables attackers to run malicious software on the VM, giving them access to VMM rights (Figure 5.13).

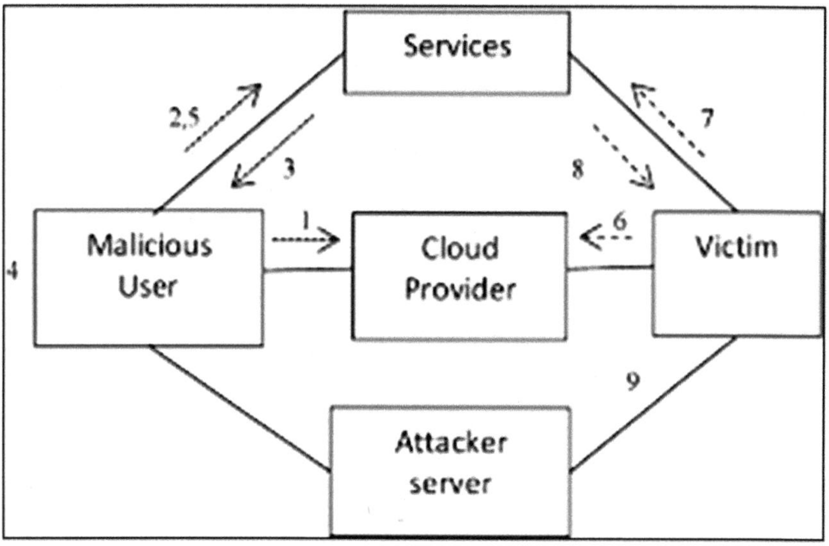

FIGURE 5.13 Malicious user attacks in the cloud.

5.8.3.3 DOS ON HYPERVISOR

A hypervisor assault is one in which the attacker takes advantage of the hypervisor, which governs numerous virtual machines on a virtual host. Malware that infects the hypervisor can impact each of the VMs that are active on the host.

The resource usage of a virtual machine may increase as a result of an infected hypervisor, which could lead to a Denial of Service (DoS) attack on the entire host or potentially multiple hosts. Since hosts are typically networked and do not necessarily require authentication for connections from other servers, they can easily infect other hosts, making the issue much more serious.

5.8.3.4 HYPERJACKING

A rootkit that has been placed on a virtual machine is used by the attacker in a hyperjacking attack to attempt to seize control of the hypervisor. If the attack is successful, the attacker will have complete control over the host and will be able to alter virtual machines' behavior, harm already-running VMs, and even launch brand-new hostile VMs.

5.8.3.5 VULNERABILITIES OF DATA

When data is stored in the cloud, there is a risk that unauthorized individuals may access the data through cloud service providers. Although many CSPs state that they do not have access to personal information, it can be difficult to fully trust their claims. Furthermore, human error can also pose a threat to cloud-based data.

5.8.3.6 MALWARE INJECTION INTO THE CLOUD

A malicious application will be created by the attacker and introduced into SaaS, PaaS, and IaaS. Once the malware has been introduced into the cloud program, it will reroute user requests to the hacker's module, resulting in the execution of harmful code. As a result, hackers can now modify data, steal information, and spy on the content.

Attacks using SQL injection and cross-site script attacks are the two types of malware injections that occur most frequently.

- **SQL Injection:** This attack targets the SQL servers of the cloud infrastructure. The attacker exploits vulnerabilities in database applications and injects malicious code. This allows them to gain unauthorized access and login credentials.
- **Cross-site Scripting:** In this attack, a cyber attacker injects malicious scripts into a vulnerable online page to gain access to the victim's web browser. The attacker may then use the authorized session cookie to access the victim's account or deceive the victim into clicking on a harmful link.

5.8.3.7 ATTACKS ON A DENIAL OF SERVICE (DOS)

DoS attacks overload the server and system, preventing users from accessing it. One drawback of DoS attacks is that they can affect multiple users by overwhelming a single cloud service. To handle the increased workload on the cloud system, additional virtual machines and services are installed, which can be harmful.

Once the cloud infrastructure starts to slow down, users will lose access to the server. Further damage can occur if hackers continue to use zombie devices or deploy attacks.

5.8.3.8 MISCONFIGURATION

Technology for cloud computing offers more than just accessibility and cooperation. Documents are also stored there. Misconfigured clouds can cause significant harm to businesses, potentially resulting in a security compromise if discovered by hackers.

Misconfiguration of the cloud refers to a flaw or hole in the user's cloud architecture that makes their data accessible. It is among the most prevalent threats to cloud security.

Misconfiguration may result from a lack of access restrictions. Businesses risk unauthorized access if they grant unrestricted access to the cloud. Attackers can easily modify or steal data stored in the cloud.

5.8.3.9 UNPROTECTED APIS

A software interface called a cloud API connects various cloud computing services, enabling the sharing of data and functionality between two programs. It is essential for businesses that utilize multiple applications, as data exchange is crucial for seamless processes.

However, if APIs are left unprotected, attackers can exploit them in various ways. One method is by taking advantage of a weak authentication feature in an API. Since the data stored in the cloud network is publicly available and easily accessible online, attackers can gain unauthorized access to it.

Another way attackers exploit APIs is by targeting the backend, which developers often overlook. Without proper authorization controls in place, hackers can modify the backend to their advantage.

5.8.3.10 UNPREDICTABLE ACTIVITY OF END-USERS

Insider attacks and breaches may result from organizations losing control over how their staff members access the cloud infrastructure. Since they already have access, insider threats would have an easy time stealing the data or information they require. No DDoS assaults, bypassed firewalls, or other means of access are required. In fact, it would be easier for the attacker.

Dealing with employees' misconduct can be challenging at times. Controlling end users can be difficult because it might be hard to hold personnel accountable for negligence or data loss. However, it is crucial to establish guidelines for monitoring, investigating, and evaluating your personnel.

5.8.3.11 MAN-IN-THE-CLOUD (MITC)

MitC hackers use flaws in the synchronization token system to intercept and modify cloud services. To provide attackers with access to subsequent synchronization with the cloud, they replace the password with a new one. Users may not be aware that their identities have been compromised, while an attacker can always revert to the previous synchronization tokens.

5.8.3.12 WEAK CRYPTOGRAPHY

Even though cloud services use cryptosystems to secure data in storage, they often rely on finite resources of entropy (such as time) to automatically create random data for data encryption. For instance, virtual computers running Linux only produce random keys from the precise moment. However, as hackers also employ advanced decoding techniques to obtain information, this might not be sufficient for reliable data encryption. Therefore, before data is transferred to the cloud, cloud engineers should consider how to safeguard the data.

5.8.3.13 NEW ATTACKS

These two categories of cyberattacks have already emerged as a fresh danger to cloud computing since they first surfaced earlier this year. Adversaries can access encrypted memory data by exploiting a design flaw in the majority of contemporary processors using malicious JavaScript code. Both methods allow attackers to read data from the kernel by compromising the separation between the operating system and applications. The fact that not every cloud user installs the most recent security fixes is a major concern for cloud developers.

5.9 DATASET SELECTED

Harmful images of 9,339 from 25 different malware families make up the Malimg [25] Dataset. The malware pictures are available in a range of sizes and lengths. Several malware families, including Rogue, PWS, Dialer, Backdoor, Worm, Worm-AutoIT, Trojan, and Trojan-Downloader, were used to create the images. Every single malware image is an 8-bit vector binary created from a PE file that has been converted into an image. The viral image sizes were altered to be entered into a CNN model. The split of families for the Malimg dataset is shown in Table 5.1. Here we abbreviate "password stealing" as "pws," "downloader" as "dl," and "backdoor" as "bd" in the dataset.

TABLE 5.1 Malware Types of Malimg Dataset

Family	Type	Samples
Adialers.C	dialer	122
Agent.FYI	bd	116
Allaple.A	worm	2,949
Allaple.L	worm	1,591
Alueron.gen!J	trojan	198
Autorun.K	worm	106
C2LOP.gen!g	trojan	200
C2LOP.P	trojan	146
Dialplatform.B	dialer	177
Donrovo.A	dl	162
Fakerean	rogue	381
Instantaccess	dialer	431
Lolyda.AA1	pws	213
Lolyda.AA2	pws	184
Lolyda.AA3	pws	123
Lolyda.AT	pws	159
Malex.gen!J	trojan	136
Obfuscator.AD	dl	142
Rbot!gen	bd	158
Skintrim.N	trojan	80
Swizzor.gen!E	dl	128
Swizzor.gen!I	dl	132
VB.AT	worm	408
Wintrim.BX	dl	97
Yuner.A	worm	800
Total	–	**9,339**

5.10 PREPROCESSING

The Malimg dataset contains images of malicious software. The raw data for each file contains the binary content encoded in hexadecimal. The goal is to convert these files into PNG images to be used as input by our CNN. We treat the series of bytes that make up a binary as if it were a grayscale PNG image. In each test, we use a configurable length with a predetermined width of 256, depending on the binary's size. Determining the type of attack when dealing with malware by simply looking at the binary malware code can be challenging. We transform a malware sample's parallel executable file into an octal decimal value ranging from 0 to 255 and store it in a decimal system vector. Regardless of the size of a malicious binary file, we choose the dimensions of a 2D matrix. Then, by repeating the grayscale signals for three slices, we transform this grayscale image into 3-channel (red (R), green (G), and blue (B)) images. Here is the result of converting 0ACDbR5M3ZhBJajygTuf.bytes from binary to PNG format. Then, we extract the relevant data to produce families or directories of normalized tensor image files. These images can be used for instruction and assessment. The pictures are cropped to the necessary 64x64 dimensions. Using the train_test_split() method, we divide the data into train and test groups in batches according to a 70–30 ratio (Figure 5.14).

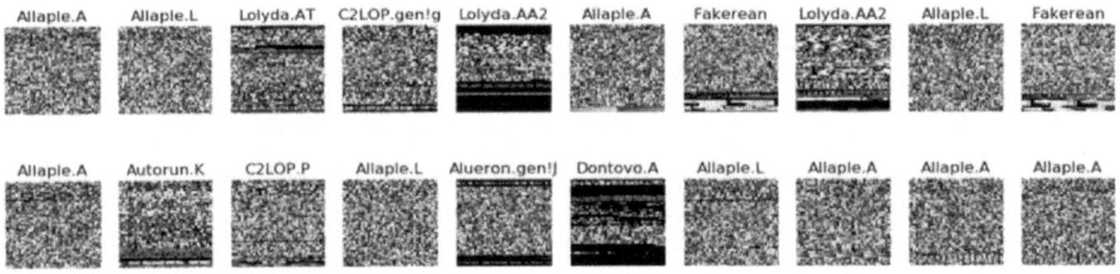

FIGURE 5.14 Malware samples.

Source: Adapted from Ref. [25]

The images need to be preprocessed to remove noise and abnormalities. Resizing the images is necessary because the image samples in the dataset have different sizes. When the dataset is not large enough, a technique called data augmentation is applied. To address this issue, data imbalances and data augmentation techniques are used when providing input data to the CNN model. This method takes the existing dataset and artificially creates a new one by transforming the input image dataset. Image data augmentation involves transformations such as rotation, shearing, zooming, flipping the images horizontally or vertically, and adjusting the brightness levels. In the experimental work, the deep learning package Image Data Generator provided by Keras is used to augment the dataset while monitoring progress. Table 5.2 shows the parameters applied to the modified dataset (Figure 5.15).

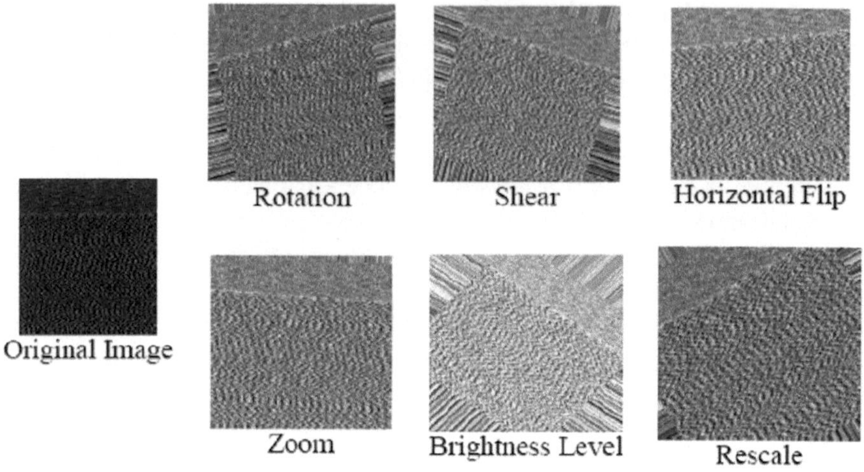

FIGURE 5.15 Malware samples after the augmentation process.

TABLE 5.2 Image Augmentation Settings of Dataset

Method	Setting	Method
Rotation Range	0.1	Rescale
Width Shift	0.1	Shear Range
Height Shift	0.1	Zoom Range
Horizontal Flip	True	Fill Mode

5.11 CONVOLUTIONAL NEURAL NETWORKS

A feed-forward neural network called a convolutional network (CNN) is biologically inspired, specifically by the structure of the visual brain of animals [26]. The most advanced neural network architecture for image classification currently is CNN. The neurons comprising CNN have biases and weights that can be learned. The three elements listed below constitute the majority of CNNs [27] (Figure 5.16):

- **Layers of Convolution:** These layers sequentially perform a set number of convolution operations (also known as linear filtering) on the image. Typically, these filters gather information from the input image regarding edges, colors, and shapes. Essentially, the filters operate on sub-regions of a picture and perform computations to produce a single value for each sub-region.

- **Pooling Layers:** Each layer is responsible for down sampling (i.e., reducing the resolution of the input nodes) the data obtained from convolution layers in order to decrease processing time and enable the computer to handle the scale of the data. This is because pooling results in fewer parameters that need to be learned in the deeper layers of the network. A commonly used pooling method that retains the maximum value in a zone is called max pooling.
- **Fully Connected Layers:** This layer classifies the output produced by the pooling and convolution layers. Each neuron in this layer is connected to every neuron in the layer above it. This type of layer is typically followed by a Dropout layer, which improves the model's ability to generalize by preventing overfitting, a common issue in the field of deep learning (Figure 5.16).

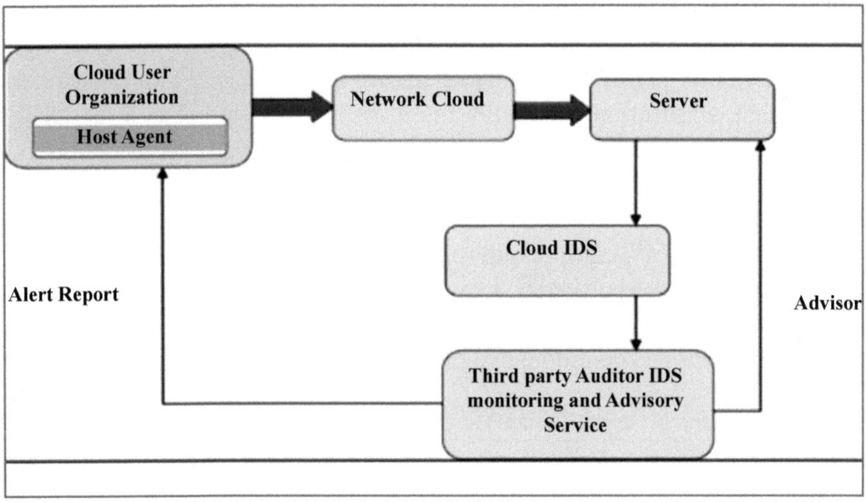

FIGURE 5.16 Cloud IDS monitoring.

5.12 FEATURE EXTRACTION USING AUTOENCODERS

The process of feature extraction involves reducing the dimensions of existing features. The feature extraction method has led to improved accuracy in detection and segmentation in various areas such as digital photos or video streams [28].

By minimizing the reconstruction layer, the auto-encoder (AE) utilizes the concept of an Artificial Neural Network (ANN) to reduce the dimensionality of the hidden layer stack. The effectiveness of the auto-encoder in extracting valuable features has been demonstrated through different variations, including Sparse Autoencoder, Variation Autoencoder, De-noising Autoencoder, and Relationship Autoencoder [29–31].

Feature extraction is one of the steps in the categorization process within the field of informatics engineering, particularly in the domain of network and information security. The increasing internet traffic in expanding networks poses a significant challenge. Additionally, the growth of IoT and cloud-based services has further intensified the connectivity.

An autoencoder, which is a type of neural network model, aims to identify a compressed representation of the input. Although they are formally trained through self-supervised learning, they are actually learned through unsupervised learning. Autoencoders are usually trained as part of a larger model that attempts to replicate the input. This is intentionally made challenging by the design of the autoencoder model, which restricts the architecture to a bottleneck at the halfway point of the model, where the data input reconstruction is performed. While there are various types of autoencoders with diverse applications, arguably the most popular one is as a model for automatic or learned feature extraction. In this scenario, once the model is fitted, the reconstruction portion can be omitted, and the model can be used up to the bottleneck. The model's output at the bottleneck is a fixed-length vector that provides a compressed representation of the input data [32].

The output at the bottleneck can then be utilized as a feature vector in a supervised learning model, for visualization, or more generally for dimensionality reduction. The model can then be fed input data from the domain.

$$Y = F_{\theta}(X) = S(WX + B_X) \tag{1}$$

$$X' = G_{\theta'}(Y) = S(W'X + B_Y) \tag{2}$$

Where the parameter f(X) is an encoded function and g(Y) is a decoded function, with the parameters W (weighting) and b (bias) for data x. S is a function activation. The decoder function g maps the hidden representations of Y back to reconstruct X.

5.13 EVALUATION METRICS

On the 64-bit Ubuntu 14.04 operating system, the TensorFlow [23] software framework is utilized for experiments. The server is a DELL R720 with 16GB of RAM and a 16-core CPU. As an accelerator, a Tesla K40m GPU from

Nvidia is used. The remaining data is training data, with one-tenth of the data being randomly chosen as test data. The cost function is cross-entropy, and the mini-batch size is 50. Tensor Flow's built-in gradient descent optimizer is used as the optimizer. Training takes 40 epochs, with a learning rate of 0.001. The following four evaluation criteria were used: F1 value, recall (R), precision (P), and accuracy (A) (F1). A classifier's total performance was assessed using accuracy. Every class of traffic's performance was assessed using precision, recall, and F1 value.

$$A = \frac{TP + TN}{TP + FP + FN + TN} \tag{3}$$

$$P = \frac{TP}{TP + FP} \tag{4}$$

$$R = \frac{TP}{TP + FN} \tag{5}$$

$$F1 = \frac{2PR}{P + R} \tag{6}$$

Where FP is the number of cases wrongly classified as X, FN is the number of occurrences incorrectly classified as Not-X, and TP is the number of instances correctly identified as X.

TABLE 5.3 Different Model Accuracies

Model	Accuracy
MLP-SVM [33]	80.46
CNN-SVM [33]	72.22
RANDOM FOREST [34]	96.26
M-CNN [34]	98.26
CNN-AE	99.11

5.14 RESULTS AND DISCUSSION

We were unable to directly use grayscale images with CNN because the input layers required the structure of an image, and the Malimg dataset is not equivalent to the ImageNet dataset (3, 224, 224). The image's Red, Green, and Blue (RGB) channels are represented by the number 3, while grayscale images require (1, 224, 224). Figure 5.17 displays the malware prediction outcomes using each of these models. The accuracy of the CNN-AE model was 99.11%, which was significantly higher. The model's highest projected accuracy after 10 runs is shown in Table 5.3. CNN-AE performed well.

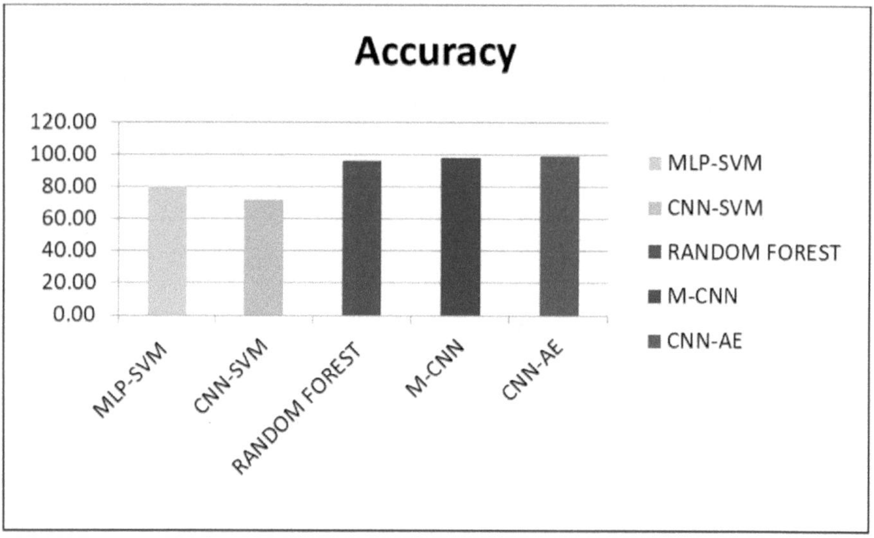

FIGURE 5.17 Model accuracies.

5.15 CONCLUSION

Deep learning techniques are being used more frequently for many applications, including cybersecurity. Thus, it is critical to determine when and which class of algorithms will produce appropriate results. We evaluate these methods for three pertinent issues in cybersecurity: spam detection, malware analysis, and intrusion detection. The automation of cloud security has several positive effects on businesses. However, compared to non-automated security, it can be more expensive and complicated. The necessary technological upgrades are expensive and time-consuming to implement. Additionally, a high level of security expertise is increasingly required to set up and manage security automation. Consequently, whether we are moving towards the cloud or are already there, a lot of antivirus products nowadays use deep learning techniques to safeguard devices against infection. When used with Windows PE binaries, deep learning architectures have shown good performance in the detection of malware. To categorize malware imagery in grayscale, we have employed CNN models. The model was successfully trained on the Malimg dataset, and the findings show that it outperforms all other compared works. To the best of our knowledge, it performs at the cutting edge in terms of malware classification on grayscale photos.

KEYWORDS

- **deep neural networks**
- **automated cyber security**
- **cloud service provider**
- **deep neural network**
- **artificial intelligence**
- **convolutional neural networks**
- **natural language processing**

REFERENCES

1. Mukkamala, A., Sung, A., & Abraham, A., (2005). Cyber security challenges: Designing efficient intrusion detection systems and antivirus tools. In: Vemuri, R. V., (ed.), *Enhancing Computer Security with Smart Technology* (pp. 125–163).
2. Martin, L., (2010). *White Paper*. http://www.lockheedmartin.com/data/assets/isgs/documents/CloudComputingWhitePaper.pdf (accessed on 27 January 2024).
3. Brooks, C., (2009). *Amazon EC2 Attack Prompts Customer Support Changes*. Tech Target. http://searchcloudcomputing.techtarget.com/news/article/0,289142,sid201_ci1371090,00.html (accessed on 27 January 2024).
4. Chen, Y., & Sion, R., (2010). *On Securing Untrusted Clouds with Cryptography* (pp. 109–114). In WPES;10.
5. Revathi, S., & Malathi, A., (2013). A detailed analysis on NSL-KDD dataset using various machine learning techniques for intrusion detection. In: *Proc. Int. J. Eng. Res. Technol.* (pp. 1848–1853).
6. Sahoo, D., Liu, C., & Hoi, S. C. H., (2017). *Malicious URL Detection Using Machine Learning: A Survey.* [Online]. Available: https://arxiv.org/abs/1701.07179 (accessed on 27 January 2024).
7. Buczak, A. L., & Guven, E., (2016). A survey of data mining and machine learning methods for cyber security intrusion detection. *IEEE Commun. Surveys Tuts.* (Vol. 18, No. 2, pp. 1153–1176). 2nd Quart.
8. Deng, L., & Yu, D., (2014). Deep learning: Methods and applications. *Found. Trends Signal Process., 7*, 197–387. [CrossRef].
9. Hinton, G., Osindero, S., & Teh, Y. W., (2006). A fast-learning algorithm for deep belief nets. *Neural Comput., 18*, 1527–1554. [CrossRef.

10. Malware Definition, (2017). https://techterms.com/definition/malware (accessed on 27 January 2024).
11. Adebayo, O. S., & Aziz, N. A., (2015). Static code analysis of permission based features for android malware classification using apriori algorithm with particle swarm optimization. *Journal of Information Assurance & Security, 10*(4), 152–163.
12. Kumar, R., Xiaosong, Z., Khan, R. U., Ahad, I., & Kumar, J., (2018). Malicious code detection based on image processing using deep learning. *Proc. of the International Conference on Computing and Artificial Intelligence* (pp. 81–85).
13. Oberheide, J., Cooke, E., & Jahanian, F., (2008). CloudAV: N-version antivirus in the network cloud. In: *Proceedings of the 17th USENIX Security Symposium (Security'08).* San Jose, CA.
14. Jon, O., Evan, C., & Farnam, J., (2007). *Cloud N-Version Antivirus in the Network Cloud.* Electrical Engineering and Computer Science Department, University of Michigan, Ann Arbor, MI 48109.
15. Matthias, S., Lars, B. A., Pablo, G., David, B., & Bernd, F., (2011). *Malware Detection and Kernel Rootkit Prevention in Cloud Computing Environments.* University of Marburg, Germany.
16. Murad, K., Shirazi, S., Zikria, Y., & Nassar, I., (2010). Evading virus detection using code obfuscation. In: *Future Generation Information Technology of Lecture Notes in Computer Science* (Vol. 6485, pp. 394–401). Springer Berlin, Heidelberg.
17. Lakshmanan, N., Sreejith, K., Gregoire, J., & Manjunath, B. S., (2011). Malware images: Visualization and automatic classification. In: *Proceedings of the 8th International Symposium on Visualization for Cyber Security* (p. 4).
18. Antonio, T., Kevin, P. M., William, T. F., Mark, A. R., et al., (2003). Context-based vision system for place and object recognition. *International Conference of Computer Vision, 3*, 153–167.
19. Qublai, K. A. M., Irfan, A., & Muhammad, Y., (2018). Cloudintell: An intelligent malware detection system. *Future Generation Computer Systems, 86*, 1042–1053.
20. Vinayakumar, R., Mamoun, A., Soman, K. P., Prabaharan, P., & Sitalakshmi, V., (2019). Robust intelligent malware detection using deep learning. *IEEE Access, 7*, 46717–46738.
21. Jiawei, S., Vargas, D. V., Sanjiva, P., Sgandurra, D., Yaokai, F., & Kouichi, S., (2018). Lightweight classification of IoT malware based on image recognition. In: *IEEE 42nd Annual Computer Software and Applications Conference* (Vol. 2, pp. 664–669). IEEE.
22. Modi, C., Patel, D., Borisaniya, B., Patel, A., & Rajarajan, M., (2013). A survey on security issues and solutions at different layers of cloud computing. *J. Super comput., 63*(2), 561–592.
23. Khalil, I. M., Khreishah, A., & Azeem, M., (2014). Cloud computing security: A survey. *Computers, 3*(1):1–35.
24. Ardagna, C. A., Asal, R., Damiani, E., & Vu, Q. H., (2015). From security to assurance in the cloud: A survey. *ACM Comput. Surv., 48*(1), 2:1–2:50.
25. Lakshmanan, N., Sreejith, K., Gregoire, J., & Manjunath, B. S., (2011). Malware images: Visualization and automatic classification. In: *Proceedings of the 8th International Symposium on Visualization for Cyber Security* (p. 4).
26. *Convolutional Neural Network*, (2017). https://en.wikipedia.org/wiki/Convolutional-neuralnetwork (accessed on 27 January 2024).
27. *Intro to Convolutional Neural Networks*, (2017). https://www.tensorflow.org/ (accessed on 27 January 2024).

28. Yuhandri, Madenda, S., Wibowo, E. P., & Karmilasari, (2017). Object feature extraction of songket image using chain code algorithm. *International Journal on Advanced Science, Engineering, and Information Technology, 7*(1), 235.
29. Meng, Q., Catchpoole, D., Skillicorn, D., & Kennedy, P. J., (2017). *Relational Auto encoder for Feature Extraction* (pp. 364–371). arXiv:1802.03145 [cs, stat].
30. An, J., & Cho, S., (2015). *Variational Auto encoder Based Anomaly Detection Using Reconstruction Probability.* Technical Report.
31. Wang, Y., Cai, W., & Wei, P., (2016). A deep learning approach for detecting malicious JavaScript code. *Security Comm. Networks, 9*(11), 1520–1534.
32. https://machinelearningmastery.com/autoencoder-for-classification/ (accessed on 27 January 2024).
33. Abien, F. A., & Francis, J. H. P., (2017). *Towards Building an Intelligent Anti-malware System: A Deep Learning Approach Using Support Vector Machine (SVM) for Malware Classification.* ArXiv, 1801.00318 [cs, stat].
34. Felan, C. C. G., Muga, II., & Felix, P., (2016). *Random Forest for Malware Classification.* arXiv preprint arXiv:1609.07770.
35. Mahmoud, K., Mrigank, R., Noman, M., Neil, B., Yang, W., & Farkhund, I. (2018). *Malware Classification with Deep Convolutional Neural Networks.* In 9th IFIP.
36. geeksforgeeks. https://www.geeksforgeeks.org/multi-layer-perceptron-learning-in-tensorflow/
37. pathmind. https://wiki.pathmind.com/restricted-boltzmann-machine

CHAPTER 6

Artificial Intelligence Models for Cyber Security

R. SWATHIKA and K. GEETHA

Department of Computer Science, Bharathiar University, Coimbatore, Tamil Nadu, India

ABSTRACT

Today's trend of digital transformation greatly improves our lifestyle and simplifies every task. However, as technology becomes more advanced, security becomes a concern. Artificial Intelligence (AI) aims to replicate human intelligence and includes applications such as voice-activated auto attendant machines, dictation systems that convert voice input to text, command-based control applications, data analytics and prediction systems, complex problem solving, decision making, and speech recognition. An AI model is a code designed to identify patterns using a set of data. It encompasses Machine Learning (ML), Deep Learning, Cognitive Science, and data analytical methods to provide accurate solutions to complex problems. AI models also utilize methodologies like robotics and Natural Language Processing (NLP), and they vary in complexity and goals.

As the volume of data increases, there is a consistent vulnerability to the resources that arises, which needs to be identified and resolved. A wide range of AI approaches are recommended for applications ranging from simple activities like following specific instructions or adjusting one's strategy for better solutions, to adopting new techniques or methods for creative solutions. AI has immense potential in Cyber Security (CS) to provide solutions for numerous threats and attacks. The system can be designed to generate alerts when threats occur and to detect new types of malware. As cyber-attacks grow in scale and complexity, AI offers security solutions to stay ahead of threats.

Leveraging Artificial Intelligence (AI) Competencies for Next-Generation Cybersecurity Solutions. Pethuru Raj, B. Sundaravadivazhagan, & V. Kavitha (Eds.)

AI frameworks should be able to identify new malware threats and attacks, enabling individuals to enhance their security by monitoring real-time activities in their daily lives through network scanning, detecting unauthorized communication attempts, unauthorized connections, usage of hostile credentials, and a higher number of brute force login attempts, among others. AI-enabled toolkits and software optimization solutions provide a range of resources to assist businesses in designing and deploying solutions rapidly, securely, and cost-effectively. This chapter discusses AI models for cyber security.

6.1 INTRODUCTION

Since artificial intelligence (AI) and cybersecurity are two distinct fields, researchers in AI have been modernizing their new logical standards through technological advancements. These advancements result in networked data in digital form that must be secured in a highly skilled manner. Figure 6.1 shows the impact of AI in cybersecurity.

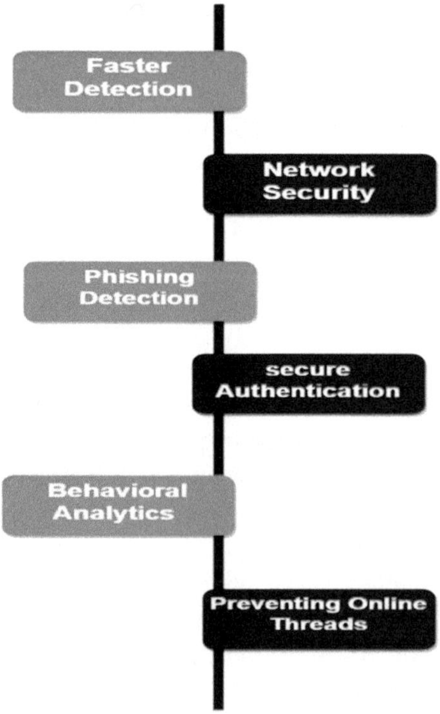

FIGURE 6.1 Imprints of AI in Cyber Security

AI has improved the traditional approach to solving problems of different kinds and accomplishes tasks that the human brain is capable of, including matching, analysis, and decision-making. AI technology is constituted with several subfields that cover almost every field of research, some of which are described in this section. Figure 6.2 illustrates a few of them [1].

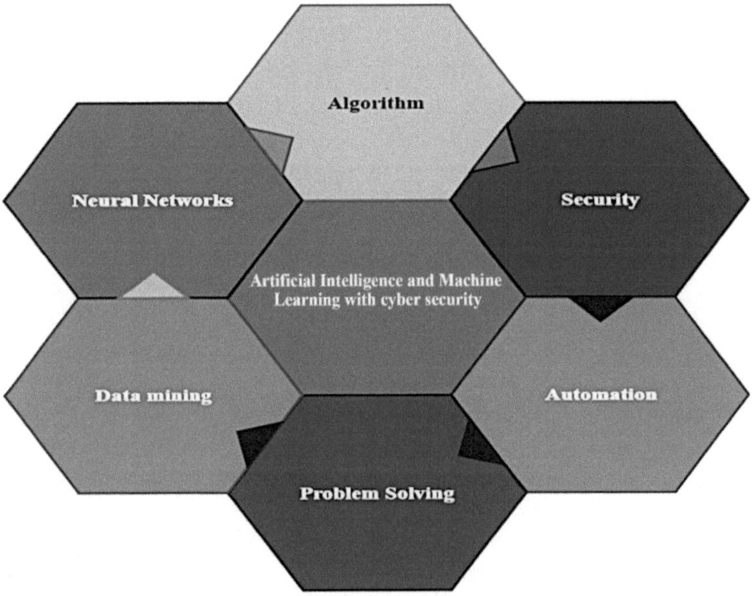

FIGURE 6.2 Applications of AI in various domains.

Currently, AI has demonstrated sufficient dependability in many applications to be used autonomously. However, human involvement is still necessary to monitor AI initiatives in automated processes. AI-based cyber security platforms can be developed using machine learning (ML) to identify or predict attacks based on anomalies detected from regular behavior [2, 3]. Once the AI model detects anomalies, humans can examine them for legitimacy or other indications of the cause.

AI is a powerful technology in applications such as traffic monitoring, reducing the need for human intervention. It can raise alarms or alerts to supervisors when anomalies are detected in regular activity. While AI has proven to be reliable enough to operate without human intervention, humans are still responsible for managing AI efforts. Analysts can analyze anomalies detected by AI to determine their authenticity and understand their cause. Although AI technology has improved its ability to detect hazards, analysts will eventually need to verify the accuracy of anomalies detected by AI systems [4, 5].

The three main goals of cybersecurity are confidentiality, integrity, and availability. Cybersecurity principles provide strategic guidance to organizations on how to protect their systems and data from cyber threats and assess the vulnerability of their systems. The four main components of these cybersecurity concepts are manage, defend, detect, and respond. In government activities, the role of cybersecurity is to identify and manage security risks. In the protected mode, it implements controls to reduce security risks. In terms of detection, it is involved in detecting cybersecurity incidents and understanding the cybersecurity events that cause these incidents. Responding to cybersecurity incidents and recovering from them are also considered principles of cybersecurity. Figure 6.3 illustrates the characteristics of cybersecurity.

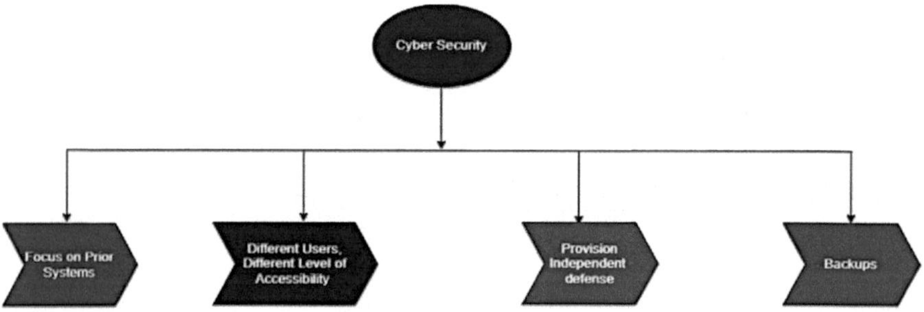

FIGURE 6.3 Characteristic of cyber security.

Cybersecurity is a field that focuses on safeguarding information in the digital world. This includes computer systems, embedded processors and controllers, and the interconnected network of information systems that make up the Internet. The phrase "cybersecurity" refers to the safety of a computer network connected to the internet. The goal of computer security is to protect valuable resources such as hardware, software, and large amounts of data. These resources are vulnerable to threats that can result in damage or impairment to an organization. Information security and system security are both part of computer security. Understanding the origins of these disciplines helps establish the framework for cybersecurity. Information security is concerned with protecting sensitive data and any information derived from it, while system security aims to ensure the proper functionality of systems [7].

The challenge of cybersecurity is to protect resources such as systems, data, and networks from these dangers. Cybersecurity employs antivirus software and intrusion detection systems (IDS) to keep computers highly secure against various types of attacks. The field has grown significantly and has successfully thwarted numerous virus attempts. Traditional security solutions for these detection systems use signatures of previous attacks [8]. By comparing current traffic with known malware and attack types stored in a database, these methods can identify and generate an alarm for any unexpected actions. However, these methods are only effective against known attacks.

We rarely witness zero-day attacks in today's world, as most attacks are now orchestrated by sophisticated software bots or firmwares. Consequently, current systems face constant threats every second. Attackers leverage tools such as AI to create swift and effective attacks. AI technologies are integrated for innovative intrusion detection, outlier identification, decryption, password cracking, and other related events.

On the other hand, cyber-attacks emerge rapidly, including antagonistic attacks, data manipulation, and other malicious activities. Interdisciplinary studies have been conducted in the past to address these issues. Advanced computing technologies like Computer Vision, Computer Image Processing, Natural Language Processing (NLP), Automated Vehicles (AV), and Robotics utilize AI, particularly deep learning (DL) models, and have shown impressive results. Neural Networks (NN) approaches excel in fitting curves in vast and high-dimensional spaces, outperforming other methods [9].

One of the issues in these latest technologies is the exponential growth of data, and the majority of networks produce significant amounts of high-dimensional data. ML techniques are employed to produce predictions based on the old data, ensuring accurate results. The fundamental benefit of an Artificial Intelligence system is its ability to function well, even in the presence of unidentified threats. In solving network-based intelligent models, predictions are made and a complex curve is fitted to distinguish between normal and malicious traffic. As a result, an alarm is triggered whenever abnormal motion is detected, different from the usual pattern. Due to the advantages of AI systems, this particular topic has been the subject of extensive research, yielding promising outcomes.

6.2 MODELING

Modeling in computer science is associated with the construction of computer programs that can simulate human intelligence. It deals with the question of how to make machines understand a problem and create intelligent decision-making with the capability of intelligent behavior.

AI modeling can be categorized into two subcategories based on the function of the model: Rule-Based and Learning-Based. In the rule-based approach, the AI model generates a pre-defined output that is based on certain rules written by the programmers. On the other hand, in the ML approach, the AI prototype is designed in such a manner that it tunes itself by learning from strategies and the data used to train the models.

In essence, an AI model uses decision-making algorithms to acquire knowledge from training and data, and then utilizes that learning to achieve predefined goals. The models are trained on a set of data to perform specific tasks, such as recognizing specific patterns in the data. A program that has been trained on a set of data to recognize specific types of patterns is called an AI model.

AI models use various types of algorithms to solve a variety of commercial problems. Different domains utilize AI models with varying levels of complexity, involving high-end technologies such as robotics, NLP, etc.

There are three kinds of AI Models

i. Artificial Narrow Intelligence (ANI);
ii. Artificial General Intelligence (AGI);
iii. Artificial super intelligence (ASI), which has superior capabilities to humans

Bots are examples of narrow AI, running automated tasks that are generally simple and monotonous in nature.

AGI, on the other hand, is a software model that represents generalized human cognition. The challenge lies in finding solutions to unfamiliar tasks. An AGI system aims to accomplish any task that a human person is capable of performing.

6.2.1 NEED FOR AI

AI is the computer simulation of human thought processes. These procedures include self-correction, inference, and learning.

The amount of work that needs to be done is increasing every day, which is why we need AI. Therefore, automating repetitive tasks makes sense. This enhances productivity while preserving the organization's human resources. Additionally, the organization can attract qualified individuals for its growth through the use of AI. Moreover, businesses today aim to automate all ordinary and regular tasks. They believe that they can automate these routine tasks using a simple application because automation is becoming more prevalent as data science advances. AI needs [18].

An AI model is a program that is trained on a set of data to identify certain patterns. AI models utilize different algorithms to solve business problems. Various fields, such as computer vision, robotics, and NLP, employ AI models of varying complexity and accuracy. ML algorithms are techniques that analyze data for pattern recognition to generate ML models. Figure 6.4 illustrates the AI [17].

6.2.2 DATA SCIENCE

Data science is the practice of developing enormous volumes of data using contemporary methods and technologies to uncover hidden patterns, learn important information, and come to business conclusions. Data science creates predictive models using powerful machine learning algorithms. Figure 6.5 showcases the data science [19].

6.2.3 MACHINE LEARNING MODELS

ASI focuses on the implementation of problems and possesses the capability to surpass human intelligence by manifesting its own cognitive skills. It includes domains where robots can think in ways that humans are not capable of.

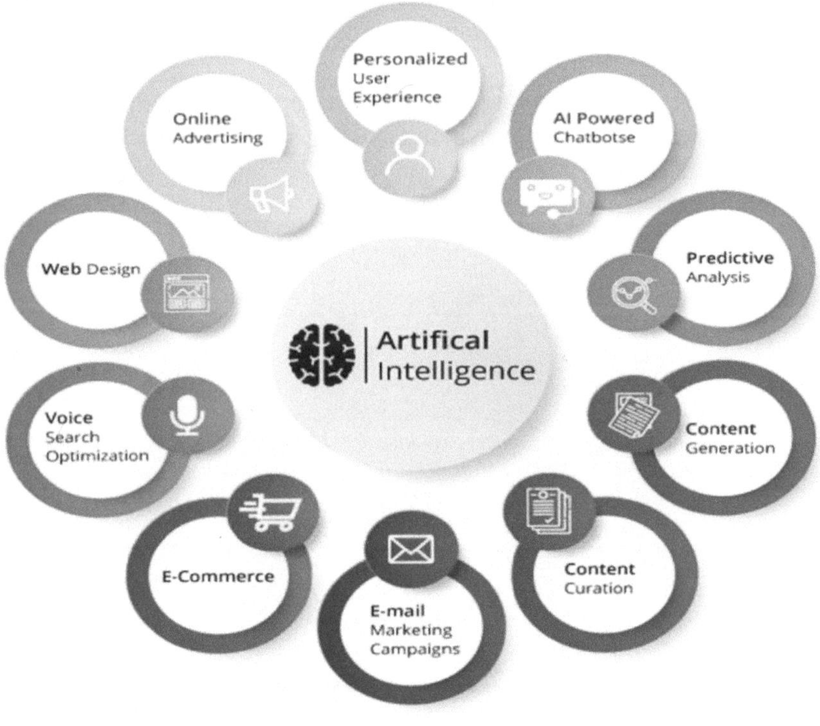

FIGURE 6.4 Artificial intelligence.

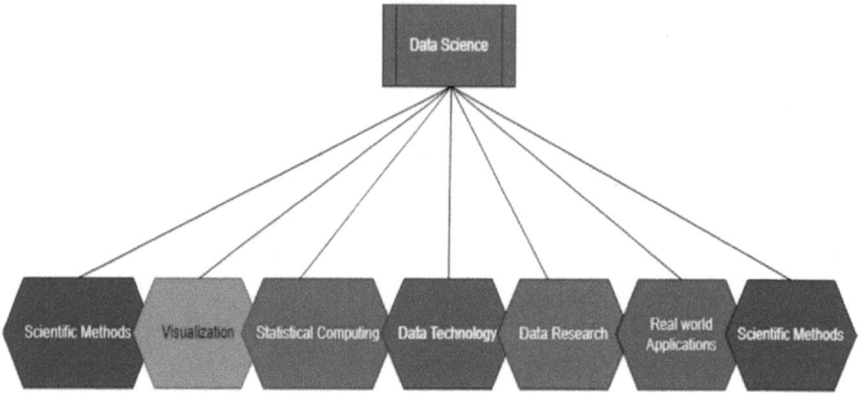

FIGURE 6.5 Data science.

However, the human brain is limited in its capacity for thought and contains only a few billion neurons.

Many people mistakenly confuse ML and AI. As ML is a part of AI, this is most likely the case. The primary distinction between the two models is that AI is focused on building machines that can mimic human thoughts, intelligence, and behaviors.

ML is concerned with giving machines the capacity to learn on their own through experience and lessons, without needing to be explicitly programmed. Figures 6.6 and 6.7 indicate the ML and AI models and ML [13].

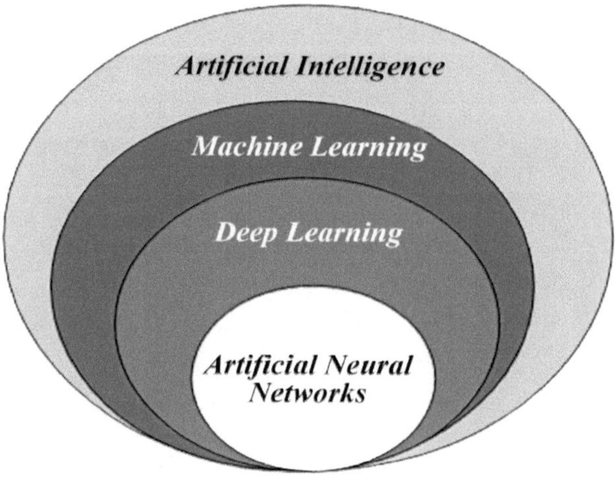

FIGURE 6.6 Models of machine learning.

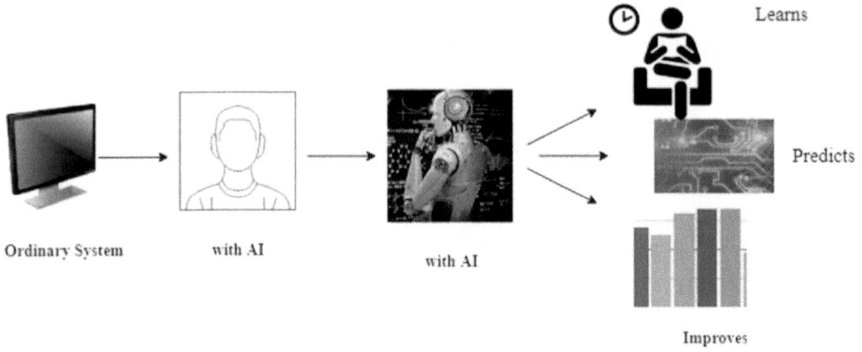

FIGURE 6.7 Machine learning.

All AI models are ML models, but all AI models will ultimately be ML models. The AI algorithms are as follows (Figure 6.8) [20]:

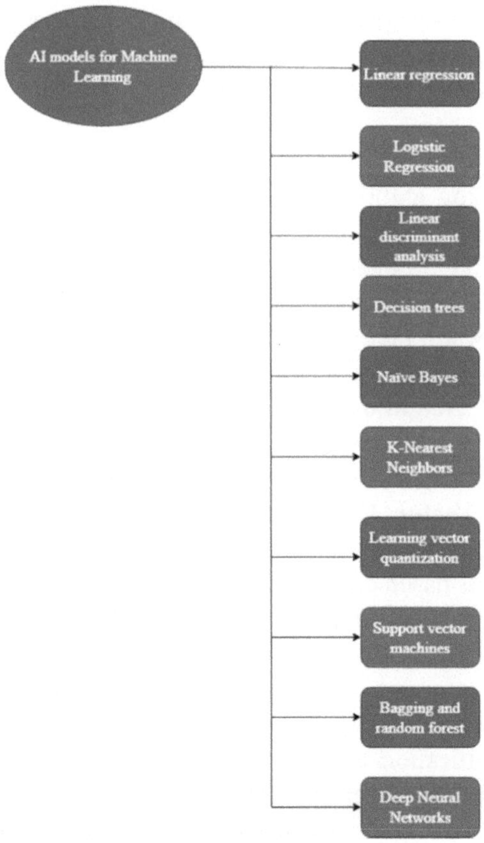

FIGURE 6.8 Models for machine learning.

- **Linear Regression:** The ML approach known as linear regression is entirely based on supervised learning. It accomplishes the task of regression by modeling a target calculation value using independent variables. Regression is commonly used to establish the relationship between variables and make estimates.
- **Logistic Regression:** Logistic regression determines the likelihood that an event, such as voting or not voting, will occur based on a set of independent variables. Since the outcome is a probability, the range of the dependent variable is from 0 to 1.
- **Linear Discriminate Analysis:** Linear discriminant analysis (LDA) is one of the often used dimensionality reduction techniques in machine learning to handle problems involving more than two classes. It is also known as Normal Discriminant Analysis (NDA) or Discriminant Function Analysis (DFA).
- **Decision Trees:** The training data is continuously split depending on a certain parameter, with descriptions of the input and the corresponding output, in a process known as supervised machine learning. Decision nodes and leaves are the two components that can be used to explain the tree.
- **Bayes Algorithm:** The Bayes algorithm, which is one of the supervised learning algorithms, uses the Bayes theorem exclusively for handling classification problems. It is primarily employed for text classification problems involving large datasets. This categorization algorithm, in comparison to others, is particularly efficient and straightforward, facilitating the creation of rapid ML models capable of generating accurate predictions.
- **K-Nearest Neighbors:** The K-Nearest Neighbors algorithm, also referred to as KNN or K-NN, is a non-parametric supervised learning classifier. It predicts and classifies how given data will be grouped.
- **Learning Vector Quantization (LVQ):** One component of the Artificial Neural Network (ANN), LVQ, is also activated by neural systems. It is based on the prototype model of the supervised learning classification algorithm and it addresses the multiclass classification problem.
- **Support vector machines:** The most successful supervised learning algorithms for support are vector machines. They are employed to solve classification and regression issues. Since the SVM method divides n-dimensional space into classes using an optimal line or decision boundary, it only needs to input new data once.

- **Bagging and Random Forest:** Before integrating the findings, a collaborative method called bagging fits several models to diverse subsets of a training dataset. A variation on bagging that also randomly chooses feature groups for each data sample is known as a random forest.
- **Artificial Neural Network and Deep Learning:** An AI called a neural network instructs computers to analyze data in a manner modeled after the human brain. Deep learning is a type of ML technique that operates with interconnected nodes in a fully connected layer structure to mimic the human brain. Figure 6.9 demonstrates the ANN network.

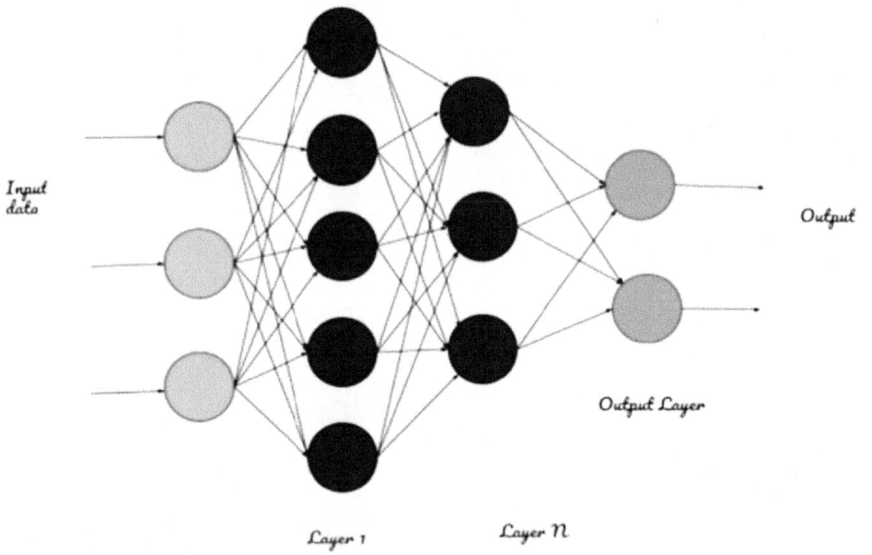

FIGURE 6.9 Artificial neural networks

- **Deep Neural Networks:** A deep neural network (DNN) is an artificial neural network (ANN) with multiple hidden layers between the input and output layers. Similar to ANNs, DNNs are capable of modeling complex nonlinear relationships. The primary objective of DNNs is to address intricate real-world problems and challenges [22]. DNNs are illustrated in Figure 6.10.

  Some other models of AI are shown in Figure 6.11. These are also used for desired output.
- **Deep Learning (DL):** DL is a method for educating computers to perform tasks that are normally performed by humans. ML, in which

a machine may learn via experience and aptitude without human guidance, is a key component of DL. DL algorithms may carry out a task repeatedly while making minor adjustments to the task each time to enhance the output, much like how people learn from their experiences. Figure 6.12 shows the concepts of deep learning [14].

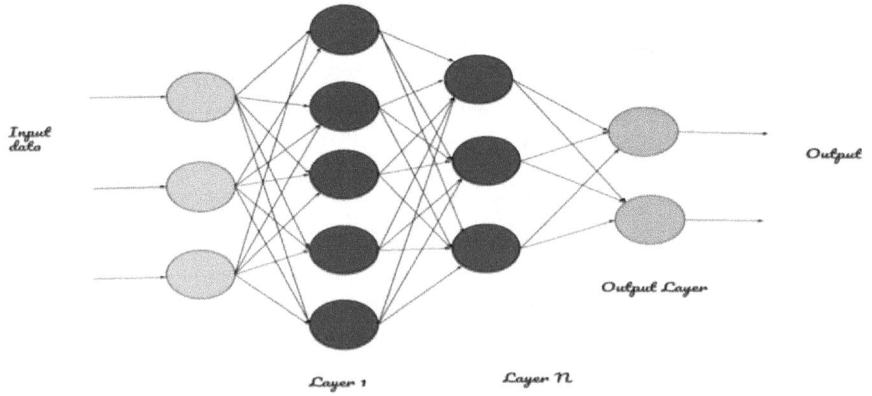

FIGURE 6.10 Deep neural networks.

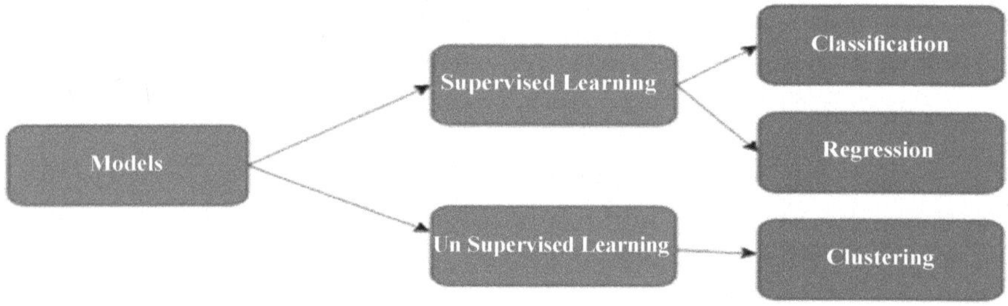

FIGURE 6.11 Some other models of AI.

DL mimics how the human brain processes information and develops designs for use in accountability. This employs the running processes of neurons and normal people's brains to find signals. By developing new broad NN and preparing with a huge extent of data, the act of NN is constantly improved. In numerous implementations, the volume of statistics to acquire from this growth in daily records formation is one reason that DL has a high proportion of debt relative to equity in cyber environments. The main use of DL over ML is its superior execution of a great amount of data [15].

6.3 AI MODELS IN CYBER SECURITY

AI can recognize cyber risks and potentially destructive behavior. AI can be particularly helpful in this case because traditional software systems cannot handle the massive number of new malware that is created every week.

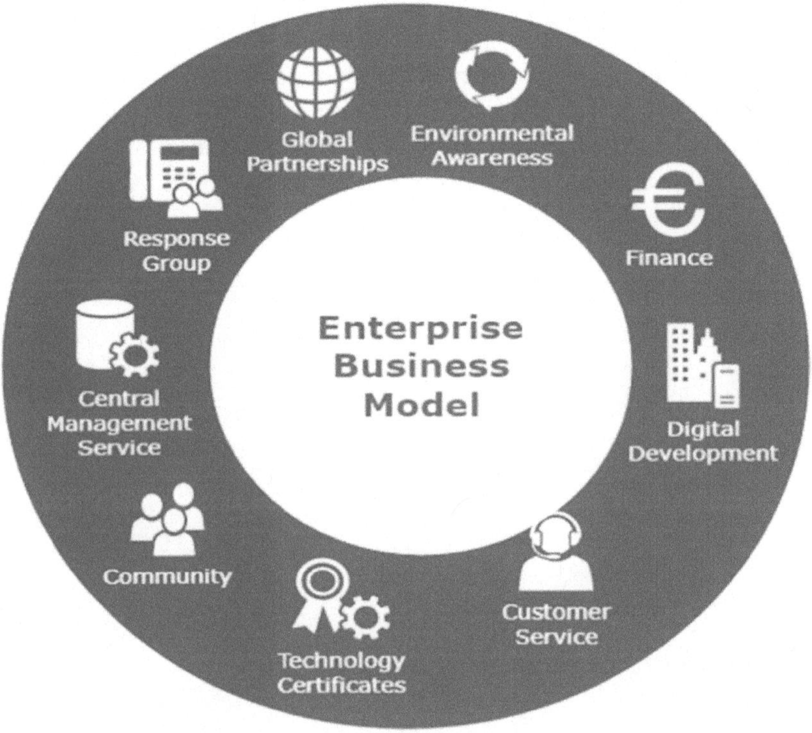

FIGURE 6.12 Deep learning.

Through the use of complex algorithms, AI systems are being trained to identify even the smallest traits of malware or cyber-attacks before they enter the system. By utilizing AI-based cyber security solutions, you can obtain the most recent information on both general and industry-specific risks. This allows you to make more informed prioritization decisions based on potential threats to your systems.

6.3.1 BATTLING BOTS

Today, a significant number of dangerous bots make up internet traffic. A serious risk that bots can present is account takeover using stolen login information, the creation of fake accounts, and data fraud. Automated threats cannot be defeated by relying solely on human intervention. Thanks to AI, we can analyze large amounts of data, and cybersecurity teams can adapt their tactics to the constantly changing environment.

6.3.2 BETTER ENDPOINT PROTECTION

AI is vital in ensuring the complete security of connections among the ever-increasing number of devices used for remote work. While the use of VPNs and antivirus software can provide protection against ransomware and other remote malware attacks, these tools often rely on signatures. Applications of AI are depicted in Figure 6.13.

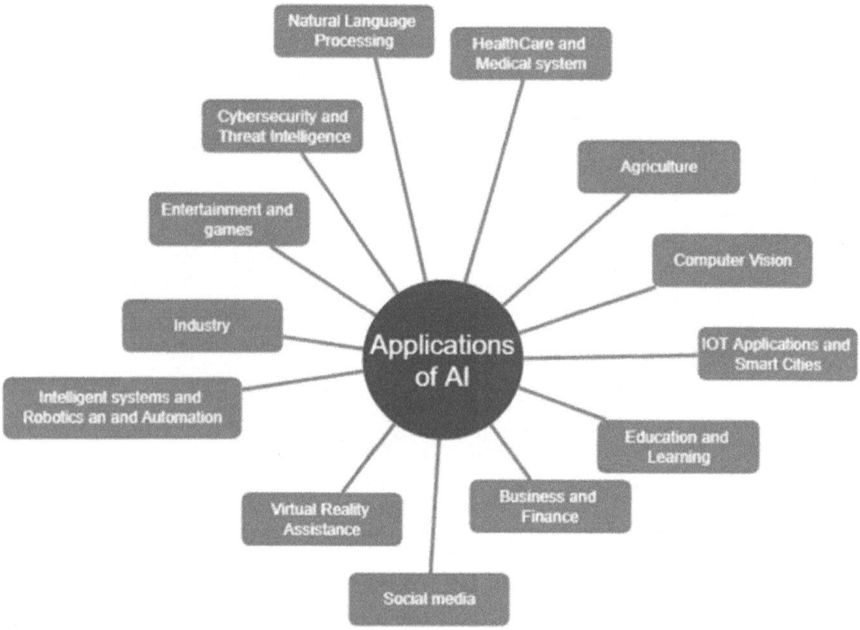

FIGURE 6.13 Applications of AI.

6.4 ATTACKS ON ML

Traditional cyber-attacks are the primary and major threat to AI models. Data, source code, and model files can all be altered, stolen, or corrupted by an accomplished hacker at any stage of the AI growth lifecycle. WWT considers every network and server architecture connected to AI systems to be vital infrastructure that requires the highest level of security. AI models can also be subjected to hacks that have been tweaked to exploit their core mathematical principles. These attacks can be divided into three categories:

i. Evasion;
ii. poisoning; and
iii. stealing.

i. **Evasion:** This requires tricking the model in a real-world setting by changing the input. A recent example in the world of cybersecurity was the bypass attack against Cylance Protect. The AV programmer could be tricked into incorrectly classifying malware by simply attaching a few phrases from a well-known computer game to the infected file. There are numerous examples of model output changing in response to slight alterations in the physical environment. For instance, applying a few sticky notes to a stop sign can make the Tesla autopilot AI misinterpret the sign, which is a particularly unsettling example [27]. Many instances have been published online. In the example below, provided by Microsoft, an AI model correctly identified the first image as a cat with 88% accuracy, while misidentifying the second image as guacamole.
ii. **Poisoning:** Attackers may find it appealing to devise strategies to skew the model by altering the data that was used to train it. Data modification, malicious data addition, or a mix of the two may be involved. The targeted dataset can contain both the training data (usually in an enterprise data lake) and the streaming real-time data that the model receives and uses as part of an update cycle. If the attacker has contact with the exercise data, they will have almost total control over the final model. Protecting the data that is saved and provided by the common security architecture is crucial. The attacker has less effect when they poison a production's input, but access is much simpler.
iii. **Stealing:** Extrapolating data from an AI model that isn't intended to be extrapolated is called stealing, also known as extraction. The basic technique is quite simple but generally hard to spot: the attacker feeds the model input, records the relevant output, and repeats. After amassing a sufficient number of records, the attacker can do offline analysis to obtain the necessary data.

6.4.1 CYBER SECURITY TOOLS THAT USE AI

AI is being used in several solutions available today to improve cybersecurity.

i. X Sophos Intercept Tool
ii. TAA tool (Symantec's Targeted Attack analytics)

iii. IBM QRadar Advisor
iv. Dark trace Antigen

i. **X Sophos Intercept Tool:** Intercept X brands practice deep learning neural networks, which mimic the human brain. Intercept X was developed using algorithms created by the US Defense Advanced Research Project Agency's first Cyber Genome Program, which was developed several years ago to detect the genetic makeup of malware and other cyber threats [23].

   Data scientists used millions of samples to train this model, giving it access to real-world feedback and a two-way exchange of threat knowledge. As a result, there are very few false-positive detections, and the detection of known and unknown malware is very accurate. Intercept X uses behavior analysis to restrict the spread of boot-record hacks and viruses.

ii. **Targeted Attack Analytics Tool:** Specific and secret threats are discovered and disclosed using the TAA tool. It combines AI and ML with the expertise, knowledge, and skills of Symantec's security researchers and specialists. The Dragonfly 2.0 campaign, which primarily targeted multiple energy corporations and aimed to infiltrate and gain access to their operational networks, was successfully thwarted by Symantec a few years ago using the TAA tool. This program focuses on identifying and analyzing abnormal activity at each network node to determine if any actions align with hidden behavior.

iii. **IBM QRadar Advisor:** This product utilizes IBM Watson technology to combat and defend against online threats, as implied by its name, and it is developed by IBM. It leverages AI to automatically scan for vulnerabilities in all exploits. By utilizing QRadar Advisor, which employs cognitive reasoning to provide valuable insights and shorten response time, security analysts can assess threat situations and minimize the risk of compromise.

   One of the key features of QRadar Advisor is automated investigation, which utilizes potential observations from incidents to analyze local data and establish a more comprehensive context. Another feature is the identification of high-priority risks, which provides crucial insights about an incident [24,25].

iv. **Dark trace Antigena:** The tool developed by Dark trace for their active and immediate self-defense, called Antigena, will expand Dark trace's sole capability to recognize and mimic virtual antibody properties that recognize and eliminate threats and infections.

Depending on the degree of the threat, it recognizes suspicious activity and reacts immediately. Antigena uses advanced machine learning technology to recognize and defend against unknown threats as they occur. Without the need for human intervention, prior knowledge of attacks, rules, or signatures, it performs the desired process. Businesses can respond to threats rapidly with the use of these automated response capabilities without needing to extend normal company working hours.

The business sector has been a prime target for cyber-attacks in recent years. A new report from Cambridge, Massachusetts indicates that manufacturing is the most targeted industry by cyber attackers. Given the significance of business production to economies worldwide, it's no wonder that cybersecurity is a top priority for manufacturers. AI can play an essential role in supporting secure business production centers and systems [26].

6.5 NETWORK SECURITY MONITORING TOOLS

These tools are used to analyze network data and detect network-based threats. Figure 6.14 illustrates the cybersecurity tools for AI.

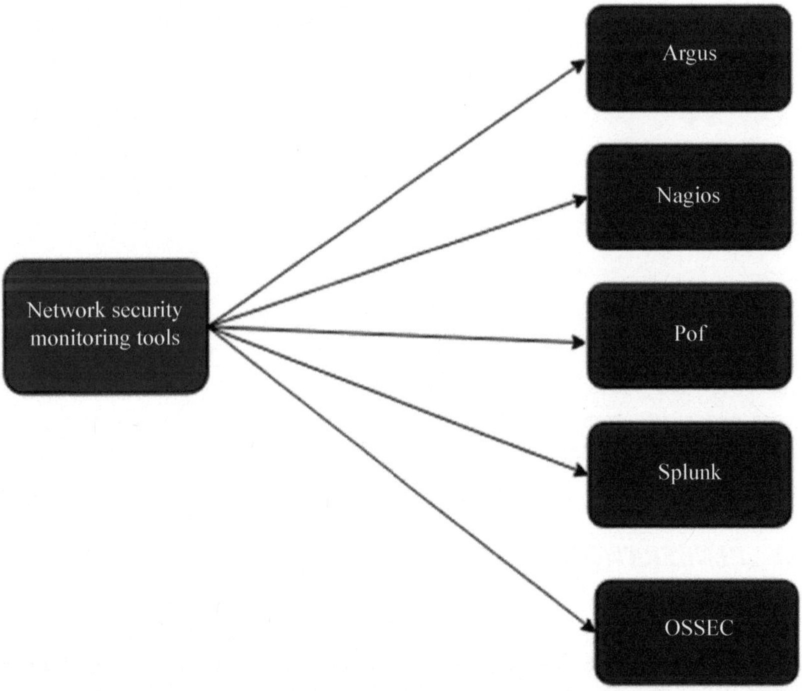

FIGURE 6.14 NSM tools.

6.5.1 ENCRYPTION TOOLS

Files are protected using encryption, which scrambles text so that only authorized users can decipher them. Figure 6.15 shows the encryption tools.

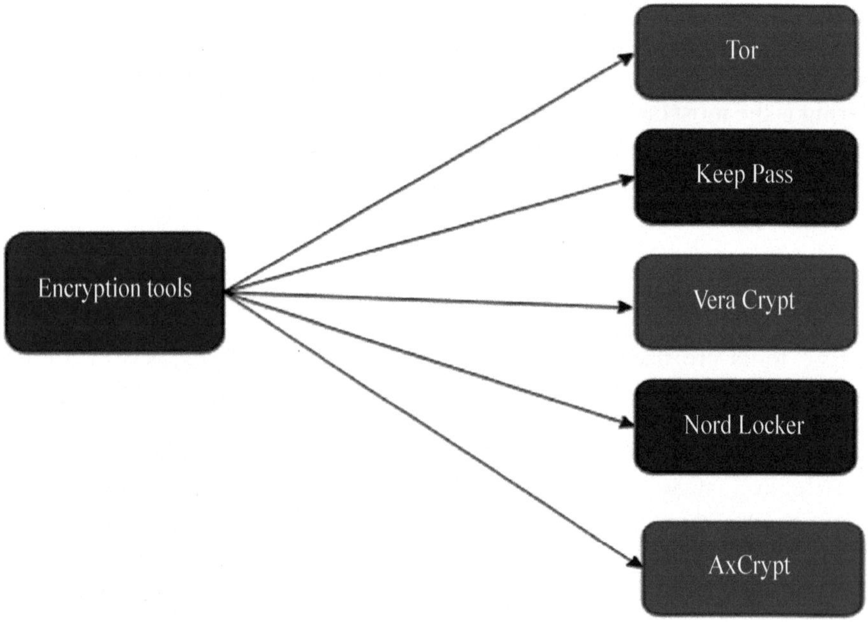

FIGURE 6.15 Encryption tools.

6.5.2 WEB VULNERABILITY SCANNING TOOLS

These software tools evaluate online applications to find security flaws such as cross-site scripting, SQL injection, and path traversal. Figure 6.16 examines the Web Vulnerability Tools.

6.5.3 PENETRATION TESTING

Penetration testing, also known as a "pen test," simulates an attack on a computer system with the purpose of evaluating the security of that system. Figure 6.17 shows the Penetration testing.

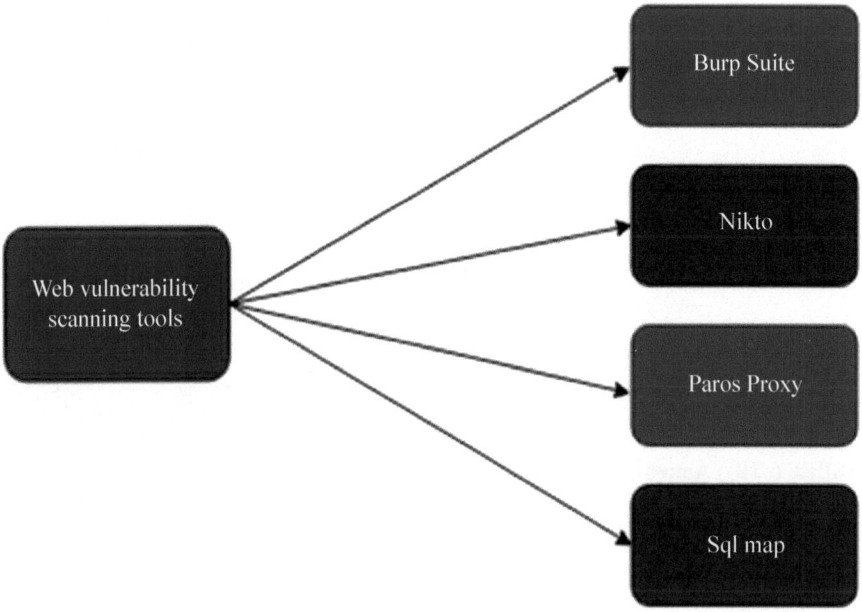

FIGURE 6.16 Web vulnerability tools.

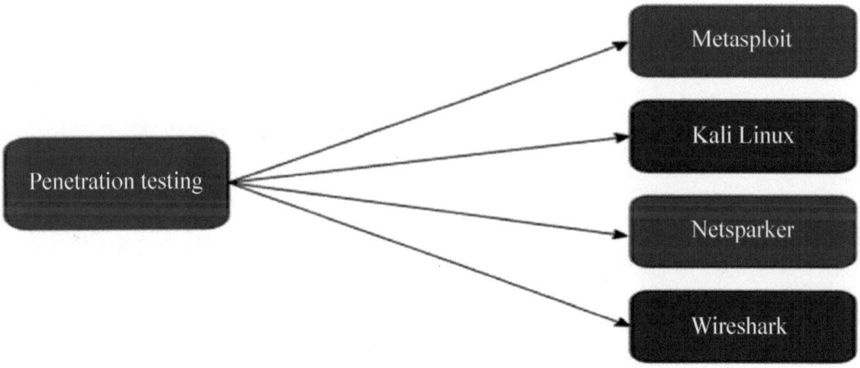

FIGURE 6.17 Penetration testing.

6.5.4 ANTIVIRUS SOFTWARE

This software program is designed to detect worms and other harmful vulnerabilities, including ransomware, worms, spyware, adware, and Trojans. Figure 6.18 illustrates some examples of Antivirus Software.

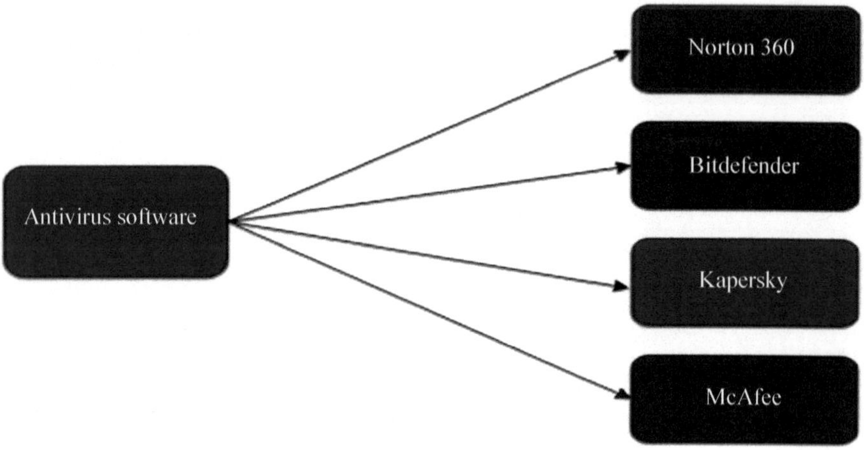

FIGURE 6.18 Antivirus software.

6.5.5 NETWORK INTRUSION DETECTION

An IDS monitors the network and system traffic, and warns the administrator if any unusual or suspicious activity is discovered. Figure 6.19 shows some examples of Network Intrusion Detection.

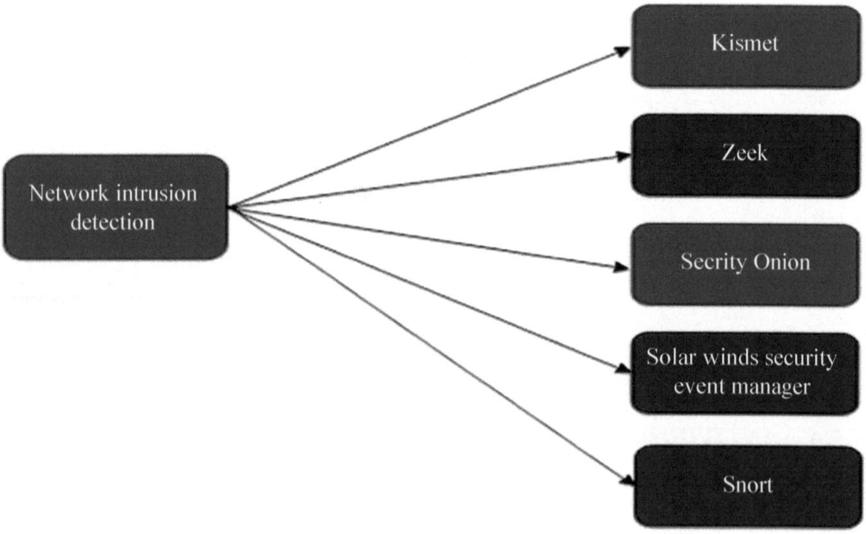

FIGURE 6.19 Network intrusion detection.

6.5.6 PACKET SNIFFERS

A packet sniffer is used to intercept, log, and analyze network traffic and data. It is also referred to as a packet analyzer, protocol analyzer, or network analyzer. Examples of such programs include Windump, tcpdump, and Wireshark.

6.5.7 FIREWALL TOOLS

Tufin, AlgoSec, FireMon, and RedSeal are some of the best firewall security management packages.

6.6 AI TECHNIQUES FOR CYBER SECURITY

A computer program called an expert system imitates human decision-making. This is the ideal illustration of a knowledge-based system. The Knowledge Base and the Inference Engine are the two subsystems that make up these knowledge-based systems. The statements and illustrations are accurate depictions of the body of knowledge. The Inference Engine is a computerized logic tool. It looks at the current state of the knowledge base, applies the necessary rules, and then adds new knowledge. The CSIA's Knowledge Base and Inference Engine, Cyber Security AI Expert System include the following elements. Figure 6.20 shows the AI expert systems.

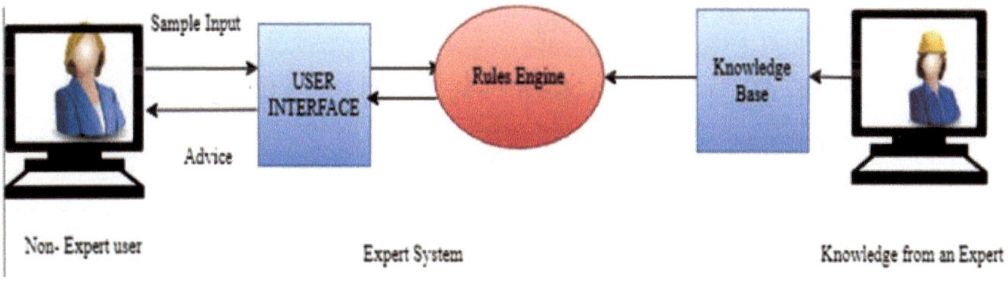

FIGURE 6.20 AI expert systems.

Features of Expert Systems:

- High-performance levels
- Easy to understand
- Utterly consistent
- Highly responsive

Constituents of Expert System
An expert system mostly contains 3 constituents:

- User Interface
- Inference Engine
- Knowledge Base

6.6.1 USER INTERFACE

The expert system communicates with the user through a user interface, receiving readable input from the user in the form of questions, and then sending those questions to the inference engine. It displays the output to the user after receiving a response from the inference engine. In other words, it is an interface that enables a user to contact an expert system even if they lack technical understanding to solve a problem.

6.6.2 INFERENCE ENGINE

The inference engine, which is the system's main processing unit, is referred to as the expert system's "brain." To make judgments or infer new information, it applies the knowledge base and inference principles. It assists in figuring out an error-free user response. The system pulls data from the knowledge base with the aid of an inference engine.

- Deterministic Inference engine
- Probabilistic Inference engine
- **Deterministic Inference engine:** It is presumed that the results of this kind of inference engine are accurate. It is based on facts and legal principles.
- **Probabilistic Inference engine:** This type of inference engine based its conclusions on uncertainty and probability.

6.6.3 KNOWLEDGE BASE

A knowledge base is a type of data repository that contains information compiled by several subject matter experts in a certain field. It is considered to be a substantial knowledge base. With a broader knowledge base, the Expert System will be more accurate.

It is similar to a database that contains information and recommendations for a specific industry or subject.

6.6.4 WHY IS EXPERT SYSTEM NEED TODAY?

The following constraints are used to develop a computer-based system for expert systems

- **No memory Limitations:** Data can be memorized both at the moment of application and stored in the required quantity. Furthermore, the ability of human professionals to recall everything at once is subject to various limitations.
- **High Efficiency:** If the knowledge base is updated with the correct information, which may not be possible for a human to do, a very efficient output is created.
- **Expertise in a Domain:** Obtaining a final output for the question is challenging due to the presence of numerous human specialists in each topic, each with distinct abilities, experiences, and expertise. However, by integrating all the facts and knowledge into the expert system, along with the knowledge acquired from human experts, it is possible to achieve an effective result.
- **Not Affected by Emotions:** Human emotions such as exhaustion, anger, depression, and worry have no impact on these systems. As a result, the performance remains unchanged.
- **High Security:** For each question, these systems offer great security.
- **Considers All the Facts:** Before responding to any inquiry, it carefully considers all pertinent data. As a result, a human expert could choose to overlook particular facts for a variety of reasons.
- **Regular Updates Improve the Performance:** If there is a problem with the expert system's output, humans can improve the system's performance by updating the knowledge base.

6.6.5 CAPABILITIES OF THE EXPERT SYSTEM

Various capabilities of an expert system are given below:

- **Advising:** It can provide advice to a human for any domain query from a certain ES.

- **Provide Decision-Making Capabilities:** I It offers the ability to make judgments in any area, such as financial decisions, decisions involving the practice of medicine, etc.
- **Demonstrate a Device:** It is effective in displaying any new items, including their features, technical details, intended uses, etc.
- **Problem-solving:** It is capable of solving issues.
- **Explaining a Problem:** Additionally, it can describe an input issue in great detail.
- **Interpreting the Input:** It can understand user input.
- **Predicting Results:** It can be applied to the outcome prediction process.
- **Diagnosis:** Because it already has several built-in medical instruments, an ES made for the medical industry may diagnose a disease without the need for additional components.

6.6.6 NEURAL NETS

Neural networks, also known as deep learning, are a cutting-edge branch of AI. They draw inspiration from the functioning of the human brain. Our brain consists of a large number of neurons, most of which are general-purpose and domain-independent. It is capable of learning from various types of data. In 1957, Frank Rosenblatt created the Perceptron artificial neuron, which paved the way for the development of neural networks. By combining with other neurons, particularly the perceptron, these networks can enhance their capabilities and address complex challenges by independently analyzing raw data.

When applied to computer protection, deep learning enables machines to determine whether a file is harmful or not without human input. Our methodology for identifying malicious attacks performs well compared to traditional machine learning methods.

6.6.7 INTELLIGENT AGENTS

Using information from the environment, input from humans, and past knowledge, an intelligent computer can perform actions or make conclusions. These programs can be used to gather information instantly according to a specified schedule or in response to human input in real time.

6.6.8 CATEGORIES OF INTELLIGENT AGENTS

The breadth of abilities and level of intellect of intelligent agents are classified into the following categories:

- **Reflex agents:** These agents do not care about the past; they only consider the present. The user initiates an event, and the agent then consults a list of pre-set rules and pre-programmed outcomes to choose how to respond. This is known as the event-condition-action rule (ECA rule).
- **Model-based agents:** Similar to reflex agents, these agents decide on a course of action, but they are more aware of their surroundings. The world model built into the internal system takes into account an agent's past.
- **Goal-based agents:** These agents add goal information or information about ideal conditions to the knowledge model that agents already have.
- **Utility-based agents:** These agents are similar to goal-based agents, but they also incorporate a utility assessment that ranks each potential scenario based on the desired outcome and selects the action that will maximize the outcome. The probability of success or the amount of resources needed are a couple of examples of grading criteria.
- **Learning agents:** Through an added learning component, these agents have the capacity to progressively advance and learn more about a specific setting over time. Feedback will be used by the learning component to determine how performance components should be adjusted to improve gradually.

6.6.9 ADVANTAGES OF AI TECHNIQUES

AI is used in several ways for cyber defense. Future systems will be more intelligent than these methods. Indeed, attackers may also employ AI in their attacks. It is obvious that the most recent advancements in ML and knowledge-understanding frameworks would improve the capability of the systems that may be deployed in terms of digital security. Figure 6.23 contains a description of the many approaches discussed in this study. Figure 6.21 shows the AI techniques.

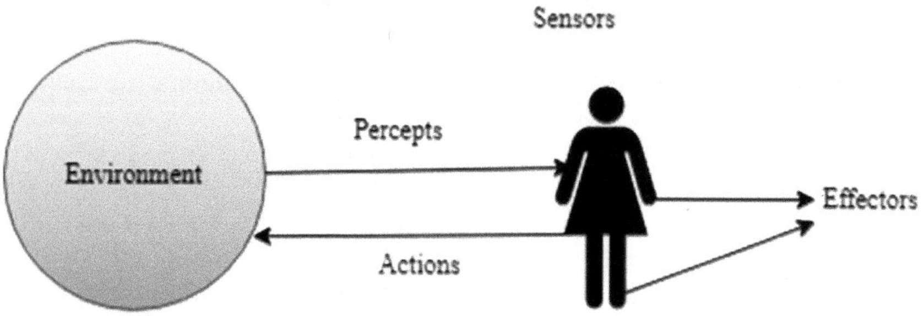

FIGURE 6.21 AI techniques.

- **Sensor:** A sensor is a device that monitors environmental changes and transmits the information to other electronic devices. An agent observes its surroundings using sensors.
- **Actuators:** Actuators are the components of machines that convert energy into motion. Actuators are the only components responsible for steering and moving a system. A rail, gear, electric motor, etc., are examples of actuators.
- **Effectors:** Effectors are objects that have an environmental influence. Effectors include things like a display screen, legs, wheels, arms, fingers, wings, and more. The benefits of AI approaches are displayed in Figure 6.22.

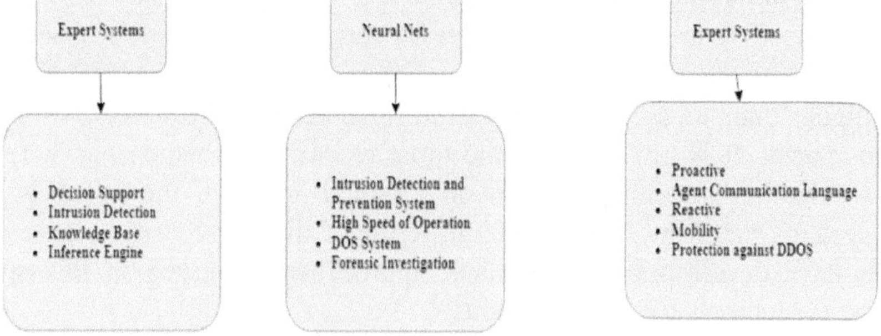

FIGURE 6.22 Benefits of AI techniques.

The results showed the potential for detecting intrusions, analyzing malware, and detecting spam using ML, deep learning, and data mining approaches for cybersecurity objectives. The ability to identify and categorize malware is now available thanks to data mining techniques, which are constantly being developed. However, it is essential to create new data mining techniques that are fast and scalable for detecting and categorizing malware.

6.7 CONCLUSION

Cybersecurity is just one of the numerous uses and benefits that AI has to offer. In today's world of rapidly evolving cyber-attacks and increasing gadgets, AI and ML can help stay up with hackers, automate threat identification, and respond more effectively than traditional software-driven or manual operations. This chapter gives a brief explanation of AI models and how they can be used in cybersecurity. It also discusses the tools used to build AI-based models and addresses attacks on ML models.

KEYWORDS

- **artificial general intelligence**
- **artificial intelligence**
- **artificial narrow intelligence**
- **cyber security**
- **natural language processing**
- **neural networks**

REFERENCES

1. Afzaliseresht, N., Miao, Y., Michalska, S., Liu, Q., & Wang, H., (2020). From logs to stories: Human-centered data mining for cyber threat intelligence. *IEEE Access, 8*, 19089–19099.
2. Laboratory of Cryptography and System Security (CrySyS), (2011). *Duqu: A Stuxnet-Like Malware Found in the Wild.* https://www.crysys.hu/publications/files/bencsath-PBF11duqu.pdf (accessed on 27 January 2024).

3. Alazab, A., Bevinakoppa, S., & Khraisat, A., (2018). Maximizing competitive advantage on e-business websites: A data mining approach. In: *2018 IEEE Conference on Big Data and Analytics (ICBDA)* (pp. 111–116). IEEE.
4. Ale, L., Sheta, A., Li, L., Wang, Y., & Zhang, N., (2019). Deep learning-based plant disease detection for smart agriculture. In: *2019 IEEE GLOBECOM Workshops (GC Wkshps)* (pp. 1–6). IEEE.
5. Allahyari, M., Pouriyeh, S., Assefi, M., Safaei, S., Trippe, E. D., Gutierrez, J. B., & Kochut, K., (2017). *A Brief Survey of Text Mining: Classification, Clustering and Extraction Techniques*. arXiv: 1707.02919 (arXiv preprint).
6. AntiyLabs, (2012). *Analysis Report on Flame Worm Samples Version 1.3.0*. Retrieved from: https://www.antiy.net/media/reports/flame-analysis.pdf (accessed on 27 January 2024).
7. Avcı, B., (2014). *Stuxnet Virüsü Nedir, Nasıl Çalışıyorve Kaynak Kodları*. Retrieved from: http://www.burakavci.com.tr/2014/06/stuxnet.html (accessed on 27 January 2024).
8. Bıçakcı, S., (2012). Yeni savaşve siber güvenlik arasinda nato'nun yeniden doğuşu, *Uluslararası İliş Kiler, 9*(34), 205–226.
9. Bıçakcı, S, (2014). NATO'nun Gelişen Tehdit Algısı: 21. Yüzyılda SiberGüvenlik. *Uluslararası İliş kiler, 10*(40), 101–130.
10. *Bilgiportal İnternet Sitesi, Virüs, Solucanve Truva Atı Nedir?* (2012). Retrieved https://www.iraj.in/journal/journal_file/journal_pdf/6-260-1465884771217-222.pdf (accessed on 27 January 2024).
11. Graham, J., & Howard, R., (2010). *Cyber Security Essentials* (pp. 198, 199). Boca Raton, Florida, ABD: Auerbach Publications.
12. Rosen, E. S., (1981). Vulnerabilities of network control protocols: An example. *Software Engineering Notes, 6*(1), 6–8. Retrieved from: https://tools.ietf.org/html/rfc789 (accessed on 27 January 2024).
13. Selvan, S., (2012). *Flame Worm – Iran Uncovers Stuxnet-Style Malware*. Retrieved from: http://www.ehackingnews.com/2012/05/flame-worm-iranuncovers-stuxnet-style.html (accessed on 27 January 2024).
14. Prajapati, P., & Shah, P., (2020). A review on secure data deduplication: Cloud storage security issue. *Journal of King Saud University-Computer and Information Sciences*.
15. Liu, Q., Li, P., Zhao, W., Cai, W., Yu, S., & Leung, V. C., (2018). A survey on security threats and defensive techniques of machine learning: A data driven view. *IEEE Access, 6*, 12103–12117.
16. Prajapati, P., Shah, P., Ganatra, A., & Patel, S., (2017). Efficient cross user client-side data deduplication in Hadoop. *JCP, 12*(4), 362–370.
17. Prajapati, P., & Shah, P., (2014). Efficient cross user data duplication in remote data storage. In: *International Conference for Convergence for Technology-2014* (pp. 1–5). IEEE.
18. Kadvani, S., Patel, A., Tilala, M., Prajapati, P., & Shah, P., (2019). Provable data possession using identity-based encryption. In: *Information and Communication Technology for Intelligent Systems* (pp. 87–94). Springer.
19. Shah, P., & Prajapati, P., (2020). Provable data possession using additive Homomorphic encryption. *Journal of King Saud University-Computer and Information Sciences*.
20. Prajapati, P., Dave, K., & Shah, P., (2020). A review of recent blockchain applications. *International Journal of Scientific & Technology Research, 9*, 897–903.
21. Xin, Y., Kong, L., Liu, Z., Chen, Y., Li, Y., Zhu, H., Gao, M., et al., (2018). Machine learning and deep learning methods for cybersecurity. *IEEE Access, 6*, 35365–35381.

22. Wei, P., Li, Y., Zhang, Z., Hu, T., Li, Z., & Liu, D., (2019). An optimization method for intrusion detection classification model based on deep belief network. *IEEE Access, 7*, 87593–87605.
23. Li, Y., Xiong, K., Chin, T., & Hu, C., (2019). A machine learning framework for domain generation algorithm-based malware detection. *IEEE Access, 7*, 32765–32782.
24. Malaiya, R. K., Kwon, D., Kim, J., Suh, S. C., Kim, H., & Kim, I., (2018). An empirical evaluation of deep learning for network anomaly detection. In: *2018 International Conference on Computing, Networking and Communications (ICNC)* (pp. 893–898). IEEE.
25. Liang, F., Hatcher, W. G., Liao, W., Gao, W., & Yu, W., (2019). Machine learning for security and the internet of things: The good, the bad, and the ugly. *IEEE Access, 7*, 158126–158147.
26. Maimo, L. F., Go´mez, A. L. P., Clemente, F. J. G., Pe´rez, M. G., & P´erez, G. M., (2018). A self-adaptive deep learning-based system for anomaly detection in 5G networks. *IEEE Access, 6*, 7700–7712.
27. Husak, M., Komarkova, J., Bou-Harb, E., & Celeda, P., (2018). Survey of attack projection, prediction, and forecasting in cyber security. *IEEE Communications Surveys & Tutorials, 21*(1), 640–660.
28. Geluvaraj, B., Satwik, P., & Kumar, T. A., (2019). The future of cyber security: Major role of artificial intelligence, machine learning, and deep learning in cyberspace. In: *International Conference on Computer Networks and Communication Technologies* (pp. 739–747). Springer.
29. Prajapati, P., Bhagat, D., & Shah, P., (2020). A review on different techniques used to detect the malicious applications for securing the android operating system. *International Journal of Scientific & Technology Research, 9*, 5255–5258.
30. Azvine, B., & Jones, A., (2019). Meeting the future challenges in cyber security. In: *Industry4.0 and Engineering for a Sustainable Future* (pp. 137–152). Springer.

CHAPTER 7

Demystifying Digital Transformation Technologies and Tools

B. SUNDARAVADIVAZHAGAN,[1] V. KAVITHA,[2] and
A. DHANU SASWANTH[3]

[1]*Department of University of Technology and Applied Sciences, Al Mussanah, Oman*

[2]*Department of Computer Science with Cognitive Systems, Sri Ramakrishna College of Arts and Science, Coimbatore, Tamil Nadu, India*

[3]*Department of Cyber Security, University of Texas at Dallas, USA*

ABSTRACT

Digital transformation technologies and tools are crucial for enhancing and advancing in every trending sector in today's world. Business is one of the fastest-growing areas in the past decade, requiring various approaches and processes to manage. Understanding market requirements, customer experiences, culture, and business procedures can all help improve the company. Digital technologies are used to build or enhance these existing functionalities, known as "digital transformation." Entrepreneurs can connect with their clients through digital technology, enriching their understanding of the market.

The four essential components of digital transformation are data, process, technology, and organizational change. Digital transformation technologies are necessary to enhance these concepts. Examples of digital transformation technologies include cloud computing, augmented reality, machine learning (ML), the Internet of Things (IoT), hyper-automation, data fabric, and artificial intelligence (AI). Digital transformation tools, such as digital accounting

Leveraging Artificial Intelligence (AI) Competencies for Next-Generation Cybersecurity Solutions.
Pethuru Raj, B. Sundaravadivazhagan, & V. Kavitha (Eds.)

tools, cloud storage, intranet platforms, communication tools, and project management tools, are used to maintain a balance with digital transformation technology and business strategy.

The primary goal of implementing digital transformation applications is to improve asset, product, and process efficiency. This increases consumer value, supports navigating through profit-growth operations, and aids in risk management. By implementing digital transformation, organizations can overcome risks such as cyber threats, data leaks, privacy concerns, operational challenges, and technological issues.

7.1 INTRODUCTION

The term "digital transformation" refers to the process of integrating new technology into a business. The ultimate goal of implementing digital transformation is to enhance creativity, efficiency, and business value. In the future, corporations will require digital transformation technology to improve their organizational structure and effectively respond to customer demands. The implementation of digital transformation tools is crucial for businesses to adapt to the ever-changing market landscape. The term "digital transformation" is also used to describe binary signals or data consisting of only ones and zeros. The concept of transformation in the digital era is not new and has been discussed since the late 1990s. Therefore, digital transformation has already become an integral part of commercial enterprises, as it has been more than 30 years since its inception [22].

This revolution began in 1948 when an American engineer and mathematician, referred to as "Dr. Claude," developed a communication theory that earned him the title of "father of contemporary digital communication and information theory." AI is a widely utilized technology that is being used to drive the digital revolution worldwide. The adoption of digital transformation has brought about changes in the way businesses operate, including overall system upkeep, culture, business operations, workflow, and procedures. In the 1980s, the expansion of digital transformation was further enhanced by the use of fiber optics. This led to the replacement of many analog signals with digital signals in various telecommunications systems, such as cable lines and cellular telephones [12].

With the widespread adoption of digital transformation by clients and industries, the digital process has become simplified, resulting in a decrease in the number of issues in client contacts. Digital firms rapidly developed their resolution capabilities and preferred to employ exclusive expert teams to form new networks and collaborations. Digital data has a significant impact on every activity and interaction of consumers in major sectors. Organizations have incorporated expanding technologies into their advancements and plans, aiming to connect clients, traders, and other investors throughout the network [6]. Commercial firms have implemented various strategies and methods to optimize the utilization of resources. They have also started prioritizing digital transformation platforms to establish a future-proof link to traditional approaches capable of performing intermediate tasks. Organizations are employing a range of digital methods to streamline commercial processes, such as accelerating production, enhancing efficiency, and gaining access to storage [8].

7.1.1 THE CONTEMPORARY CHARGE OF DIGITAL TRANSFORMATION

In the current landscape of digital transformation, it is expected that outdated standards and underperforming 500 firms will be replaced in the near future. "Digital transformation" has become a prominent topic, with DX taking the lead in terms of innovation and leaving traditional approaches behind [7].

More than 40% of total expenses are allocated to digital transformation, with technology costs exceeding 3 trillion dollars. The "Global 2000" companies have come to realize that their ability to deliver digitally transformed products, experiences, and services is crucial for their economic success. Approximately 30% of these companies will need to make a low-cost investment equivalent to at least 10% of their earnings. By 2021, the integration of digitalization into everyday life will be evident through the prevalence of tenders, expenditures, AI, bots, and other digital technologies. Additionally, one of the seven digital titans, including Tencent, Alibaba, Facebook, Google, Amazon, Apple, and Baidu, will account for 20% of all events. If 60% of foreign spending amounts to 2 trillion dollars, the compound annual growth rate of spending from 2016 to 2021 will be 16% [19] (Figure 7.1).

7.1.2 ROLE OF DIGITAL TRANSFORMATION TECHNOLOGIES IN BUSINESS

For example, by checking the supply chain, IoT sensors are utilized to trace the shipment of food from the farm to the trade outlet. This stage displays the transportation routes as well as the food's condition. Every second will be meticulously documented. As a result, a possible risk is reduced. Automated trust can assist manufacturers in securing a large number of contractors quickly, as well as increasing transparency for consumers, investors, and regulators [13]. The concepts of augmented reality and mixed reality are employed in work assistance as well as data training to enhance the physical environment for users. Additionally, the virtual reality approach enables practice of soft skills in addition to work skills. Extended reality is mostly employed in a variety of applications that combine AI with mobile devices and Internet of Things (IoT) headgear. These concepts help oil rig personnel work safely on intricate operations [10].

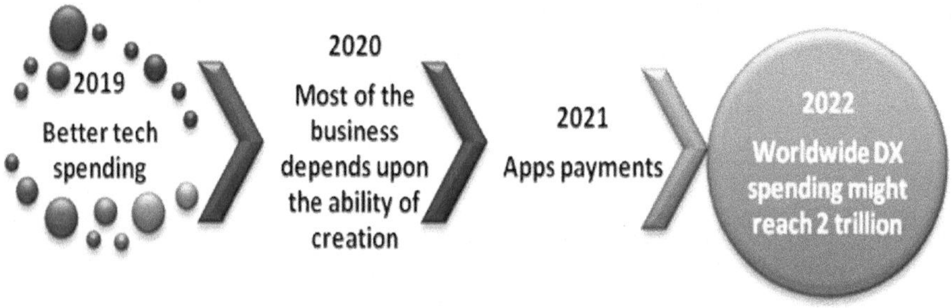

FIGURE 7.1 Digital transformation from 2019 to 2022.

Virtual interfaces are primarily used to detect and interpret emotions by analyzing behaviors, whole-body movements, and brain waves, rather than relying on speech and text like traditional interfaces. Users can interact with the digital environment through augmented workflows, which combine contextual user data and information collected through the interface. In order to achieve innovative and successful outcomes, people must naturally engage with technology [17]. Intelligent automation enables systems to learn from past decisions and patterns, allowing people to focus on more important tasks. Data is collected from various sources, including machine data, human input, IoT sensors, analytics, feedback, and historical transactions. Utilizing this information improves prediction and maintenance [22].

The digital world has been utilized to facilitate test scenarios and gain new insights without requiring individual participation. Digital reflections refer to virtual representations of real objects. By using virtual replications or digital reflections, one can test activities against real-time challenges to forecast or guide rapid outcomes and business decisions. On a larger scale, hyperlinked networks enable machine-to-machine interactions. Hyper connectedness is leveraged to bring together information, products, and people in unconventional ways, revolutionizing society and business. The main objective is to connect critical assets, such as people and locations, to create agile and adaptable enterprises [9].

When digital transformation technologies are employed in the workplace, there are quite a few advantages. The use of the right tools for effective workflow can assist an organization in boosting its productivity as a result of digital transformation. They can automate manual processes. Profitability and efficiency will improve for businesses that employ digital transformation. With digital transformation technologies, businesses become more nimble. Businesses may increase their market speed by enhancing their agility with the aid of software development. This digital transformation promotes digital culture and increased cooperation, as well as assisting the organization in moving forward digitally. Customers will have higher expectations of digital experiences than organizations, and digital transformation may help them meet those expectations. Understanding customer wants and developing company plans both help to deliver better customer insights. Because the majority of the activities are automated, resource management will be simplified, resulting in effective resource management. Digital transformation allows for improved data collection since all manual processes are transferred to digital duties, ensuring data confidentiality and privacy [11].

7.2 LITERATURE REVIEW

Digital transformation technologies are used by a large number of organizations (Table 7.1).

TABLE 7.1 Global Expenditure with Future Prediction

Years	Expenditure
2017	0.96
2018	1
2019	1.18
2020	1.31
2021	1.5
2022	1.8
2023	2.1
2024	2.4
2025	2.8

Since 2017, global investment in data transformation services and technology has increased, with projections for the future (Figure 7.2).

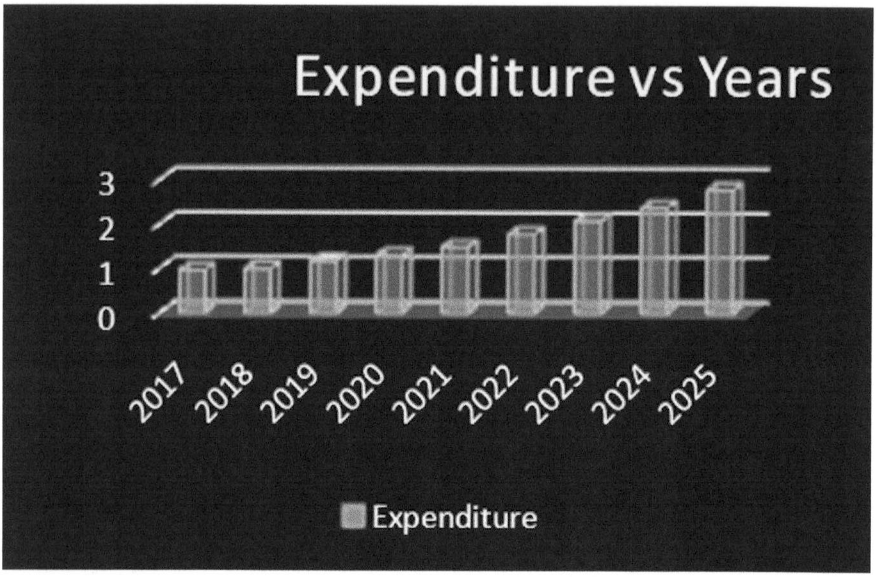

FIGURE 7.2 Global expenditure with future prediction.

According to a survey conducted by Statista, spending on "digital transformation technologies" is projected to exceed 2.1 trillion by 2023. The spending values in US dollars are listed in Figure 7.2 [3], along with the corresponding years.

7.3 DIGITAL TRANSFORMATION TECHNOLOGIES

Digital transformation is the use of digital technology across all areas of a business, requiring a shift in how individuals do their jobs and deliver value to customers. As a result, it is an enlightening modification that requires organizations to constantly challenge their position, experiment, and adapt to be comfortable with failure (Figure 7.3).

When PricewaterhouseCoopers (PwC) examined over 200 emerging technologies to identify those that will have the greatest commercial impact across industries, they referred to them as the "Essential Eight" [20].

FIGURE 7.3 The fundamental eight technologies for digital transformation.

7.3.1 THE TOP MOST DIGITAL TRANSFORMATION TECHNOLOGIES

Technology is constantly evolving these days, with progressive change leading to a significant rise in the rate of advancement. Many things have changed as a result of COVID-19, despite the fact that every new technology and trending technology evolves with time. This epidemic paved the path for contactless services, and many organizations began inventing innovative methodologies to accelerate their advancement. The IoT, blockchain, artificial intelligence with machine learning (AIML), and 5G are some of the developing technologies that will shape the next era [10].

1. **Internet of Things:** Many objects may now be accessible via networks such as Wi-Fi and Bluetooth. Household appliances, automobiles, banks, workplaces, and personal gadgets are already connected through the Internet, making the IoT the future. The IoT platform enables the connection of physical objects and customers through a digital link. IoT not only facilitates connectivity, transmission, and sensing, but it also recognizes the value of data in the big data universe [22].
2. **Block Chain:** The blockchain is often stored as a distributed, self-contained database of digital transactions. The main benefit of blockchain in digital technology is that it allows for the establishment of trust in every peer-to-peer transaction without the need for an intermediary to audit the transactions. People and service providers can have a high level of trust utilizing blockchain. Blockchain allows all users to access a virtual database within a network, and each user can retain a local copy of the database. Because the blockchain is integrated with cryptographic mechanisms, it is difficult to follow and moderate the transaction. This is how blockchain provides security and trust as a bonus feature. Smart contracts are used in a variety of commercial processes.
3. **Artificial Intelligence and Machine Learning:** This decade has witnessed the emergence of AI alongside ML. However, due to its increasing trend, it is still regarded as a cutting-edge technology. AI has revolutionized voice and image recognition, relevant recommendation apps, and popular social media platforms for smartphones, among other domains. AI assists in uncovering patterns and predicting demand in various sectors such as marketing, banking, and healthcare [13]. ML is a subset of AI that creates a significant demand for skilled professionals across diverse industries. According to a recent estimate, by 2025, ML, automation, and AI will encompass 9% of all occupations.
4. **5G:** The primary goal of this technology is to improve connection, reliability, and customer experience by increasing bandwidth. With the help of secure video conferencing and messaging with higher quality and speed, the 5th generation also allows people to work with wireless data from anywhere in the world [16].

7.3.2 TRANSFORMATION OF INDUSTRIES TO DIGITALIZATION

Digital technology is extremely important for all types of organizations, from micro-enterprises to major corporations. The messages from numerous publications, research, and panel discussions on how businesses can become more relevant and competitive are growing stronger and more evident as the world undergoes increasing digital transformation. Similarly, digital technology has revolutionized the way businesses engage with their consumers. In the past, businesses would rely on consumers to discover them through personal contact or meetings, but the landscape has changed, and customers are increasingly using technology to find what they need. Many sectors, including agriculture, education, healthcare, information technology, and transportation, have experienced significant advancements as a result of digitization.

1. **Agriculture:** Agriculture is one of the most vital industries on the planet because food preservation is impossible without it. Digitalization has brought significant changes to agriculture. In this sector, digital technologies such as the IoT, AI, 5G, and blockchain play a critical role in increasing production. IoT devices are used in agriculture to monitor farms in real time, with cameras connected to the network and suitable auditing work being completed. AI and ML applications can be used to forecast demand and sales in the agriculture sector. Additionally, AI and ML can be used to assess and identify products based on their quality, size, and weight.
2. **Education:** Students' community learning experiences are enhanced by digital technology. Specifically, 5G plays a critical role in education. There was no way to study during the pandemic, but 5G enabled students to connect with their teachers from the comfort of their homes. Students began to develop a capability known as "self-learning." Nowadays, they have access to information on any subject. Educational sites and applications have started incorporating augmented reality and virtual reality concepts to make topics more engaging for students. This experiment has resulted in a significant increase in the use of digital technology, leading to improved learning productivity and efficiency. Additionally, learning technology and tools have enhanced critical thinking capacity, which is the fundamental level of reasoning skills [21].
3. **Healthcare:** Digital technology is also being used to assist the healthcare industry in improving its efficiency and efficacy. This

remains true as health technology enables service providers to enhance quality, reduce costs, increase access, eliminate inefficiencies, and personalize medication delivery for patients. Technologies like augmented reality, virtual reality, 5G, blockchain, and AI are being utilized to enhance the well-being of the healthcare sector. The healthcare industry is also seeking assistance from digital technology to enhance efficiency and performance, allowing providers to improve quality, lower costs, increase access, eliminate inefficiencies, and customize drug administration for patients. The healthcare industry is benefiting from technologies such as augmented reality, virtual reality, 5G, blockchain, and AI.

4. **Transportation:** Nowadays, traveling the world without a guide is simple; individuals simply take their smartphones with them and explore the globe. Anyone with a smartphone can access any area thanks to the network. These smartphones use GPS for location, allowing users to find their current location as well as their destination. Digital technology is used in automobiles to provide not only location but also AI and ML capabilities. For drivers, AI provides real-time notifications and driving analyses. Many apps are being developed that use ML to identify driver sleepiness and prevent accidents [20].

7.4 HEALTHCARE INDUSTRY WITH DIGITAL TECHNOLOGIES

In the year 2022, COVID-19 is present worldwide. As a result of this pandemic, several sectors have undergone significant transformations, with the healthcare industry experiencing the most notable changes. The healthcare sector is actively exploring new ideas to enhance services and provide higher quality care through the integration of digital technology. With the increasing number of patients, delivering effective and timely treatment has become increasingly challenging for healthcare facilities. In this critical situation, digital transformation plays a vital role in preserving medical strategies.

Through the collaboration of various digital transformation technologies such as AI, IoT, 5G, augmented reality, virtual reality, and mixed reality, healthcare centers are striving to offer better and faster treatment to patients. Additionally, there is a new term called XR, which stands for extended reality.

Organizations that provide healthcare services vary in pace throughout the world, based on political agendas, regulatory frameworks, individual organizations, and distinct digital transformation goals, among other factors. By adopting IoT and cloud computing, healthcare is attempting to monitor distant health.

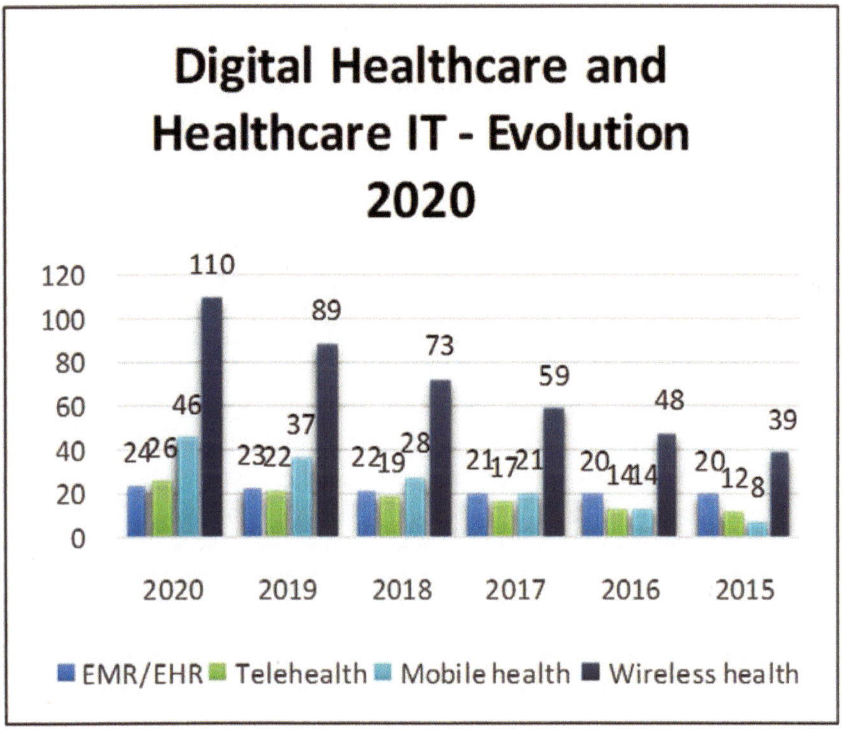

FIGURE 7.4 Graph of digital healthcare evolution.

Wireless health is maintained through the use of wireless technologies for medication and treatment, remote monitoring and diagnostics, and m-Health over mobile health, which is managed by mobile devices such as laptops, smartphones, and even wearables, as shown in Figure 7.4. In 2020, all forms of healthcare services, including wireless health, telehealth, electronic health records, and electronic medical records, will be available compared to previous years. The electronic health record, also known as the electronic medical record or patient chart, is a digital recording of a patient's medical history. According to the forecast, wireless health is maintaining a peak every year, among other techniques. In contrast, telehealth is maintaining the lowest except in the year 2020. Before the evolution of mobile health, the second technology was electronic records, but it gradually lost its growth when mobile health became common. In short, wireless health has shown vigorous growth within five years. This wireless health includes digital technologies such as IoT, AI, augmented reality, virtual reality, ML, cloud computing, etc.

1. **AI in Healthcare Industry:** Essentially, digital healthcare utilizes technology to enhance patients' wellness and health by leveraging data obtained from them. The latest techniques, such as monitoring devices, daily medications, wearable devices, health-related emails, electronic medical records, and health management software, are employed, eliminating the need for patients to manually provide medical data. Digital healthcare, combined with AI, is used to improve individual health. Examples include digital or virtual visits, digital check-in for appointments, digital insurance cards, and more. AI can perform various tasks in different therapeutic settings. It allows physicians to be supported, and patients can consult with doctors using augmented reality technology. AI can assist with illness detection, antibiotic resistance prevention, risk prediction using electronic health records, enhanced cancer therapies, expanded services, prevention of fraudulent billing, and more.
2. **IoT in Healthcare Industry:** The Internet of Things (IoT) is gaining popularity worldwide. Applications such as telemedicine and tele health care technology are being adopted in the healthcare industry. This integration of systems, smart devices, and various technologies has a significant impact on healthcare businesses. The use of IoT in healthcare has resulted in several advantages, including improved data analysis, enhanced patient experience, better medicine management, end-to-end connectivity, monitoring and alerting, and more. Health insurance firms, hospitals, clinics, patients, and physicians all utilize IoT technologies. Health insurers collect data from health monitoring devices with transparency and patience. IoT devices gather health-related data, while integrated filtering devices in clinics and hospitals enable real-time monitoring of health equipment. Individuals can also use IoT-enabled gadgets like wireless heart rate monitoring cuffs and fitness bands for personalized attention, particularly beneficial for senior patients. By incorporating IoT into the healthcare industry, physicians become more attentive and actively engage with their patients.

3. **5G in Healthcare Industry:** Many healthcare sectors have increased their investments in digital technology to improve the healthcare industry as a result of COVID-19. By 2021, the global 5G network is projected to have advanced by 39%. IoT devices, also known as connected devices, are one of the primary drivers of healthcare growth. Due to the minimal latency of 5G networks, high-speed data transfer is enabled, which will help save money by 2030. The 5G network allows specialists and patients to stay connected, and large files (such as medical data) can be quickly shared. Consequently, 5G facilitates telemedicine, improving the quality of telehealth while reducing costs over time. The 5G network also promotes remote surgery, where a hospital surgeon wears a headset with a video feed that is broadcast to a remote site through 5G, allowing the remote expert to receive advice from the in-hospital surgeon. Healthcare providers can enhance their performance, capacity, reliability, and security by utilizing a 5G network. WebRTC offers video, audio, and data transfer capabilities for telemedicine.
4. **AR/VR in the Healthcare Industry:** The capabilities of augmented reality and virtual reality have opened up plenty of possibilities in the healthcare sector. From telehealth apps to surgery, augmented reality and virtual reality may improve the healthcare industry. Virtual environment is the most widely used technology. This is exemplified by surgeons using mixed reality headsets like Microsoft's HoloLens 2. This headset allows surgeons to gather information from the headset while performing surgery, enabling them to use both hands simultaneously. Other physicians can also monitor the procedure through the helmet-mounted video view. This technology also assists nurses in locating veins for blood draws. The IoT concept and AI are both crucial for the advancement of augmented reality. There are several benefits to adopting augmented reality and virtual reality in healthcare; for instance, when a child needs a stitch removed, a virtual reality headset can be provided to the child, allowing them to experience the content from the headset instead of enduring the pain.

7.5 MANUFACTURING INDUSTRY WITH DIGITAL TECHNOLOGIES

The main goal of implementing digital technologies in the manufacturing industry is to improve decision-making capabilities, adaptability to changing customer demands, operational efficiency, cost reduction, manufacturing quality improvement, and revenue generation. Other advantages include customer centricity, better data utilization, improved outsourcing, enhanced processes, and increased innovation.

According to market predictions, AI is expected to expand at a rate of $17.2 billion by the year 2025, as shown in Figure 7.5. The key focus should be on avoiding losing the competition. To achieve this, the industry will need to adopt a predictive approach, which relies on machine learning (ML) and AI. Some significant advantages of utilizing ML in conjunction with AI include predictive or preventive maintenance, defect detection, supply chain optimization, and generative design. AI is deployed when humans are no longer able to perform certain tasks, and it employs various ML methods to solve such problems. Machines are already capable of delivering high-quality work, surpassing human capabilities in terms of efficiency and precision.

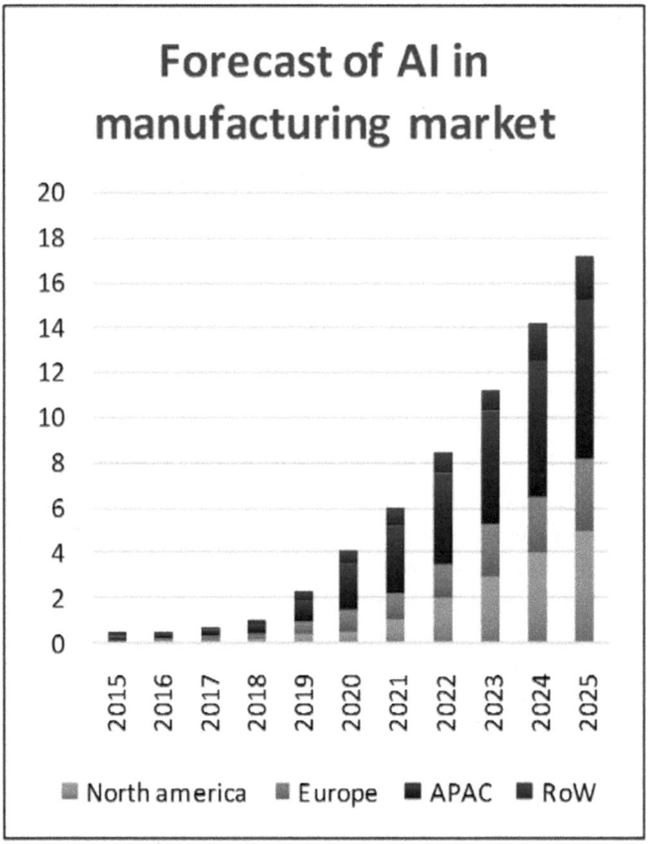

FIGURE 7.5 Graph of AI in the manufacturing market.

1. **IoT and IIoT in the Manufacturing Industry:** Using IoT-enabled cloud technologies, everyday processing utensils are kept in excellent working order. The Industrial Internet of Things (IIoT) is a component of IoT that focuses on manufacturing-related applications. The production process of the business is monitored in real-time, enabling data-driven decisions and manufacturing analytics. Real-time monitoring allows industries to reduce potential risks, enhance energy efficiency, improve working conditions, and increase worker productivity, among other benefits. This enhanced IoT can offer various advantages to manufacturers, including future insights, new functionalities, and services, among others. The application of industrial technology to IoT can improve asset quality, corporate processes, operational performance, safety and risk management, worker experience, and logistics, among other aspects. The manufacturers' objective is to deliver high-quality products with higher profit margins. The IoT opens up new possibilities for preparing and enabling new revenue streams.
2. **AR/VR in the Manufacturing Industry:** The application of augmented reality allows workers or employees to expedite collaboration operations and enhance data-driven decision-making. Employees can utilize augmented reality glasses to comprehend information and perform tasks more efficiently and effortlessly. By employing augmented reality and virtual reality applications such as maintenance, inventory management, and rapid prototyping, augmented reality and virtual reality have the potential to boost global revenue by 2030. Different machinery requires distinct maintenance processes, and providing "maintenance instructions" accessible through augmented reality glasses simplifies significant maintenance tasks. Certain procedures in inventory management are not meant to be fully automated. Enabling augmented reality and virtual reality in the workplace can assist human workers by providing information on the locations of items. Rapid prototyping is achieved by enabling manufacturers to visualize the virtual representation of the product.

7.6 RETAIL INDUSTRY WITH DIGITAL TECHNOLOGIES

The retail business is one in which products and services are consumed directly by customers. The broad company approach that uses digital technology to maximize accuracy and efficient production for consumers is known as retail digital transformation. To successfully adopt digital transformation in the retail industry, retailers must have a thorough understanding of all aspects of their operations, including customer service, inventory planning, pricing, and sourcing. Customer interaction or communication between the customer and the brand, risk detection or evaluating recent expenditures to avoid fraud and hazards, demand forecasting or predictive analysis of customer purchases, and forecasting future performance are the basic units for digital technology in the retail industry. Data analytics is used to get a better picture of risks and losses.

1. **AI in the Retail Industry:** AI would instruct retailers to provide perfect customer service in various circumstances, including automatic tracking of customer mood for better personalized responses. AI can enable chatbots to provide 24/7 customer service. AI would handle a large number of customers politely. AI also offers visual search, allowing customers to upload a picture and search for a similar product. This eliminates the need for a product description; customers can make a purchase based on an uploaded image. For the customer service team, AI-driven technology for mood tracking helps retailers interact with customers based on their mood. AI tracks the customer's mood. AI also provides virtual dressing rooms, reducing congestion in physical dressing rooms at clothing stores. AI technology positions products on customer images for better observation of color, size, and style.
2. **5G in the Retail Industry:** Through personalization, retailers can enhance the customer experience. 5G is defined as the next generation of mobile internet, combining various mechanisms and technologies. In the retail industry, there are four main benefits of 5G: next-generation digital commerce, advanced analytics, in-store augmentation, and consumer interaction (Figure 7.6).
3. **IoT in the Retail Industry:** The IoT enables customers to have a more personalized purchasing experience, which helps companies enhance sales, loyalty, and customer satisfaction. The IoT can be utilized to reduce waiting times at stores. Customers who are next in line can engage in any helpful activities provided by the store. Smart shelves are used in the IoT to manage products, and sensors track the items in real time. Various IoT development platforms, such as IBM IoT Foundation, Android Things, Oracle IoT, AWS IoT Platform, and Google Cloud IoT Core, can be utilized to deliver these services. The integrated sensors in shopping carts allow businesses to visualize

the flow of sales, which helps them improve their operations. These carts are equipped with sensors that use communication protocols to monitor and record cart movements and distances.

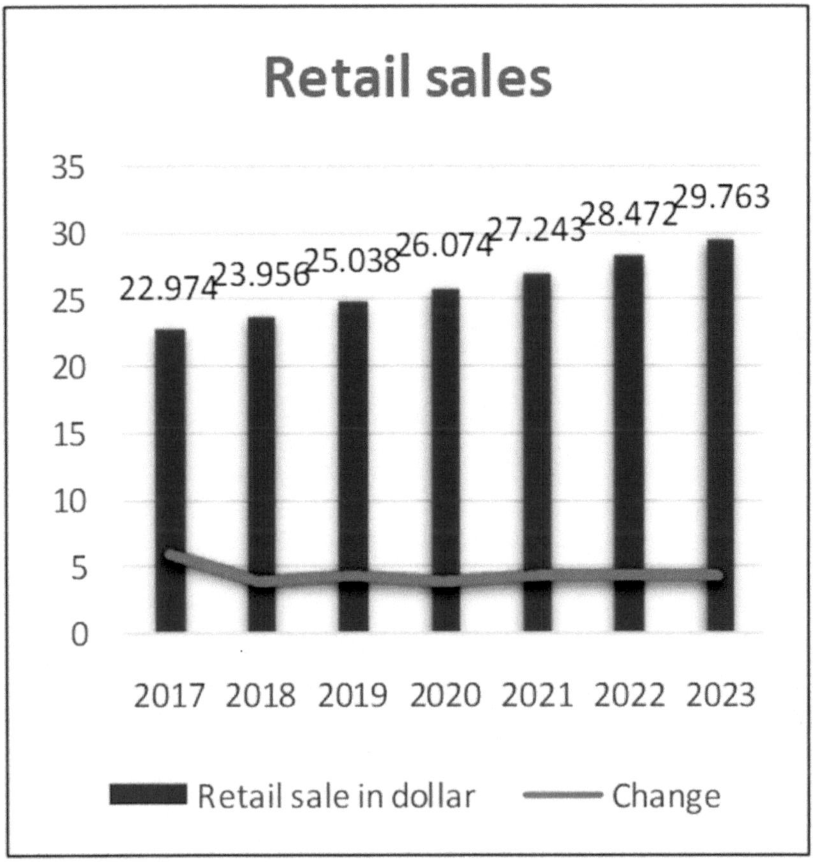

FIGURE 7.6 Graph of global retail sales.

7.7 COMPARISON OF VARIOUS INDUSTRIES (Table 7.2)

The analysis was based on the business of the current decade. With the help of data analytics, it is predicted that the retail industry will be the most utilized industry in the future compared to others. According to information obtained from Splunk in January 2020, 21% of retail IT and enterprise managers internationally reported that their establishment currently uses 5G

TABLE 7.2 Comparison Between Artificial Intelligence, IoT, IIoT, and AR/VR

Industries	AI	IoT/IIoT	AR/VR
Healthcare	Artificial Intelligence proposes an array of improvements for medicinal care over experimental decision-making techniques. Accomplishment of AI associated with ML in the business can enable physicians, and doctors to accompany patients with more defined and accurate data, permitting them to expand extraordinary insights into diagnostics, care processes, treatment variability, as well as patient conclusions.	IoT has been established to impact the medicinal care Industry through smart connected devices, systems, and things that are proposed by billions of users to leverage statistics and facilitate them to formulate timelier, specific, and contextualized decisions. IoT in the medicinal care industry has opened doors for numerous prospects.	VR & AR in medicinal care has opened the door for the prosperity of innovative prospects. These technologies recommend workable clarifications to several of the mainly critical challenges of the medicinal care system and power users with an embarrassment of assorted opportunities in areas like wide-ranging diagnostics and medical training.
Manufacturing	In the present day, top-level administration of the production industry is empowering AI to enrich equipped effectiveness, efficiency, and performance. From gaining momentous cuts in downtime to producing better-designed products and humanizing the safety of employees, AI is accurately evolving manufacturing.	The most outstanding and well-known IoT use cases recline in operations, asset management, and workforce management. For example, manufacturers can ascertain anticipatory maintenance programs with real-time monitoring; and enhance power efficiency and functioning conditions through intellectual air management, risk management, employee efficiency, and more.	AR facilitates the workers to pace up the assemblage process and enrich decision-making. Using AR glasses, human resources can see statistics, similar to assembly guiding principles, or a successive number of components, predictable on the real part, facilitating quicker and easier work procedures.
Retail	Artificial intelligence is an expression that is frightening approximately in various industries, but most people do not fully grasp what it denotes. AI means various counts of technologies, together with machine learning as well as predictive analytics that can assemble progression	The workflow and structural design of IoT under different retail use cases using Azure service are expressed below. The developer might use several IoT development platforms similar to the AWS IoT platform, Google Cloud IoT Core, Oracle IoT, Android Things, or	Color to texture, wrapping to manufactured goods labels – who thought inspection of all these and more accurate from the couch comfort can get further 'real,' well beyond 360° images?

TABLE 7.2 *(Continued)*

Industries	AI	IoT/IIoT	AR/VR
	and analyze troves of statistics, and use that information to envisage, forecast, notify, and assist retailers make precise, data-driven business decisions.	IBM IoT Foundation Device Cloud. The reference has to be taken from Azure IoT Services and describe the use cases utilizing Azure IoT Suite.	

services, and 63% predicted that their business would adopt 5G services in the future. Figure 7.7 illustrates the current and future industry forecasts.

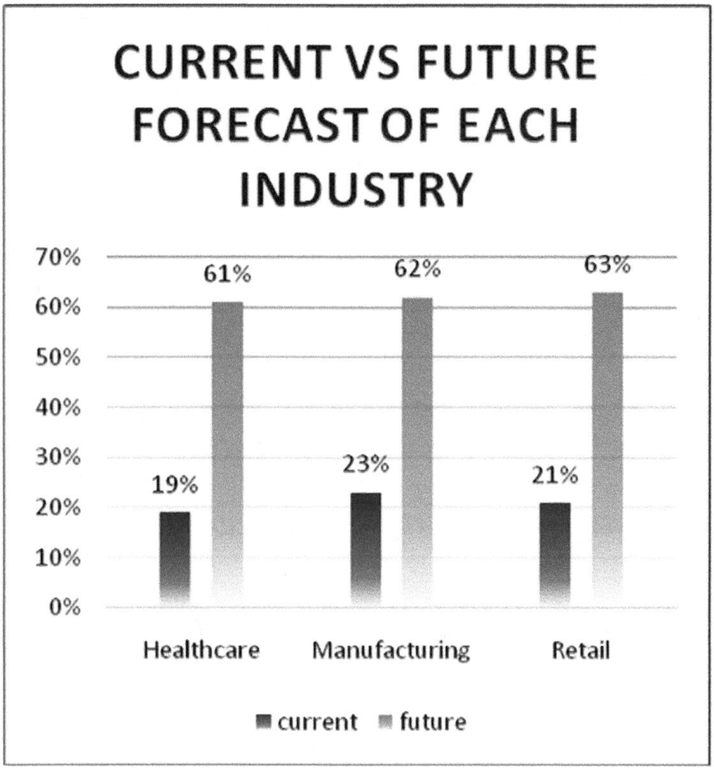

FIGURE 7.7 Forecast vs current usage.

7.8 VARIOUS APPLICATIONS FOR DIGITAL TRANSFORMATION

Figure 7.8 shows some different kinds of digital transformation technologies.

1. **Automating Trust:** To establish trust, the section on automating trust utilizes three technologies from the "Essential Eight technologies," namely IoT, AI, and blockchain. These technologies are employed to verify the authenticity of data, authenticate identities, and enable secure collaborative communications [13].

   Integrated machinery can provide means to establish trust in digital, human assets, and physical assets. When combined, "IoT and blockchain" will create an indisputable supply chain, ensuring that consumers receive reliable products that have not been tampered with or altered during the process.

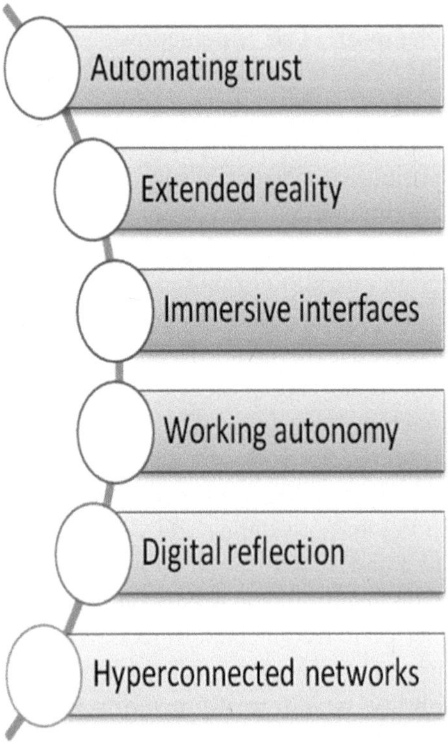

FIGURE 7.8 Various digital transformation applications.

2. **Extended Reality (XR):** XR has the most important propositions in the entertainment industry. In addition to that, cell phones and extended reality equipment can provide uninterrupted statistics about the globe. In the commercial field, augmented reality and mixed reality are widely applied to relieve work and improve client experience in real-time. Virtual reality supports users in gaining perfect experience in job skills techniques and soft skills techniques. Extended reality always incorporates AI in every application, such as IoT headsets or other devices, ensuring safe operation [10].
3. **Working Autonomy:** The term "Automation" encompasses the process of transforming mental procedures into digital procedures through the integration of automation, intelligent systems, and robotics.

"Working autonomy" refers to the range in which mechanization has evolved from individual explanations to providing a comprehensive workflow of mechanizations and integrating disparate systems to deliver fully automated end-to-end systems.

4. **Digital Reflection:** The interactions, interdependent physical processes, and complex processes are virtually represented. This is known as digital reflection. The ability to digitize the world has significantly increased in recent decades. This significant improvement can be attributed to the use of analytical tools, IoT, and simulation modeling. Without human intervention, the digital world can evolve, which aids in foreseeing upcoming scenarios and conducting testing. Digital reflection refers to virtual replicas of physical objects such as human societies, places, processes, and products. In human societies, people have started experimenting and experiencing simulations to gain deeper knowledge about the life cycle and the environment [7].
5. **Hyper-Connected Networks:** The Internet of Things (IoT) and the infrastructure of networks are used to process information at a tremendous speed, enabling proper connectivity between a vast number of individuals, systems, and devices. The deep interactions between automated systems and human beings contribute to the attainment of connectivity and processing speed. The combination of a large number of IoT-connected objects with cloud computing and 5th generation mesh networks results in low latency, high speed, and improved stability, which greatly benefits the growth of large-scale industries [21]. With increased utilization of the IoT concept, connections with AI-based responses and predictions will be established, eliminating the need for cloud computing in certain scenarios [9].

7.9 CONCLUSIONS

To summarize, digital transformation is a broad concept that offers numerous benefits in the corporate world. Each year, digital transformation contributes to the progress of emerging businesses. This digital revolution represents a significant trend in the corporate landscape. By achieving this trend through digital transformation, risks can be minimized. Various fundamental principles in digital technologies can be utilized to fulfill the responsibilities of an entrepreneur. Digital transformation encompasses cutting-edge concepts such as IoT, AI, robots, augmented reality, drones, blockchain, virtual reality, and 3D printing.

As the digital transformation era advances, it adapts to the changing times, whether consciously or unconsciously. People are comfortable with digital transformation technologies because, compared to the past, utilizing such technologies in complex industries is relatively simple and error-free. There are several barriers to digital transformation, such as the absence of a digital transformation mentor and employees lacking product placement skills. However, these barriers can be overcome by enhancing skills and technical expertise.

KEYWORDS

- **artificial intelligence**
- **blockchain technology**
- **COVID-19**
- **digital transformation**
- **digital transformation technologies**
- **industrial internet of things**
- **internet of things**

REFERENCES

1. Irena Sajovic & Bojana Boh Podgornik, (2022). "Bibliometric Analysis of Visualizations in Computer Graphics: A Study," SAGE Open 1–17.
2. Kokkinakos, P., Markaki, O., Koussouris, S., & Psarras, J., (2016). Digital transformation: Is public-sector following the enterprise 2.0 paradigm? In: *Digital Transformation and Global Society* (pp. 96–105). Springer International Publishing, Cham.
3. Van, E. N. J., & Waltman, L., (2014). Visualizing bibliometric networks. In: Ding, Y., Rousseau, R., & Wolfram, D., (eds.), *Measuring Scholarly Impact* (pp. 285–320). Cham: Springer International Publishing.
4. Dremel, C., Wulf, J., Herterich, M., Waizmann, J., & Brenner, W., (2017). How AUDI AG established big data analytics in its digital transformation. *MIS Q. Executive 16*(2), 81–100.
5. Lei, Z., & Jing, Y., (2016). Study on human resource reform in the digital transformation. In: *Proceedings of the 2016 Joint International Information Technology, Mechanical and Electronic Engineering; AER – Advances in Engineering Research* (Vol. 59, pp. 471–477).
6. Richert, A., Shehadeh, M., Willicks, F., & Jeschke, S., (2016). Digital transformation of engineering education. *Int. J. Eng. Pedagogy, 6*(4), 23–29.

7. Kaivo-oja, J., Roth, S., & Westerlund, L., (2017). Futures of robotics. Human work in digital transformation. *Int. J. Technol. Manage., 73*(4), 16–205.
8. Kohli, R., & Johnson, S., (2011). Digital transformation in latecomer industries: CIO and CEO leadership lessons from Encana oil & gas (USA) Inc. *MIS Q. Executive 10*(4), 141–157.
9. Abrahamson, E., (1996). Management fashion. *Acad. Manag. Rev., 21*(1), 254–285.
10. Lucas, H. C. J., Agarwal, R., Clemons, E. K., El Sawy, O. A., & Weber, B., (2013). Impactful research on transformational information technology: An opportunity to inform new audiences. *MIS Q., 37*(2), 371–382.
11. Walsh, M. N., & Rumsfeld, J. S., (2017). Leading the digital transformation of healthcare. *J. Am. Coll. Cardiol., 70*(21), 2719–2722.
12. Haux, R., (2006). Health information systems – past, present, future. *Int. J. Med. Inform., 75*(3, 4), 268–281.
13. Sharma, D., Singh, A. G., & Bajaj, R., (2019). Evolution from ancient medication to human-centered healthcare 4.0: A review on health care recommender systems. *Int. J. Commun. Syst.*, e4058.
14. Webster, J., & Watson, R., (2002). Analyzing the past to prepare for the future: Writing a literature review. *MIS Q., 26*(2), 3.
15. Peek, S. T. M., Wouters, E. J. M., Van, H. J., Luijkx, K. G., Boeije, H. R., & Vrijhoef, H. J. M., (2014). Factors influencing acceptance of technology for aging in place: A systematic review. *Int. J. Med. Inform., 83*(4), 235–248.
16. Rojas, E., Munoz-Gama, J., Sepúlveda, M., & Capurro, D., (2016). Process mining in healthcare: A literature review. *J. Biomed. Inform., 61*, 224–236.
17. Allahyari, M., et al., (2017). A brief survey of text mining: Classification, clustering, and extraction techniques. In: *Proceedings of KDD Bigdas* (p. 13).
18. Yu, D., Xu, Z., Pedrycz, W., & Wang, W., (2017). Information sciences 1968–2016: A retrospective analysis with text mining and bibliometric. *Inf. Sci. (Ny), 418, 419*, 619–634.
19. Van, E. N. J., & Waltman, L., (2011). Text mining and visualization using VOS viewer. *ISSI Newsl., 7*(3), 50–54.
20. Bosilj, V. V., Ivančić, L., & Suša, V. D., (2018). A preliminary literature review of digital transformation case studies. In: *ICMIT 2018: 20th International Conference on Managing Information Technology* (pp. 737–742).

CHAPTER 8

Big Data Analytics in Cyber Threats and Artificial Intelligence

V. KAVITHA,[1] B. SUNDARAVADIVAZHAGAN,[2] and K. GOWRI[1]

[1]*Department of Computer Science with Cognitive Systems, Sri Ramakrishna College of Arts and Science, Coimbatore, Tamil Nadu, India*

[2]*Department of University of Technology and Applied Sciences, Al Mussanah, Oman*

ABSTRACT

The data is available everywhere, from the universe to the tiniest microbes. As the usage of techniques and entities increases, the quality of the data also improves. However, maintaining such a huge amount of data becomes a challenge. This challenge led to the term "big data" being coined in 2005. According to a survey report on big data analytics (BDA) and cyber security, it has been declared that 62% of cyber attackers are involved in hacking, 51% are connected to malware attacks, and 43% are using social media platforms to gain unauthorized access to data.

Having large and meaningful data for processing and problem-solving is crucial. This process of studying vast amounts of distinct data sets using advanced techniques is known as BDA. In today's cyber world, where everyone is using networks and data daily, cyber threats are becoming more prevalent. Detecting vulnerabilities is essential for data security. BDA can be utilized as an innovative approach to address these threats by implementing a simple methodology.

Leveraging Artificial Intelligence (AI) Competencies for Next-Generation Cybersecurity Solutions. Pethuru Raj, B. Sundaravadivazhagan, & V. Kavitha (Eds.)

In this research paper, the Louvain clustering algorithm is used to predict various types of cyber attacks. The goal of this technique is to easily identify existing attacks through periodic audits.

8.1 INTRODUCTION

Big data analytics (BDA) is the practice of employing advanced analytical strategies to large and diverse datasets that consist of structured, semi-structured, and unstructured data from various sources, ranging in size from terabytes to zettabytes. The term "big data" is used to describe databases that are too large or complex for traditional relational databases to handle and process with minimal latency. Big data possesses one or more of the following characteristics: high volume, high velocity, or high variety [1]. The term "Big Data" is used to refer to datasets that are extremely large or complex, surpassing the capabilities of conventional data processing software.

The main distinction between traditional data and big data lies in terms of volume, velocity, and variety. Volume refers to the amount of data that has been generated; velocity refers to the speed at which the data is being generated, and variety encompasses different types of structured and unstructured data. The application of BDA in cyber-security is of utmost importance.

By analyzing data from networks and computers, analysts can extract valuable network information. This analysis enables decision-makers to make more informed choices, identifying necessary actions and providing recommendations for policies, strategies, events, and tools related to network operations.

8.1.1 BIG DATA

Big data is a collection of large amounts of uncategorized and categorized information. This information is complex and expansive, making it difficult for humans to organize properly. Fortunately, there is software available now to manage this type of information. People can utilize this data for decision-making and predictions. All big datasets have three main characteristics, known as the 3 Vs:

1. **Volume:** Big data contains millions of formless, low-density data points. Businesses can store data in dozens of terabytes to hundreds

of petabytes. This type of data is curated by data scientists using appropriate software.

2. **Velocity:** It refers to the direct grouping and application of big data. Big data is collected and processed rapidly to provide the most up-to-date results in various big data software systems, capturing data in real-time.
3. **Variety:** Big data encompasses diverse and disparate types of data within a single unstructured database. Traditional database management systems, on the other hand, only handle structured and relational data, lacking the ability to manage unstructured data that is organized using Big Data Analytics (BDA).

The concept of big data originated in the 1960s and 1970s. However, during that time, storing such large amounts of data was not feasible due to the sheer volume. The practical application of big data emerged from industries like manufacturing and social media platforms such as YouTube and Facebook, where data is generated on a daily basis.

In 2005, data handling software like MySQL and Hadoop were introduced, enabling better storage and analysis of datasets. This marked the beginning of a significant development in data handling techniques, including platforms like Apache, Hadoop, and Apache Spark, which are widely used for managing big data today.

Big data continues to advance, with many companies recognizing the benefits of predictive analytics. It plays a crucial role in influencing the Internet of Things (IoT) and cloud computing, allowing for the collection of vast amounts of data and the application of machine learning (ML) to create more accurate models.

8.1.2 CYBER SECURITY THREATS

A cyber security threat is a malicious act that compromises or steals data and disrupts digital existence in general. Cyber threats encompass computer viruses, data breaches, Denial of Service (DoS) attacks, and other attack vectors [2]. The objectives of cyber-attacks are to cause damage, disruption, unauthorized access to computer networks, intellectual property, or theft of information technology. Cyber threats can originate from trusted insiders within an organization or from external parties in remote locations.

Cyber threats are generated by various threat actors, including terrorist groups, organized crime organizations, hacktivists, hackers, natural disasters, and unintentional actions by authorized users. The number of attack vectors has significantly increased in recent times due to the growing global connectivity, widespread use of cloud computing services, and outsourcing by organizations. Overall, cyber security is of utmost importance to safeguard all types of information from damage and theft. This includes Personally Identifiable Information (PII), Protected Health Information (PHI), sensitive data, and government and industry information systems.

8.1.3 CYBER THREAT INTELLIGENCE

Gartner's definition of "Threat Intelligence" is confirmation based on knowledge about previous or emerging menaces or hazards to assets. Cyber threat intelligence provides an enhanced knowledge of cyber threats and allows the world to conduct research on various kinds of cyber threats in a timely and accurate manner [3]. Cyber threat intelligence is enhanced through a cyclic task called the intelligence cycle. Some activities are treated as a protocol in the intelligence cycle, such as planning the data collection, enhancement, and preparation of the evaluated report based on reliable information. The main purpose of threat intelligence is to demonstrate the risks that firms face from external threats, such as zero-day threats and APT. Threat intelligence includes in-depth knowledge about specific threats, such as the attacker's identity, motivation, and capabilities. Threat intelligence offers enhanced insight into the threat landscape with its recent techniques, tactics, and procedures [4].

8.2 LITERATURE REVIEW

Big data applications are used in business, technology, and also in medical fields. The future of business and technology is based on big data. Some examples of big data usage are:

- **Product Decision Making:** Big data is used by businesses like YouTube and Flipkart to develop products based on future trends. They analyze old product performance to determine what consumers want, and they can also use pricing data to control prices and sell more to their target customers.
- **Testing:** Big data can analyze a large amount of sensor reading, bug reports, and hardware specifications. This analysis can help companies

and individuals improve their sensor or hardware performance and also assist in maintaining user credentials.

- **Marketing:** Dealers analyze big data from previous marketing efforts to gather information about product sales, demand, and customer preferences.
- **Healthcare:** Medical professionals utilize big data to determine appropriate medications for various illnesses. By analyzing previous cases, they can match the right medicine to the corresponding illness. Additionally, they take into account the potential side effects of these medications. For example, if many individuals experience headaches after taking a certain medication, it is recorded as a potential side effect in the data.
- **Customer Experience:** Big data is used to gain customer insights and product reviews after the launch of a product. This includes analyzing customer knowledge, product reviews, social media references, online reviews, and feedback on product videos in order to obtain a more comprehensive review.
- **Machine Learning:** Big data is mainly used in machine learning (ML) and artificial intelligence (AI). ML is now used everywhere, such as Amazon, Netflix, and YouTube, where they predict how it works and what a customer wants. You know that every person has recommendations based on previous data, common purchase items, and session items [7].

8.2.1 HOW BIG DATA HELPS IN CYBER SECURITY?

Enhancing cyber security should be a top priority for many organizations in the digital age, as the lack of proper tools can lead to disastrous consequences. This is why more and more companies around the world are utilizing big data to defend against potential hacker attacks.

However, it is important to focus on achieving optimal cyber security in big data, as hackers can potentially gain unauthorized access to data processed using big data technology. Therefore, it is clear that big data has its advantages and disadvantages. Big data is arguably the best way forward in terms of cyber security, as detecting threats at the earliest possible opportunity is now easier than ever before [8].

Cybercriminals are using increasingly sophisticated big data strategies, making it more difficult to protect our laptops and the internet. Traditional preemptive security tools and technologies used for data mining and deterring cyber-attacks are simply not sufficient for many organizations, especially those dealing with large volumes of data. This is why cyber security professionals are turning to big data analytics. Big data analytics allows companies to identify potential threats and attack patterns by analyzing the data leading up to the attack. Intelligent big data analytics enables experts to develop a predictive model that generates alerts at the exact moment it detects a potential entry point for a cyber security attack.

AI and revolutionary device-learning models can play a pivotal role in the establishment of such mechanisms. Analytics-based solutions enable organizations to anticipate and prepare for potential events in their processes. Big data analytics can assist in monitoring a large set of customer and system activities to deter any chances of cyber attacks. This approach plays a crucial role in detecting and preventing data breaches and other similar cyber attacks.

Furthermore, security professionals can automate their processes to expedite the recovery process and combat data breaches when a Cyberattack occurs. Enterprises can leverage data from various monitoring tools such as OSSEC, Nagios, Splunk, etc. Big data analytics can address this issue by mechanizing processes at scale. The Intrusion Detection Systems can be enhanced comprehensively with real-time analytics for the detection and prevention of malicious activities encountered in a system. Such systems eliminate the risk before a cybercriminal gains unauthorized access to the network.

For instance, employers scan associations with different databases from proxy logs, secure and appropriate domains, and monitor the overall health of the organization's internet.

8.3 BIG DATA ANALYTICS IN CYBER SECURITY

Big data is arguably the most effective method when it comes to cyber security, as detecting threats at the earliest possible opportunity is now easier than ever before.

The big data methods used by cybercriminals are becoming increasingly difficult to defend against. Traditional preemptive security tools and techniques used for data mining and preventing cyber-attacks are no longer sufficient for many businesses, especially those that handle large volumes of data. This is why big data analytics (BDA) has become increasingly utilized in the field of cyber security.

BDA allows organizations to identify potential threats and attack patterns by analyzing the data related to the actions leading up to an attack. Intelligent BDA enables experts to develop predictive models that generate alerts at the precise moment when an entry point for a cyber security attack is detected. Artificial intelligence (AI) and advanced machine learning (ML) models can play a crucial role in the development of such systems. Analytics-based solutions enable businesses to forecast and prepare for potential events in their operations.

BDA can contribute to the 24-hour care of a large set of actions by operators and systems to deter any possibilities of cyber-attacks. This method plays a crucial role in the detection and prevention of data breaches and other similar cyber-attacks. Additionally, security experts can automate their processes to expedite the retrieval process and combat data breaches when a cyber-attack occurs (Figure 8.1).

i. **Malware Detection:** Malware detection plays an important role in computer systems and network security. Although numerous machine learning models have been used in the detection of malicious binary programs, the performance of unified strategies has not been extensively examined. Additionally, the vast amount of malware has made it a big data challenge, requiring security investigators and practitioners to deploy big data technologies to acquire, store, analyze, and visualize malware information.
ii. **Forensic Analysis:** The meaning of forensic analysis can be described as a thorough process of observing, examining, and documenting the motive, progression, and conversations related to a security incident or violation in accordance with state and federal laws. Forensic analysis is commonly employed in court investigations, particularly in criminal investigations. It encompasses a wide range of analytical techniques and expertise.
iii. **Threat Hunting:** Cyber threat hunting is an active security exploration through networks, endpoints, and datasets to pursue hateful, doubtful, or dangerous activities that have evaded detection by existing tools. Therefore, there is a distinction between cyber threat discoveries and cyber threat hunting. Threat discovery is a somewhat passive approach to monitoring statistics and systems for potential security issues, but it is still crucial and can assist in threat detection. Practical cyber threat hunting approaches have evolved to utilize innovative threat intelligence and information correlation to identify and classify potential threats in the event of an attack.

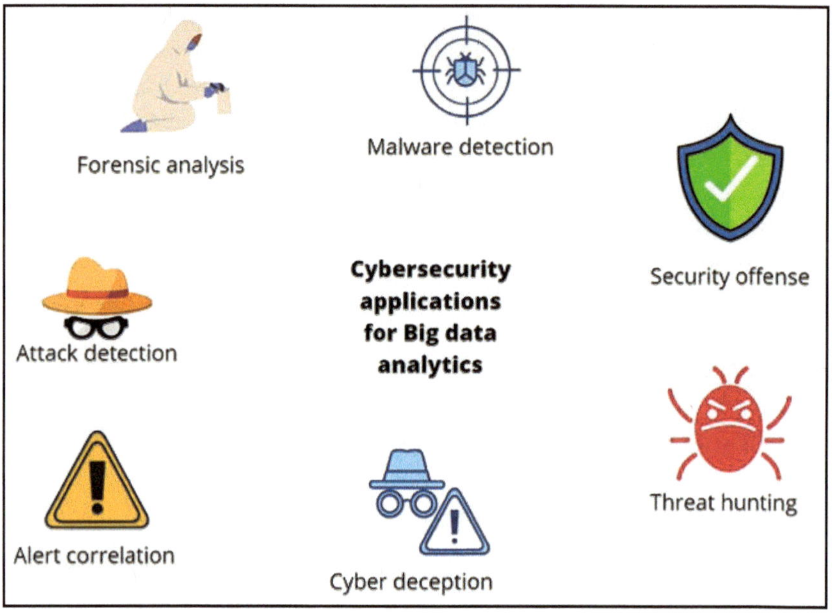

FIGURE 8.1 The major applications of big data analytics using cyber security.

iv. **Security Crimes:** Cybercrimes are illegal activities carried out in a sophisticated manner, where computers are either used as tools or targets, or both. Cybercrime often includes illegal access to computers, data manipulation, and virus/worm attacks.

8.4 COMMON TYPES OF CYBER THREATS (FIGURE 8.2)

i. **Malware:** Malware is a software program that is used for adware and viruses. Malware is triggered when a person clicks on a hyperlink, which results in the installation of malicious software.
ii. **Emotet:** Emotet is a sophisticated, interconnected financial Trojan that primarily functions as a downloader for various financial Trojans. It remains one of the most costly and destructive malware.
iii. **Denial of Service:** A distributed denial of service (DDoS) attack does the same thing, however, the assault is created from a network of computers. A botnet is a form of DDoS where tens of thousands and millions of systems may be infected with malware and controlled by a hacker, according to Jeff Melnick of Netware, an information technology security software company. Botnets, sometimes referred

to as zombie systems, target and overwhelm a target's processing capabilities. Botnets are located in various physical locations and are difficult to trace.

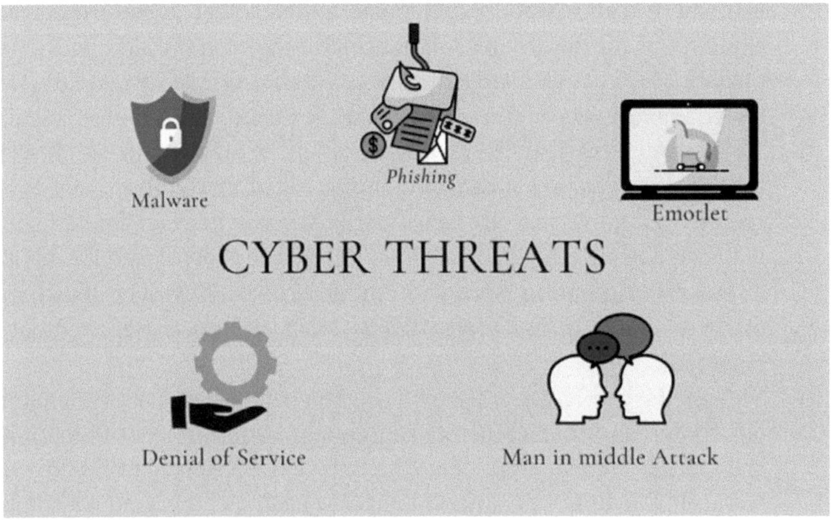

FIGURE 8.2 Some common types of cyber threats.

iv. **Man in the Middle:** "Man-in-the-Middle" (MITM) hackers insert themselves into a two-party transaction. By intercepting the traffic, they can intercept and steal data. Attacks often occur when a user connects to an insecure public Wi-Fi network.

v. **Phishing:** Phishing attacks use fake messages, such as emails, to deceive the recipient into opening them and carrying out secret commands, such as providing a credit card number. The objective is to obtain sensitive information like credit card and login data, or to install malware on the target's system.

8.5 MAJOR CHALLENGES

i. **Unaware Employee:** Employees with a purpose to spot and save protection cracks will need fundamental training in the unique approaches that cyber security threats can present themselves. For the most part, this includes spam, phishing, malware, ransomware, and social engineering.

ii. **No Periodic Updating:** Updating, however, is not to be overlooked. These updates are provided for surprisingly useful purposes. Occasionally, they include improvements or popular developments. However, more frequently, they fix risky bugs in the operating system and, more importantly, patch vulnerabilities that cyber-attacks can exploit. It is for this last reason, especially, that you should update your operating system as soon as possible when prompted to do so. Even better, make sure you have set up your device to update automatically.

iii. **No Strong Firewall:** Firewalls, sometimes referred to as network security devices, are a central component of any enterprise's cyber security strategy. A strong set of perimeter and internal firewalls in a network can help block malicious traffic and slow down the spread of attacks originating from within the network. However, to achieve the best results from firewall deployment, effective management is essential.

iv. **Storage Issue:** The rapid growth of data has placed greater demands on storage and management. Big data storage refers to the storage and management of large-scale datasets while ensuring reliability and scalability. We will examine key issues such as big storage systems, distributed storage systems, and large data storage strategies. On one hand, the storage infrastructure needs to provide a reliable storage service; on the other hand, it should offer an efficient access interface for querying and analyzing large amounts of data.

v. **Vulnerable Networks:** Network protection vulnerabilities are strengths or flaws within the system's software, hardware, or administrative procedures. Network vulnerabilities can be both non-physical and physical. If a virus or malware downloads into the operating system, it can potentially infect the entire network. Improperly installed hardware or software, operating systems or firmware that have not been updated.

8.6 REMEDIES OF THE LIMITATIONS

i. **Educate the Employee:** When a cyber threat occurs, it is crucial for the victims to be aware of what is happening. If some attacks are taking control of the company's database, the employees must know the attack scenario. Educating employees is a significant solution for this purpose. By educating them about common threats, the company can prevent the attacks to a certain extent.

ii. **Perfect Audit:** Auditing is a lawful and steady process. The SOC analyst must be appointed for the auditing. This person will perform surveillance of the company to prevent cyber-attacks. Through proper auditing, one can assess the possibilities of cyber threats and attacks [13].

    The company may have weak techniques to counter the attacks and attackers. Initially, the company must enhance the strength of its techniques. By strengthening the techniques, the algorithm will have a higher chance of working efficiently, which in turn leads to the anticipation of cyber threats [14].

iii. **Port Scanning:** To ensure the stability and security of the systems, it is necessary to recruit an expert in the field. A white hat hacker is preferred for this task, as they can perform authorized attacks to identify vulnerabilities. Port scanning is a useful technique to identify active ports and prevent potential attacks. Any unnecessary active ports can be deactivated and protected from malicious activities.

As part of the decision-making process, port scanning can be implemented to identify vulnerabilities. These vulnerabilities can then be compared with relevant threats and appropriate countermeasures can be applied. This research work also discusses the importance of auditing.

8.7 METHODOLOGY

Anyone with a physical or remote connection to the assets of a specific concern can be exposed to the concept of cyber risk. For example, trusted staff members of the organization may accidentally misplace reliable information, casual workers may be unaware of the company's procedures and policies, or malicious insiders with authorized access to critical information and systems.

Due to the aforementioned reasons, the concerned organization will face the challenge of cyber threats. It is crucial to predict and address these threats within a short timeframe, as failure to do so can result in significant damage to the entire system. To handle cyber-attacks, a combination of conventional and innovative techniques is required. Among these techniques, the Louvain Clustering technique has been chosen to apply the dataset of cyber-attacks for faster attack prediction. The outcome of this process will help minimize cyber-attacks and prevent extensive damage.

The Orange tool has been selected for its user-friendly interface and drag-and-drop functionality. Orange simplifies the implementation of various methodologies such as the SVM algorithm, KNN, Random Forest algorithm, Louvain clustering algorithm, etc. [12] (Figure 8.3).

FIGURE 8.3 The picture illustrates the workflow from the orange tool with the Louvain clustering algorithm.

The dataset is a CSV file downloaded from Kaggle.com. Initially, the CSV file import icon is set to load the dataset from that CSV file. In the next step, the Louvain clustering algorithm is applied. One can easily join the scatter plot to the clustering algorithm and view the plotted graph without any trouble. This graph will be displayed in a separate window called a scatter plot. In this window, the x-axis and y-axis can be selected using a combo box, and sorting can also be done using colors [11].

8.7.1 PROPOSED METHODOLOGY

The Louvain clustering algorithm is a clustering method used to propose and predict groups in large networks. Cyber-attacks come in various kinds and patterns, and it is crucial to identify and prevent them to avoid significant and lasting damages. Discovering and predicting these cyber-attacks requires a specific machine learning technique. Therefore, the Louvain clustering algorithm is applied to minimize the vast number of different attacks [9]. The Louvain clustering algorithm offers numerous advantages, including:

- Louvain clustering has a sufficient number of manageable parameters
- Requires no prior assumption on the graph
- Works on data points connected by a graph
- Used for complex networks
- Using the Louvain clustering algorithm can learn how groups of nodes are partitioned.

8.7.1.1 LOUVAIN ALGORITHM

The Louvain algorithm is very straightforward and well-designed. This algorithm achieves excellent results, such as modularity, through two simple phases: local moving of nodes and aggregation of the network. In the first phase, specific nodes are moved to the corresponding community that maximizes the quality function. In the second phase, a combined network is created based on the separation obtained in the previous phase. Each community in this separation becomes a node in the aggregate network. These two phases are repeated until the quality function can no longer be improved. Typically, the Louvain algorithm starts with a "singleton partition," where each node is in its own community. However, it is also possible to start the algorithm with a different partition. To discover enhanced and refined partitions, multiple iterations of the algorithm can be performed, using the partition obtained in one iteration as the starting point for the next iteration (Figure 8.4).

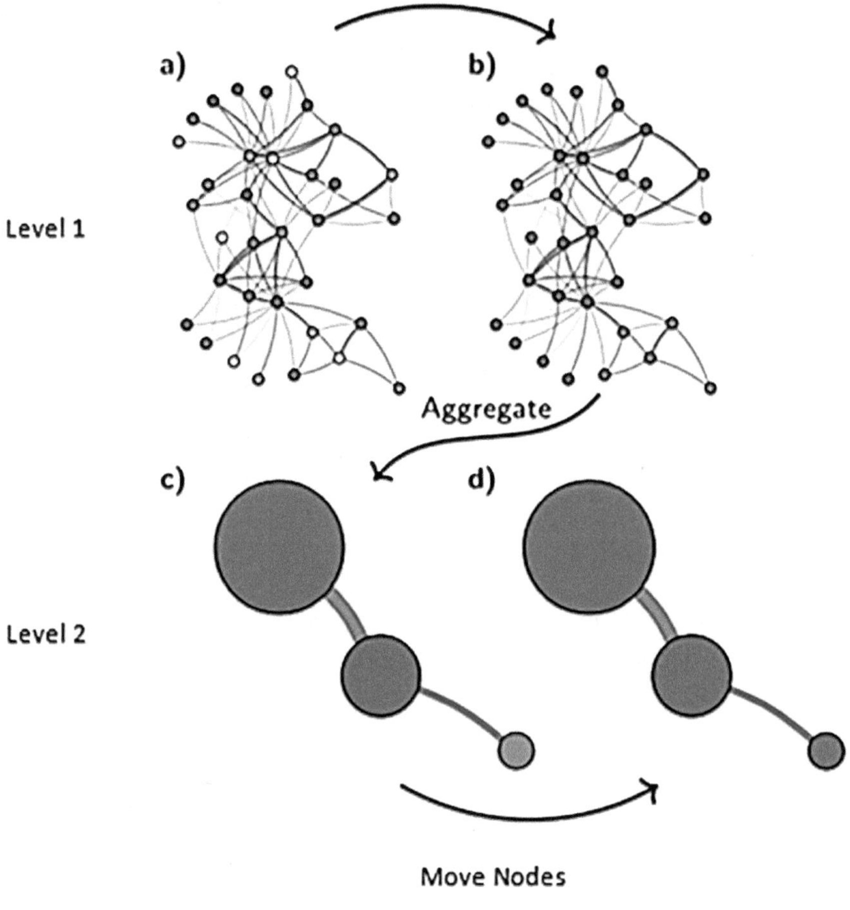

Move Nodes

FIGURE 8.4 The working structure of the Louvain algorithm.

Source: Reprinted from Ref. [18] https://creativecommons.org/licenses/by/

8.7.1.2 BADLY CONNECTED COMMUNITIES

The Louvain algorithm can potentially identify subjectively imperfect connected communities. In particular, Louvain may discover societies that are internally disconnected. This means that an element of a disengaged community can easily connect with another element through a path outside the society. Essentially, the issue of disjointed societies is not merely a theoretical concern.

The Louvain algorithm begins with a singleton division, where each node is in its own community. The algorithm then moves individual nodes from one group to another in order to uncover separations. Based on these separations, an aggregate network is created. The algorithm continues to move individual nodes within the aggregate network. These steps are repeated until satisfactory improvements are achieved.

Interestingly, iterating the algorithm amplifies the challenge, despite enhancing the overall process. According to the Louvain algorithm, a node may be moved to different groups even if it serves as a connection between various elements of its previous community. Removing a specific node from its previous community results in the separation of the former group. Other nodes in the previous community will then be moved to different communities. However, the remaining nodes may still be sufficiently connected to each other within the community, despite the fact that the society has become disconnected.

Noticeably, the Louvain algorithm may uncover the most severe cases of disconnected communities. It may encounter various challenges that can lead to the discovery of weakly associated communities. Generally, Louvain is capable of handling poorly connected communities that differ from the well-known challenge of achieving a promised range of modularity. Due to the range of resolution, modularity may cause smaller communities to be merged into larger ones.

Consequently, the issue that Louvain addresses is independent of the problem of decision boundaries with poorly connected communities. In the case of modularity, communities may have significant substructure along with the decision boundary, which is limited by Louvain. Although the Louvain algorithm may seem to perform exceptionally well in discovering high-quality partitions, in its standard form, it guarantees that it will not create divisions that cannot be combined. This indicates that the communities are well separated. As the Louvain algorithm iterates, the quality of the divisions will continuously improve until the algorithm reaches a point where further enhancements are not possible.

8.7.1.3 STEPS TO FOLLOW THE LOUVAIN ALGORITHM

Step 1: N nodes in a graph network are mentioned. Assigned the weight of each edge to begin with every node is assigned to the various partitions or communities. Hence,

No of Partitions = Number of Nodes N

Step 2: For each neighbor, the node is verified using the entire modularity amplified by shifting the partitions. Node i is shifted to a specific partition j. None of the observed gains are positive, so I will retain the current partition label.

Step 3: Continuously, step 2 is repeated for all the nodes in sequence. Furthermore, this iteration continues until no further improvement in modularity can be achieved. The suggestion is to reach a local maximum modularity, beyond which it would not be feasible to separate. It is highly likely that the nodes will be visited multiple times to measure the change in modularity by moving them to different partitions nearby.

Step 4: Step 2 involves recreating a novel network by collectively gathering the nodes labeled in a similar community during the initial process. The weights of the edges among the different new communities are established by accumulating the weights of the edges from each node in a particular community to another.

The Louvain algorithm successfully optimizes modularity. Any supervised or unsupervised ML algorithm requires an optimization procedure to choose the convergence criterion. In the scenario of community detection, modularity is one of the most appropriate optimization metrics. The quality of the communities, referred to as partitions, is calculated through the modularity of the partition. Modularity M is defined by the formula shown below.

$$M = \frac{1}{2m}\sum\nolimits_{i,j}\left[X_{i,j} - \frac{k_i k_j}{2m}\right]_{\delta(a_i a_j)} \quad (1)$$

where; $X_{i,j}$ is the weight of the edge among 'i' and 'j;' $K_{i,j}$ is the sum of weights of the vertex attached to the vertex I, referred to as the degree of the node; a_i, a_j are the community for the assigned vertex 'i;' $\delta\,(a_i a_j)$ is 1 when x = y, and otherwise it is 0; and $M = \left(\frac{1}{2}\right)\sum\nolimits_{ij} A_{ij}$ means the number of links.

8.8 RESULTS

The Louvain clustering algorithm is applied to the dataset. In this dataset, the periodic audit is performed on all kinds of ports [10] (Table 8.1).

The dataset of the Attack.csv file contains the following attributes: port numbers, flow duration, flow packets, and attack. This dataset will be considered and enhanced with the Louvain algorithm. Various attribute combinations are applied with the clustering algorithm, resulting in multiple clustering combinations.

TABLE 8.1 The dataset "Attack.csv" contains four distinct columns: Port, Flow Durations, Flow Packets, and Attack.

Port	Flow Duration	Flow Packets	Attack
7973	244117	20.48198	0
8080	1.19E+08	0.690214	0
5355	64527484	0.805858	0
5353	86799103	2.788047	0
3298	85	23529.41	1
1124	17	647058.8	0
465	1901599	21.03493	1
445	10964463	5.38102	0
443	174669	91.60183	1
389	21	238095.2	1
139	41415	1810.938	1
137	3892118	84.78674	1
123	68016491	0.058809	1
80	67469	118.573	1
53	47749	41.88569	1
42	17	764705.9	0
21	2	1000000	0
135	60	33333.33	0
1900	6027484	14.59979	1
139	249	16064.26	0

In the result section, the Louvain clustering algorithm plots are illustrated. The clustering technique provides a clear analysis that will be used to accurately report the exact scenario of the cyber-attack. Figures 8.5–8.8 explain various parameters of cyber-attack comparison, such as the flow duration and port, flow packets, and port, using various graph representations.

8.8.1 FLOW DURATION VS PORT

Figure 8.6 explains the comparison among the two consecutive parameters of flow duration and ports of the cyber-attack datasets. The flow duration is measured in minutes. The largest value of flow duration is 120, according to the graph. The most vulnerable ports are less than 1,000 ports, such as port numbers 465, 443, 389, 139, and 137. These ports are used for udp, smtp, http, or https, idp. Figure 8.5 also provides further explanation.

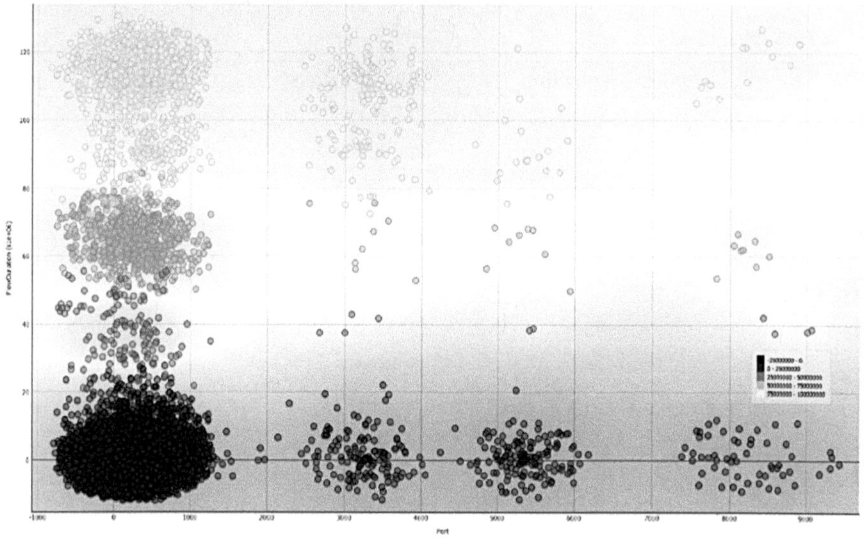

FIGURE 8.5 Louvain clustering algorithm for flow duration and ports.

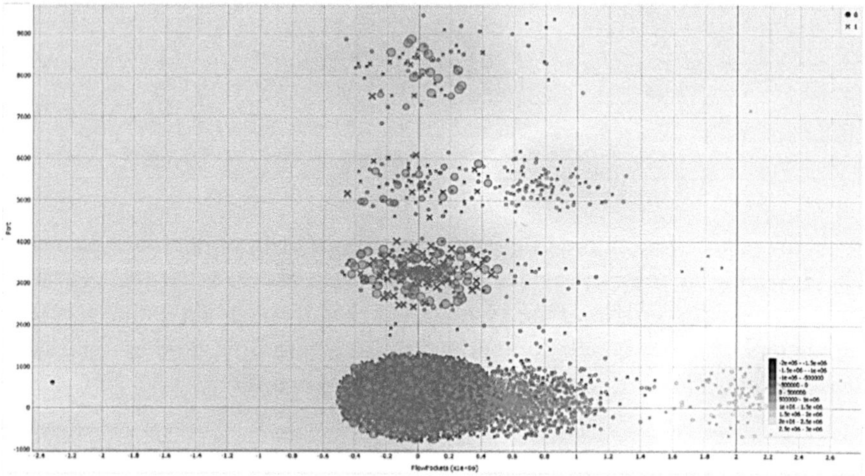

FIGURE 8.6 Louvain clustering algorithm for flow duration and ports.

8.8.2 FLOW PACKETS VS PORT

These plotted attacks are denoted specifically, as each type of attack has been marked with different shapes. There are mainly two possibilities: true or false. The occurrence of the attacks is illustrated here. Flow packets range from 0 to huge numbers monthly. The thread found flow packets ranging from -0.4 to 1. During data transmission, the data is sent through several packets. The vulnerability occurs in the packets which are plotted above.

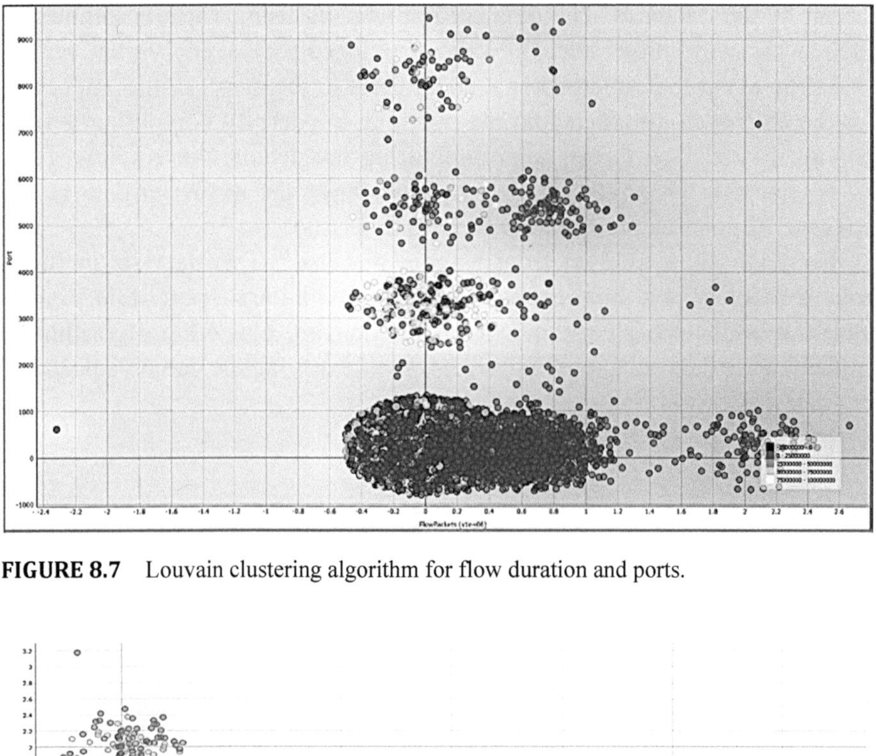

FIGURE 8.7 Louvain clustering algorithm for flow duration and ports.

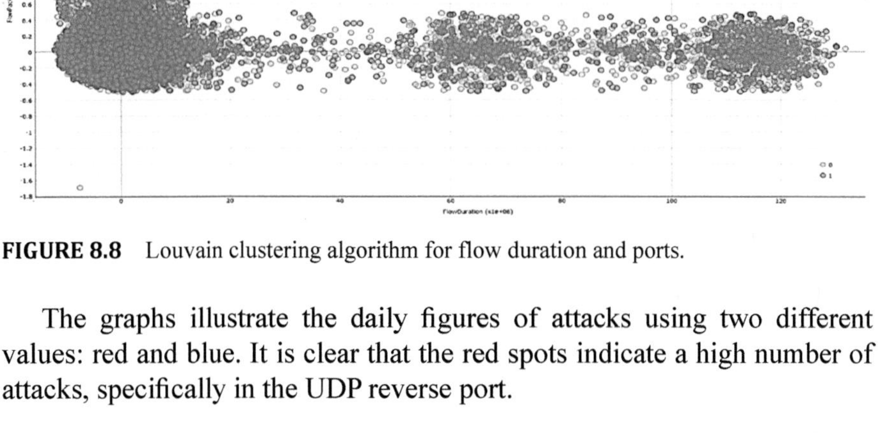

FIGURE 8.8 Louvain clustering algorithm for flow duration and ports.

The graphs illustrate the daily figures of attacks using two different values: red and blue. It is clear that the red spots indicate a high number of attacks, specifically in the UDP reverse port.

8.9 CONCLUSIONS

The most valuable thing is data. Every person is trying to protect this data from unauthorized access. These unauthorized access attempts occur due to threats to the database. The data set is protected using various techniques such as password protection, firewalls, privileged access, encryption, etc.

In the event that an attacker sends a view to the system, it is important for the employee to understand the potential impact. By comparing it with previously occurred threats and configuring the details, the employee can assess the risk. However, challenges arise when the historical data is not updated. Regular audits can help address this issue.

For analysis, a data set from Kaggle was used. This data set includes columns such as port number and attack type. With the support of the orange tool, the predictive algorithm called the Louvain clustering algorithm is applied to the data set to predict threats. By identifying the valuable port, the problem can be verified.

KEYWORDS

- **big data analytics**
- **machine learning**
- **man-in-the-middle**
- **personally identifiable information**
- **protected health information**

REFERENCES

1. Aarushi, A., Harshit, M., & Dayanand, W. J., (2017). Big data analytics in cyber security. *International Journal of Engineering Research and Technology.*
2. Rohit, K., Anvesh, B. V., & Ranjith, R K., (2019). Cyber security. *Journal of Business and Public.*
3. Roji, K., & Sharma, G., (2019). *Cyber Security Challenges and Big Data Analytics.* Crimson Publishers.
4. Pelin, A., Bharat, B., & Rohit, R., (2019). Big data analytics for cyber security. *Hindawi Security and Communication Networks.*

5. Iqbal, H. S., Kayes, A. S. M., Shahriar, B., Hamed, A., Paul, W., & Alex, N., (2020). Cyber security data science: An overview from machine learning perspective. *Journal of Big Data.*
6. Murad, A. R., Mohd, A. M., & Anazida, Z., (2017). Big data analytics adoption for cyber security: A review of current solutions, requirements, challenges, and trends. *Journal of Information Assurance and Security.*
7. Athor, S., & Andri, A., (2019). Cyber risk prediction through social media big data analytics and statistical machine learning. *Journal of Big Data.*
8. Liyuan, S., Hongyun, Z., & Chao, F., (2021). Data security governance in the era of big data status, challenges, and prospects. *Data Science and Management.*
9. Pasquale, De. M., Emailio, F., Glacomo, F., & Alessandro, P., (2011). Generalized Louvain method for community detection in large networks. *IEEE Proceedings.*
10. Sayan, G., Mahantesh, H., Antonino, T., Ananth, K., Hao, L., Chavarria-Miranda, D., Arif, K., & Assefaw, H. G., (2018). Distributed Louvain algorithm for graph community detection. *IEEE International Parallel and Distributed Processing Symposium.*
11. Naoto, O., Hiroshi, T., & Mary, I., (2016). A simple acceleration method for the Louvain algorithm. *International Journal of Computer and Electrical Engineering.*
12. Siti, H. H A., Zuraida, A. A., Norhazwani, M. Y., Nurul, H. M. Z., Nurul, A. H., Mohd, F M., Siti, A. A., et al., (2021). Comparison between Louvain and Leiden algorithm for network structure: A review. *Journal of Physics: Conference Series*. IOP Publishing.
13. Ashwini Sheth, Sachin Bhosale, & Farish Kurupkar (2021). *Research Paper on Cyber Security.* Contemporary Research in India (ISSN 2231-2137).
14. Jabber, A., & Quddus, T., (2021). COVID-19 pandemic: A new era of cyber security threat and holistic approach to overcome. *IEEE Asia-Pacific Conference on Computer Science and Data Engineering (CSDE).*
15. Vemula, G., et al., (2015). Big data analytics for detection of frauds in matrimonial websites. *International Journal of Computer Science Engineering and Technology (IJCSET)* (Vol 5, No. 3, pp. 57–61).
16. Richard, A. D., (2002). Insurance fraud. *The Journal of Risk and Insurance, 69*(3), 271–287.
17. Bresfelean, V. P., Mihaela, B., Nicolae, G., & Calin-Adrian, C., (2007). Data mining clustering techniques in academia. In: *ICEIS* (No. 2, pp. 407–410).
18. Traag, V. & Waltman, L. & van Eck, Nees Jan. (2019). From Louvain to Leiden: guaranteeing well-connected communities. Scientific Reports. 9. 5233. 10.1038/s41598-019-41695-z.

CHAPTER 9

Cyber Security-Based Solutions for IoT Edge Devices: Critical Issues and Services

M. R. SUNDARAKUMAR

School of Computing Science and Engineering, Galgotias University, Greater Noida, Utter Pradesh, India

ABSTRACT

In the modern digital era, industries and companies face a major challenge in handling millions of devices in a network while ensuring security. Edge computing is a problematic concern for the mobile and IoT application environments as it involves handling more devices connected at the edges. Providing security for the entire cluster of devices at the cluster edges is complicated in the cloud computing domain. Although few classical approaches and technologies are handled at the edge devices, providing security among all devices during frequent data transactions is less reliable. Providing security among the devices that are connected through IoT is difficult due to the huge number of volumes. Quality of Service (QoS) and vulnerability issues are the major problems of the edge devices connected to IoT. Accurate results and less latency require a common approach or technology to apply to these situations. Cybersecurity is a modern technique that provides security to edge devices connected to the cloud environment with the help of cryptographic principles. The vulnerabilities of IoT edge devices are controlled using cybersecurity since firmware updates are often lacking. When data is stored in a data warehouse in a cloud environment, data distribution among plenty of devices can be handled with cybersecurity techniques. Cybersecurity is leading the world with its features, and future developments in cybersecurity will reveal more standard policies and standards for cloud storage principles.

Leveraging Artificial Intelligence (AI) Competencies for Next-Generation Cybersecurity Solutions. Pethuru Raj, B. Sundaravadivazhagan, & V. Kavitha (Eds.)

9.1 INTRODUCTION

9.1.1 IOT DEVICES AND SENSORS

In the present scenario, the devices used in various real-world applications are accessed and controlled in an IoT (Internet of Things) environment. This involves handling sensors and actuators, which are the cause of all applications running behind the industries. Handling a massive number of devices collected together in clusters is challenging, but the edge devices used to communicate between the clusters play a predominant role in cloud computing. However, vulnerabilities have affected the functionalities of the devices, particularly at the edges of the clusters, due to the inefficiency of handling huge volumes of devices simultaneously. This leads to less latency and low performance during the transmission of data consistently. Most devices are controlled using sensors, which control the applications either manually or electronically based on the user applications. Different types of sensors and actuators are used in devices for their applications, as shown in Figure 9.1, which denotes the IoT devices and sensors used in various applications.

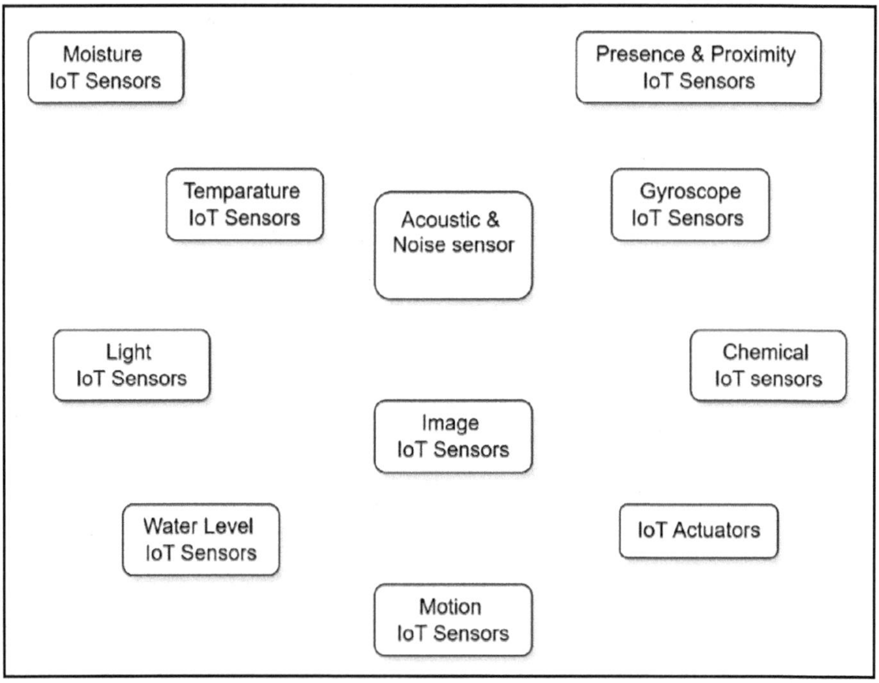

FIGURE 9.1 IoT devices and sensors.

9.1.2 IOT EDGE DEVICES

Edge computing is an innovative approach that involves running applications on mobile and IoT devices. This technique enables the handling of large amounts of data at data centers while consuming it at the edges of the network. The edge computing platform manages the deployment of data centers to retrieve the entire data amount over the network. With increasing concerns about latency and bandwidth issues, edge computing is a highly promising technology that offers solutions to these problems. In cloud computing, a data center is the foundation of the network, which stores data from several resources connected to the internet. The integration and upgrade of these data centers will significantly impact cloud storage for real-time global applications. Although various mechanisms have been developed to address these issues, the solutions are not yet optimal. Figure 9.2 illustrates the IoT edge devices and their functionality.

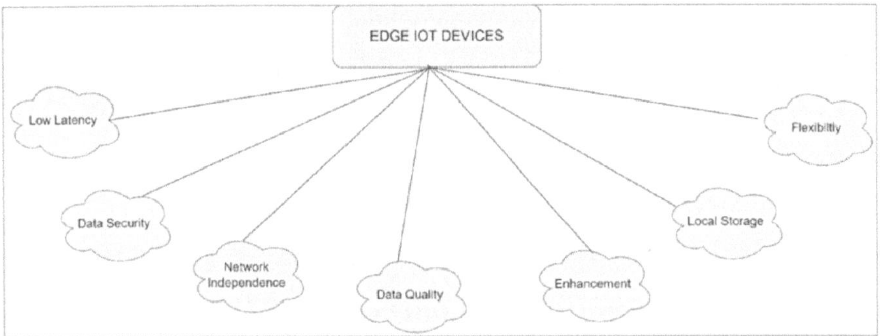

FIGURE 9.2 IoT edge devices.

9.1.3 IOT EDGE COMPUTING

Edge computing refers to accessing multiple data centers from various cloud environments on the same platform connected via the internet. Devices with IoT sensors are connected to the nodes and accessed by other devices in real-world applications. The nodes are connected through a wired connection which ensures fast latency and bandwidth. However, as the number of nodes connected over the network increases, the bandwidth may become limited. With an increasing number of devices connected at the edge of the network, their access to the huge storage decreases. Data is generated from multiple sources worldwide, but the storage capacity is limited, leading to a high retrieval process. The speed and accuracy of data retrieval may vary depending on the occurrences of data located in storage. (See Figure 9.3 for a visual representation.)

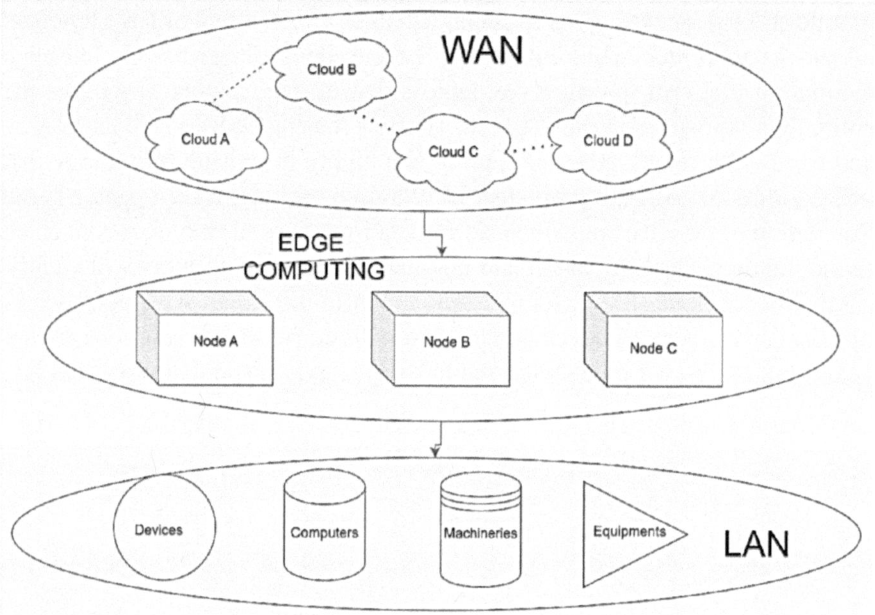

FIGURE 9.3 IoT edge devices.

9.1.4 IOT EDGE COMPUTING SECURITY AND CHALLENGES

The devices that are connected to the edges of the network can be categorized based on their level of security, which enables them to provide simulated access during critical situations. To improve the overall performance, it is essential to strengthen the security at the edges of the network. There are various aspects that typically define security in edge computing.

- To ensure security, various parameters are taken into consideration such as encrypted tunnels, access control, policies, standards, and firewall configurations.
- The network layer is important for accessing data from the outside world through its applications.
- Edge computing is not centralized access control, but it poses critical challenges for providers in implementing early threat detection.
- Patches have been created and upgraded from the existing system and then implemented to keep devices in an updated condition at all times.

- Maintenance of vulnerabilities in typical places and also discovering new ones.
- Outdated software usage, weak log credentials, domain-specific problems, and platform-oriented issues are also causes of security weaknesses in edge computing.

Figure 9.4 denotes the edge computing security over IoT applications.

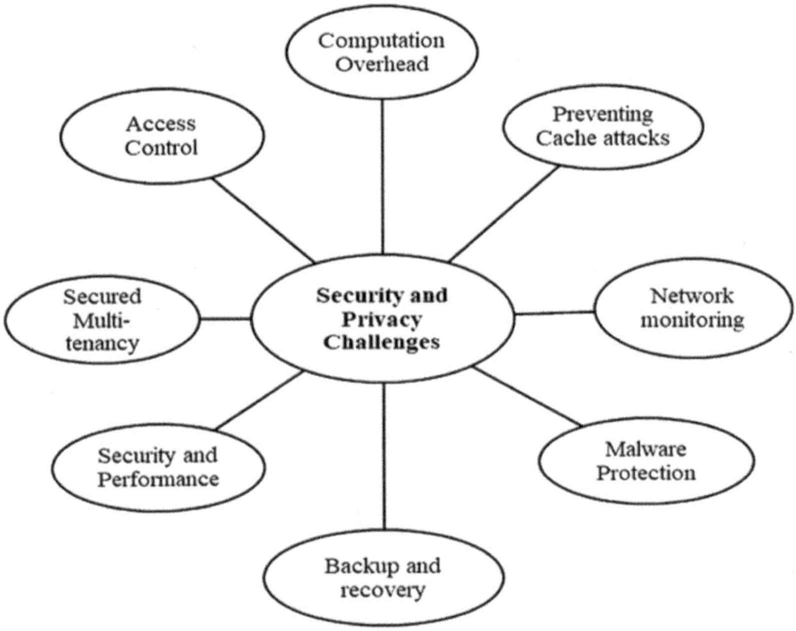

FIGURE 9.4 Edge computing security issues.

Source: Reprinted with permission from Ref. [49]. Copyright © 2019, Springer Nature Switzerland AG

9.1.5 IOT EDGE COMPUTING ARCHITECTURE

The challenges faced in IoT edge computing devices are mainly in scalability, reliability, transmission delay, interoperability, and availability. These challenges can be addressed with the help of IoT sensors and devices used at edge networks, which provide reduced latency, longer battery life, low bandwidth requirements, system resilience, and new architecture. This enables the handling of cloud data centers with continuous data analytics. To address security issues, a separate gateway is created to intelligently process real-time and short-term data storage over the network. Smart IoT devices play a crucial role in developing this new architecture, with the middle layer consisting of analytics and processing components. Decisions are made at the edge layer computer and then processed in the cloud storage. The architecture used for the IoT edge computing process.

9.1.6 IOT EDGE COMPUTING – REAL-LIFE USE CASES

Depending on the number of devices increased in real-world applications to perform edge computing operations, classical cloud computing has several issues in configuring policies, standards, and data security threats. The storage places of cloud data centers were wasted due to the low usage of connected resources. However, edge computing was introduced because it provides decentralized data processing, which helps industries with low dependencies. After edge computing comes into the industries, it will give advantages to the industries in terms of privacy and data policies, robust applications, low operational costs, reliability, scalability, and minimal latency. The use cases were taken at different scenarios in various domains for developing real-world applications with these new interventions, and the results were analyzed to compare the performances. Most of the use cases have been considered in healthcare sector devices, autonomous vehicles, retail industries, smart security solutions, and video conferencing facilities. Describes the sectors that are used for edge computing.

9.1.7 IOT EDGE COMPUTING – SECURITY SOLUTIONS

The solutions for the security challenges and issues at the cloud edge devices while storage plays a major role in the entire system. Either may be a tool, policy, standard, or protocol, and several parameters have been taken into consideration.

- Administrative rights have been granted to individuals responsible for maintaining databases and storage in the cloud.
- Updates and patches of the software and hardware should be properly done.
- Accounts, passwords, and authentication must be highly confidential.
- Monitoring the regular activities of the network devices connected through the internet.

- Providing strengthened security like cyber security for the entire storage system.
- Modern tools and algorithms used for improving security systems.
- Different networks like LAN, WAN, MAN, and wireless networks can be isolated with their ranges and provide security using standards.
- Adopted techniques can be integrated with the existing systems.
- Remote assistance and control devices should be avoided in certain cases.

9.2 CYBERSECURITY

Cybersecurity is a field that aims to protect large volumes of data using standardized methods and principles. It ensures the security of cloud storage systems by implementing diplomatic strategies and enables secure sending and receiving of digital information. Despite the several principles applied to run the cybersecurity system on a large scale, security remains the most significant factor for clusters in a cloud computing environment. There are eight primary factors considered for cybersecurity features and characteristics, which include:

1. Identification and Asset Management
2. Risk factor
3. Accessibility control
4. Handling of threats
5. Security standards
6. Business Principles and Recovery Methods
7. Events Occurred
8. Awareness of security management.

9.2.1 IDENTIFICATION AND ASSET MANAGEMENT

Minimizing the exposure of data in a network is crucial to prevent security concerns. Attackers can quickly intrude on a network system, but this method of minimizing data exposure can help to deter them. It involves identifying all devices connected to the network and keeping a log of them in a secure repository. However, ensuring robustness of this method can be challenging due to software platforms and their accessibility. Hardware components are also crucial in developing security measures to prevent third-party intrusion into the network.

9.2.2 RISK FACTORS

During the identification process, it was found that there were unsecured routers, servers, access points, and workstations due to poor security measures. This poses a risk and we are taking all necessary steps to remove it using the latest techniques available. Risk assessment is used to identify potential vulnerabilities in the network and assess the risks associated with them without compromising the security model. It involves checking the entire system, including processes, applications, and functions, to identify any threats and calculate the risks associated with data access. Risk assessment is particularly important in a cloud computing environment where multiple systems are connected to the network.

9.2.3 ACCESSIBILITY CONTROL

In this session, we can discuss the entry of authorized persons and their limitations over the network. Data can be classified based on the client's requirements, whether it is stored on a single server or distributed on a centralized server. People who work for a company/industry can access this data, but outsiders also have permission to access it. This decision was made in accordance with policies and standards. Access control mechanisms are classified into wired, wireless, physical, and network access control (NAC) tools to develop accessibility methods. Some users have full permission to handle the data, while others have limited permissions due to this method. Least privilege data is more secure when compared to prior data because of its necessity.

9.2.4 HANDLING OF THREATS

During this session, vulnerability and patch updating requirements were discussed in light of the threats that have arisen with the edge computing method. To ensure network security, frequent scanning of the entire network and sometimes backing up the entire data is necessary to identify most of the vulnerabilities. This can be managed using tools and patches for upgrading. Patches provide solutions for the identified threats without creating new issues. Therefore, interactions between software and hardware need to be regularly updated with the current techniques to maintain network security.

9.2.5 SECURITY STANDARDS

Administrative controls are provided to authenticated users and various policies have been created to enable 24x7 monitoring of the process. However, this does not grant full access to the data, particularly server data, and it is managed in the cloud. While creating standards, the main issues to consider are network security with multi-factor authentication (MFA), segmentation, endpoint security, and content filtering. Different protocols are used on the entire network system, and other factors can affect it.

9.2.6 BUSINESS PRINCIPLE AND RECOVERY METHOD

Different types of data are stored in various clusters on a cloud. These data can be classified and accessed on the cloud using different principles and recovery methods in the event of a disaster such as floods, earthquakes, power grid failures, and so on. Quality of Service (QoS) and Denial of Service (DDoS) are two random events that can occur, affecting the effectiveness of data delivery to clients. If this module fails, it will not recover any data, and the cost of alternative options will also increase.

9.2.7 EVENTS OCCURRED

Numerous events and records are generated daily, resulting in a large volume of data that exceeds the cluster's size. All event-generated data has been allocated resources that are accessible to administrative personnel. The steps of data classification in cloud storage include preparation, identification, containment, eradication, recovery, and lessons learned.

9.2.8 AWARENESS OF SECURITY MANAGEMENT

The system will provide clients with various tools like education, training, assessments, workshops, and other ICT tools to increase cybersecurity awareness. This will benefit society as a whole by disseminating cybersecurity knowledge to everyone. Figures 9.5 and 9.6 depict cyber attacks and ways to protect against them.

- Malware
- Botnets
- Virus, Trojan
- Spam
- Phishing
- Malicious Codes
- Web-based Attack

FIGURE 9.5 Various kinds of attacks in Cybersecurity

FIGURE 9.6 How to protect against cyber security attacks.

Source: Reprinted from Freepik https://www.freepik.com/free-vector/protect-against-cyber-attacks-infographic_8268976.htm.

9.3 CYBERSECURITY CHALLENGES AND SOLUTIONS

IoT cyber security attacks are handled with the help of several security principles that ensure secured hardware and software parameters. Initially, users configure the elements presented in the IoT environment according to their requirements, and these elements are accessed through the gateway to the net network. The connectivity between these two networks is established using any of the current communication techniques such as Bluetooth, Wi-Fi, LORA,

ZigBee, or any of the latest techniques. The data is stored in cloud clusters for analytics and extraction. Modern tools such as Hadoop, SPARK, and their ecosystems are used for analytics to extract data from the huge storage. Since the storage is platform-independent, it must be distributed among the internet resources connected to the large network. Finally, real-world applications can be sorted out to produce accurate results of data in any dimension using CRM, ERP, SCM, and PLM. These techniques are used to produce accurate values for queries written in any of the latest languages, such as Java or Python. In this scenario, three principles are used to provide security against cyber attacks:

- Secure Device;
- Secure Communication;
- Secure Cloud; and
- Secure Life cycle Management.

9.3.1 SECURED DEVICE

Device intelligence is a new concept in cyber security that involves gathering accurate values from devices across a network. This involves updating device configurations to provide more precise predictions in real-world applications. At the edges of the cloud, data is either processed manually or policies are set to predict values. Cyber security principles are used to provide security across the network, with the latest techniques employed to improve the speed of data retrieval.

9.3.2 SECURED COMMUNICATION

Devices are connected to the edges of cloud clusters and equipped with high-configuration nodes to process data. Communication between the clusters is managed through messages that are passed via brokers in the cloud. Brokers and agents in edge computing are responsible for facilitating communication between different clusters that are connected to the internet.

9.3.3 SECURED CLOUD

Cloud security involves identifying the right users who can access the cloud storage for data processing. Authentication is used to ensure that only the right users can access and transmit data between clusters. There are certain principles that apply to cloud user authentication, which assign rules and limitations across all connected clusters. Encryption techniques are used to encrypt messages during data transmission, and common keys are shared between sender and receiver. This ensures that hackers cannot gain access to the system during data transmission. If hackers gain access to the system at the edge devices, they may receive false information due to encryption techniques. Various standards and policies are available for industries and companies to maintain data security.

9.3.4 SECURED LIFE CYCLE MANAGEMENT

The process of managing the secure life cycle was successfully completed with the help of remote control of devices at their edges. However, not updating devices at regular intervals led to drastic changes in the cloud systems. Integrating firmware, malware, antivirus, middleware, and firewall principles will help improve the security system on the larger network. The cyber-attacks on IoT edge devices.

9.4 CONCLUSION AND FUTURE DEVELOPMENT

In the future, the development of cybersecurity on IoT edge devices will be implemented using modern technologies such as ML, artificial intelligence, deep learning, and big data processing. Most of these techniques deal with huge data storage in cloud-based storage, along with their security principles and standards. Although several policies and standards have been put in place to control the entire cloud storage, the accuracy of these remains low when compared with other techniques. Big data and other frameworks are used to predict data security at edge devices, but due to the high volume, the results deviate from the original prediction. Algorithms have been developed to predict accurate values, but the speed of data retrieval and bandwidth of data processing remain slow in the larger network. Cybersecurity provides encryption standards for security, but when hackers identify the credentials of the users, it becomes easier for them to hack the data. Reduced vulnerabilities, QoS, DDoS, and improved encryption standards are used to improve cybersecurity attacks.

KEYWORDS

- **cybersecurity**
- **denial of services**
- **IoT edge computing**
- **multi-factor authentication**
- **quality of service**

REFERENCES

1. Makrani, H. M., He, Z., Rafatirad, S., & Sayadi, H., (2022). Accelerated machine learning for on-device hardware-assisted cybersecurity in edge platforms. In: *2022 23rd International Symposium on Quality Electronic Design (ISQED)* (pp. 77–83). IEEE.
2. Hazra, A., Alkhayyat, A., & Adhikari, M., (2022). Blockchain-aided integrated edge framework of cybersecurity for internet of things. *IEEE Consumer Electronics Magazine*.
3. Nakkeeran, M., & Mathi, S., (2022). A generalized comprehensive security architecture framework for IoT applications against cyber-attacks. In: *Artificial Intelligence and Technologies* (pp. 455–471). Springer, Singapore.
4. Timko, A. M., (2022). *Cybersecurity of Internet of Things Devices: A Secure Shell Implementation.* https://urn.fi/URN:NBN:fi:amk-202205139197.
5. Do, X. T., & Le, K. H., (2022). Towards remote deployment for intrusion detection system to IoT edge devices. In: *Recent Advances in Internet of Things and Machine Learning* (pp. 301–316). Springer, Cham.
6. Ghimire, B., & Rawat, D. B., (2022). Recent advances on federated learning for cybersecurity and cybersecurity for federated learning for internet of things. *IEEE Internet of Things Journal*.
7. Manokaran, J., & Vairavel, G., (2022). An empirical comparison of machine learning algorithms for attack detection in internet of things edge. *ECS Transactions*, *107*(1), 2403.
8. Molokomme, D. N., Onumanyi, A. J., & Abu-Mahfouz, A. M., (2022). Edge intelligence in smart grids: A survey on architectures, offloading models, cyber security measures, and challenges. *Journal of Sensor and Actuator Networks*, *11*(3), 47.
9. Tom, R. J., John, V. M., & Renukadevi, G., (2022). A Secure and reliable IoT-edge framework with blockchains for smart Micro Grids. In: *Proceedings of International Conference on Deep Learning, Computing, and Intelligence* (pp. 651–669). Springer, Singapore.
10. Kaushik, K., (2022). Blockchain enabled artificial intelligence for cybersecurity systems. In: *Big Data Analytics and Computational Intelligence for Cybersecurity* (pp. 165–179). Springer, Cham.
11. Sangwan, U., & Chhillar, R. S., (2022). Comparison of various classification techniques in cyber security using IoT. *International Journal of Intelligent Systems and Applications in Engineering*, *10*(3), 334–339.

12. Alkhayyat, A., & Adhikari, M. (2022). *Blockchain-aided Integrated Edge Framework of Cybersecurity for Internet of Things.*
13. Attkan, A., & Ranga, V., (2022). Cyber-physical security for IoT networks: A comprehensive review on traditional, blockchain and artificial intelligence-based key-security. *Complex & Intelligent Systems*, 1–33.
14. Rani, D., Gill, N. S., & Gulia, P., (2022). Classification of security issues and cyber-attacks in layered internet of things. *Journal of Theoretical and Applied Information Technology*, *100*(13).
15. Alterazi, H. A., Kshirsagar, P. R., Manoharan, H., Selvarajan, S., Alhebaishi, N., Srivastava, G., & Lin, J. C. W., (2022). Prevention of cyber security with the internet of things using particle swarm optimization. *Sensors*, *22*(16), 6117.
16. Pandey, U., & Chaurasia, B. K., (2022). *IoT Edge-Based Framework for Sybil and Buffer Overflow Detection.* https://doi.org/10.21203/rs.3.rs-1693583/v1.
17. d BH, T., & Lehara, F. (2022). Cyber Security Perspective of Top Future Technologies. *Building Cyber Resilience against Hybrid Threats, 6.*
18. Dutta, N., Jadav, N., Tanwar, S., Sarma, H. K. D., & Pricop, E., (2022). *Cyber Security: Issues and Current Trends*. Springer.
19. Radanliev, P., & De Roure, D., (2022). Advancing the cybersecurity of the healthcare system with self-optimizing and self-adaptative artificial intelligence (part 2). *Health and Technology*, *12*(5), 923–929.
20. Gupta, L., Salman, T., Ghubaish, A., Unal, D., Al-Ali, A. K., & Jain, R., (2022). Cybersecurity of multi-cloud healthcare systems: A hierarchical deep learning approach. *Applied Soft Computing*, *118*, 108439.
21. Vermesan, O., & Bacquet, J., (2022). *Internet of things–the Call of the Edge: Everything Intelligent Everywhere*. CRC Press.
22. Almaiah, M. A., Hajjej, F., Ali, A., Pasha, M. F., & Almomani, O., (2022). A novel hybrid trustworthy decentralized authentication and data preservation model for digital healthcare IoT based CPS. *Sensors*, *22*(4), 1448.
23. Reddy, Y. H., Ali, A., Kumar, P. V., Srinivas, M. H., Netra, K., Achari, V. J., & Varaprasad, R., (2022). A comprehensive survey of internet of things applications, threats, and security issues. *South Asian Res. J. Eng. Tech.*, *4*(4), 63–77.
24. Stead, M., Pilling, F., Gradinar, A., & Ian, F., (2022). *Sustainable X Secure Edge: Design Guidelines for Future Data-Driven Edge-IoT Devices and Services.*
25. Nukavarapu, S. K., (2022). *Machine Learning (ML)-Assisted Tools for Enhancing Security and Privacy of Edge Devices.*
26. Attkan, A., & Ranga, V., (2022). Montasari, R., & Jahankhani, H., (2021). *Artificial Intelligence in Cyber Security: Impact and Implications*. Springer International Publishing.
27. Minh, Q. N., Nguyen, V. H., Quy, V. K., Ngoc, L. A., Chehri, A., & Jeon, G., (2022). Edge computing for IoT-enabled smart grid: The future of energy. *Energies*, *15*(17), 6140.
28. Mazher, N., Alhadaad, M., & Shagdar, O., (2022). *A Brief Summary of Cybersecurity Attacks in V2X Communication.*
29. Abdullahi, M., Baashar, Y., Alhussian, H., Alwadain, A., Aziz, N., Capretz, L. F., & Abdulkadir, S. J., (2022). Detecting cybersecurity attacks in the internet of things using artificial intelligence methods: A systematic literature review. *Electronics*, *11*(2), 198.
30. Alawida, M., Omolara, A. E., Abiodun, O. I., & Al-Rajab, M., (2022). A deeper look into cybersecurity issues in the wake of COVID-19: A survey. *Journal of King Saud University-Computer and Information Sciences.*

31. Khoso, F. H., Arain, A. A., Lakhan, A., Kehar, A., & Nizamani, S. Z., (2021). Proposing a novel IoT framework by identifying security and privacy issues in fog cloud services network. *Int. J., 9*, 592–596.
32. Demir, A. K., & Alam, S., (2021). Advancing artificial intelligence-enabled cybersecurity for the internet of things. In: *Handbook of Research on Advancing Cybersecurity for Digital Transformation* (pp. 118–143). IGI Global.
33. Zhao, Y., Cheng, G., Duan, Y., Gu, Z., Zhou, Y., & Tang, L., (2021). Secure IoT edge: Threat situation awareness based on network traffic. *Computer Networks, 201*, 108525.
34. Ahmed, M., & Haskell-Dowland, P., (2021). *Secure Edge Computing: Applications, Techniques, and Challenges*. CRC Press.
35. Vermesan, O., & Bacquet, J., (2022). *Internet of Things – the Call of the Edge: Everything Intelligent Everywhere*. CRC Press.
36. Moseley, R., (2021). *Advanced Cybersecurity Technologies*. CRC Press.
37. Karimipour, H., & Derakhshan, F., (2021). *AI-Enabled Threat Detection and Security Analysis for Industrial IoT*. Springer International Publishing.
38. Hasan, A. M., Rahim, M. S., Ahmed, S., & Levula, A., (2021). Usage of blockchain for edge computing. In: *Secure Edge Computing* (pp. 261–277). CRC Press.
39. Chen, X., Susilo, W., & Bertino, E., (2021). *Cyber Security Meets Machine Learning*. Singapore: Springer.
40. Memos, V. A., Psannis, K., & Lv, Z., (2022). A Secure network model against bot attacks in edge-enabled industrial internet of things. *IEEE Transactions on Industrial Informatics*.
41. Salwan, G., Kaushal, P., Gupta, S., & Nayak, P., (2021). A robust cyber security: Challenges and opportunities. *Artificial Intelligence, Machine Learning, and Data Science Technologies*, 283–297.
42. Maleh, Y., Baddi, Y., Alazab, M., Tawalbeh, L., & Romdhani, I., (2021). *Artificial Intelligence and Blockchain for Future Cybersecurity Applications* (Vol. 90). Springer Nature.
43. Kamhoua, C. A., Kiekintveld, C. D., Fang, F., & Zhu, Q., (2021). *Game Theory and Machine Learning for Cyber Security*. John Wiley & Sons.
44. Bernabe, J. B., & Skarmeta, A., (2022). *Challenges in Cybersecurity and Privacy-the European Research Landscape*. CRC Press.
45. Goutam, R. K., (2021). *Cybersecurity Fundamentals: Understand the Role of Cybersecurity, Its Importance and Modern Techniques Used by Cybersecurity Professionals (English Edition)*. BPB Publications.
46. Gupta, B. B., & Dahiya, A., (2021). *Distributed Denial of Service (DDoS) Attacks: Classification, Attacks, Challenges, and Countermeasures*. CRC press.
47. Chatterjee, D., (2021). *Cybersecurity Readiness: A Holistic and High-Performance Approach*. SAGE Publications.
48. Khanna, K., Estrela, V. V., & Rodrigues, J. J. P. C., (2021). *Cyber Security and Digital Forensics: Proceedings of ICCSDF 2021* (Vol. 73). Springer Nature.
49. Bragadeesh, S.A., Arumugam, U. (2019). A Conceptual Framework for Security and Privacy in Edge Computing. In: Al-Turjman, F. (eds) Edge Computing. EAI/Springer Innovations in Communication and Computing. Springer, Cham. https://doi.org/10.1007/978-3-319-99061-3_10

CHAPTER 10

A Structural Framework for Modeling Multi-Stage Network Attacks

S. SUGANYADEVI[1] and B. MAHALAKSHMI[2]

[1]*Department of Computer Science, Sri Krishna Arts and Science College, Coimbatore, Tamil Nadu, India*

[2]*School of Computer Studies, Rathnavel Subramaniam College of Arts and Science, Coimbatore, Tamil Nadu, India*

ABSTRACT

Protection software is a vital feature of computer/network systems. It is used to restrict unauthorized access. Data safety is important to prevent a program from accessing data that is not part of its memory area. This ensures that any infection or defect within a program cannot spread to other programs or the operating system. When code attempts to access memory it does not own, a hardware fault called a segmentation fault or storage violation exception occurs, resulting in the termination of the offending program. Additional memory protection methods, such as randomizing the layout of address spaces and encrypting executable data, are used for data security.

10.1 INTRODUCTION

Computer security is the protection of computing structures against unauthorized access. It is essential for defending personal and sensitive information [2, 3, 5]. The Computer Security Act of 1987 requires organizations to protect sensitive systems, practices, and provide PC protection training, as shown in Figure 10.1.

Leveraging Artificial Intelligence (AI) Competencies for Next-Generation Cybersecurity Solutions. Pethuru Raj, B. Sundaravadivazhagan, & V. Kavitha (Eds.)

10.2 COMPUTER SECURITY

IT security helps the organization achieve its goals. Computer security safeguards an organization's data, systems, and intellectual property, protecting vital assets. Additionally, secure computing is fundamental to good management. An organization's information and computer systems are valuable resources in furthering its goals. Ensuring their safety is also crucial for the company's cash, property, or personnel.

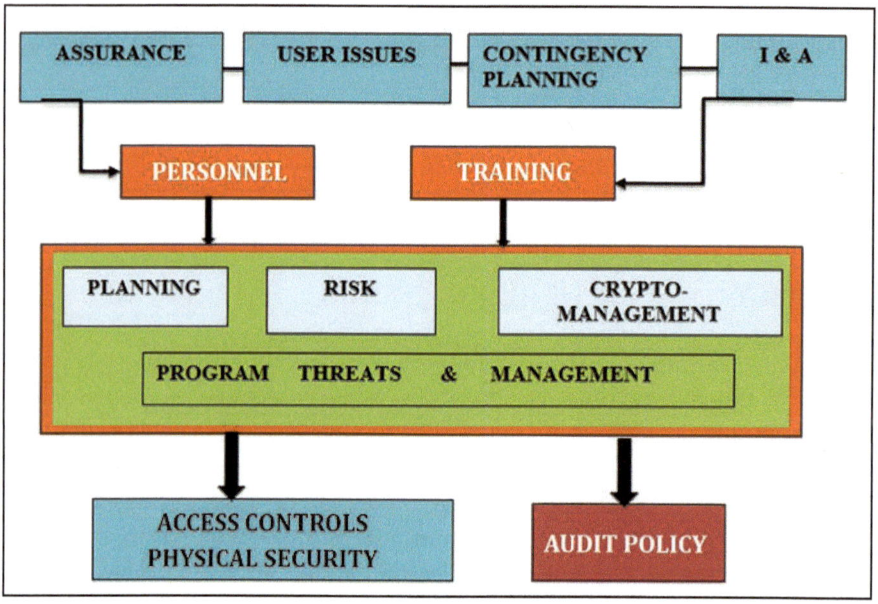

FIGURE 10.1 Computer security.

When an organization's data and computers are connected to other networks, management becomes accountable for events beyond the company. Consequently, management may need to [4].

- Understand what famous stage or safety for external system(s)
- Are you trying to ensure that the exterior gadget provides sufficient protection for the organization's needs?

The cost of controls should not exceed the predicted benefits. Therefore, it is important to carefully evaluate the monetary and nonmonetary costs and advantages of security measures. The level of security implemented should align with the resources invested and the significance placed on the computer systems, as well as the nature, likelihood, and consequences of potential security breaches. The required level of protection may vary for different computer systems in the future.

In terms of cyber safety, clear roles and responsibilities should be established for the owners, providers, and clients of computer systems. The scope of an employee's obligations may be limited to the company or extend beyond it. If a system has users outside of the organization, the owners have a contractual obligation to ensure those users have confidence in the system's security by disclosing information about the existence and general scope of security measures. Therefore, a comprehensive and integrated strategy is essential for computer security. Providing adequate computer security requires a holistic approach that considers factors both within and beyond information technology security. The information lifecycle refers to the process of how information changes over time. Dependencies between security measures are instances of unlinked dependence.

It is essential to regularly re-evaluate computer security. Computers and their environments are constantly changing. The security needs for a system must adapt to the ever-evolving state of its components and the people who use them, the data and information contained within, and the threats to which those systems are vulnerable. In a social context, there are limits to computer security. Multiple factors, including societal issues, may hinder security's ability to support the organization's objectives. For instance, conflicts may arise in the workplace between the need for security and the right to personal privacy. However, security can also contribute to achieving societal goals and enhancing people's information integrity.

10.3 GOALS OF MUTUAL SECURITY

Computer systems are susceptible to threats that can cause significant losses [6]. This damage can range from compromising database integrity to fires that can destroy entire computer centers. To mitigate the risks associated with operating a data system, managers and users need to understand the vulnerabilities of the system and the threats that can exploit them. Understanding the risk environment enables the system administrator to implement the most appropriate security measures. In certain cases, managers may find it more reasonable to accept the expected losses.

10.3.1 ERRORS AND OMISSIONS

The presence of errors and omissions jeopardizes the integrity of data and systems. All users who produce or modify data contribute to this problem, not just data entry assistants performing hundreds of daily transactions. Many programs do not have enough quality assurance checks, especially user-made ones for PCs. Users, data entry clerks, system operators, and programmers often make errors that contribute directly or indirectly to security vulnerabilities. Malfunctions can occur at any time during the life cycle of a system. Robert Courtney, a computer security expert and former member of the Computer Systems Security and Privacy Advisory Board, undertook a long-term analysis of economic losses associated with computers and found that 65% of those losses were due to human mistakes and omissions. "Bugs," as errors in software development are commonly known, can have varying degrees of impact. Installation and maintenance issues are other sources of security problems. In 1988, the President's Council on Integrity and Efficiency (PCIE) conducted an assessment that revealed that all 10 mainframe computer locations it examined had made installation and maintenance mistakes that left the systems vulnerable to serious security breaches.

10.3.2 FRAUD AND THEFT

Fraud and theft can be "automated" using computers, and innovative approaches can be used to steal from online services. Assuming that minor inconsistencies may not be scrutinized, some people may use computers to skim small sums of money from many financial accounts. Theft of computer hardware and software is a possibility, in addition to the use of technology in fraud and theft. The Computer Crime Unit of the United States Department of Justice has claimed that "insiders constitute the biggest threat to computer systems." Both insiders and outsiders can commit fraud and theft in a computer system [9, 10]. Authorized users of the victim computer system are more vulnerable to attack since insiders have access to and experience with the system. Ex-employees might pose a risk to a company because of the sensitive information they know how to access.

10.3.3 INTERRUPTION

The downsizing of public and private sector organizations has resulted in a group of individuals who possess organizational knowledge and practical access to laptop computers. These employees are familiar with their employer's computer systems and applications, collectively understanding the potential impact of certain actions that could cause damage, mischief, or sabotage. While the number of incidents involving employee sabotage is smaller compared to theft cases, the frequency of such incidents can still be relatively high [7].

Some common examples of computer-related employee sabotage include:

- Destroying hardware or facilities;
- Planting logic bombs that destroy;
- Programs or data;
- Entering data incorrectly;
- "Crashing" systems;
- Deleting data;
- Holding data hostage; and
- Changing data.

10.3.4 PHYSICAL-INFRASTRUCTURE SUPPORT

Strength disasters (outages, spikes, and brownouts), loss of communications, water leaks, sewer issues, a lack of transportation services, fires, floods, civil unrest, and strikes are all examples of the loss of supporting infrastructure.

10.3.4.1 HACKERS

Malicious hackers, often referred to as crackers, gain unauthorized access to systems. The group may consist of both external individuals and insiders. The expansion of government and business networks has been linked to an increase in hacking. According to a 1992 study of a single Internet site, attacks occur at least three times a day [8].

When assessing the threat posed by hackers, it is important to consider both past and potential future harm. Although the costs of cyber-attacks are currently smaller compared to losses from insider theft and sabotage, the issue of hackers is widespread and severe. Hostile hacker activity commonly targets the automated public telephone system. Despite more well-known and dangerous threats, the hacker threat often receives more attention. The Computer Crime Unit of the U.S. Department of Justice provides three reasons for this. Firstly, the threat posed by hackers has only recently come to light. Secondly, while discipline mechanisms can help, firms have long been concerned about internal threats. Lastly, the issue lies in the fact that these protections are not sufficient.

10.3.4.2 INTELLIGENT ENGINEERING

Economic espionage is another term for state-sponsored spying on businesses abroad. The term "industrial espionage" refers to the illegal practice of obtaining sensitive information from one company or government agency and using it to benefit another. Industrial espionage can be committed by businesses looking to gain a competitive advantage in the market or by government officials trying to support their domestic industries. Computer security plays a crucial role in protecting against these threats due to the widespread use of computers in processing and storing information. However, it is unlikely to deter authorized personnel from selling this data. According to a study funded by the American Society for Industrial Security (ASIS), the theft of confidential company information increased by 260% between 1985 and 1992. The most damaging types of compromised data were the value of the information, industrial advancements, and product development and specification details.

10.3.4.3 CODE OF CRACKERS

Cracker code refers to various "uninvited" software such as Trojan horses, worms, logic bombs, and viruses. Sometimes mistakenly associated only with personal computers, malicious code can attack other platforms [11-13]. A 1993 study of viruses found that while the number of recognized viruses is increasing exponentially, the variety of virus incidents is not. They discovered that viruses are becoming more prevalent, but only "gradually."

Actual costs attributed to malicious code have resulted primarily from device outages and the time spent by workers in repairing the systems. Nonetheless, these expenses can be significant.

Malicious Software: A Few Key Terms:

- **Virus:** A code phase replicates by attaching copies of itself to existing executables. The reproduction of the virus occurs when a person executes the new host program. The virus can also include an additional "payload" that triggers when specific prerequisites are met.
- **Trojan Horse:** A software that performs a desired task but also includes unexpected (and unwanted) functions. This application could be altered to randomly delete one of the users' files whenever they use

a useful feature (such as editing), but these deletions are unforeseen and certainly not desired.

- **Worm:** A self-contained, self-replicating program that does not require a host program. The application creates a copy of itself and motivates it to execute. Worms typically use network services to propagate to other host systems.

10.4.3.4 INTELLIGENCE AGENCIES

Intelligence agencies from other countries may pose a threat. While economic espionage is a concern, unclassified systems could also be targeted by foreign intelligence agencies seeking to further their agendas. Examples of non-classified material that could be of interest include travel plans of high-ranking officials, civil defense and emergency preparations, manufacturing technology, satellite data, personnel and payroll data, law enforcement, investigation, and security files.

10.4.3.5 SECURITY ADMINISTRATION

The invasion of one's privacy can come from a variety of directions. Government employees have been accused of selling people's personal information to private investigators and other "information brokers" in many instances. For example, in 1992, the Justice Department announced that it had arrested more than two dozen individuals for allegedly buying and selling data from Social Security Administration computers. The investigation revealed that Social Security Administration employees had unrestricted access to over 130 million employment records. Another investigation found that 5% of IRS employees in a specific region had accessed the tax returns of individuals they were familiar with or had a personal interest in. While some employees may have filed tax returns dishonestly, the vast majority did so in good faith.

10.4.3.6 PROGRESS OF THREAD

Structured approaches to identifying, quantifying, and addressing threats are essential parts of the development process:

- Just like spacing and designing; and
- Just like coding and testing.

One technique is presented here. There are others (e.g., OCTAVE).The Threat Modeling Process is:

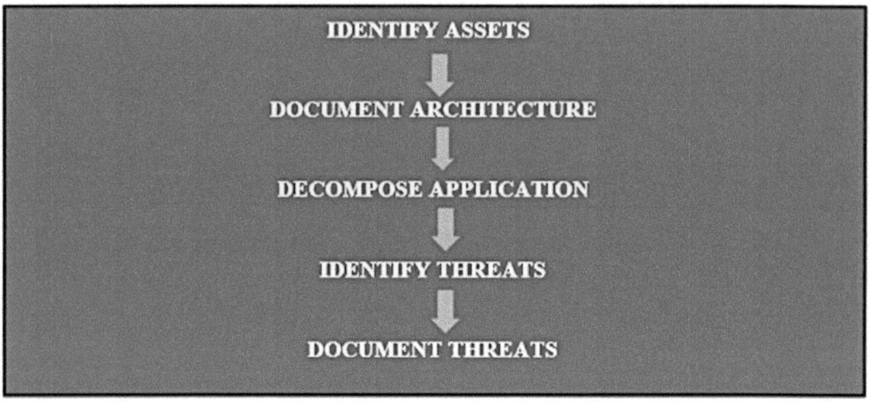

FIGURE 10.2 Threat modeling process.

- **STRIDE**

  As described step in Figure 10.3.

- **Documenting Threats**

  Document threats using a template (Table 10.1).

TABLE 10.1 Threat

Theft of Auth Cookies by Eavesdropping on Connection	
Threat target risk	Connection between brewers and web server risk management
Attack techniques	Attack uses a sniffer to monitor traffic
Countermeasures	Use SSL/TLS to encrypt traffic
Theft of Auth Cookies via Cross-Site Scripting	
Threat target	Vulnerable application code
Risk	Support Management
Attack techniques	Attack sends emails with malicious links to users
Countermeasures	Validate input: HTML encodes the output

- **Rating Threats**

  Simple model

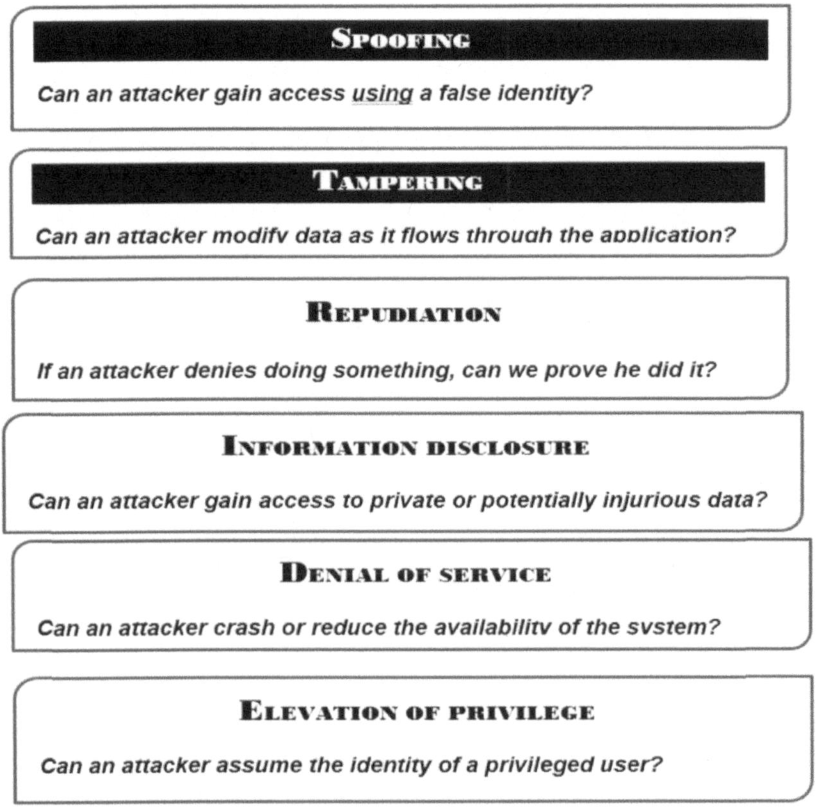

FIGURE 10.3 STRIDE.

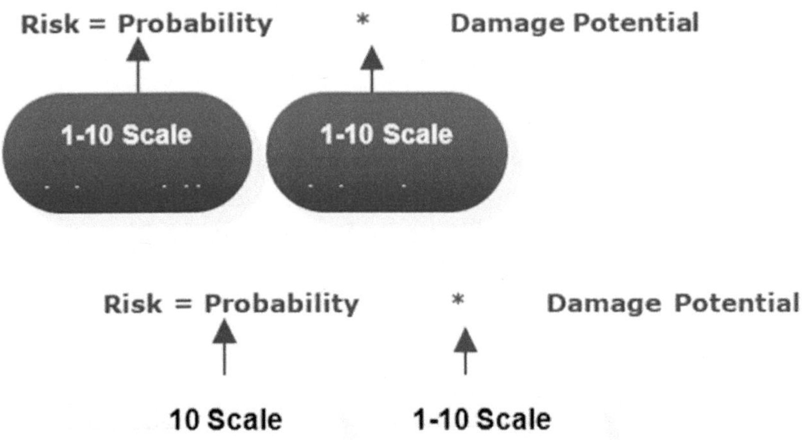

FIGURE 10.4 DREAD role.

Example:

Threat	D	R	E	A	D	Sum
Auth cookie theft (eavesdropping)	3	2	3	2	3	13
Auth cookie theft (XSS)	3	2	2	2	3	12

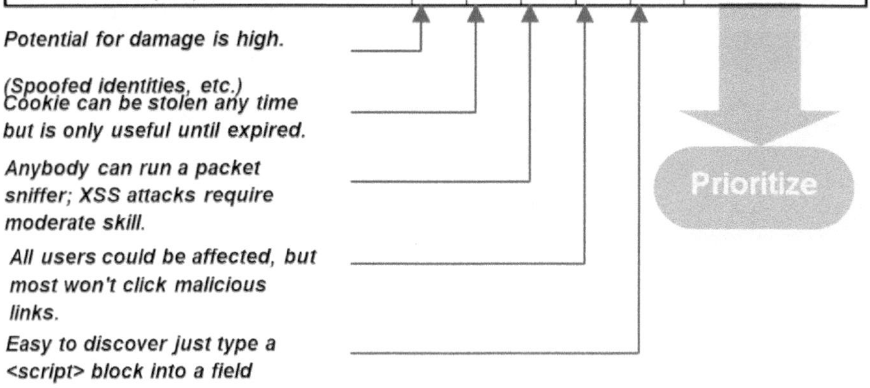

FIGURE 10.5 Thread model.

10.5 PROTECTION FROM CRYPTOGRAPHIC

As a subfield of mathematics, cryptography conceals information via secret codes. It is a vital component of computer security because of the invaluable protection it provides. For instance, cryptography can help ensure data secrecy, integrity, electronic signatures, and sophisticated user authentication [14, 15] (Figures 10.6 and 1.7).

10.5.1 CRYPTOGRAPHY TOPICAL

Algorithms and keys are the backbone of cryptography. When applied to a given input value, a cryptographic algorithm produces a new value that meets predetermined security requirements. Cryptography can be broken down into two distinct subfields: secret-key (or symmetric) systems and public-key (or asymmetric) systems (also called asymmetric systems).

- **Secret Key Cryptography:** For data encryption and decryption in secret key cryptography, both the sender and the receiver use the same key. The key in secret key cryptography must remain secret, as the name suggests. The best-known secret key scheme is the Data Encryption Standard (DES), issued by NIST as the Federal Information

Processing Standard (FIPS). The use of cryptography with a secret key dates back millennia. Early variants merely swapped the printed characters to obscure the message.

- **Secret Public Key Cryptography:** Modern public key cryptography relies on complex mathematical techniques. In contrast to secret key cryptography, which only requires knowledge of a single key, public key cryptography requires knowledge of two keys, one for each party. Public key cryptography is helpful in situations when the communication parties cannot be trusted or do not have a shared key.
- **Secret Hybrid Cryptographic System:** A computer system can use secret and public key cryptography in a complementary approach, with each type serving a purpose to reap both benefits while avoiding the drawbacks. Public key systems are used for automated key distribution, while secret key systems encrypt large amounts of data simultaneously.
- **Secret Key Escrow:** Since cryptography can provide highly secure encryption, it can hinder the government's efforts to conduct lawful electronic surveillance. NIST has issued the FIPS for encrypted data. As part of a federal government policy called "voluntary key escrow," portions of the decryption keys are distributed to escrow authorities. To decrypt the encrypted data, possession of both keys is necessary.
- **Secret Encryption:** Encryption is one of the most efficient and low-cost methods for protecting sensitive information. Decryption is the process by which this is accomplished in reverse. Plaintext refers to the original, readable data before encryption, while cipher text refers to the resulting, unreadable message. It is unnecessary to prevent leakage of the cipher text after data has been encrypted. Cryptography uses encryption to ensure that only authorized parties can decode encoded messages or data. Encryption does not prevent eavesdropping but prevents the interceptor from reading the encrypted message. Using an encryption algorithm, plaintext (the original message or data) is transformed into cipher text (which must be decrypted before it can be read) in an encryption scheme [16]. A pseudo-random encryption key generated by an algorithm is used in an encryption scheme for practical reasons. Messages encrypted with a well-designed encryption scheme can be decrypted without the key, but only with time, effort, and computing power. The message is encrypted and can only be decrypted by the intended recipient using the key provided by the sender [17].

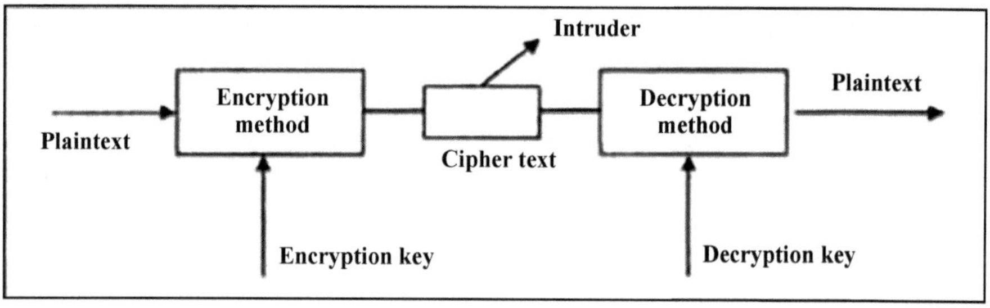

FIGURE 10.6 Encryption process.

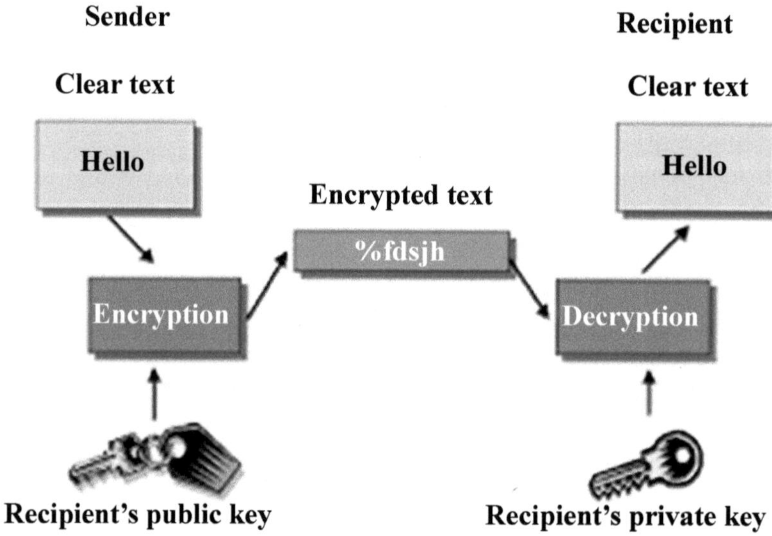

FIGURE 10.7 Encryption model thru key.

10.6 AUTHENTICATION SECURITY

Several innovative authentication procedures are predicated on cryptography. Instead of sending secret information over a public network, authentication can be accomplished by proving familiarity with a cryptographic key, such as a human with a fingerprint (Figure 10.8).

A person's identity can only be trusted after it has been confirmed through authentication. In other words, authenticating someone entails checking their identity with multiple sources (e.g., password). The Greek word authentic, from which we get the word "authentic," means "genuine" [18].

FIGURE 10.8 Create a Security key for Authentication.

10.7 MESSAGE AUTHENTICATION CODES (MAC)

In cryptography, a Communication Authentication Code (MAC) is a short piece of statistics used to affirm a message and to furnish integrity and authenticity assurances on the message [19]. Authenticity guarantees confirm that the message came from its purported source, whereas integrity guarantees uncover any tampering, whether unintentional or purposeful (Figure 10.9).

- Message authentication is concerned with:
  - Protecting the integrity of a message
  - Validating the identity of the originator
  - Non-repudiation of origin (dispute resolution)
- Will consider the security requirements and then three alternative functions used:
  - Hash function
  - Message encryption
  - Message authentication code (MAC)
- A small fixed-sized block of data
  - Generated from message + secret key
  - MAC = C(K,M)
  - Appended to the message when sent

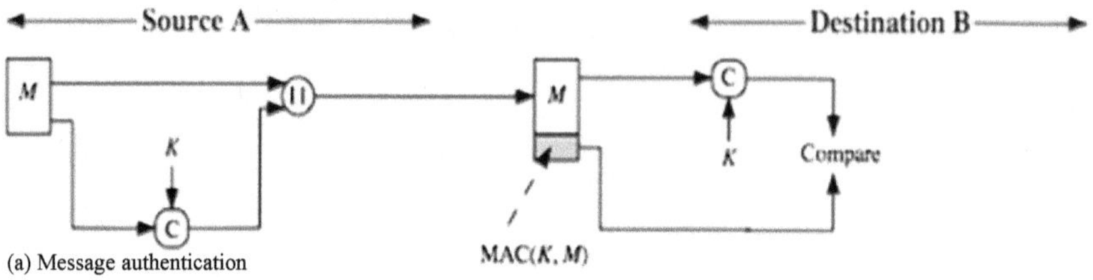

FIGURE 10.9 Processing of MAC.

- The MAC provides authentication and can also use encryption for secrecy
  - o Generally use separate keys for each
  - o Can compute MAC either before or after encryption
  - o Is generally regarded as better done before
- Why use a MAC?
  - o Sometimes only authentication is needed
  - o Sometimes authentication persists longer than the encryption (e.g., archival use)
- MAC is not a digital signature

i. MAC Properties

- MAC is a cryptographic checksum

$$MAC = C_K(M)$$

  - o Condenses a variable-length message M
  - o Using a secret key K
  - o To a fixed-sized authenticator
- It is a many-to-one function
  - o potentially many messages have same MAC
  - o but finding these needs to be very difficult

ii. Requirements for MACs

- Taking into account the types of attacks need the MAC to satisfy the following:
  1. Knowing a message and its MAC, it is infeasible to find another message with the same MAC.
  2. MACs should be uniformly distributed
  3. MACs should depend equally on all bits of the message

iii. Security of MACs

It handles the following attacks

- Brute-force attacks exploiting
  - o Strong collision resistance hash has cost $2^{m/2}$.
- 128-bit hash looks vulnerable, 160-bits better
- MACs with known message-MAC pairs
  - o Can either attack keyspace (cf key search) or MAC
  - o At least a 128-bit MAC is needed for security
- Cryptanalytic attacks exploit structure
  - o Like block ciphers want brute-force attacks to be the best alternative
- More variety of MACs so harder to generalize about cryptanalysis

10.7.1 KEYED HASH FUNCTIONS AS MACS (HMAC)

Keyed-hash message authentication codes (HMACs) are a type of Message Authentication Code (MAC) that are calculated using a cryptographic hash function and a secret cryptographic key. They can be used to verify the authenticity and integrity of a communication, similar to any other MAC. HMAC can be implemented using various cryptographic hash functions such as MD5 or SHA-1, resulting in HMAC-MD5 or HMAC-SHA1 respectively. The security of HMAC depends on the quality of the key, the size of the hash result, and the security of the underlying hash function.

An iterative hash algorithm divides a message into predetermined-sized chunks for compression. The hashing algorithms MD5 and SHA-1 use blocks of 512 bits. HMAC produces an output of the same size as the underlying hash function, which can be trimmed if necessary.

The use of HMACs is standardized and generalized by FIPS PUB 198. In 1996, Mihir Bellare, Ran Canetti, and Hugo Krawczyk defined HMAC and provided its analysis in RFC 2104. They also introduced a less commonly used variation called NMAC. HMAC-SHA1 and HMAC-MD5 are utilized in protocols like IPSec and TLS (Figure 10.10).

i. Definition:

KeyedHash = Hash (Key | Message)

ii. Points:

- Use, without modifications, hash functions

- Allow for accessible replaceability of embedded hash function.
- Preserve real ordinary overall performance of hash characteristics barring large degradation.
- Use and deal with keys effortlessly.
- Have an especially understood cryptographic assessment of authentication mechanism strength.

Hash function on the message:

$$HMAC_K(M)=Hash[(K^+ XOR\ opad) \| Hash[(K^+ XOR\ ipad) \| M)]$$

Here, where K^+ is the key padded out to size, opad, ipad are specified padding constants

- Overhead is just 3 more hash calculations than the message needs alone
- Any hash function can be used. Example MD5, SHA-1, RIPEMD-160, Whirlpool

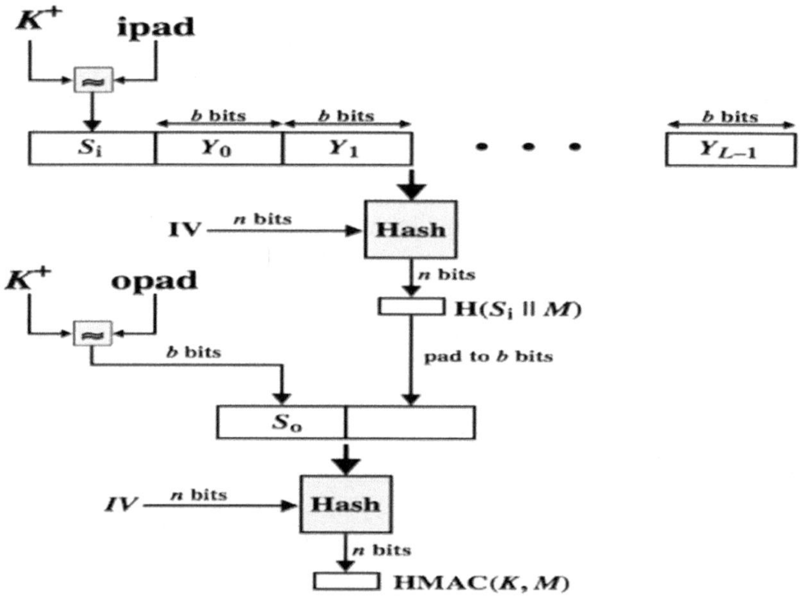

FIGURE 10.10 HMAC–process flow.

iii. HMAC Security:

Proved security of HMAC relates to that of the underlying hash algorithm:

- brute force attack on the key used.

- birthday attack (but since keyed would need to see a large number of messages).
- choose the hash function used based on speed versus security constraints.

10.7.2 USING SYMMETRIC CIPHERS FOR MACS

Using Symmetric Ciphers can utilize block cipher chaining mode and utilize the final block as a MAC. Data Authentication Algorithm (DAA) is a widely used MAC based on DES-CBC, with IV=0 and zero-padding of the last block when encrypting a message. The use of DES in CBC mode is either the final block as the MAC or the leftmost M bits (16M64) of the final block. The remaining MAC is now considered too small for security (Figure 10.11).

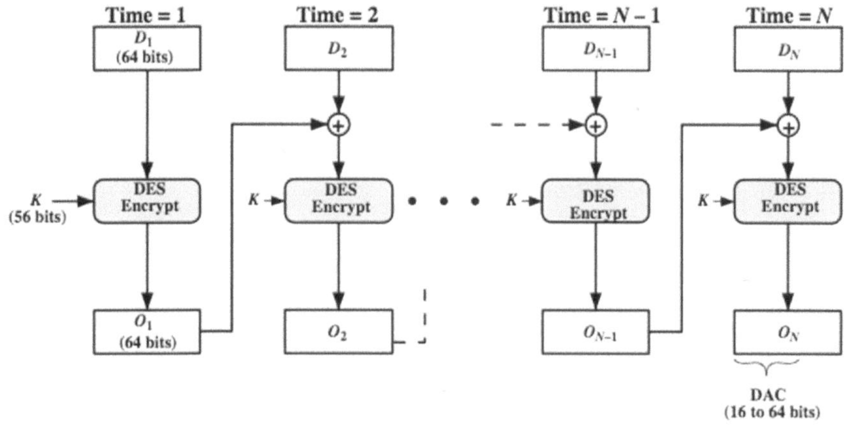

FIGURE 10.11 Data authentication algorithm.

As previously seen, the DAA (CBC-MAC) is used in government sectors and industry. However, it has a message dimension limitation. It can overcome the usage of two keys & amp; padding. Thus, forming the Cipher-based Message Authentication Code (CMAC) adopted through NIST SP800-38B.

The structure of Stallings CMAC is depicted in Figure 10.12. The size of the underlying cipher blocks is utilized. The message is divided into n blocks m1·· Mn, padded if necessary. The algorithm uses an n-bit regular key k1 or k2 (depending on whether the message was padded or not) and an okay-bit encryption key. The key size (ok) for AES can be 128, 192, or 256 bits, while for triple DES it is 112 or 168 bits. According to the text [19], the constants k1 and k2 can be derived from the actual key ok by encrypting zero and multiplying it in gf(2n).

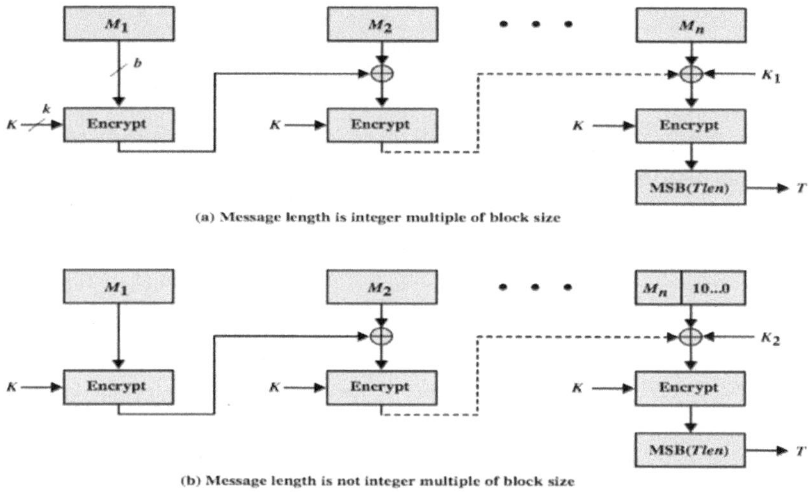

FIGURE 10.12 Cipher-based message authentication code processing.

i. Authenticated Encryption

It is aimed to simultaneously protect the confidentiality and authenticity of communications, and they are often required, but usually separate. The procedures are:

- Hash-then-encrypt: E(K, (M || H(M))
- MAC-then-encrypt: E(K2, (M || MAC(K1, M))
- Encrypt-then-MAC: (C=E(K2, M), T=MAC(K1, C)
- Encrypt-and-MAC: (C=E(K2, M), T=MAC(K1, M)

Its decryption /verification is honest, but security vulnerabilities with these types of.

10.7.3 CCM–COUNTER WITH CIPHER BLOCK CHAINING-MESSAGE AUTHENTICATION CODE

The CCM consists of:

- NIST standard SP 800-38C for WIFI
- Variation of encrypt-and-MAC approach

- Algorithmic ingredients
- AES encryption algorithm
- CTR mode of operation
- CMAC authentication algorithm
- Single key used for both encryption & MAC

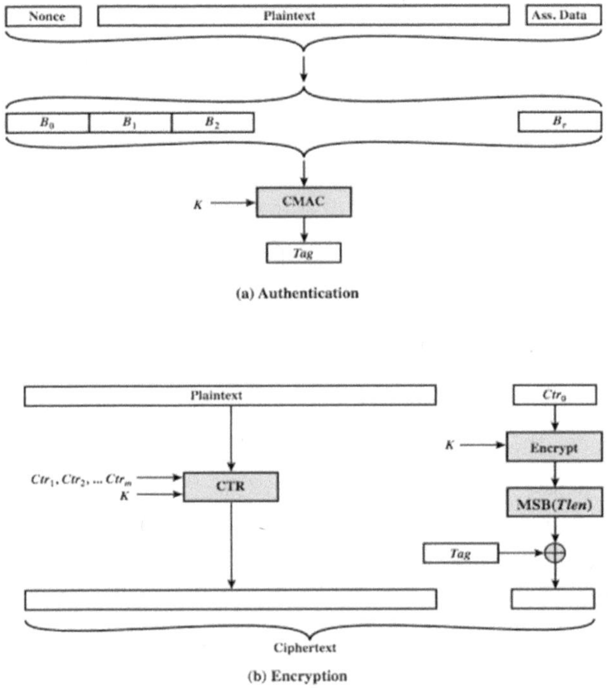

FIGURE 10.13 The CCM operation.

The CCM procedure is represented in Figure 10.13. The nonce, associated data, and plaintext are all necessary for authentication. The data is shown as a sequence of blocks, designated B0 through Br. The nonce and the lengths of the N, A, and P elements can be deduced from the formatting bits of the first block. Following that are zero or more blocks with the letter A, and then zero or more blocks with the letter P. The CMAC method takes the created block sequence as input and generates a MAC value of length Tlen, where Tlen is less than or equal to the block length (Figure 10.13(a)). Encryption requires the generation of a nonce-independent sequence of counters. Using the unique counter Ctr0, the authentication tag is encrypted in CTR mode.

The tag is combined with the Tlen most significant bits of the output through XOR to produce a hidden tag. The plaintext is encrypted using CTR mode with the remaining counters. The resulting ciphertext is created by merging the encrypted plaintext and encrypted tag. CCM is an algorithm with a high degree of complexity. Two full passes must be made through the plaintext to generate the MAC value and encrypt the plaintext. If forced to choose between the nonce length and the tag length, the specification details impose an unnecessary constraint.

Moreover, keep in mind that in the CTR encryption mode, the encryption key is used twice: once to generate the tag and once to encrypt the plaintext plus tag. It is unclear whether or not these additional layers of complexity improve the algorithm's safety. After looking at the algorithm from two perspectives, both agree that CCM offers a safe solution.

10.7.4 GALOIS/COUNTER MODE (GCM)

The NIST standard SP 800-38D is parallelizable. Here, the message is encrypted using a variant of CTR. The cipher text is multiplied with the key and length over (2^128) to generate an authenticator tag. It also has a GMAC MAC-only mode.

Two functions:

1. GHASH – a keyed hash function
2. GCTR – CTR mode with incremented counter

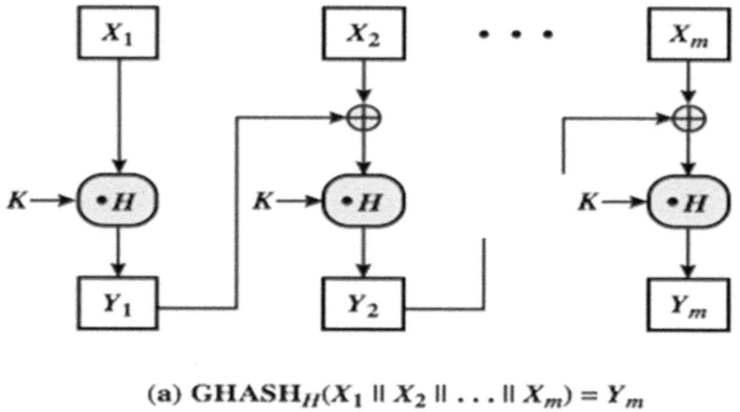

FIGURE 10.14 GCM phenomena.

Ghash(x) takes and processes the hash key h and a bit string x, where len(x) = 128 m bits for a positive integer m, to generate a 128-bit Mac price. The function can be specified as depicted in Stallings Figure 10.14(a). The GHASH *H(X)* function can be expressed as:

$$(X_1 \bullet H^m) \text{ XOR } (X_2 \bullet H^{m-1}) \text{ XOR. XOR } (X_{m-1} \bullet H^2)$$
$$\text{XOR } (X_m \bullet H) \text{ where } \bullet \text{ designates multiplication in } GF(2^{128})$$

This method has practical performance implications. If the identical hash key is used to authenticate multiple messages, the values h2 and h3 can be pre-calculated once for use with every authenticated message [20, 21]. Additionally, the blocks of the facts to be authenticated (x1, x2, ... xm) can be processed in parallel due to the unbiased computations (Figure 10.15).

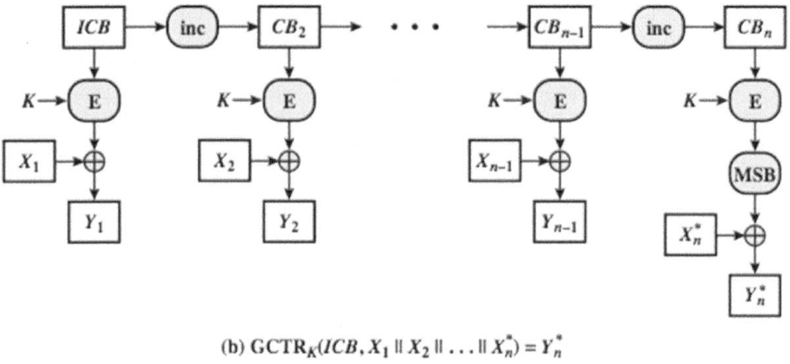

FIGURE 10.15 GCM function.

The function Gctrk(icb, x) takes an input of a mystery key, denoted as 'okay', and a bit string x of arbitrary length. It produces a ciphertext y with a bit period of len(x) (x). The characteristic can be designated as illustrated in Stallings' parent 12. In this case, the feature inc32(s) increments the rightmost 32 bits of s by 1 mod 232, while leaving the other bits untouched. It is worth noting that the counter values can be quickly created and the encryption approaches can be carried out simultaneously.

The entire authenticated encryption function can be defined using the diagram in Figure 10.16 from Stallings. The inputs include the plaintext P, the secret key K, the initialization vector IV, and any other authenticated data A. To generate the hash key, we first encrypt a sequence of zeros with the secret key K. Then, the pre-counter block is created using the IV (J0).

For example, if the IV is 96 bits in length, the pre-counter block is built by appending the padding string 0 || 1. If the IV is not 96 bits, it is padded with zeroes (or left unpadded) until the length of the string is a power of two (128 bits), which is the block size. The pre-counter block is generated by first applying the GHASH function to the original text, then appending 64 zeroes to the end, followed by the 64-bit representation of the length of the IV.

Therefore, GCM is built upon the CTR mode of operation and expands it by adding a MAC that authenticates the message and some additional data that can only be validated. The hashing function only uses multiplications in a Galois field to generate a hash. Since multiplication is a straightforward operation that can be efficiently implemented in hardware within a Galois field, it was chosen to be used. The most effective mode of operation for high-speed packet networks is CTR-based authenticated encryption. Utilizing GCM mode is secure. Thus, the information will not be lost.

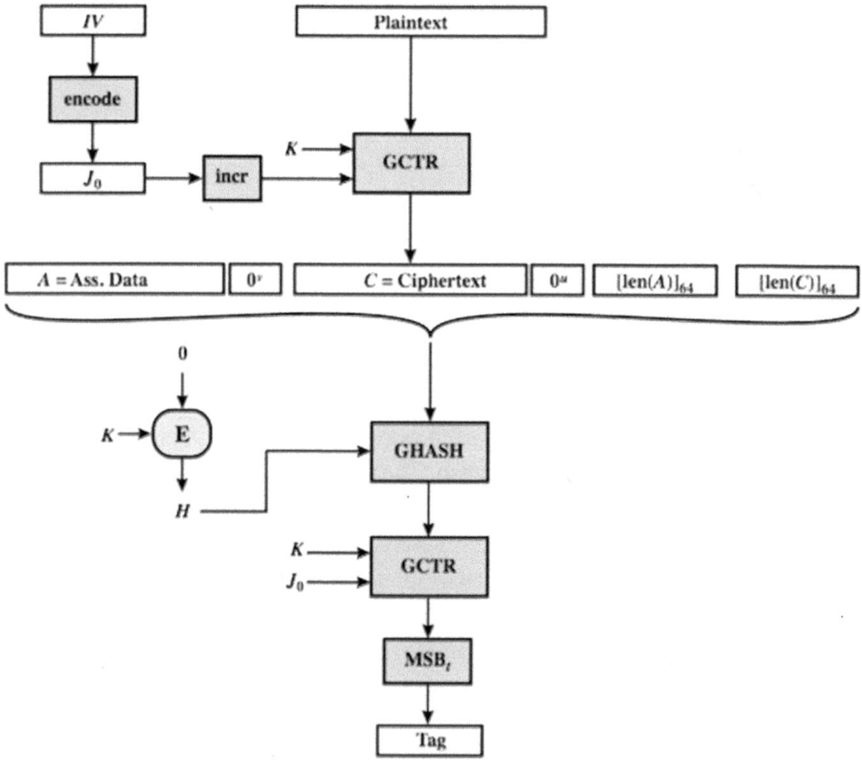

FIGURE 10.16 GCM process flow.

10.8 SUMMARY

Application security is a software engineering challenge for engineers to ensure that the system averts attacks. Information security is enhanced by the operating system, browsers, email clients, document management systems (DMS), middleware supporting remote computing and database access, and libraries of reusable components. Since its components are well understood, most attacks target system infrastructures, such as web browsers. In terms of systems administration, a secure network reinforces the system against intrusion.

KEYWORDS

- **cipher-based message authentication code**
- **computer security**
- **cryptography and authentications**
- **data authentication algorithm**
- **Galois/counter mode**
- **information security**
- **message authentication codes**
- **president's council on integrity and efficiency**

REFERENCES

1. Ahamad, S., Nair, M., & Varghese, B., (2013). A survey on cryptocurrencies. In: 4th *International Conference on Advances in Computer Science, AETACS* (pp. 42–48). Cite seer.
2. Alessandretti, L., ElBahrawy, A., Aiello, L. M., & Baronchelli, A., (2018). Anticipating cryptocurrency prices using machine learning. *Complexity*.
3. Alexander, C., & Dakos, M., (2020). A critical investigation of cryptocurrency data and analysis. *Quantitative Finance, 20*(2), 173–188.
4. Al-Yahyaee, K. H., Mensi, W., Ko, H. U., Yoon, S. M., & Kang, S. H., (2020). Why cryptocurrency markets are inefficient: The impact of liquidity and volatility. *The North American Journal of Economics and Finance, 52*, 101168.
5. Ardia, D., Bluteau, K., & Rüede, M., (2019). Regime changes in bitcoin GARCH volatility dynamics. *Financ. Res. Lett., 29*, 266–271.
6. Aspris, A., Foley, S., Svec, J., & Wang, L., (2021). Decentralized exchanges: The "wild west" of cryptocurrency trading. *Int. Rev. Financ. Anal., 77*, 101845.

7. Bartolucci, S., Destefanis, G., Ortu, M., Uras, N., Marchesi, M., & Tonelli, R., (2019). *The Butterfly Effect: Impact of Development Practices on Cryptocurrency Prices.*
8. Baur, D. G., Dimpf, T., & Kuck, K., (2018). Bitcoin, gold and US dollar-a replication and extension. *Financ. Res. Lett., 25*, 103–110.
9. Bauriya, A., Tikone, A., Nandgaonkar, P., & Sakure, K. S., (2019). Real-time cryptocurrency trading system. *Int. Res. J. Eng. Technol. (IRJET) 06.*
10. Bedi, P., & Nashier, T., (2020). On the investment credentials of bitcoin: A cross-currency perspective. *Res. Int. Bus. Financ., 51*, 101087.
11. Cross, J. L., Hou, C., & Trinh, K., (2021). Returns, volatility, and the cryptocurrency bubble of 2017–18. *Econ. Model,* 105643.
12. Kakushadze, Z., (2018). Crypto asset factor models. *Algorithmic Finance (Preprint)*, 1–18.
13. Kumar, D., & Rath, S., (2020). Predicting the trends of price for Ethereum using deep learning techniques. In: *Artificial Intelligence and Evolutionary Computations in Engineering Systems* (pp. 103–114). Springer.
14. Luu, D. H. T., (2019). Spillover risks on cryptocurrency markets: A look from VAR-SVAR granger causality and student copulas. *Journal of Risk and Financial Management, 12*(2), 52.
15. Ma, F., Liang, C., Ma, Y., & Wahab, M., (2020). Cryptocurrency volatility forecasting: A Markov regime-switching Midas approach. *J. Forecast, 39*(8), 1277–1290.
16. Manahov, V., (2021). Cryptocurrency liquidity during extreme price movements: Is there a problem with virtual money? *Quant Finance, 21*(2), 341–360.
17. Zha, Q., Kou, G., Zhang, H., Liang, H., Chen, X., Li, C. C., & Dong, Y., (2020). Opinion dynamics in finance and business: A literature review and research opportunities. *Financial Innovation, 6*(1), 1–22.
18. Zhang, W., & Li, Y., (2020). Is idiosyncratic volatility priced in cryptocurrency markets? *Res. Int. Bus Financ., 54*, 101252.
19. Zhang, S., Yao, L., Sun, A., & Tay, Y., (2019). Deep learning-based recommender system: A survey and new perspectives. *ACM Comput. Surv., 52*(1). https://doi.org/10.1145/3285029.
20. Zhengyang, W., Xingzhou, L., Jinjin, R., & Jiaqing, K., (2019). Prediction of cryptocurrency price dynamics with multiple machine learning techniques. In: *Proceedings of the 2019 4th International Conference on Machine Learning Technologies* (pp 15–19). ACM.
21. Zhou, H., & Kalev, P. S., (2019). Algorithmic and high frequency trading in Asia-Pacific, now and the future. *Pac. Basin Financ. J., 53*, 186–207.

CHAPTER 11

Artificial Intelligence Techniques for Malicious Domain Detection in Cyberspace

A. SALEEM RAJA,[1] B. SUNDARAVADIVAZHAGAN,[2] G. PRADEEPA,[1] R. JUSTIN,[1] and R. VINODINI[3]

[1]*IT Department, College of Computing and Information Sciences, University of Technology and Applied Sciences, Shinas, Sultanate of Oman*

[2]*IT Department, College of Computing and Information Sciences, University of Technology and Applied Sciences, Al-Musannah, Sultanate of Oman*

[3]*Little Angel Institute, Karur, Tamil Nadu, India*

ABSTRACT

The World Wide Web provides a platform for global information sharing, and people have everyday access to a variety of services. The internet is used by businesses as a low-cost medium for communication and brand promotion. This excessive use of the internet also allows cybercriminals to operate. Cybercriminals employ a range of techniques to bypass web security. Malicious code embedded in web pages steals sensitive information from victims and exploits vulnerabilities in their systems for later use. Nowadays, malicious domain names are generated by algorithms and URL shortening services, which are exploited to evade the security features of web filters and firewalls. Once victims click on these URLs, fraudsters steal their data or install malware on their computers for further access. Malicious URLs are typically spread through email, SMS, and social media platforms

Leveraging Artificial Intelligence (AI) Competencies for Next-Generation Cybersecurity Solutions. Pethuru Raj, B. Sundaravadivazhagan, & V. Kavitha (Eds.)

like Facebook and Twitter. It is necessary to utilize AI-based approaches to proactively identify such harmful domain names. For automated classification, researchers have increasingly used machine learning (ML) and deep learning approaches. Despite the publication of various research articles on this topic, the problem remains unsolved due to factors such as constantly evolving attack vectors and a lack of adaptable solutions. This chapter provides a comprehensive overview of the research domain.

11.1 INTRODUCTION

The Internet and the World Wide Web have dramatically transformed our world and way of life. We now rely on various online applications in cyberspace to carry out our daily tasks. This digital landscape has made communication, business, learning, and other activities more convenient. However, the widespread use of the internet has also led to an increase in criminal activity in cyberspace. Cybercriminals tirelessly search for vulnerabilities and employ various tactics to deceive individuals into revealing sensitive information. Malicious websites serve as a valuable tool for these criminals to carry out illegal operations. These websites are intentionally designed to cause harm, often used for phishing and malware activities. Phishing sites often disguise themselves as legitimate websites to deceive internet users into divulging important information. Their primary objective is to steal data such as personal information, credit card details, and security passwords. If an employee falls victim to a rogue website, it can expose the company to data theft or ransomware attacks. Malware, on the other hand, is malicious software used to destroy data, take control of devices, or gain unauthorized access to networks and monitor device activities. Victims are often unaware that malware is being installed on their devices through malicious websites, making it difficult to detect. According to the APWG's most recent Phishing Activity Trends Report [1], the first quarter of 2022 witnessed a record-breaking 1,025,968 phishing attacks, making it the worst quarter to date, as shown in Figure 11.1.

Figure 11.2 shows that during the first three months of 2022, 107,848 different consumers were protected from at least one instance of malware attempting to steal money from their bank accounts by using Kaspersky solutions [2].

It is essential to implement Artificial Intelligence (AI)-based solutions in order to limit or prevent access to malicious websites. AI is a broad domain that includes different technologies such as machine learning (ML), deep learning, natural language processing (NLP), and computer vision, among

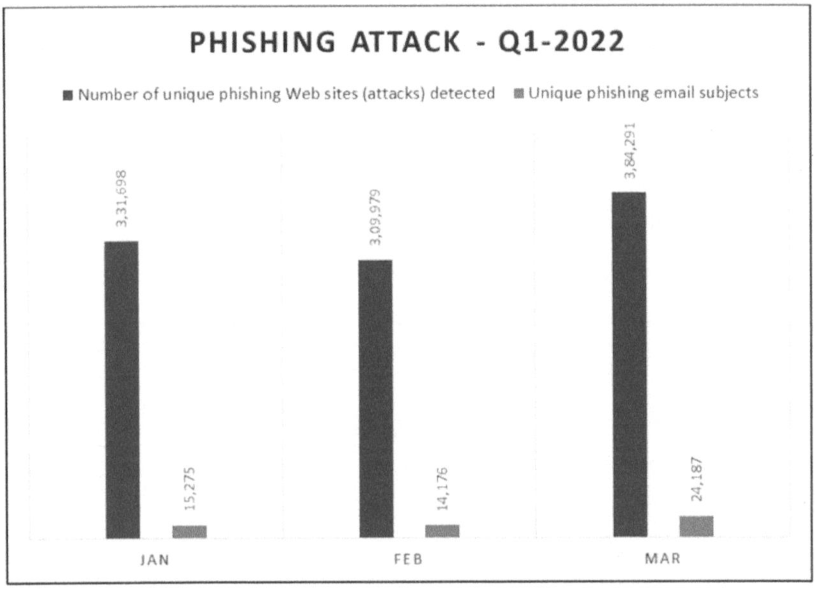

FIGURE 11.1 APWG Phishing attack report.

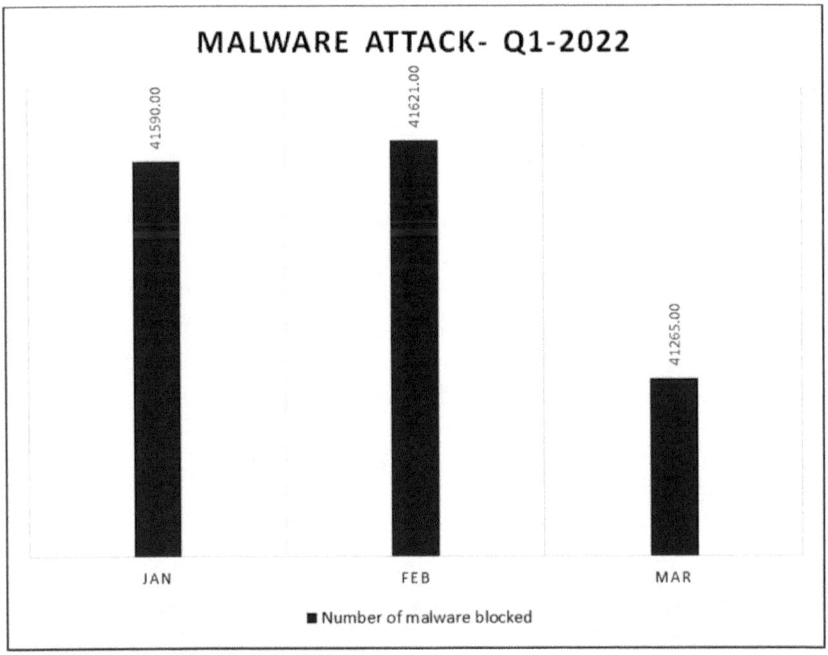

FIGURE 11.2 APWG phishing attack report.

others. Figure 11.3 depicts the technologies under AI. Researchers have increasingly utilized ML and deep learning approaches to automatically detect malicious websites. This chapter presents different approaches for detecting malicious websites and the most utilized algorithms in ML and deep learning with recent research works.

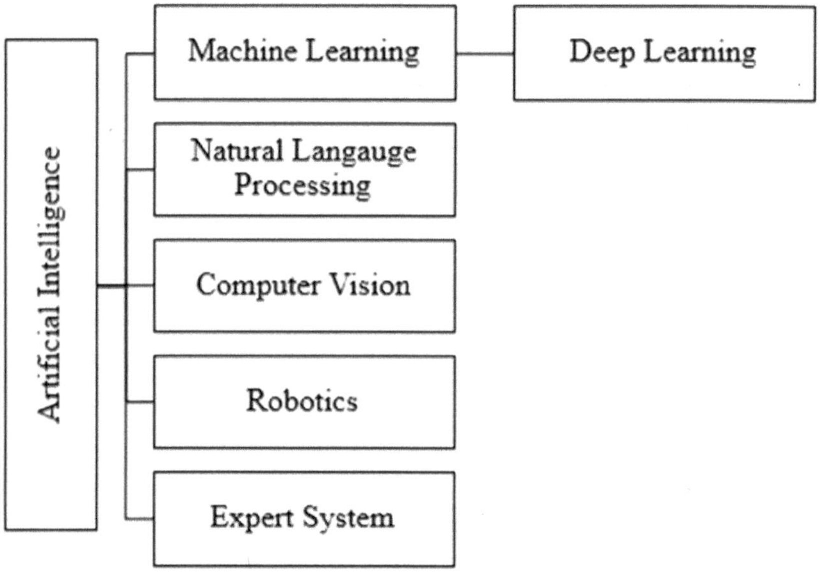

FIGURE 11.3 Technologies in AI.

The remaining sections are structured as follows. Section 2 outlines the background information about URL and content analysis. Section 3 covers the different approaches to malicious domain detection in cyberspace with recent research works. NLP Techniques in malicious domain detection are explained in Section 4. Section 5 presents the ensemble approach. The conclusion includes a summary of the chapter.

11.2 BACKGROUND

Understanding the structure of the Uniform Resource Locator (URL) and web page technology is essential for detecting or classifying malicious and benign websites. The URL is the way to address resources on the internet. The URL consists of various parts as listed in Table 11.1.

http://my.tut. edu: 80/python/index. jsp? un = xyz#id

TABLE 11.1 Anatomy of a URL

Elements of URL	Description
http	HTTP stands for protocol. It determines how to access the server
my.tut. edu	A website can be located on the internet by its distinctive reference, known as its domain name.
80	The port number is always required but rarely seen in URLs. The default port number for HTTP is 80.
python/index. jsp	A file or directory on the web server is normally what the path refers to.
un = xyz	A URL's query string contains informational snippets called parameters.
#id	An internal page reference, often known as a named anchor, is a fragment.

Understanding JavaScript, Cascading Style Sheets (CSS), and HTML are typically the three primary languages required to develop a website or an application. Code flaws or vulnerabilities in the design elements could allow hackers to insert malicious code into a legitimate web page. Some websites contain malicious code because they were created with harmful intentions. Malicious code usually collects user data from the browser and downloads malware as soon as the user accesses such pages. Therefore, it is necessary to examine the content of the web page for better detection accuracy. The contents of the webpage are analyzed using two different methods: static code analysis and dynamic code analysis [3], as shown in Table 11.2.

TABLE 11.2 Code Analysis Method

Method	Description
Static code analysis	Download the source code of the webpage for analysis
Dynamic code analysis	Examine the code of the webpage while it is running

Static code analysis is safer than dynamic code analysis. It is necessary to extract the text features while processing either a URL string or the content of a web page. The term "feature extraction" is used to describe the procedure of reducing unstructured data to a set of quantifiable characteristics that may then be processed further. Figure 11.4 shows the types of features extracted from URL text.

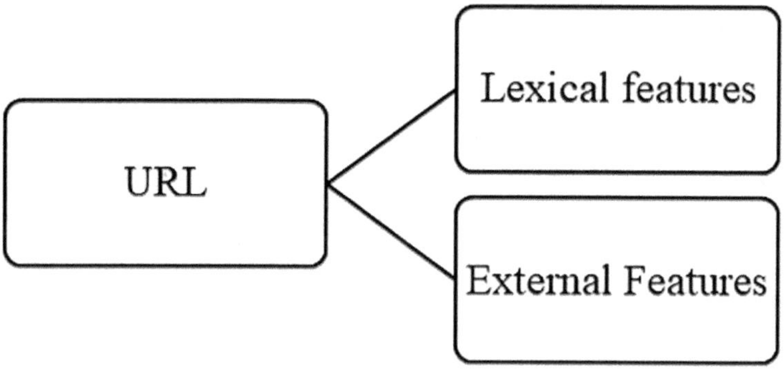

FIGURE 11.4 Types of features in URL.

- **Lexical Features of URL string [4]:** Characters, words, special characters, numbers, URL parts, etc., can all be identified by the text's lexical properties. The primary benefit of focusing solely on URL text is an increase in processing speed.
- **External Features of the URL [5]:** The external features of a URL refer to the characteristics that are obtained from external servers or sources, resulting in additional processing costs. These features can include DNS information of the web page, the page's rank, and more. Figure 11.5 illustrates the various types of features that can be extracted from webpage content.

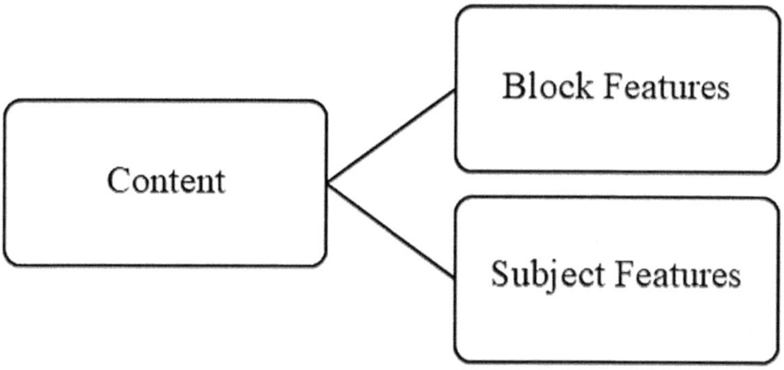

FIGURE 11.5 Types of features of web page content.

- **Block Features [6]:** The basic building blocks of webpage elements are considered here, which include tags, elements, and attributes of HTML, JavaScript, etc.

- **Subject Features [7]:** The web page's subject features include the webpage's textual and visual features.

11.3 APPROACHES TO DETECT MALICIOUS WEB PAGES

Numerous strategies have been presented by researchers to identify malicious websites. These methods fall into four broad categories, as illustrated in Figure 11.6.

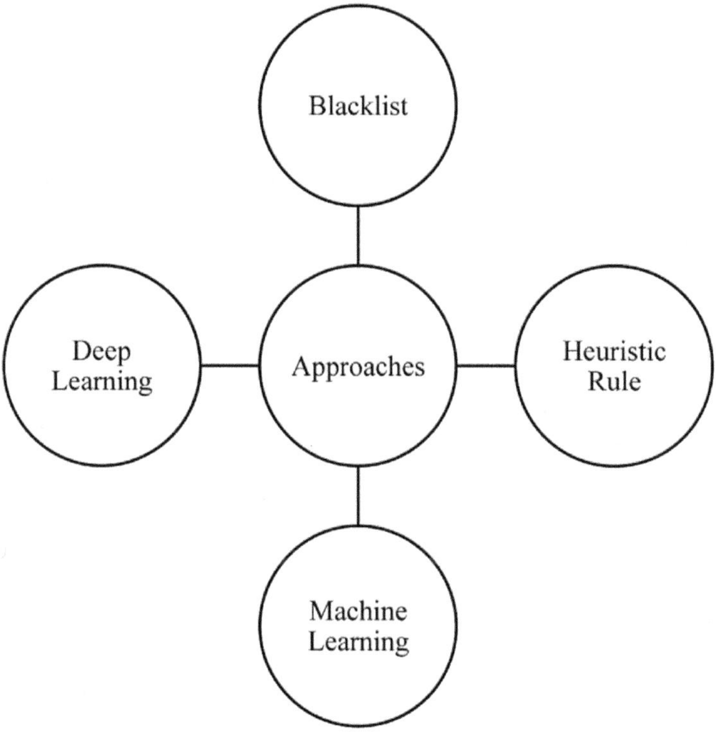

FIGURE 11.6 Different approaches to detect malicious web pages.

The blacklist approach [8] is based on a list of detected bad URLs, either manually or automatically. Blacklisting services are integrated into popular browsers, mobile apps, and search engines to block access to harmful URLs. Examples of such tools include Microsoft's Smart Screen and Google's Safe Search. However, maintaining and updating a blacklist with a large number of entries is an almost impossible task. Without regular updates, newly added malicious URLs cannot be identified by the system. The next approach is rules-based [9], which involves using various characteristics of domain names, web page content, and additional information like DNS information and third-party ranking services. Unlike blacklisting, this approach applies to a wider range of contexts. Rami et al. [17] identified essential features of phishing websites and heuristic rules for classifying benign and phishing URLs, including URL-based features, irregular features, page content-based features, and domain-based features. Hafiz et al. [18] provided a list of 47 prevalent and significant URL lexical features from multiple datasets, including count features, Bow, length feature, IP features, and frequency features. Additionally, it verifies if the URL is attempting to download or retrieve a file with specific extensions like .exe or .zip. Saleem et al. [19] proposed heuristic rule-based malicious URL detection with URL and server-based features.

In a third method, ML techniques, a self-learning methodology, are used to train a model using a diverse dataset of malicious and legitimate URL attributes for phishing detection. Classifying URLs is a binary classification problem. Multiple models (algorithms) exist for classification. The classification algorithms that are used most frequently in ML are presented in Table 11.3.

TABLE 11.3 Classification algorithms in Machine Learning [10]

1. Regression
It is a model that can make predictions. In order to make a prediction, it establishes a connection between the dependent and independent variables. Examples include linear regression (LR) and logistic regression (Log R).
2. Support Vector Machine (SVM)
It is a classification model used to determine the decision boundary (hyperplane) to appropriately classify the data points.
3. K-Nearest Neighbor (KNN)
It is a type of classification model used to group data points. The data points that form a cluster are located close to each other.
4. Decision Tree (DT)
It belongs to the category of supervised learning algorithms and is typically applied to classification-related problems. It applies to dependent variables of both categorical and continuous types. Using this approach, the population is split into homogeneous groups. Groupings are made as distinct as possible using the most important attributes/independent variables.
5. Random Forest (RF)
It's a classification model. The term "forest" is used to describe a collection of decision trees. RF uses a collection of decision trees in order to significantly improve the accuracy of classification.

TABLE 11.3 *(Continued)*

6. Naïve Bayes (NB)
Naive Bayes is a probability model in which it is assumed that the likelihood of the occurrence of specific features within a class is independent of the likelihood of the occurrence of other features.
7. Boosting
Boosting combines the predictions of numerous base estimators to improve the resilience of a single estimator. It takes several predictors that are either weak or average and combines them into one robust prediction. Examples include Gradient Boosting (GB), AdaBoosting (AB), and XGBoosting (XGB).

The following metric is used to evaluate the performance of different machine learning methods. In Table 11.4, we can see how the metrics are derived using the confusion matrix [12].

TABLE 11.4 Confusion Matrix

		Predicted	
Actual		Positive (1) (Malicious)	Negative (0) (Benign)
	Positive (1) (Malicious)	TP	FN
	Negative (0) (Benign)	FP	TN

T= True, F=False, P=Positive, N=Negative

1. **Accuracy (acc):** Indicates how many accurate predictions were made compared to the total number of predictions.

$$acc = \frac{TP + TN}{TP + TN + FP + FN}$$

2. **Precision (Pr):** Ratio of true positive predictions to the total number of positive predictions.

$$Pr = \frac{TP}{TP + FP}$$

3. **Recall (Rc):** Ratio of correct positive predictions to the total number of actual positives.

$$Rc = \frac{TP}{TP + FN}$$

4. **F1-Score (FS):** The harmonic means of precision and recall.

$$FS = \frac{2\,x\,Pr\,x\,Rc}{Pr + Rc}$$

There are numerous research papers on malicious domain detection using the ML approach. Multiclass categorization was proposed by Dhamaraj et al. [20] to identify malicious URLs. The system utilizes a total of 183 features, including 63 lexical features, 34 content-related features, and 18 host-based features. SVM and multiclass CW learning were two of the methods experimented with. A total of 49,935 URLs were collected from sources such as PhishTank, Alexa, and jw SpamSpy, consisting of 26,041 benign, 8,976 phishing, 11,297 malicious, and 3,621 spam URLs. The average prediction accuracy for the test was 98.44%. Patgiri et al. [21] conducted an experimental study on the most popular ML classifiers, including SVM and Random Forest. The extraction of URL lexical and host-based features from the dataset was examined at ratios of 60:40, 70:30, and 80:20. According to the results, the random forest classifier outperformed the SVM classifier significantly. The accuracy results of the random forest were more stable compared to the SVM accuracy results. McGahagan et al. [22] presented a system with two sets of features (26 and 50) for detecting malicious websites. System performance is assessed by various ML algorithms such as Logistic Regression (LR), K-Nearest Neighbor (kNN), Random Forest (RF), Gradient Boosting (GB), Ada Boost (AB), and Extra Tree (ET). Random Forest and Extra Tree provide 90% accuracy. Manan et al. [23] have listed key features of malicious websites, especially content-based features including HTML and JavaScript. Javad et al. [24] presented a malicious web page detection system with a set of content-based features (HTML, JavaScript, JQuery, and XSS attack features). Different sources of the dataset are used for the experiment, including MDL, DNS-BH, Zeus domain block list, CLEAN-MX real-time database, and Malc0de blacklist. The experimental result shows that the C4.5 Tree algorithm gives 97.61% accuracy on the proposed set of features. Jaynish et al. [25] described a variety of features derived from the URL. These features include those associated with the address bar, abnormal features, features based on HTML-JavaScript, and domain-based features. To appropriately classify the URL, heuristic rules are built with various weights. Vahid et al. [26] used 30 different features, including lexical, webpage-based, and server-based features. Data were collected from Phishtank. Different ML algorithms were used for the experiment, including LR, RF, DT, Ada boost, KNN, SVM, Gradient boost, and XGBoost algorithms. The XGBoost algorithm yielded better results. Pradeepa et al. [27] proposed a lightweight method for malicious URL classification, which includes lexical, web page-based, bag of words, and based features. The dataset was taken from Phistank and Kaggle. The results showed that RF achieved 99% accuracy. Chaiban et al. [28] investigated the influence of feature sources on malicious website detection, including Lexical features from the URL, content features, host features, and embedding features. The Singh dataset was used for experiments. The results showed that XGBoost achieved 97% accuracy with JavaScript features. Malak et al. [29] demonstrated a system for detecting phishing that uses ML and deep learning. Two different datasets (the UCI Phishing dataset and the Kaggle dataset) were used to extract hybrid features. For this experiment, ML and DL models were employed. The results indicated that the random forest algorithm offered higher accuracy. Rong et al. [30] proposed malicious content analysis using a static and dynamic method. The static analysis uses URL lexical features, HTML, and JavaScript features. The dynamic analysis examines the shell codes during web page execution. Saleem et al. [31] employed 27 lexical characteristics of the URL to train the ML system for appropriate classifications. Malicious URL detection systems based on rules and ML use different domain features.

Existing research shows that feature engineering is the key to the performance of an ML model. It is a complex process that requires a deep understanding. Additionally, there is a possibility that the vast potential of the dataset may not be fully utilized by the limited number of features being used [11]. To tackle these challenges, researchers employ deep learning algorithms that harness the potential of the training datasets. Deep learning is a multilayer neural network consisting of one input layer, one or more hidden layers, and one output layer. Each layer is composed of nodes, also known as neurons, which are stacked together in the layers as illustrated in Figure 11.7.

Input data is sent to each node. The inputs are multiplied by arbitrary weights, the calculation's outcomes are obtained, and then a bias is added. Finally, activation functions, which are nonlinear, are ultimately employed to choose which neuron to fire. Figure 11.8 shows a single-layer perceptron. Table 11.5 shows the most popular models of neural networks used to solve classification problems, and Table 11.6 outlines the common activation functions used by DNNs.

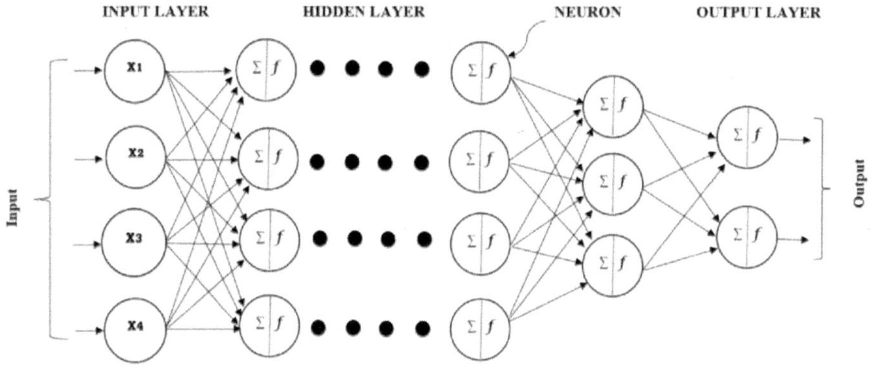

FIGURE 11.7 Deep neural network with N hidden layers.

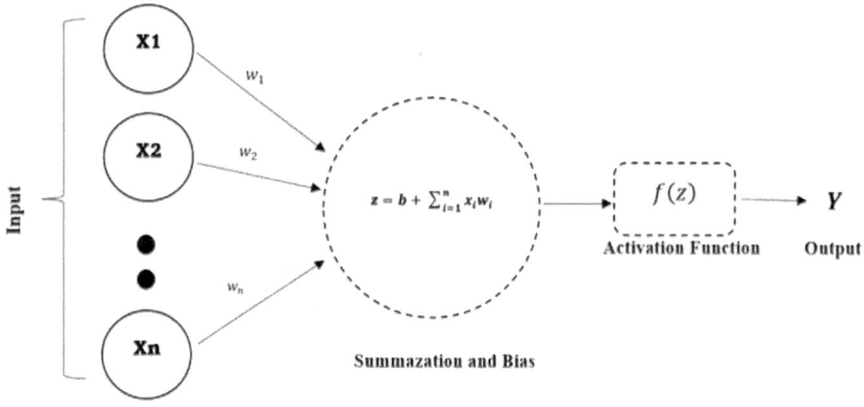

FIGURE 11.8 Single layer perceptron.

TABLE 11.5 Deep Learning Model for Classification

1. Convolution Neural Network (CNN) [12]
An artificial neural network consisting of one or more convolutional layers is known as a convolutional neural network (CNN). This type of network is commonly used for image processing, classification, segmentation, NLP, speech recognition, and other applications that involve auto-correlated input. Convolution entails sliding a filter over an input.
2. Recurrent Neural Network (RNN) [13]
RNNs gather dynamic information from sequential data through recurrent connections between hidden layer nodes. They are capable of categorizing sequential data and maintaining a contextual state. The RNN is commonly employed in contexts involving sequences, such as movies composed of individual image frames, audio recordings composed of individual clips, and textual passages composed of individual words.

TABLE 11.5 *(Continued)*

3. Long Short-Term Memory Networks (LSTMs) [14]
LSTM, a type of recurrent neural network, has more memory efficiency than traditional recurrent neural networks. LSTMs are superior at recognizing patterns. When the LSTM traverses each hidden layer, vital information is preserved and unnecessary information is eliminated from every cell.
4. Gated Recurrent Unit (GRU) [15]
GRU is a popular type of recurrent neural network. It requires less memory and is quicker than LSTM. However, LSTM is more accurate when working with datasets that contain longer sequences.

TABLE 11.6 Common Activation Function Used in Deep Learning Model

1. Sigmoid
Sigmoid is a classification function. The function accepts any input value and returns values between 0 and 1. If the input value is larger (more positive), the output value will be closer to 1.0, while if the input value is smaller (more negative), the output value will be closer to 0.0.
2. Tanh
Tanh refers to the activation function of the hyperbolic tangent. The function receives real numbers and produces a value between –1 and 1. If the input is larger (more positive), the output will be closer to 1.0. If the input value is smaller (more negative), the resulting value will be closer to –1.0.
3. ReLU
The rectified linear activation function, or ReLU, overcomes the limitations of the sigmoid and tanh activation functions. It addresses the vanishing gradient issue but introduces the problem of dead units. The ReLU function returns 0 if the input value is negative; otherwise, it returns the input value.
4. SoftMax
It is a type of sigmoid function used for multi-class classification. In multiclass classification, it returns a probability distribution over target classes.

Recent studies have utilized deep learning algorithms to identify malicious domains. Xiang Dong et al. [37] employ a character-level approach and an improved deep learning model for detecting malicious domain names. To achieve the highest level of detection accuracy, the enhanced deep learning model combines CNN, TCN, and LSTM. The accuracy ratings obtained are 99.76% for CNN+LSTM and 98.81% for CNN+TCN. Malak et al. [38] have introduced a phishing detection system based on machine learning and deep learning. Hybrid features were extracted from two distinct datasets. This experiment included a variety of ML models. The results indicate that the random forest algorithm offers an accuracy of 96.53%. Eint et al. [39] proposed a method for detecting phishing URLs by utilizing the domain and path features of the URL. The performance of the proposed method is evaluated using balanced (D1) and imbalanced (D2) datasets. Keras and ELMO word embedding are employed to encode the text. The results show that the LSTM network achieves an accuracy of 97% with D2.

11.4 NATURAL LANGUAGE PROCESSING (NLP) IN MALICIOUS DOMAIN DETECTION

NLP is broadly defined as the automated manipulation of natural languages, such as speech and text, by computer programs [32]. Computers are only capable of understanding binary code; they cannot comprehend human languages like Hindi or English. NLP enables a computer to comprehend English and Hindi. In comparison to other approaches, NLP with ML and NLP with Deep Learning receive the most attention from researchers due to their processing capabilities. NLP preprocessing techniques help convert text into numerical vectors. Vectorization is the process of transforming a dataset's vocabulary words or tokens into a numerical vector. The input to ML and deep learning models is comprised of these vectors. The following are a few of the typical vectorization techniques:

11.4.1 BAG OF WORDS (BOW)-COUNT VECTORIZER

The bag-of-words model [33] models text using ML methods. The BoW model is a method of vectorization that uses numerical feature vectors to represent text. The BoW model keeps track of the word count for each document it occurs in. Each vector serves as a feature column for the ML model. Let's consider an example with two documents.

- document_1 = "This is a good tip for life. I will not miss this anymore."
- document_2 = "This is not at all good advice."

Step 1: Create a corpus that includes all the documents.

- Corpus = "This is a good tip for life. I will not miss this anymore. This is not at all good advice."

Step 2: Feature selection is the subsequent task which selects the unique words from the corpus.

Features_bag = ['This,''is,''a,''good,''tip,''for,''life,'.,"'I,''will,''not,' 'miss,''this,''anymore,''This,''is,''not,''at,''all,''good,''advice']

Step 3: Usually, the stop words are removed from the features. So the final features set consists of unique words.

Features_bag =['good,''miss,''life,''advice,''tip,''anymore']

Step 4: Assign a number to each unique word in the Feature_bag.
{'good': 2, 'miss': 4, 'life': 3, 'advice': 0, 'tip': 5, 'anymore': 1}

Step 5: Vectorization of the documents as in Table 11.7

TABLE 11.7 Result of the Count Vectorizer

document	good:2	Miss:4	Life:3	advice:0	tip:5	anymore:0
1	1	1	1	1	1	1
2	1	0	0	1	0	0

Issues:

- New words in sentences increase vector length.
- The vectors would have many 0s, generating a sparse matrix.
- Both the word order and the grammatical structure of the sentences are being disregarded.

11.4.2 N-GRAM

An N-gram [34] consists of a collection of N words, phrases, or sentences that are all interconnected. The N-gram can be composed of large blocks of words or smaller groups of syllables. N-gram models are created by counting the occurrences of word sequences in a text corpus and estimating their probability of appearing. This aids in the learning of new words and enhances the understanding of the document by the bag-of-words model. N-gram models can be applied at either the character or word level. Consider an example document and its N-Grams in Table 11.8.

document_1 = "Reehan is a successful businessman."

TABLE 11.8 N-Gram for the Document_1

N value	Generated N-gram
1	['Reehan,''is,''a,''successful,''businessman']
2	['Reehan is,''is a,''a successful,''successful businessman']
3	['Reehan is a,''is a successful,' a successful businessman']

Issue:

- The N-gram model has a problem with capturing long-distance context.

11.4.3 TERM FREQUENCY – INVERSE DOCUMENT FREQUENCY (TF-IDF)

Term frequency-inverse document frequency is a statistical method for determining the significance of a word within a collection of documents. Multiplying the total number of times a word appears in a document (TF) and the word's inverse document frequency across a collection of texts produces this result (IDF) [35]. This approach creates an association between words and sentences, but it only examines them from the standpoint of frequency.

Term Frequent (TF) is a way to measure how often a phrase or term (t) appears in a document (d).

$$tf_{t,d} = \frac{n_{t,d}}{Number\ of\ terms\ in\ the\ document}$$

The term "t" appears in the text "d" n times. Consider an example of documents (document_1, document_2, document_3). The corresponding TF is depicted in Table 11.9, IDF is in Table 11.10 and finally, TF-IDF is shown in Table 11.11.

Consider an Example:

- document_1: Good boy
- document_2: Good Girl
- document_3: Boy Girl Good.

TABLE 11.9 TF for Three Documents

	TF		
Word	**Document 1**	**Document 2**	**Document 3**
Good	1/2	1/2	1/3
Boy	1/2	0	1/3
Girl	0	1/2	1/3

The IDF (Inverse Document Frequency) is a metric used to evaluate the information quality of a term. For common words like stop words, the IDF is low. The IDF measures the importance of a word.

$$idf_t = log\frac{number\ of\ documents}{number\ of\ documents\ with term\ 't'}$$

TABLE 11.10 IDF for the Three Documents

	IDF
Word	**IDF**
Good	Log(3/3)=0
Boy	Log(3/2)=0.176
Girl	Log(3/2)=0.176

Next, calculate the TF-IDF score for each word in the corpus. Words with higher scores are more important:

$$(tf_{idf})_{t,d} = tf_{t,d} * idf_t$$

TABLE 11.11 TF*IDF for the Three Documents

		TF*IDF	
	Good	**Boy**	**Girl**
Sentence 1	1/2 * 0 = 0	1/2 * 0.176 = 0.088	0
Sentence 2	1/2 * 0 = 0	0	1/2 * 0.176 = 0.088
Sentence 3	1/3 * 0 = 0	1/3 * 0.176 = 0.058	1/3 * 0.176 = 0.058

Issue:

- TF-IDF is based on the bag-of-words (BoW) model; therefore, it does not account for text position, semantics, or co-occurrences across documents. Due to this limitation, TF-IDF is primarily useful as a lexical feature.

11.4.4 WORD EMBEDDING

Word embedding [36] is a type of word representation that bridges the gap between human and machine understanding of language. It involves acquiring text representations in an n-dimensional space, where words with similar meanings have similar representations. Word embedding is a technique for encoding words and texts using dense vector representations. Popular word embedding algorithms include Global Vectors (GloVe), Word2Vec, and Keras Word Embedding.

Many deep learning-based detection systems utilize NLP techniques for text processing. Lakshmanarao et al. [40] proposed a malicious website detection system that employs NLP techniques such as count vectorizer, TFIDF vectorizer, and hash vectorizer for feature extraction. Various ML-based classifiers were tested on the Kaggle dataset, and the results showed that the RF algorithm achieved 97.5% accuracy. Bocheng et al. [41] presented a deep learning-based malicious detection system using NLP. Experiments were conducted with machine and deep learning algorithms, and the LSTM model with attention mechanism achieved 98.9% accuracy. Routhu et al. [42] proposed a phishing detection method using the text of webpages. Around 10,000 URLs, including phishing and benign ones, were used for the experiment. The results of the experiments showed that the multimodal domain-specific text achieved 99.34% detection accuracy. Zhang et al. [43] presented a system that extracts different features from URLs, as well as content, visual similarity, and reputation-based features, and applies NLP techniques for text processing. The results showed that the AdaBoost plus word embedding method achieved 99.8% accuracy.

11.5 ENSEMBLE APPROACH

Some machine learning classifiers perform well in a specific dataset but perform poorly in other datasets due to certain statistical properties or hyperparameter configurations. This can involve trial and error and requires more time. To address this issue, ensemble learning refers to algorithms that combine the predictions from two or more models. The base machine learning classifiers in an ensemble classifier are referred to as the ensemble members. Ensemble members can be of the same type or different types. The prediction of each ensemble member is combined using a statistical mean, mode, or any sophisticated method as shown in Figure 11.9. The ensemble classifier outperforms the base machine learning classifier in terms of performance and data generalization. Combining multiple base machine learning classifiers increases accuracy while reducing the bias and variation of the individual classifier. It is also a robust model that can handle various types of datasets.

The ensemble technique has been widely utilized by researchers in recent years. Ubing et al. [44] described a phishing detection system with an ensemble hard (majority) voting classifier. The UCI ML phishing dataset was utilized. The dataset included 30 distinct features, including URL and content-based features, and an ensemble learning model was constructed by combining several ML models (GNB, Support Vector Machine (SVM), kNN, LR,

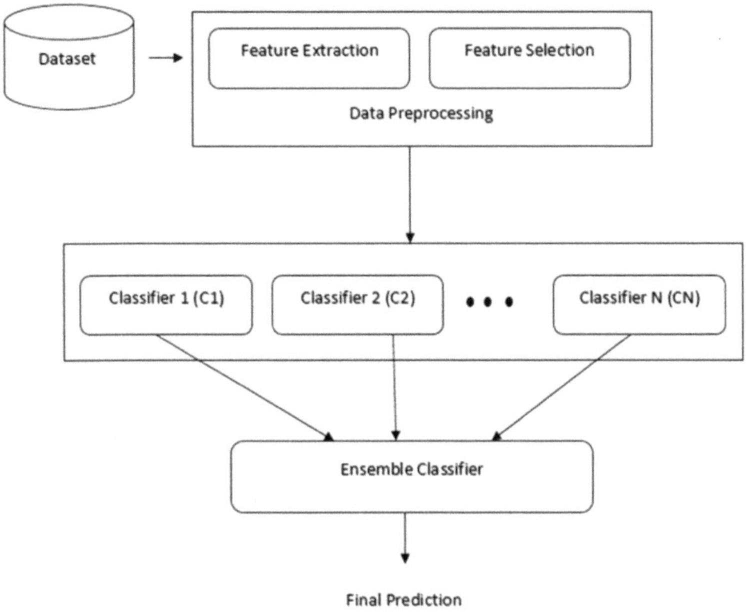

FIGURE 11.9 Ensemble classifier.

multilayer perceptron NN, GB, and RF classifiers). The results demonstrate that the proposed model provides 95% accuracy. A voting classifier developed by Basit et al. [45] was used to categorize the phishing URLs. The experiment used a dataset from the UCI ML repository with 11,055 URLs and 30 features. An ensemble learner was created by combining three ML classifiers (ANN, KNN, and C4.5 decision tree classifiers). According to the experimental findings, the KNN+RF classifier had an accuracy of 97.33%. Vishva et al. [46] presented a phisher fighter system for phishing URL detection. The proposed ensemble model was trained and tested using various datasets, including the UCI, LIAR, and Phish tank datasets. The URL and content of the web pages were analyzed. For URL analysis, 30 URL features from the UCI dataset were used. For content analysis, the LIAR dataset and textual contents were processed using NLP techniques. The Phish Tank dataset was utilized for testing. Different ML models (LR, RF, SVM, SGD, and NB) were used to select the best two models for the final classification. The results showed an accuracy of 90.68% in the proposed system [47] [48]. Mohammed et al. developed a phishing website detection method using an ensemble stacking classifier. A genetic algorithm (GA) was used to fine-tune the hyperparameters of the chosen models (RF, AB, XGB, bagging, GB, and LGBM). The three best models were selected for the stacking ensemble classifier. Three datasets from UCI and Mendeley were used for the experiment, and the results showed accuracies of 97.16%, 98.58%, and 97.39 for dataset1, dataset2, and dataset3, respectively [49]. Altyeb presented a detection method for phishing websites using an ensemble weighted soft voting classifier that combines several ensemble classifiers such as RF, AB, XGB, bagging, GB, and LGBM. The UCI dataset was used for the experiment, and the results showed 95% accuracy [49]. Ammara et al. demonstrated a stacking model for phishing website detection using a UCI dataset. Stacking the highest-performing base ML classifier resulted in a better-performing model. Combining RF, NN, and the bagging classifier yielded an accuracy of 97.4% [50]. Ehsan et al. used hybrid features with popular ML models such as RF, AB, GB, and XGB, and achieved better results than existing models [51].

11.6 CONCLUSION

Cybersecurity has become a major concern in today's business world. Attackers use different techniques to deceive internet users into performing illegal actions, such as collecting sensitive data, spreading malware, and using the victim's system for further attacks. Most of these attacks are spread through SMS, email, and social media posts. Malicious links are embedded in messages or websites, leading users into the attacker's trap. As a result, recognizing malicious domain names has become a significant concern. Given the ever-changing attack vectors, innovative and practical approaches are needed to timely detect malicious web pages. This chapter provides an in-depth discussion of various methods that can be utilized in the detection of malicious domains.

KEYWORDS

- **artificial intelligence**
- **cascading style sheets**
- **K-nearest neighbor**
- **machine learning**
- **natural language processing**
- **support vector machine**
- **uniform resource locator**

REFERENCE

1. https://docs.apwg.org/reports/apwg_trends_report_q1_2022.pdf(accessed on 27 January 2024).
2. https://securelist.com/it-threat-evolution-in-q1-2022-non-mobile-statistics/106531/ (accessed on 27 January 2024).
3. Hajian, Vafaei, M., Tayarani, & Sadrnezhad, (2017). Analyzing new features of infected web content in the detection of malicious web pages. *The ISC International Journal of Information Security, 9*(2), 161–181.
4. Cho Do, X., Hoa, D. N., & Tisenko, V. N., (2020). Malicious URL detection based on machine learning. *International Journal of Advanced Computer Science and Applications (IJACSA), 11*(1).
5. Pradeepa, D., (2022). Review of malicious URL detection using machine learning, soft computing for security applications. *Advances in Intelligent Systems and Computing* (Vol. 1397, pp. 97–105). Springer.
6. Chaiban, S., & Soliman, S. L., (2022). Investigating the influence of feature sources for malicious website detection. *Applied Science, 12*.
7. Sahoo, L., & Hoi, (2019). Malicious URL Detection Using Machine Learning: A Survey. https://doi.org/10.48550/arXiv.1701.07179.
8. Prakash, K., & Kompella, G., (2010). *PhishNet: Predictive Blacklisting to Detect Phishing Attacks* (pp. 1–5). IEEE INFOCOM.
9. Saleem, Pradeepa, & Arul, (2022). Mudhr: Malicious URL detection using heuristic rules based approach. *AIP Conference Proceedings, 2393*, 020176.
10. Aurélien, G., (2017). *Hands-On Machine Learning with Scikit-Learn and TensorFlow.* O'Reilly Media.
11. Jason, B., (2016). Deep Learning with Python: Develop Deep Learning Models on Theano and Tensor Flow Using Keras. Ebook, Edition v1.7, Chapter 4, 6, 7.
12. https://towardsdatascience.com/an-introduction-to-convolutional-neural-networks-eb0b60b58fd7 (accessed on 27 January 2024).
13. https://forum.huawei.com/enterprise/en/what-is-recurrent-neural-network/thread/738873-895(accessed on 27 January 2024).
14. https://machinelearningmastery.com/gentle-introduction-long-short-term-memory-networks-experts/ (accessed on 27 January 2024).
15. https://towardsdatascience.com/understanding-gru-networks-2ef37df6c9be(accessedon 27 January 2024).
16. https://www.geeksforgeeks.org/activation-functions-neural-networks/(accessed on 27 January 2024).
17. Rami, M., & Fadi, T., & McCluskey, L. (2015). *Phishing Website Features*. https://eprints.hud.ac.uk/id/eprint/24330/6/MohammadPhishing14July2015.pdf (accessed on 27 January 2024).
18. Hafiz, M. J. K., Niyaz, Q., Devabhaktuni, V. K., Guo, S., & Shaikh, U., (2019). Identifying generic features for malicious URL detection system. In: *IEEE 10th Annual Ubiquitous Computing, Electronics & Mobile Communication Conference (UEMCON)* (pp. 0347–0352). New York City, NY, USA. doi: 10.1109/UEMCON47517.2019.8992930.
19. Saleem, Pradeepa, & Arulkumar., (2022). Mudhr: Malicious URL detection using heuristic rules-based approach. *AIP Conference Proceedings* (Vol. 2393, No. 1).

20. Dharmaraj, P., & Jayantrao, P., (2018). Feature-based malicious URL and attack type detection using multi-class classification. *The ISC International Journal of Information Security, 10*(2), 141–162.
21. Patgiri, R., Katari, H., Kumar, R., & Sharma, D. (2019). Empirical study on malicious URL detection using machine learning. In: Fahrnberger, G., Gopinathan, S., & Parida, L. (eds.). Distributed Computing and Internet Technology. ICDCIT 2019. *Lecture Notes in Computer Science, vol. 11319*. Springer, Cham. https://doi.org/10.1007/978-3-030-05366-6_31.
22. McGahagan, Bhansali, Pinto, & Cukier, (2019). A comprehensive evaluation of webpage content features for detecting malicious websites. *Latin-American Symposium on Dependable Computing* (pp. 1–10).
23. Manan, Ahmed, & Kahar, (2018). Characterizing current features of malicious threats on websites. *Intelligent Computing & Optimization* (Vol. 866). Springer.
24. Javad, Majid, Mohammad, & Zohre, (2017). Analyzing new features of infected web content in detection of malicious web pages. *The ISC International Journal of Information Security (IseCure)* (Vol. 9, No. 2, pp. 161–181).
25. Jaynish, (2018). *Design and Implementation of Heuristic Based Phishing Detection Techniques*. Master's degree thesis, University of Victoria.
26. Vahid, Mohammad, & Izadi, (2020). *Phishing Detection Using Machine Learning Techniques*. Sharif University of Technology, Tehran, Iran.
27. Pradeepa, & Devi, (2022). Lightweight approach for malicious domain detection using machine learning. *Scientific and Technical Journal of Information Technologies, Mechanics, and Optics, 22*(2), 262–268.
28. Chaiban, Sovilj, Soliman, Salmon, & Lin, (2022). Investigating the influence of feature sources for malicious website detection. *Applied Sciences, 12*(6).
29. Malak, & Samitha, (2022). Phishing attacks detection using machine learning and deep learning models. *International Conference on Data Science and Machine Learning Applications (CDMA)* (pp. 175–180).
30. Rong, Yan, Jiefan, & Binbin, (2017). Detection of malicious web pages based on hybrid analysis. *Journal of Information Security and Applications, 35*, 68–74.
31. Saleem Raja, Vinodini, & Kavitha, (2021). Lexical features based malicious URL detection using machine learning techniques. *Materials Today: Proceedings, 47*(1), 163–166. https://doi.org/10.1016/j.matpr.2021.04.041.
32. https://machinelearningmastery.com/natural-language-processing/ (accessed on 27 January 2024).
33. Ghaleb, Alsaedi, Saeed, Ahmad, & Alasli, M., (2022). Cyber threat intelligence-based malicious URL detection model using ensemble learning. *Sensors, 22*, 3373.
34. Shreyas, & Nisha, (2018). N-gram assisted YouTube spam comment detection. *Procedia Computer Science, 132*, 174–182.
35. Chan, (2018). *NtMalDetect: A Machine Learning Approach to Malware Detection Using Native API System Calls.* arXiv:1802.05412v2 [cs.CR].
36. Crişan, Florea, Halasz, Lemnaru, & Oprisa, (2020). Detecting malicious URLs based on machine learning algorithms and word embeddings. In: *IEEE 16th International Conference on Intelligent Computer Communication and Processing (ICCP)* (pp. 187–193).
37. Xiang, D., Hao, J., Feng, C., Jian, BaoShan, Bao, P.,& Qi, T., (2022). A malicious domain detection model based on improved deep learning. *Safe and Fair Machine Learning for Neuroscience.*

38. Malak, S., (2022). Phishing attacks detection using machine learning and deep learning models. *International Conference on Data Science and Machine Learning Applications (CDMA)* (pp. 175–180).
39. Eint, H., (2021). *Phishing URL Detection Using Information-Rich Domain and Path Features*. DEIM.
40. Lakshmanarao, Babu, & Bala, (2021). Malicious URL detection using NLP, machine learning and FLASK, international conference on innovative computing. *Intelligent Communication and Smart Electrical Systems (ICSES)* (pp. 1–4).
41. Bocheng, Xianang, & Pengxiang, (2021). Malicious URL detection system based on LSTM and attention mechanism. *Journal of Physics: Conference Series, 2037.*
42. Routhu, Amey, & Alwyn, (2022). Application of word embedding and machine learning in detecting phishing websites. *Telecommunication Systems, 79*, 33–45.
43. Zhang, Zeng, Jin, Yan, & Geng, (2017). Boosting the phishing detection performance by semantic analysis. *International Conference on Big Data* (pp. 1063–1070).
44. Ubing, Jasmi, Abdullah, Jhanjhi, & Supramaniam, (2019). Phishing website detection: An improved accuracy through feature selection and ensemble learning. *International Journal of Advanced Computer Science and Applications, 10*(1).
45. Basit, Zafar, Javed, & Jalil, (2020). A novel ensemble machine learning method to detect phishing attack. *IEEE 23rd International Multitopic Conference (INMIC).*
46. Vishva, & Aju, (2021). Phisher fighter: Website phishing detection system based on URL and term frequency-inverse document frequency values. *Journal of Cyber Security and Mobility, 11*(1), 83–104. River Publisher.
47. Mohammed, Faisal, Zeyad, Badiea, Tawfik, Mohammad, Abdulrahman, & Talal, (2021). An optimized stacking ensemble model for phishing websites detection. *Electronics, 10*(11).
48. Zeyad, Badiea, Mohammed, Faisal, Tawfik, Mohammad, Abdulrahman, & Talal, (2022). Phishing websites detection by using optimized stacking ensemble model. *Computer Systems Science & Engineering, 41*(1).
49. Altyeb, (2017). Phishing websites classification using hybrid SVM and KNN approach. *International Journal of Advanced Computer Science and Applications, 8*(6).
50. Ammara, Hikmat, Tassawar, Nazish, Farah, Almas, & Maryam, (2020). Phishing web site detection using diverse machine learning algorithms. *The Electronic Library, Emerald Publishing Limited, 38*(1), 65–80.
51. Ehsan, Mohammad, & Mauro, (2022). *An Adversarial Attack Analysis on Malicious Advertisement URL Detection Framework.* arXiv:2204.13172v1.

CHAPTER 12

Malicious Domains Detection Based on Machine Learning Techniques

D. SASIREKHA, HANNAH VIJAYKUMAR, A. PARAMESWARI, and K. MAHESWARI

Department of Computer Science, Anna Adarsh College for Women, Chennai, Tamil Nadu, India

ABSTRACT

This research refers to a study on information security, focusing on phishing-type attacks. The main objective is to describe the primary risks associated with phishing attacks for individuals and organizations, as well as to indicate the best practices to mitigate the occurrence of such attacks. Phishing websites are web pages that imitate banking services, currency transactions, and other financial operations that require authentication or registration of sensitive information. These websites are developed by attackers with the intention of obtaining user credentials. In this study, a phishing website detector was developed using a classifier trained with various page characteristics. The purpose was to predict whether a page is a phishing website or not. It is believed that the integration of this tool into browsers, firewalls, and email services can significantly reduce the number of victims of this type of fraud. The final model achieved an accuracy of 95% and an F1 measurement of 95%.

12.1 INTRODUCTION

We have witnessed a revolution in the last two decades: the popularization of the Internet has changed almost all human relationships, especially the way most people work. An office worker guides their daily activities and

Leveraging Artificial Intelligence (AI) Competencies for Next-Generation Cybersecurity Solutions. Pethuru Raj, B. Sundaravadivazhagan, & V. Kavitha (Eds.)

conducts their projects through the exchange of messages. These messages communicate more ways to buy and sell his projects. The first threats to computer security were created, but different ways of protecting systems also emerged, creating layers that make attacks difficult. This made attackers focus their efforts on the most vulnerable link in the chain: people. Deceiving people has been a tactic used for a long time and continues to be prevalent today, especially with the numerous channels through which messages can be received, such as SMS, WhatsApp, social networks, forums, and more. The list goes on and on.

As technology advances, new solutions to everyday problems emerge, along with new ways for attackers to target individuals and companies. Currently, the most commonly used technique is phishing, where people are lured into scams executed in various ways and with varying levels of technological sophistication. In this paper, we will describe these attacks, present methods to identify them, and highlight the most effective ways to prevent them.

To address this issue, our work aims to develop a tool that can predict the legitimacy of a webpage given its URL. This will be achieved through a machine learning (ML) model trained using techniques [12, 13].

12.2 INFORMATION SECURITY

The main objective of information security is to prevent the leakage and manipulation of data and systems by unauthorized third parties.

The main pillars of information security are availability, integrity, and confidentiality. These pillars must be maintained. In addition, other aspects that must be taken into account include authenticity, responsibility, and reliability.

Availability ensures that the respective information within the system functions smoothly. The requested data must be retrieved correctly at the desired time, which means that systems must be protected from failures.

In terms of integrity, it is necessary for data to circulate or be stored securely, without interference from external parties that could compromise it, leading to risks of loss or damage to the integrity of the data.

Thus, in terms of integrity, the accuracy of information and the fluidity of secure data storage are indispensable issues when it comes to information security.

Confidentiality ensures that data is accessible only to authorized individuals within an organization, preventing confidential information from being stolen by systems through cyberattacks and other means.

12.2.1 PHISHING

Phishing is a fraudulent attempt to obtain sensitive information, such as login and password, from users of a system [1, 2]. This is usually done through a fake website, which mimics a legitimate one, where users enter their data believing they are on the real domain.

12.2.2 MAIN TYPES OF PHISHING

Examples of some types of Phishing are:

12.2.2.1 SCAM

Attempts occur where criminals try to deceive the user into disclosing personal information such as bank account numbers, passwords, and credit card numbers by opening malicious links or files.

12.2.2.2 BLIND PHISHING

The most common of all phishing attacks is characterized by the large number of emails sent with random content, relying solely on "luck" to determine whether and which user will fall into the trap.

12.2.2.3 SPEAR PHISHING

This type of attack is already targeted at a specific group. It can be against government officials, users, or customers of a company, and even against a specific person. Its purpose is to gain access to the database in order to obtain confidential information, as well as confidential or financial files.

12.2.2.4 CLONE PHISHING

This method clones an original site to attract users. The focus of this type of attack is to obtain valuable data from users, which they believe is on the original site, and end up providing this data. The most common data turns out to be credit card data. In the end, the user is redirected to the original site and ends up not even realizing that they have fallen into a scam.

12.2.2.5 WHALING

This type of attack is focused on individuals with high executive power, such as the president of a corporation, and it is carried out on behalf of the company they work for. These attacks are often associated with judicial intimidation or internal corporate notifications. In addition to employing tactics similar to spear-phishing, which targets important individuals, this type of attack utilizes methods such as fraudulent communications that appear to originate from a seasoned or highly influential person within the organization. This adds an additional element of social engineering to the scam, as employees typically do not refuse requests from individuals they perceive as important.

12.2.2.6 VISHING

In this type of phishing attack, voice mechanisms are used, which can be accompanied by a Short Message Service (SMS) stating that the victim's card has been blocked. The victim is then instructed to call a specific number to request its release. Criminals utilize Voice over Internet Protocol (VoIP) to easily conceal the caller's identity.

12.2.2.7 PHARMING

These attacks involve DNS. Attackers poison the DNS to target users on a large scale. When users try to access a specific website by typing the Uniform Resource Locator (URL), they are redirected to a fake page created for the attack.

12.2.2.8 SMISHING

A type of phishing focused solely on SMS is known as smishing. These messages typically aim to embarrass the user with false debts or manipulate them into making impulsive decisions based on emotions, such as promises of sweepstakes, prizes, or receiving a large sum of money.

12.2.2.9 SOCIAL ENGINEERING

Social engineering is the art of exploiting human psychology, rather than hacking techniques, to gain access to buildings, systems, or data. For example, instead of finding software vulnerabilities, a social engineer might call an employee and pretend to be someone from IT support, trying to trick the employee into revealing their password [3, 4]. Table 12.1 summarizes and characterizes the main types of phishing.

TABLE 12.1 Types of Phishing

Type of Phishing	Where it Occurs	Main Focuses
Scam	Malicious links and files	Final user
Blind phishing	Malicious links and files with random shots	Final user
Spear phishing	Malicious links and files	End-users (civil servants, business customers)
Clone phishing	cloned sites	companies and banks
Whaling	Malicious links and files	Final user
Vishing	Phone calls and SMS	High executive power user
Pharming	Link redirecting to a fake website	Final user
Smishing	SMS	End user companies
Social engineering	interpersonal attacks	Final user

12.2.3 BEST PRACTICES ADOPTED

In addition to always staying informed about the types of phishing, you should also be aware of the following prevention methods (Table 12.2):

TABLE 12.2 Best Practices Adopted

Best Practices	How to Use Them
Analysis of Email	Always check the sender and content of emails.
Use of antivirus	Always keep your antivirus on the latest update updated.
Using two-step verification	With two-step verification, the protection of your account or files will be doubled.
Firewall	It will keep your computer safe from Phishing attacks with other types of possible security attacks.
Personal information	Never share personal information with strangers, especially passwords.
self-study	Stay informed about new types of attacks for prevention.
Information security training	With formal courses, to update skills and acquire new information

In the next section, we will develop a phishing website classifier.

12.3 METHODOLOGY

This section presents the main steps of the method developed to detect phishing websites.

12.3.1 FEATURE EXTRACTOR

Often, phishing pages have characteristics and patterns that distinguish them from legitimate pages [5, 6–17]. Therefore, by extracting various attributes or elements from a page (which we will call characteristics), its legitimacy can be classified.

Thus, a program was developed capable of extracting 24 characteristics of a page from its URL. Some of them are directly related to the URL string itself, such as its length or the use of certain characters, while others are related to its content, such as the use of pop-ups or cookies [15–17].

It is important to note that not all feature extractor functions were used in the final version of the work, as many were discarded after being evaluated (see Results and Discussion).

The characteristics from a tool, described below, were selected for the final model:

The characteristics from a to w, as described above, were presented [5, 6], all of which are originally binary or ternary (indicating malicious, suspicious, or legitimate behavior). The characteristic related to the length of the URL, for example, was originally ternary, receiving a certain label if its number of characters fell within pre-established values. As this practice is not very robust, we replaced it with the actual extracted value itself, which, in the example above, is the number of characters.

12.3.2 DATABASE

Pages such as PhishTank provide URLs of phishing pages, while other services like Umbrella provide lists of supposedly legitimate websites. In this way, a total of 7,120 URLs were categorized as either phishing or legitimate, in equal proportion.

Furthermore, the uniqueness of domains was ensured in the selection of these websites to avoid dataset bias. It was also verified that the URLs returned a valid response.

Ch.	Specification	Explanation	Example	Value
a)	IP address in the URL	Malicious websites may not have a domain and can only be accessed from their public IP address.	https://222.222.222.222/s cam.html.	0 or 1
b)	At sign-in in the URL	Many browsers ignore content preceding the at sign. Attackers use this technique to trick their targets into believing they are accessing a legitimate domain.	https://paypal.com@peyp al.com takes you to the peypal.com website	0 or 1
c)	"//" in the URL	Similarly, to the above feature, in some browsers, the "//" set outside the transfer protocol context (http/https) causes the text that precedes it to be ignored.	https://paypal.com//peypa l.com takes you to the paypal.com website	0 or 1
d)	Hyphen in the URL	Hyphens are rarely used in legitimate URLs. Phishing sites, however, use this character to give the domain legitimacy.	http://login-bradesco.com.br	0 or 1
e)	Use of HTTPS	HTTPS is a transfer protocol that establishes a secure connection between server and client. It therefore lends credibility to a page and is most commonly used on legitimate pages.	https://abc.com.br uses HTTPS, unlike http://abc.com.br.	0 or 1
f)	Favicon from the domain itself.	Favicon is the icon, usually with the page logo, displayed in the browser tab. Favicons loaded from a domain other than the one accessed may indicate that the page is not legitimate.	–	0 or 1
g)	Use of pop-up	Pop-ups are windows created by the website accessed and, currently, it is not so common to use them on legitimate websites. This is mainly due to the fact that browsers have been creating obstacles to their use, due to their ability to open domains other than the one accessed. On the other hand, this function can be convenient for attackers.	–	0 or 1

(Continued Table)

Ch.	Specification	Explanation	Example	Value
h)	Size of the URL.	Long URLs may contain malicious content or be used to hide the phishing domain.	https://paypal.com/xxxxqq q/content/0018a 2125ec66 86c9fa0f24ab7312716/?us er=&_ verify? service=mail &data:text/html;charset= utf-8;base64,PGh0bWw+DQ o8c3R5bGU+IGJv ZHkge yBtYXJnaW46IDA7IG92 ZXJmbG93 OiBoaWRkZ W47IH0gPC9zdHlsZT4N CiAg PGlmcmFt	Number of characters present in the URL
i)	Number of own links	Own links are those that point to the domain they are on. On average, legitimate pages have more self-links.	–	Number of own links.
j)	Number of open ports.	A page server, if properly configured, should only have ports 80 (HTTP) or 443 (HTTPS) open. Other open ports can pose a threat to the user as they allow the server to use non-essential services.	Connection is required to ports 21 (FTP), 22 (SSH), and 3306 (MySQL) of the page's IP address, in order to identify whether they are open. Other ports are not tested so as not to make the feature calculation time too high.	Number of open ports among 21, 22, and 3306.
k)	Age of certificate.	Phishing websites are usually active for a short time, as after being notoriously identified as malicious, they are deactivated. So, when they have a certificate, it is usually recent. Legitimate sites, however, may have had a certificate for years.	–	Age of certificate
l)	Age of domain	As mentioned, phishing sites tend to be active for a short time, so the average age of their domains is lower than that of legitimate sites.	–	Domain age
m)	Domain registration time	As phishing websites tend to be active for a short time, the attacker usually contracts the domain for a short period, while legitimate websites usually make contracts lasting several years, to avoid renewing them frequently.	–	Domain registration time

(Continued Table)

Ch.	Specification	Explanation	Example	Value
n)	Alexa ranking	Alexa Ranking provides a list where millions of websites are categorized by a combination of their number of hits and unique users. On average, legitimate websites rank better than phishing websites.	–	Position in Alexa Ranking
o)	PageRank	The PageRank algorithm provides an integer value from 0 to 10 that measures the importance of a page, according to the number and quality of links on the web that point to it.	–	The value assigned by PageRank.
The following characteristics, after being evaluated, were not used in the final model:				
p)	HTTPS simulation.	Phishers can put "https" at the beginning of the domain to simulate that the site uses HTTPS.	–	0 or 1
q)	Sending information by email	Malicious websites can send user-submitted credentials directly to an attacker's email. Implementation: search for mailto: (HTML) and mail() (PHP) methods.	–	0 or 1
r)	Use of URL shortening services	Attackers can use shortening services to hide suspicious URLs. This increases the chance of Internet users accessing them when they find them, thus being redirected to the phishing page. Implementation: The URL is compared to a list of shortening services.	–	–
s)	The right mouse button is disabled	Phishers can disable the use of the right mouse button on the site so that the user has difficulties in getting their malicious source code. Implementation: search in the source code for "event.button==2" (JavaScript), which disables the right button.	–	0 or 1

(Continued Table)

Ch.	Specification	Explanation	Example	Value
t)	Use of iframe	iFrame displays another page within the current one. It can be used for harmful purposes such as loading malicious websites without the user's consent.	–	0 or 1
u)	Analysis of anchors.	In HTML, anchors are elements that, when clicked, lead to other points on the same page or different pages (links). A high rate of anchors that lead to other domains or that lead nowhere can indicate malicious intent.	–	Percentage of suspicious anchors as described.
v)	Number of redirects.	Phishing websites may have a higher average amount of redirects when accessing them than legitimate websites.	–	Number of redirects.
w)	Number of subdomains	In the URL, subdomains precede a website's domain. In general, legitimate pages do not have more than one, besides the "www."	http://www.hostpoint.ch.17a902ef.tcorner.com, whose domain is tcorner, has 3 subdomains, in addition to "www."	The number of subdomains present in the URL, other than "www."
The following test feature was also introduced:				
x)	Random value	Characteristics are used to evaluate others since effective characteristics should not have less or equal importance to them. It is not introduced in the final model.	–	0 or 1, determined at random.

Next, features were extracted from the obtained pages using the program described in the previous section. At the end of the process, a spreadsheet was generated where each line contains the characteristic values and the label (legitimate or phishing) of a given site.

12.3.3 CLASSIFIER MODEL TRAINING

To train the phishing URL classifier model, the dataset obtained in the previous step was divided into 25% for testing and 75% for training. The procedure was applied in a stratified manner, ensuring that both sets had an equal distribution of phishing and legitimate data.

Next, a pipeline was created to normalize the data provided to it, regardless of whether it was for training, testing, or real use. Two normalization techniques, MinMax [3] and Z-score [4], were tested for this purpose. The classifier algorithms evaluated were Decision Tree, Logistic Regression, Support Vector Machine (SVM), and Random Forest [5]. Ultimately, the Random Forest algorithm with Z-score normalization was selected.

For each algorithm, a set of possible values for its hyperparameters was chosen. A grid search [6] with 5-Folds cross-validation [7] was then performed to determine the optimal hyperparameters. Thanks to the pipeline, each fold had separate normalization for its training and testing. The trained model with the best cross-validation result was obtained at the end of the process.

Then, the test set (which has not been used yet) is applied to the newly trained model to acquire the values of its accuracy, precision, recall, and F1 measure [8, 9].

By repeating the process described in the two previous paragraphs for different classification algorithms and normalization forms, the best combination was obtained.

Finally, the impacts of removing certain sets of features from the dataset were evaluated to determine which set should compose the final model. This evaluation was performed by analyzing the dataset and calculating the Gini Importance [15] of the characteristics.

12.3.4 USE OF THE TRAINED CLASSIFIER MODEL

To classify a site that has not been seen before using the trained model, its characteristics are extracted and analyzed. The trained model then predicts whether the site is legitimate or not.

12.4 RESULTS AND DISCUSSION

In this section, the results obtained from the application of the proposed method are presented and discussed.

12.4.1 FIRST EXPERIMENT

12.4.1.1 SELECTION OF BINARY CHARACTERISTICS

After preparing the dataset, as described in Section 3.2, its binary characteristics were analyzed to verify if they produced the expected result. Thus, Table 12.3 was generated, in which true positives correspond to the characteristics that demonstrated the expected result for phishing pages. On the other hand, false positives are the characteristics that present a value opposite to that expected on legitimate pages and should occur on malicious pages.

TABLE 12.3 Count of True Positives and False Positives for Each Binary Characteristic

Feature	True Positives	False Positives
IP in URL	81	0
shortening service	16	6
@ in the URL	90	0
// in the URL	51	0
use of hyphen	502	122
Usage of HTTPS	2,775	772
Favicon from own domain	3,247	2,224
HTTPS simulation	3	0
Sending information by email	352	428
The right mouse button is disabled	6	0
Use of popup	310	77
iframe usage	1	1

The characteristics marked in shade in Table 12.3 were removed from the dataset as they did not yield the expected result. In the case of "sending information by e-mail," the opposite outcome was observed, with more false positives than true positives. For the other shaded characteristics, only a few true positives were obtained, making them of little use in identifying the legitimacy of a web page. Consequently, these aforementioned characteristics were discarded and are no longer utilized in the study.

12.4.1.2 SELECTION OF THE TRAINING ALGORITHM

Using the features not eliminated in the previous step, 10 models were trained for each algorithm proposed in Section 3.3.

TABLE 12.4 Accuracy of the Best-Trained Model for Each Algorithm

Algorithm	Random Forest	Decision Trees	Logistic Regression	Support Vector Machine
accuracy	95.03%	87.13%	78.6%	45.08%

As detailed in Table 12.4, greater accuracy was achieved with the Random Forest algorithm, which was used in all subsequent stages of the work.

12.4.1.3 ANALYSIS OF THE IMPORTANCE OF CHARACTERISTICS

For the best model obtained in the previous step, the importance of each characteristic was calculated, as shown in Table 12.5.

TABLE 12.5 The importance of Gini, in descending order, of the characteristics of the best trained model

Feature	Importance
URL Size	38.4%
Page Rank	12.5%
Open doors	10.7%
Alexa ranking	10.4%
certificate age	4.8%
Usage of HTTPS	4.7%
Domain registration time	4.3%
number of sub domains	3.9%
Domain age	3.4%
Number of redirects	1.9%
Anchor Analysis	1.6%
Number of own links	1.4%
Favicon from own domain	0.5%
use of hyphen	0.3%
Use of popup	0.3%
// in the URL	0.2%
IP in URL	0.1%
@ in the URL	0%

According to Table 12.5, the characteristics related to the length of the URL showed much greater importance than the others, accounting for 38.5%. There is a significant discrepancy even in relation to the second most important characteristic, which has an importance of 12.6%. Additionally, it represents more than a third of the total importance of the 18 characteristics.

Therefore, it can be concluded that there is bias. Consequently, in practice, long legitimate URLs analyzed by this model are erroneously classified as phishing, regardless of the value of the other characteristics.

It has been identified that the bias stems from the generated dataset, where several phishing URLs contain paths unlike legitimate URLs. For instance, an example address with a path is https://atagucsea.com/wp-content/themes/twentyfifteen/cloud9/gucemail/. Without any path, it would be https://atagucsea.com. Therefore, the current conjecture of the legitimate part of the dataset does not accurately represent a real scenario, where Internet users navigate through different paths of a given page.

12.4.2 SECOND EXPERIMENT

12.4.2.1 DATASET BIAS CORRECTION

In order to mitigate the bias reported in the previous experiment, a new dataset was generated with legitimate URLs containing paths. This was done through a search for links in the HTML of the pages. As a result, the average size of these URLs increased, and consequently, the value of the associated characteristic.

12.4.2.2 SELECTION OF NON-BINARY CHARACTERISTICS

Next, the average of the values of the non-binary characteristics of the dataset was calculated. This allowed for the identification of characteristics with results outside the expected range, similar to what occurred with the binary characteristics in the previous experiment.

The characteristics shaded in Table 12.6 presented a close average value for phishing and legitimate sites and, therefore, are not able to provide evidence of the origin of the page. Thus, they were removed from the dataset and were no longer used in the work.

TABLE 12.6 Average of Non-binary Characteristic Values for Phishing Sites and Legitimate Sites

Feature	Phishing Sites	Legitimate Sites
URL size	64.74	47.11
Number of subdomains	0.28	0.28
Certificate age (days)	120.99	387.01
Domain registration time (days)	427	913.93
Open doors	1.46	0.16
Anchor analysis	42.43	41.98
Redirects	0.37	0.49
Domain age (days)	1863.91	4292.09
Alexa ranking	1103538.47	680310.14
Page rank	0.76	3.62
Number of own links	3.94	16.44

12.4.2.3 INTENSIVE TRAINING

Using only the characteristics retained after the two selection processes (Sections 4.2.1 and 4.2.2), 500 models were trained.

In addition to evaluating the models based on their accuracy, precision, and recall, the F1 measurement was introduced and reported in Table 12.7.

TABLE 12.7 Results of the model with greater accuracy, after intensive training.

	Precision	Revocation	F1-Score	Accuracy
Legitimate	93%	96%	95%	94.38%
Phishing	95%	93%	94%	

From the precision results, information is obtained that when a sample is classified as legitimate, this classification is correct 93% of the time. Classifications like phishing are correct 95% of the time.

From the recall, 96% of legitimate websites are recognized as such. For phishing, the figure is 93%.

It is observed that the best accuracy obtained among the 500 trained models (94.38%) is lower than that obtained in the first experiment (95.03%, Table 12.4), for 30 models. However, the accuracy shown in Table 12.5 is closer to the real one, since the model is no longer biased.

To demonstrate the last statement, the amounts of the model in question were calculated. It was found that the importance of the URL size feature, originally biased, decreased from 38.5% to 18.2%, as shown in Table 12.8.

TABLE 12.8 Importance, in Descending Order, of the Model Characteristics

Feature	Importance
URL Size	18.1%
Alexa ranking	17.1%
Page Rank	15.7%
Open doors	13.2%
certificate age	9.4%
Domain age	7.6%
Usage of HTTPS	5.8%
Domain registration time	6.0%
Number of own links	4.7%
Favicon from own domain	1.2%
use of hyphen	0.6%
Use of popup	0.6%
IP in URL	0.2%
@ in the URL	0.2%
// in the URL	0.1%

Note that, although the URL size feature continues to be the most important, its discrepancy in relation to the second place has significantly decreased. In the previous experiment, the difference between the two was 25.9%, while in the present experiment, it is 1%.

On the other hand, the last ones are of little importance. In the following experiment, the impact of removing them from the dataset will be analyzed (Section 4.3.3).

12.4.3 THIRD EXPERIMENT

12.4.3.1 STANDARDIZATION

As a first step of the experiment, the normalization process was introduced. As mentioned in section 3.3, different normalizations were applied to the training, cross-validation, and testing stages. This was done to prevent data from one stage interfering with the others, which could introduce bias.

To evaluate the performance of the MinMax and Z-score algorithms, 20 models were trained for each algorithm. Additionally, the same number of models were generated without normalization for comparative purposes.

TABLE 12.9 Accuracy of the Best Model Trained with Each Normalization Algorithm

Normalization	Maximum Accuracy
MinMax	93.74%
Z-Score	93.75%
-	93.14%

Better accuracy was achieved when the normalization process was applied, and as a result, it was integrated into the training of the subsequent steps.

However, both normalization algorithms yielded the same accuracy for their best models. Therefore, it is currently not possible to exclude the use of either one.

12.4.3.2 PRINCIPAL COMPONENT ANALYSIS

In view of the characteristics with little importance presented in Table 12.6, Principal Component Analysis (PCA) [11] was performed. This procedure aims to reduce the dimensionality of the model, that is, to reduce the number of features.

20 models were trained with PCA for each normalization algorithm. For MinMax, it was possible to obtain dimensionality reduction while preserving 99% of the original variance of the characteristics. For the Z-score, the minimum variance for which there was a reduction was 97%.

TABLE 12.10 Preservation of the original variance of the characteristics, number of reduced dimensions, and accuracy

Normalization	Variance	Reduction	Maximum Accuracy
Z-score	96%	1	91.84%
MinMax	98%	2	89.37%

All values are relative to the best PCA-trained model of each normalization algorithm. For Z-score, there was a dimensionality reduction of 1, while for MinMax, the reduction was of 2 dimensions. Better accuracy was obtained for PCA with Z-score; however, the value (91.85%) is lower than that obtained without PCA (93.76%, Table 12.9), for the same amount of trained models (Table 12.11).

TABLE 12.11 Importance of the Characteristics of the Best Trained Model, in Descending Order

Feature	Importance
Alexa ranking	19.4%
URL Size	19.1%
Open doors	13.1%
Page Rank	12.7%
certificate age	9.2%
Domain age	7.3%
Domain registration time	6.4%
Usage of HTTPS	5.1%
Number of own links	4.5%
random value	1.1%
Favicon from own domain	0.9%
use of hyphen	0.5%
Use of popup	0.3%
IP in URL	0.2%
@ in the URL	0.2%
// in the URL	0.2%

It is concluded, therefore, that the PCA technique did not prove to be a viable option for feature selection. Thus, an alternative method was studied and described in the following section.

12.4.3.3 ANALYSIS OF CHARACTERISTICS WITH LOW IMPORTANCE

Feature X was introduced (Section 3.1) and was not used until then. This characteristic randomly assigns a value of 0 or 1 to each sample in the dataset. Therefore, we will utilize it during training to identify characteristics that are less significant than yours. It is understood that these characteristics, at least individually, have minimal consistent impact on the identification of the origin of websites, as they are less important than a randomly assigned characteristic.

Training a model with characteristic x, the following values were obtained:

Analyzing Table 12.11, the characteristics of low importance are identified (below the random characteristic): "// in the URL, @ in the URL, IP in the URL, Use of pop-up, Use of hyphen, Favicon from the domain itself."

Having accomplished its purpose, the random characteristic x is no longer used in any subsequent step.

Then, the removal of each of the unimportant features was tested separately, as well as the removal of all of them concomitantly. For this, 5 model structures were created; each with a specific dataset separation into training, testing, and cross-validation. In addition, they have a specific Random Forest building. Thus, for each combination of normalization and characteristics, it was possible to create 5 models that were consistent across experiments.

The sets of features and normalizations marked in the shade in Figure 12.1 showed the best results. In particular, the use of MinMax without feature removal and with pop-up feature removal achieved the best accuracy (92.3%). Next, 100 models were trained for each set highlighted in Table 12.12.

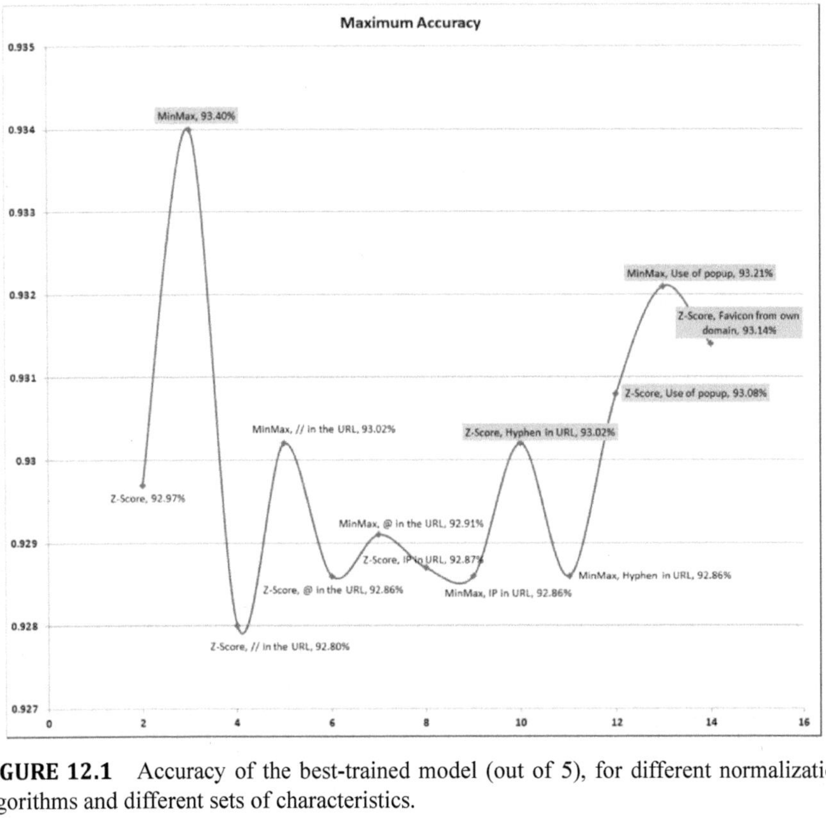

FIGURE 12.1 Accuracy of the best-trained model (out of 5), for different normalization algorithms and different sets of characteristics.

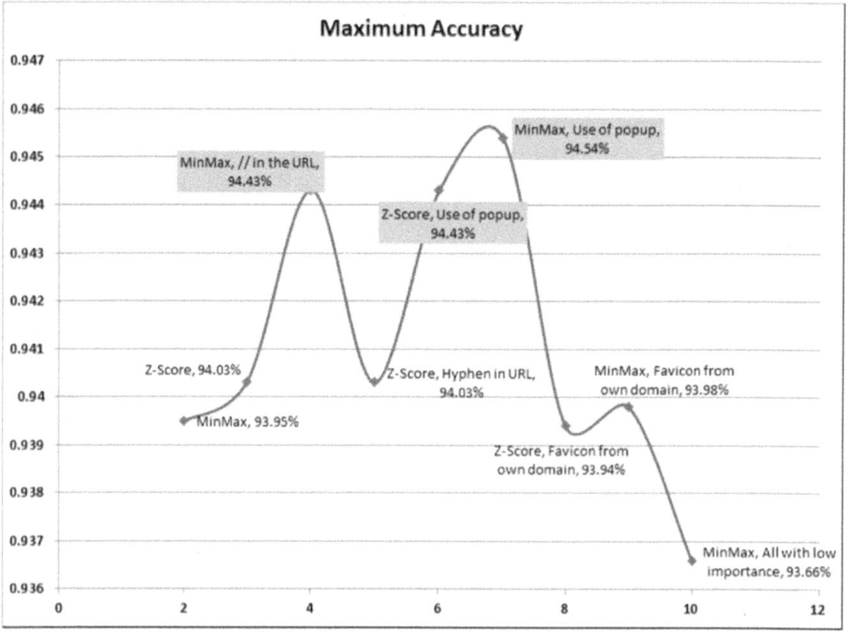

FIGURE 12.2 Accuracy of the best-trained model (out of 100), for different normalization algorithms and different sets of characteristics.

In Figure 12.2, the sets of characteristics and normalizations that presented the best accuracy are highlighted in shaded, after training 100 models for each. Specifically, these results were superior to the best-unbiased model generated thus far in the study (Table 12.7), achieving an accuracy of 94.38% after 500 training sessions.

It is also worth noting that the worst result was obtained when all 6 less important characteristics were removed. One possible explanation is that while their individual importance may be low, certain combinations involving these characteristics may be significant in identifying the origin of a website.

12.4.3.4 FINAL MODEL

Finally, 1,000 models were trained for the sets shaded in Table 12.12. Models without feature removal were also trained.

After intensive training on the 6 cases presented in Table 12.12, we achieved better accuracy (95%) for normalization by Z-score and without feature removal. Therefore, in the final model, we kept the features with low importance. This indicates that the selection of binary and non-binary features performed in the first two experiments (Sections 4.1 and 4.2) was sufficient.

TABLE 12.12 The accuracy of the best-trained model (out of 1,000), for different normalization algorithms and different sets of characteristics

Normalization	Features(s) removed	Maximum Accuracy
-	-	94.37%
MinMax	-	94.82%
Z-Score	-	95.01%
MinMax	// na URL	94.71%
Z-Score	Uso de *pop-up*	94.43%
MinMax	Uso de *pop-up*	94.54%

TABLE 12.13 Precision, Recall, and F1 Measurement of the Final Model (Shaded in Table 12.12)

	Precision	Revocation	Measure F1
Legitimate	95%	98%	96%
Phishing	97%	94%	96%

The worst accuracy in Table 12.12 also occurred for a case in which all characteristics were maintained, but without normalization, revealing the importance of this procedure.

From the results of the accuracy of the best model, we can conclude that when a sample is classified as legitimate, this classification is correct 94% of the time. Classifications as phishing are correct 96% of the time.

From the recall, 97% of legitimate websites are recognized as such. For phishing, the value is 93%.

In the end, 15 characteristics were kept, namely Alexa Ranking, URL Size, Open ports, Page Rank, Age of certificate, Domain Age, Domain registration time, Use of HTTPS, Number of own links, Favicon from the domain itself, Hyphen usage, Use of popup, IP in the URL, @ in the URL, and // in the URL.

12.4.3.5 CLASSIFIER AND PERFORMANCE

Figures 12.3 and 12.4 are examples of the classifier's response, which displays: the result of each characteristic (in red if it has a typical phishing value), the time used to classify the URL, and the prediction. Please note that in the case where the prediction is phishing (Figure 12.2), several characteristics are displayed in red in the output.

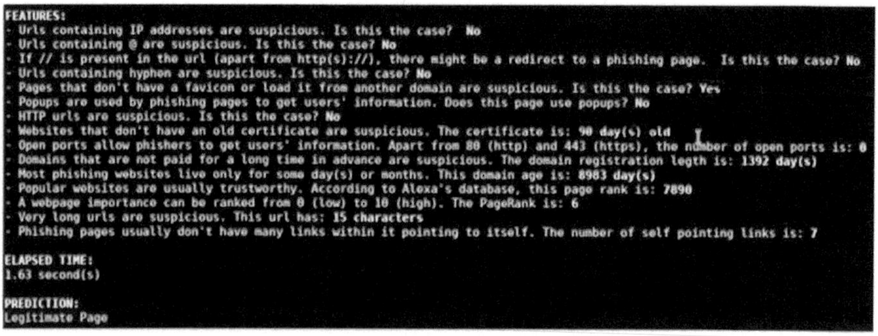

FIGURE 12.3 Classifier response, when receiving legitimate URL as input.

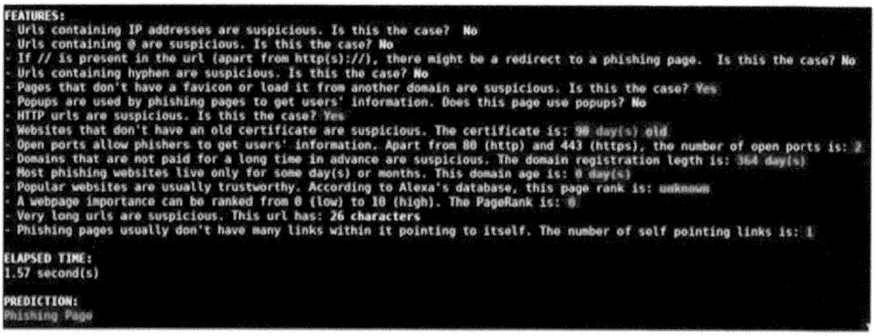

FIGURE 12.4 Classifier response, when receiving a phishing URL as input.

The program utilizes threads to calculate features that require internet access, such as verifying the age of the domain or its certificate. The time taken by these features practically contributes to the total time of the URL ranking. For broadband with a download speed of 240 Mbps, the sorting time averaged just over 1.5 seconds. However, in times of low connectivity, this value can be significantly higher, highlighting the direct dependence of a good connection for fast classification. On the other hand, the processing power required to run the classifier is always low. For an Intel i5-4200M computer (2.5 GHz, 64-bit, 4 cores), only 0.1% of the CPU was utilized. As for memory, only 128 MB were used.

12.5 CONCLUSIONS

In this work, a phishing website classifier was developed and trained as an example. It is capable of correctly classifying a significant number of domains, achieving an accuracy and F1 measure of 95%.

With a good internet connection, the time required to rank a URL is about 1.5 seconds. Therefore, by using the tool in conjunction with browsers, for example, it would be possible to prevent fraud. Sufficient time would be available to classify the pages accessed by Internet users and, in the case of phishing, to alert them before they enter sensitive information.

In general, applications such as antivirus programs rely on public lists of phishing URLs reported by their victims to protect their users. The alternative studied here, which alerts Internet users about accessed fake websites, could also automatically report them to these lists. This would eliminate the need for victims to report malicious domains.

The developed work has thus demonstrated the effectiveness and validity of AI techniques in the field of information security, particularly in the prevention of phishing attacks.

KEYWORDS

- **machine learning**
- **pattern recognition**
- **principal component analysis**
- **short message service**
- **uniform resource locator**
- **voice over internet protocol**

REFERENCES

1. Alkawaz, M. H., Steven, S. J., & Hajamydeen, A. I., (2020). Detecting phishing website using machine learning. In: *2020 16th IEEE International Colloquium on Signal Processing & Its Applications (CSPA)* (pp. 111–114). 10.1109/CSPA48992.2020.9068728.
2. Al-Ahmadi, S., (2020). PDMLP: Phishing detection using multilayer perceptron. *Int. J. Netw. Secur. Appl. (IJNSA), 12*, 1–14.

3. Ma, D., Zhang, S., Kong, F., & Fu, Z., (2021). Malicious domain name detection based on Doc2vec and hybrid network. In: *IOP Conference Series: Earth and Environmental Science* (Vol. 693). IOP Publishing: Bristol, UK.
4. Behnke, M., Briner, N., Cullen, D., Schwerdtfeger, K., Warren, J., Basnet, R., & Doleck, T., (2021). Feature engineering and machine learning model comparison for malicious activity detection in the DNS-Over-HTTPS protocol. *IEEE Access, 9*, 129902–129916.
5. Mohammad, R. M., Thabtah, F., & Mccluskey, T. L., (2013). Predicting phishing websites based on self-structuring neural network. *Neural Computing and Applications, 25*, 443–458.
6. Mohammad, R. M., Thabtah, F., & Mccluskey, T. L., (2015). An assessment of features related to phishing websites using an automated technique. In: 7th *International Conference for Internet Technology and Secured Transactions* (pp. 492–497).
7. Von, N. J., (1928). Zur theorie der gesellschaftsspiele. *Mathematische Annalen, 100*, 295–320.
8. Kreyszig, E., (1979). *Advanced Engineering Mathematics* (4th edn., p. 880). Wiley.
9. Russel, S., Norvig, P., (2020). *Artificial Intelligence: A Modern Approach* (4th edn., pp. 657, 684, 692, 697). Pearson.
10. Shekar, B. H., & Dagnew, G., (2019). Grid search-based hyper parameter tuning and classification of microarray cancer data. In: *Second International Conference on Advanced Computational and Communication Paradigms* (ICACCP) (pp. 1–8).
11. Lachenbruch, P. A., & Mickey, M. R., (1968). Estimation of error rates in discriminant analysis. *Techno metrics, 10*, 1–12.
12. Van, R. C. J., (1979). *Information Retrieval* (2nd edn.). Butterworth-Heinemann.
13. Chinchor, N., (1992). MUC-4 evaluation metrics. *Fourth Message Understanding Conference* (pp. 22–29).
14. Louppe, G., Wehenkel, L., Sutera, A., & Geurts, P., (2013). Understanding variable importance's in forests of randomized trees. In: *26th International Conference on Neural Information Processing Systems* (Vol. 1, pp. 431–439).
15. Hotelling, H., (1933). Analysis of a complex of statistical variables into principal components. *Journal of Educational Psychology, 24*, 417–441.
16. Sahingoz, O. K., Buber, E., Demir, O., & Diri, B., (2019). Machine learning based phishing detection from URLs. *Expert Systems with Applications, 117*, 345–357.
17. Benavides, E., Fuertes, W., Sanchez, S., & Sanchez, M., (2020). Classification of phishing attack solutions by employing deep learning techniques: A systematic literature review. In: *Developments and Advances in Defense and Security* (pp. 51–64). Springer, Singapore.

CHAPTER 13

Use of Artificial Intelligence and the Thread Detection Method in Cyber Security

S. KARTHIYAYINI,[1] KARTHIKEYAN,[2] and B. AYSHA BANU[1]

[1]Department of Information Technology, Mohamed Sathak Engineering College, Tamil Nadu, India

[2]Department of CSE (AI&ML), Vardhaman College of Engineering, Hyderabad, Telangana, India

ABSTRACT

The "ordinary" cyberattacks have evolved into "intelligent" cyberattacks since cybercriminals have been unable to avoid this challenge. AI and online safety Advertise Conspire is an organization that makes a difference when it comes to identifying, describing, and thwarting cyber threats to maintain data privacy. People's increasing awareness of security issues, advancements in data innovation, improvements to police work methods, and the volume of information amassed from numerous sources have all prompted firms to implement reliable and effective Cybersecurity measures. The motivations behind these cybercriminals include political rivalry, rivals seeking an advantage and damaging the reputation of others, global data theft, and radical non-secular clusters. The majority of cyber-attacks are targeted. In the modern day, the issue of Cybersecurity has become more problematic. Information breaches, ID theft, captcha-cracking, and similar incidents are widely reported, causing concern among the public and corporations. Designing effective controls and methods and implementing them with extreme precision to address cyber-attacks and violations has proven to

Leveraging Artificial Intelligence (AI) Competencies for Next-Generation Cybersecurity Solutions.
Pethuru Raj, B. Sundaravadivazhagan, & V. Kavitha (Eds.)

be an ongoing challenge. There are various approaches to leveraging AI to enhance organizational cybersecurity. Intrusion detection systems can compare an AI-derived representation of "normal" organizational behavior to identify anomalies that indicate malicious activity resulting from a breach.

13.1 INTRODUCTION

The rapid assessment of cyber threats presents significant challenges to cybersecurity. The integration of artificial intelligence into cybersecurity systems helps mitigate the ever-evolving cybersecurity risks faced by multinational corporations. With the increasing computing power, storage capabilities, and data collection, deep learning and Intelligent Systems (AI) are being applied across various industries and applications [1]. The vast amount of data generated cannot be effectively managed by humans alone. Cybersecurity encounters numerous complex issues, including intrusion detection, privacy protection, proactive defense, anomaly detection, advanced threat detection, and more. Moreover, new variations of threats continue to emerge and spread. Artificial Intelligence differs from psychology in its focus on computation, and it differs from computer science in its emphasis on activity, thought, and perception. Businesses acknowledge the need for reliable and advanced cybersecurity measures due to the growing awareness among individuals, advancements in information technology, improvements in intelligence and law enforcement services, and the increasing volume of information gathered from various sources [2]. The internet is transforming how people work, learn, and live today. It also enables the integration of social life and the web, which increases security risks in various ways. The current focus is on developing the capability to detect cyber attacks and network threats, particularly those that have already been observed. The process of implementing cyber defensive measures and protocols to protect information, programs, servers, and system infrastructure from unauthorized access or modification is known as cybersecurity [3]. The majority of our computer systems and organizational infrastructure rely on web interfaces. In order to safeguard information, foster growth in enterprises, and maintain security, virtually all corporations, governments, and even individuals have come to depend on cybersecurity.

13.2 ARTIFICIAL INTELLIGENCE

AI with cybersecurity is a technique for analyzing multiple types of risk data and the connections between risks in your project data frameworks to identify unanticipated types of attacks. By comparing client behavior throughout an environment to that in a comparable environment, artificial intelligence (AI) security solutions are adapted to distinguish between "safe" and "malicious" behavior. This process is often referred to as "unsupervised learning" where the system makes patterns without human supervision. For some AI platforms, like Vectra, "deep learning" is another key application for detecting malicious behaviors. Cybersecurity includes IT asset inventory, risk exposure, controls effectiveness, breach risk prediction, incident response, and improved communication around Cybersecurity within the organization. With their ability to quickly analyze millions of events and identify various types of threats, AI technologies can help enhance security (Figure 13.1).

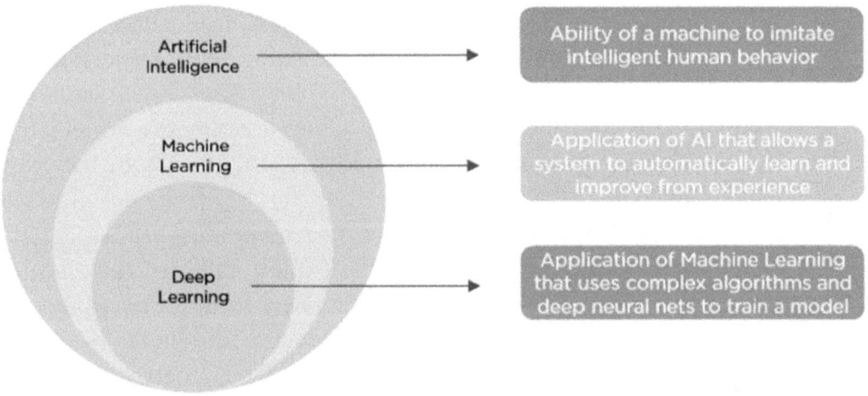

FIGURE 13.1 Connection between deep learning, machine learning, and artificial intelligence.

13.3 DETECTING NETWORK INTRUSIONS WITH AI

Accessing computer networks without the owner's consent is known as network intrusion. Both physical and logical intrusions are potential. Physical intrusion contrasts with logical intrusion, which involves an intruder attempting to enter your computer system through a network [4]. A deliberate act of using network properties or endangering the communication or its information is known as a "network intrusion."

The six steps of network intrusion are as follows:

1. **Information Gathering about the Target:** During this phase, intruders gather data on every aspect of the target. They attempt to learn about the benefits and drawbacks of their target, including email addresses, details on open-source projects, network information, etc. They also strive to understand all of the network's device functionalities in order to plan an attack effectively. Additionally, they spend time searching for weak spots that could be exploited.
2. **Inceptive Exploitation:** Activities related to intrusion are concerning at this phase. Networks can be exploited by infecting websites that the victim frequently visits. There are additional methods to jeopardize the network and gain more control over it, such as through SQL injection, phishing, and watering hole attacks. Intruders patiently execute these exploitations to minimize the chances of detection.
3. **Establish Persistence:** Intruders continue to engage in nefarious behavior while avoiding the victim's awareness. Undetected exploitation operations are rapidly increasing, including script inspection and key discovery.
4. **Malware Installation:** This stage involves creating the initial work by introducing malware into the network, ranging from the least dangerous to the most dangerous.
5. **Penetrating Deeply into the Network:** As intruders gradually gain access to the entire network, they can exploit anything they desire. Once they achieve their goal, they disconnect from the network.
6. **Elimination of Fragments:** Before leaving the network, some hackers erase all evidence of their activity out of concern that their breach will be found.

13.4 MACHINE LEARNING

Machine learning (ML), on the other hand, is less demanding to define: It is an AI technique that assigns the task of learning to a computer, using algorithms to discover patterns within a dataset [5]. One type of ML, called deep learning, is effective at solving a variety of problems. The term "deep learning" is often used interchangeably with "neural networks" because it learns using a similar mechanism to natural neural systems. Cybersecurity groups are using ML to detect and mitigate advanced threats more efficiently in this rapidly evolving landscape. ML can make Cybersecurity less complex, more proactive, less expensive, and ultimately more effective. ML has numerous applications in Cybersecurity, including detecting cyber threats, enhancing existing antivirus programs, combating cybercrime, and improving businesses' AI capabilities, among others (Figure 13.2).

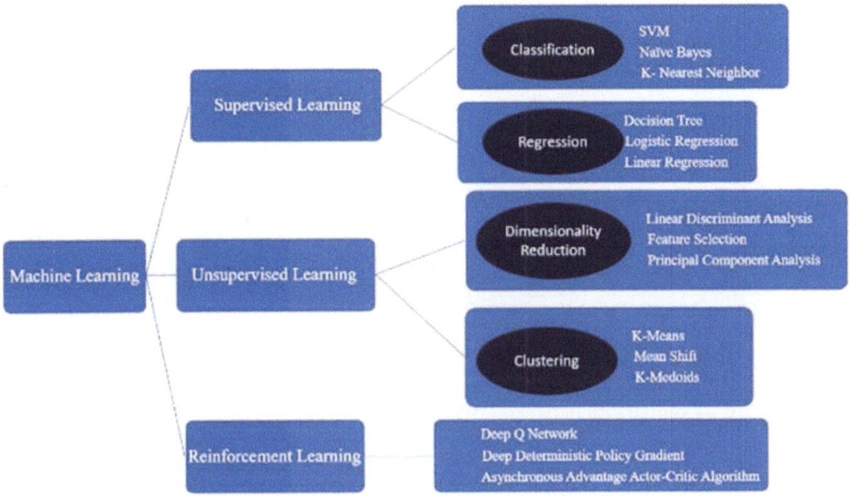

FIGURE 13.2 Methods and algorithms for machine learning.

13.5 SYSTEM FOR DETECTING INTRUSIONS

When a software application is used for intrusion detection, it is referred to as an Intrusion Detection System (IDS). Intrusion detection involves monitoring network traffic and computer events to detect unexpected occurrences. IDS, used in network security, can identify potential hazards, interruptions in services, unauthorized access, or data loss. During IDS testing and training, a graphical user interface may be provided, allowing users to access various functions. In Figure 13.3, both a Network-Based Intrusion Detection System (NIDS) and a Host-Based Intrusion Detection System (HIDS) are deployed based on the specific activity. For example, HIDS investigates host computer events, while NIDS inspects packets captured by system devices such as routers. A hybrid detection method combines the best aspects of both environments [6].

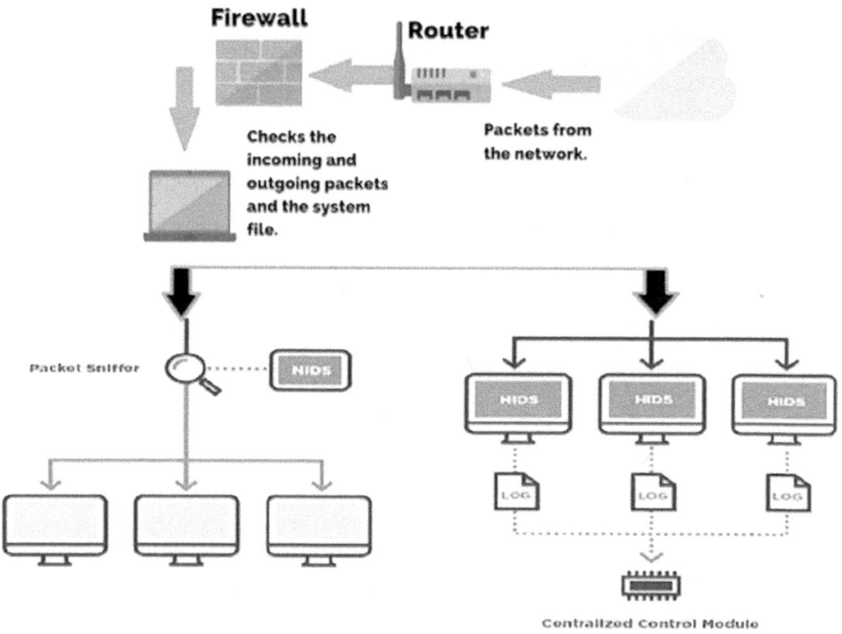

FIGURE 13.3 Comparison between HIDS and NIDS.

13.5.1 APPROACH TO AN INTRUSION DISCOVERY SYSTEM

According to Table 13.1, there are two categories of intrusion detection techniques: Incongruity Discovery Methods and Misappropriation Discovery Methods.

- **Analysis of Anomaly:** According to this model, distinct abnormal traffic can be identified from steady circulation with a high likelihood, while exact irregular traffic has a low likelihood. Algorithms for anomaly identification based on statistical learning and unsupervised learning can recognize uncommon and unrecognized attacks.
- **Detection of Misuse:** This method utilizes a signature-based approach. By leveraging recognized attack signatures, detection can occur when threats are monitored by an IDS. This supervised learning-based method can identify unethical or suspicious behaviors that could be employed to counteract similar assault behaviors.

TABLE 13.1 Compares Various Techniques to Intrusion Detection Systems

	Variance Detection	Misappropriation Detection
Detection of occurrences	Recognized and unidentified attacks	Only recognized attacks
Discovery performance	High untrue alarm	Little false alarm
Attack contextual data essential	No, rely on data for a portion of the feature design.	Yes, be contingent on information for all detections
Discovery of efficiency	Depending on the difficulty of the model	High, converse a connection to a data repository.
Apprise essential	No requirement for warnings	Yes, requires apprises

13.5.2 RECENT WORKS ON INTRUSION DISCOVERY SYSTEMS

The most effective IDS has been developed using AI and ML techniques and procedures in a number of applications that attempt to continuously recognize evolving threats and attacks. AI was used to provide greater accuracy and reduce the false alarm rate after ML served as the fundamental framework for classification.

Figure 13.4 illustrates the key contrast between the two renowned techniques, which depends on how the system acquires information. Deep learning relies on artificial neural network connections that train data without much human input, while the performance of machine learning depends on how the data is taught [7]. Here and in Table 13.2, we list other distinctions between these two domains.

Dependencies on data: This measure indicates the amount of data available. Classical machine learning, which is based on rules, performs better with smaller datasets. On the other hand, deep learning performs better with larger datasets, as more data is needed for precise clarification and comprehension.

Processing parameter is a technique used to simplify complex data and facilitate the implementation of learning algorithms. In other words, the feature procedure is utilized to categorize and identify features in new information. Unlike machine learning, where the expert must determine the necessary representations, deep learning algorithms automatically identify the representations.

Interpretability is the ability of a model to understand human language. An interpretable model can be comprehended without the use of additional techniques or instruments. However, it is challenging to define how deep

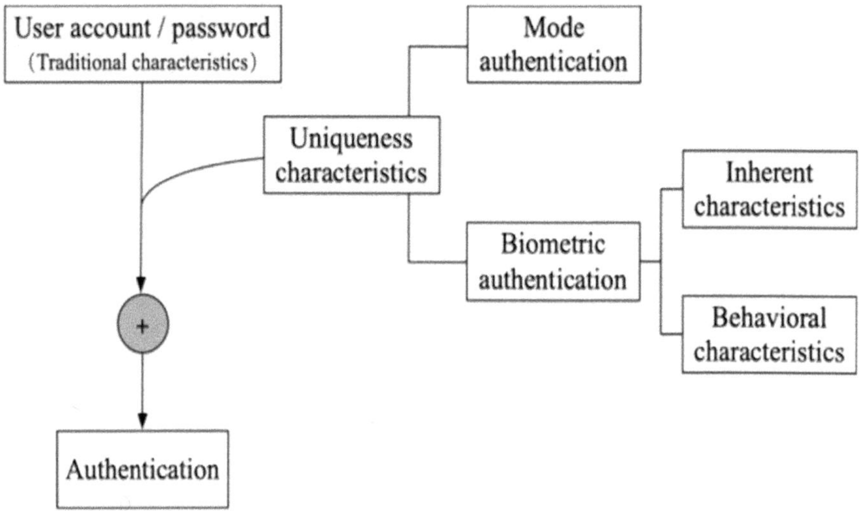

FIGURE 13.4 AI process.

learning represents neurons and how the layers cooperate, making it difficult to explain how the outcome was reached.

In traditional machine learning, the problem is divided into smaller parts that are solved one at a time until the final answer is obtained. Deep learning, on the other hand, aims to fully solve the problem.

13.6 ROLE OF ARTIFICIAL INTELLIGENCE IN CYBER SECURITY

13.6.1 CYBER SECURITY POLICIES

A few straightforward but effective rules can be followed to guarantee the availability, confidentiality, and integrity—the three key objectives of cybersecurity.

1. **Prioritize Prior Systems:** The most challenging aspect of maintaining resource availability, confidentiality, and integrity is focused on the most critical systems, which are provided with the strongest defenses. Meanwhile, less critical systems are protected using alternative methods.
2. **Various People, Various Degrees of Accessibility:** Different users should have varying levels of access to the data, and the data

TABLE 13.2 Assessment of Two Cases

	Machine Learning	**Deep learning**
Effort	Data in More	countless data
Yield	Mathematical standards	Mathematical ideas, manuscript, noises
Component necessities	Low-end machines like CPU	Machines with GPU
How it works	Various algorithms are employed to learn from historical data	To interpret relationships and features, neural networks are employed to transfer the input via processing layers.
Human Intervention	Requires human intervention a lot	Does need much human intervention
How its managed	In order to interpret relationships and features, neural networks are employed to transfer the input via processing layers. Data scientists instruct the algorithms to look at particular dataset variables.	Once the procedure begins, the algorithms will work independently to examine the data.
Dataset size	Works healthy with the small-medium data Item	Active fine through a big data Item
Number of layers	A short system with only one hidden unit, input, and output	input, output, and at least three hidden layers make up a complex system.
Features	Manual credentials of the structures	Involuntary documentation of the significant structures
Execution Time	Insufficient time or hours	Rare periods or weeks
Exercise Period	Extended period	Brief period
Conclusion	Computers make a judgment using historical data.	Machines make decisions using an artificial neural network.
Hyperparameter tuning	Competence of change is partial	It can be tuned in many ways
Implementations	Prediction and simple requests	Complex applications

TABLE 13.2 *(Continued)*

	Machine Learning	Deep learning
Problem-solving	Separated into sub-problem	Solve the whole problem, end to end
Interpretability	Some algorithms' results are simple to understand, like DT, while others, like SVM, are more complex.	Difficult to understand
Authority	Low processing power	Requires high execution influence
Procedures Number	Various	Insufficient
Exactness	Fewer correctness than DL	Higher accuracy than ML

accessible to each user should depend on their user types. This pertains to the minimum necessary permissions for a specific role. Consequently, any change in privileges will directly correspond to a modification in responsibilities.

3. **Establishment of Self-governing Défense (Protocols):** Using multiple authentication protocols instead of just one is a much better idea. The primary objective is to increase the effort required for an attacker by requiring them to complete multiple tasks to bypass several layers of defense, significantly reducing the likelihood of successful cyberattacks.
4. **Holdups:** Failures can occur, but preparing for their impact can reduce significant damage to the system, network, or individual. This is a highly effective approach that is implemented in various fields [8].

13.6.2 CONTEMPORARY WARFARE (WARFARE IN CYBERSPACE)

In the history of the world, fighting has never been a pleasant experience. Despite the fact that none of the people participating in these cruel operations ever found them to be profitable, they continue today and now utilize cutting-edge technology. There is no universally agreed-upon definition of cyber warfare; Richard A. Clarke describes it as "activities by a nation-state to access another nation's systems or networks with the intention of causing damage or disruption." These days, online is where combat trains.

Threats that could result in cyber warfare:

a. Espionage is the act of spying on people for military or political purposes.
b. Sabotage: Intentional destruction of something, sometimes for political motives.
c. Cyber propaganda is the passing of information in ways that significantly affect the interests of the public. All of these activities are now carried out online, giving rise to cybercrimes, which develop over time and eventually result in cyber terrorism. It is possible to solve the challenging problem of countering cyber terrorism by either reducing cybercrime or implementing highly protected systems. The aforementioned technologies can be used to create intelligent systems that can protect against them.

13.6.3 DETECTING THREADS

AI can be used to identify online risks and potentially harmful activity. This is an area where AI can help because traditional program frameworks can't keep up with the massive volume of current malware produced each week. AI frameworks can now recognize malware, execute design recognition, and recognize even the tiniest behaviors of ransomware attacks as soon as they reach the system by using sophisticated calculations. By sifting through publications, news, and assessments of cyber dangers, analysts can learn about an organization's vulnerabilities using natural language processing made possible by AI [10]. Additionally, ML can be used to detect any deviation from the norm or potential vulnerabilities in the midst of "regular" activity and alert users to the possibility that their critical information may have been compromised. With the right systems in place, your team won't even realize they're there; instead, you'll be able to take swift action to ensure the security of your sophisticated infrastructure.

13.6.4 PASSWORD PROTECTION AND AUTHENTICATION

In some cases, passwords serve as the barrier between our accounts and cybercriminals. While advancements like biometric verification are a step in the right direction, they have limitations and can still be vulnerable to hacking. AI analyzes and evaluates personal variables during the login process to generate a risk score for the situation. For instance, a user accessing from specific IP addresses or attempting to log in during late hours may indicate a potential risk. The AI continuously develops and refines algorithms to identify elements that indicate an attack, employing various methods to solve the problem and comparing the results with the dataset. Eventually, it discovers a set of algorithms that enable accurate prediction of threats most of the time.

13.6.6 BREACH RISK PREDICTION

AI-based security frameworks can analyze and learn from verifiable behavior and logs to predict how and where an organization is likely to be breached. In effect, businesses can use these insights to allocate resources and assets in vulnerable areas, effectively enhancing the firm's cyber resilience. Apart from detecting new types of malicious activity and programs, AI-based network monitoring tools can track user activity in real time to establish a baseline behavior profile and respond to anomalies [11].

13.6.7 BEHAVIOR ANALYTICS

Another critical utilization of AI in Cybersecurity begins with its capacity to analyze behavior. This means that ML algorithms can analyze how you use your device and web platforms to learn about and provide an example of your behavior. Nuances can incorporate everything from your typical login times and IP addresses to your typing and browsing patterns. If the AI algorithms detect any unusual activities or behavior that deviates from your normal patterns, it can flag it as suspicious or even block the user. AI can provide deep CRM analytics and, more importantly, insights on customer behavior, expectations, preferences, and desires. If implemented correctly, AI can empower companies to offer the right products to the right customers at the right time (Figure 13.5).

13.6.8 IDENTIFY CYBER ATTACKS

Distinguishing dangers, fraudulent assaults, and other pernicious exercises is one of the key areas in which Artificial Intelligence advances can add value.

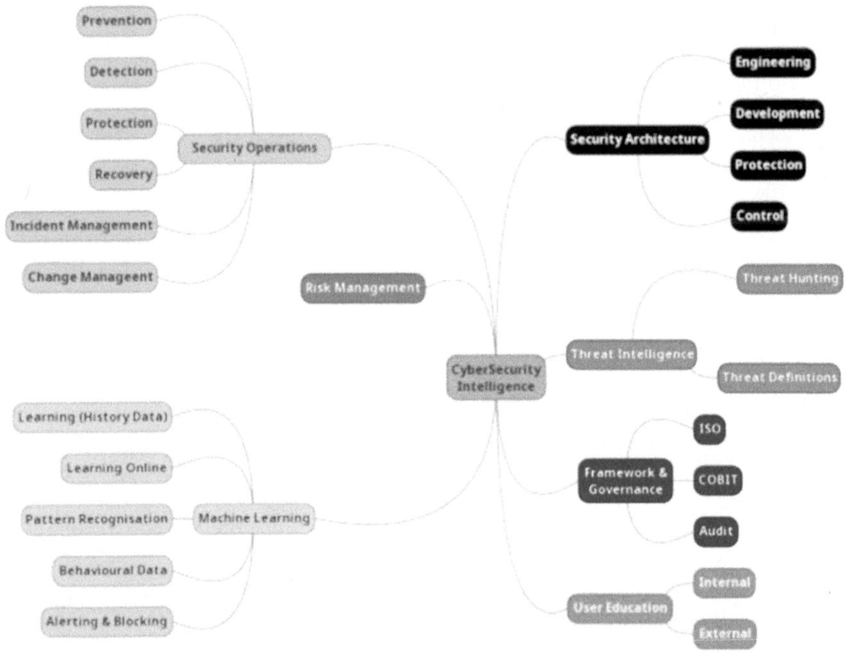

FIGURE 13.5 Detection process in AI.

Conventional systems often lack the advanced tech capabilities required to manage and monitor large volumes of data and the quality of transactions and malware being generated, making the application of AI to systems and processes the most reasonable option. AI service providers are integrating the best ML algorithms, NLP applications, and Deep Learning techniques to provide solutions that have secure algorithms to detect viruses and malware, often by running pattern recognitions in programs [12]. These solutions are being tested, with AI algorithms being trained to identify even the smallest malware and ransomware attacks within the system. In some cases, the algorithms can even predict future attacks and take necessary action for prevention.

13.6.9 CONTEMPORARY CYBERSECURITY CHALLENGES

1. The more systems we safeguard, the more of a shared responsibility cybersecurity becomes. After gaining a basic understanding of how much the cyber world impacts every entity, from individuals to organizations to national and international levels, one can observe that technologies such as Artificial Intelligence (AI), Machine Learning (ML), Blockchain, Deep Learning, Big Data Analysis, Data Science, IoT, etc., present challenges in terms of users' online security. But why do these technologies compromise security? In this field, significant technical work is being done to create goods and services for humanity. However, in the long term, economic benefits and convenience come at the expense of privacy and protection. With the increasing dependence on the Internet, sophisticated digital systems, and cloud computing, which are rapidly advancing technologies with serious security flaws, the frequency and variety of Cyberattack have also increased [13]. The lack of trustworthy and standardized computing infrastructures has made it difficult to mitigate threats. Challenges can manifest in various forms, have diverse root causes, and exist on different scales. Let's examine them from various perspectives.
2. **Innovations:** As technology advances daily, cybercriminals and other predators are constantly looking for weaknesses. The continually increasing prevalence of cybercrime will result in the creation of 3.5 million new jobs in the cybersecurity sector. Due to developments in new technology, security needs have increased by an astounding 350 times in the last five years.
3. In addition to technologies, one of the obstacles to cybersecurity is users on a large scale. Users' inclination to act irresponsibly and,

occasionally, their ignorance of risks associated with the resources they utilize are some common causes of this. Furthermore, we cannot ignore some users' involvement in criminal behavior as well as a few negligent high-ranking individuals. Many times, a single user click is all it takes to start a tragedy.

4. **Financial Outlay:** Technological, product, and service developments are simply one factor to consider when assessing the cybersecurity landscape. Too often, we overlook another important factor, which is the price of these advancements. People and companies tend to invest in cybersecurity only when they experience online attacks. The antiquated strategy of considering future risks when investing in cybersecurity attracts cyber criminals.

13.7 AI APPEARS TO SAVE

It seems strange to continue reading stories with headings like "What Has AI Become?" given the increasing importance of intelligent robots in various aspects of our lives, including the military, medical, economic, financial, and political fields. This is a scenario that Turing had predicted would occur: machine intelligence spreading throughout our information-based economy and becoming so familiar and well-integrated that it goes unnoticed.

As a result of the growing usage of mobile, cloud, and networked computers, cyberattacks have increased in size and variety. With numerous interconnected devices, fraudsters have more access points to exploit. Moreover, these access points are often insecure. The expansion of the Internet of Things (IoT) has led to a significant rise in cyberattacks. The news and media frequently discuss cyber fraud [14].

It is necessary to extract threat attributes from threats, encode threat attributes into software to detect threats, and identify threats using conventional cybersecurity procedures. Modern cyberattacks are far more sophisticated than earlier methods. Cybersecurity and AI have never before had any connection. But as time goes on, their differences get fuzzier. An excellent illustration of the connection between AI and cybersecurity is the Completely Automated Public Turing Test to Tell Computers and Humans Apart (CAPTCHA). The user must enter the letter while it is hidden or otherwise distorted in this test. Conventional cybersecurity techniques are commonly referred to as "Signature-based procedures." After a quick review of signature-based strategies, our discussion will center on ML-based techniques.

13.7.1 TECHNIQUES USING SIGNATURES

Information security methods known as "signature-based techniques" involve comparing a malware instance's signature, which is a byte sequence of code, to a database of malware signatures. This allows for the recognition of malware or cyberattacks. The database contains "blacklists" of programs that are believed to be harmful. Signature-based techniques rely on the idea that malicious software can be identified using these signatures, also known as malicious patterns. However, this approach fails when new malware or attacks are discovered that do not have recognized patterns or signatures. Unfortunately, the current environment does not support these signature detection methods. While still useful at the novice level, signature-based detection has several drawbacks:

1. **Vulnerable to Circumvention:** They are well-known because signature patterns are used to construct signatures for malware or attacks. They can be easily deceived by hackers who employ techniques such as code reordering and adding no-ops (obfuscation).
2. **Zero-Day Attacks:** Signature-based malware detection systems are created using existing malware as a foundation. However, they are unable to identify new and unidentified malware or even known malware variants. As a result, they lack accurate signatures and are ineffective in recognizing polymorphic malware. Therefore, signature-based detection does not provide security against zero-day vulnerabilities. Each strain of malware requires a new signature in a signature-based detector, resulting in an exponential growth of the signature library.

13.8 AI-BASED TOOL IN CYBER SECURITY

- **CyberSecTK:** It is an available Python library for creating the toolkit required for preparing and quality extraction of cyber-security-related information. It aims to bridge the gap between strategies and cyber security. It may be a suite of program modules, data sets, and instructional exercises supporting investigation and education in cyber security and defense required for cyber-security experts to develop a strategy from scratch.
- **Cognito by Vectra:** It is an AI system that detects and responds to attacks within the cloud, data center, IoT, and enterprise networks. Some of the benefits of using the Vectra Cognito platform include

automated threat detection, empowering threat hunters, providing visibility across entire deployments, and more.

- **Defplorex:** It could be an ML toolkit for large-scale e-crime forensics. It could be an adaptable toolkit based on open-source libraries for analyzing millions of compromised web pages. It utilizes ML and visualization techniques to transform unstructured data into meaningful high-level representations.
- **Sophos' Intercept X tool:** It may be a cyber-security tool that coordinates with a deep learning neural network. This tool works by transforming endpoint security from a reactive to a proactive approach, providing protection against both known and previously unseen threats. Sophos Intercept X employs a comprehensive defense-in-depth strategy for endpoint security, rather than relying on a single security method. The features of this tool include implementing data execution prevention, stack pivoting, and heap spraying, among others.

13.9 THE SCOPE AND FUTURE ASPECTS

Researchers predict that by 2023, almost all new software applications and products will have AI capabilities. This will inevitably alter how we work, live, and conduct business. Despite being in its infancy, AI has already demonstrated its limitless potential for carrying out tasks precisely and successfully in various areas, including healthcare, education, manufacturing, retail, and cybersecurity.

Like everything else, AI has two sides. Concerns have been raised about the potential harmful applications of AI. However, discussions on AI still emphasize its benefits. According to research published in The Guardian, "we foresee that the increasing use of AI systems will contribute to the expansion of existing risks, the introduction of new threats, and a shift in the conventional nature of threats."

Undoubtedly, AI has the potential to significantly enhance cybersecurity if deployed and educated properly. It can provide real-time defense against cyberattacks while using fewer resources. As cyber threats constantly evolve, machine learning techniques can quickly identify new patterns that are challenging for human analysts to gather and evaluate. Human analysts can then focus on analyzing the results and developing innovative tactics to proactively combat criminals, thanks to the deep analysis capabilities of machine learning. The use of deep learning and machine learning in defense systems will undoubtedly raise the bar for cybersecurity.

13.10 CONCLUSION

The development of AI has become a necessary innovation for advancing the efforts of human data security groups. Since individuals are no longer able to adequately secure the active project assault surface, AI provides much-needed inspection and risk-distinguishing evidence that can be used by cybersecurity specialists to reduce breach risk and advance security posture. AI is capable of identifying cybersecurity risks and alerting experts to the need for quick action to address them. In light of the current environment, the use of AI is thus developing in several data innovation fields, such as information security, software testing, and cybersecurity. Utilization of artificial intelligence tools in cybersecurity is diverse and promising. However, implementing all the clusters of rebellious, noteworthy speculations and time may be required. Nevertheless, very few systems within the world require a full stack of protections as not all computer programs are debilitated equally, and not all computer programs compromise sensitive data.

KEYWORDS

- **artificial intelligence**
- **completely automated public Turing test to tell computers and humans apart**
- **host-based intrusion detection system**
- **intrusion detection system**
- **machine learning**
- **network-based intrusion detection system**

REFERENCES

1. Lalitha, R., (2021). AI vs. machine learning vs. deep learning. *Artificial Intelligence (AI)*, 73–82. https://doi.org/10.1201/9781003005629-3.

2. Huang, J., (2021). Construction of evaluation system platform of university work in data mining technology. *Cyber Security Intelligence and Analytics*, 759–762. https://doi.org/10.1007/978-3-030-69999-4_104.
3. Kirti, R. B., Harsh, S., & Neha, K., (2020). *Chapter 79 The Role of Artificial Intelligence in Cyber Security*. IGI Global.
4. Putra, A. S., & Surantha, N., (2019). Internal threat defense using network access control and intrusion prevention system. *International Journal of Advanced Computer Science and Applications, 10*(9). https://doi.org/10.14569/ijacsa.2019.0100948.
5. Yang, H., (2022). Transforming pharma with data science, AI, and machine learning. *Data Science, AI, and Machine Learning in Drug Development*, 1–17. https://doi.org/10.1201/9781003150886-1.
6. Kerri, J., (2019). *Polymer-COF Hybrid Combines the Best of Both Worlds* (Vol. 97, No. 29, pp. 139–149). C&EN Global Enterprise. https://doi.org/10.1021/cen-09729-scicon2.
7. Ghavami, Peter (2020). Machine learning, deep learning—Artificial neural networks, 2019. *Big Data Analytics Methods*, 155–184. https://doi.org/10.1515/9781547401567-009.
8. Vargas, J., Goel, A., Stoner, T., & Fowler, K., (2019). Heart failure hospitalist can reduce length of stay, can reduce readmission, and can reduce cost. *Journal of Cardiac Failure, 25*(8), S121. https://doi.org/10.1016/j.cardfail.2019.07.345.
9. Ramegowda, D., & Lin, M., (2021). Can learning-based hybrid DVFS technique adapt to different Linux embedded platforms? In: *2021 IEEE Smart World, Ubiquitous Intelligence & Computing, Advanced &; Trusted Computing, Scalable Computing &; Communications, Internet of People and Smart City Innovation (Smart World/SCALCOM/UIC/ATC/IOP/SCI)*. https://doi.org/10.1109/swc50871.2021.00032.
10. Temechu G. Zewdie (2020). IoT security and the role of Ai/Ml to combat emerging cyber threats in cloud computing environment, (2020). *Issues in Information Systems*. https://doi.org/10.48009/4_iis_2020_253-263.
11. General Chairs, (2021). *2021 IEEE Real-Time Systems Symposium (RTSS)*. https://doi.org/10.1109/rtss52674.2021.00005.
12. Narayan, V., & Shaju, B., (2022). Malware and anomaly detection using machine learning and deep learning methods. *Research Anthology on Machine Learning Techniques, Methods, and Applications*, 149–176. https://doi.org/10.4018/978-1-6684-6291-1.ch010.
13. Wei, X., & Xiu, L., (2020). An all-digital highly programmable TAF-DPS based true random number generator working on principles of frequency-mixing and frequency-tracking. In: *2020 IEEE International Symposium on Circuits and Systems (ISCAS)*. https://doi.org/10.1109/iscas45731.2020.9180408.
14. Djenna, A., & Eddine, S. D., (2018). Cyber-attacks classification in IoT-based-healthcare infrastructure. In: *2018 2nd Cyber Security in Networking Conference (CSNet)*. https://doi.org/10.1109/csnet.2018.8602974.
15. Asmaa, H., Teddy, S. G., Mohamed, H. H., Murad, H., Mira, K., & Robiah, A., (2022). Machine learning and deep learning approaches for cyber security: A review. *IEEE Access*.

CHAPTER 14

The Impact of AI on Cybersecurity in Sustainable Universal Healthcare: Current Developments, Issues, and Future Prospects

A. KALAIVANI[1] and M. JAITHOON BIBI[2]

[1]*Department of Computer Technology, Nallamuthu Gounder Mahalingam College, Pollachi, Tamil Nadu, India*

[2]*Department of Computer Technology, Sri Ramakrishna College of Arts and Science, Coimbatore, Tamil Nadu, India*

14.1 INTRODUCTION

Cyber security is the technique used to prevent any unwanted access to computer systems, networks, and applications. Artificial Intelligence (AI) is now being used by more complex cyber attacks to circumvent cyber defenses. AI has often been used to continually manage and secure health system Internet-of-Things (IoT) sensors and Networked Embedded Infrastructure gadgets as they interface with and disassociate via health centers [1]. The Cyber-Physical System (CPS) is an adaptive system made up of physical and cybernetic parts that is operated and managed by an AI system.

Healthcare is essential for a successful life. The widespread perception is that a smart health sector could be the answer to making healthcare delivery and access simpler and more affordable. A substantial amount of research is being done to incorporate the application of smart healthcare for maintaining patients' wellness with specific chronic medical conditions [2]. It is made up of many different sensors, accelerometers, and wireless devices to gain a comprehensive view of different populations [3]. In addition to these,

Leveraging Artificial Intelligence (AI) Competencies for Next-Generation Cybersecurity Solutions.
Pethuru Raj, B. Sundaravadivazhagan, & V. Kavitha (Eds.)

numerous other types of equipment are utilized to monitor and analyze the vast amounts of health records. The main four aspects of the ecosystem are the end users (patients or medical professionals), sensors that gather and transfer data, the communication system over which the data transfers, and the application [4]. To ensure the success of this smart healthcare ecosystem, many people involved in the medical field are required. Despite the advantages, the platform as a whole is still developing and faces a number of challenges. Numerous research projects have been carried out to address the main difficulties encountered in the ecosystems for smart healthcare.

The healthcare delivery system is becoming increasingly automated as a result of the usage of smart medical equipment, which makes it a favorable field for numerous cyberattacks. To highlight potential assaults and abnormalities and their tracking with machines, a modest survey research on deep learning-based security systems (IDS) was conducted [5]. According to this, a different research article focuses on the security flaws of distinct architectural layers [6]. To illustrate the security concerns from node to ecosystem management of privacy, trust, and accountability, a systematic study technique was used [7]. Additionally, these smart gadgets transmit data to the cloud through a third-party consultant. The issues associated with the data stored in the cloud are both technical and non-technical. The main issues with electronic healthcare records (EHR), the clouds, and big data for healthcare were identified by Yuan et al. [8]. The issues in the areas of pervasive healthcare safety and privacy, e-health cloud, and healthcare big data have been the subject of several research papers [7, 9, 10]. However, the sustainability of this ecosystem will ultimately depend on a smart healthcare ecosystem made up of multiple involved parties and addressing the major concerns of each contributor's opinions.

IoT technology is playing a significant role in healthcare in the recent smart health scientific breakthrough because of its capability to predict, prevent, and carefully control the spread of viral illnesses such as the coronavirus.

By leveraging artificial intelligence (AI) technology and offering a better perception of wellness data to allow for more individualized health coverage, Intelligence Digitalization has grown in popularity based on IoT medical practices. In addition to providing real-time decisions, it can handle robust processing of massive data from internet of things sensing devices. Establishing a secure and affordable healthcare system has received relatively little attention, despite the impressive strides made in research on intelligence and data protection for patient monitoring. Many key aspects of our healthcare system could undergo a revolution. The success can be attributed to the AI-driven Internet of Things (AI-IoT). Figure 14.1 showcases automation innovation for safe, long-term preventative screening.

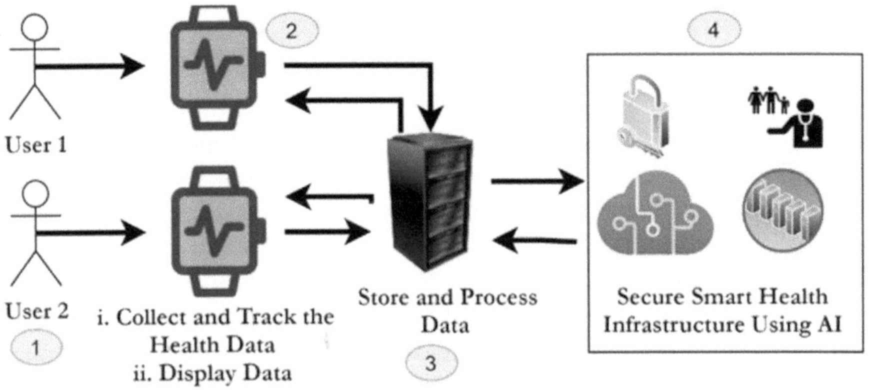

FIGURE 14.1 Automation innovation for safe, long-term preventative screening.

The importance of sustainable transition in health systems, population safety, and medical fields has been acknowledged by numerous organizations [11]. The National Information Science Education and Domestic Production Strategic Framework has been launched by the Intelligent Global Communications Investigation and Advancement initiative. This approach has made it clear how important it is to work together in the domains of public health, behavioral and social science, architecture, mathematics, and analytics to explore the crucial creativity needed to improve the quality of medical system services [12]. This incorporation is now possible thanks to recent exciting advances in pattern recognition, cognitive computing, reinforcement learning, somewhat improved cloud platforms, and the accessibility of fresh data samples.

In order to increase the accuracy of data implications, revolutionary techniques can aid in the creation of software applications for analyzing multidimensional and multiphase medical and personal health information. Spontaneously and logically gathering, joining, analyzing, and understanding data from public sources, strategies, and schemes, data science primarily focuses on technical and scientific advances made by interdisciplinary teams. Additionally, this comprehensive and intelligent data collection will aid in improving healthcare services. Among the challenges are several issues with data collection, coordination, synthesis, and representation from multiple sensory platforms, data management, and other consumer as well as medical devices. Several fundamental issues, such as standardized data interoperability, interoperability and reuse, data preprocessing, efficiency, unpredictability quantitative determination, high reliability, simulation models and analysis, confidentiality, together with relevant conceptual and computer security, are at the heart of these challenges.

EHR data can be used to create predictive results with uncertainty for testing and validation, as well as to improve the predictability of model development and simulation techniques. This could be the subject of comprehensive research [12]. According to the Medical Council, "Internet and telecommunications services in healthcare, including disease monitoring and supervision, education, and investigation" are important. Scientists also believe that "Smart Healthcare" is a combination of data and applications that promote health initiatives, online healthcare management, and associated AI and data web harvesting. These methods can increase security and accuracy in health informatics and personalized healthcare. Despite the growing popularity of deep learning, investigators have only occasionally used this method to forecast results from multimodal health data. They prefer to use an algorithm based on statistical techniques and multiple regressions to create healthcare predictions [13–15]. The authors of this chapter emphasize the importance and challenges of using AI to address cybercrime problems that endanger the privacy, accuracy, and accessibility of data for affected health services during current trends.

14.2 USING CRYPTOGRAPHY TO MONITOR REMOTE WELLBEING HEALTH

Medicaid organizations state that cryptography is a top priority in their review. The healthcare sector has been and continues to be one of the most frequently targeted sectors by hackers. Unfortunately, the issue is only getting worse. Breaches do occur due to the high value of digital evidence. Additionally, cybersecurity threats are constantly evolving. Hackers are now employing more sophisticated and automated attacks, utilizing pattern recognition and intelligence systems. By leveraging modern social engineering techniques, cybercriminals can piece together information from social media to create highly accurate profiles of individuals or the data that healthcare organizations possess. The ongoing digital transformation in the healthcare sector is further complicating Cyber security requirements, resulting in increased complexity, greater usage of connected devices, a diminishing perimeter, and all of this happening at an accelerating pace.

14.2.1 MEDICAID NETWORK CRYPTOGRAPHY IN THE PHASE OF DIGITAL TRANSFORMATION

Organizations in the healthcare industry are implementing cutting-edge technologies, including cloud, mobile, IoT, big data, and sophisticated analytics. These new solutions add a new layer of complexity that necessitates a different approach to security. The majority of businesses have long distinguished between resources and users inside and outside the network. However, when healthcare organizations adopt digital transformation, they become more receptive to interactions with other relevant parties in integrated care/collaborative care environments, such as patients, partners, employees, government health insurance authorities, and other providers. Users can now access resources from anywhere, not just inside the network perimeter. Surgeons and primary care doctors from a separate network may share information. Hospitals may continue to remotely evaluate patients even after they are released.

Business workflows, which continually evolve with technologies, are inherently associated with risk. Healthcare organizations must incorporate security into their procedures from the beginning. Security should be a core business value that guides every decision made by healthcare organizations. As healthcare organizations modernize their operations digitally, it becomes crucial to build a secure connected ecosystem. To enable scalability, acceleration, intelligence, and automation in this ecosystem, the four primary security disciplines of digital certificates, intrusion prevention, vulnerability assessment, and authentication and authorization must be elevated to a new level. The vision and mission of the firm and the broader business virtual world should align with this plan.

Distributed network security should consist of the following key components:

- **Terminal security:** To control vulnerabilities on a per-device basis.
- **Verification and validation:** To verify users' identities and restrict access, the solution should include user authentication capabilities and granular security controls. These features will ensure that users have the appropriate level of access to the required information.
- **File encryption exchange:** Protocols and the usage of encryption should protect data transmission.
- **Common elements:** The solution should use network segmentation to isolate individual systems and containerization to prevent threats

from spreading rapidly across systems once they have gained access to the network.

- **Trusted administration:** Healthcare institutions require a risk management strategy that considers the possibility of users and resources residing outside the network's perimeter. Knowing what constitutes regular and abnormal conduct on the network is one method to build trust in such a setting. A baseline for typical network activity can be created using AI-powered technologies for identification and authorization management. These tools assess each user and device, as well as their application, security, and quality of service requirements, to define a standard baseline behavior. AI can then immediately detect any unusual occurrences or behaviors, such as a surveillance system accessing an EMR or EHR, and provide analytics to pinpoint the reason for the change.

An intelligent network administration system, using a software-defined framework, can provide the necessary speed to keep up with the digital transition.

The healthcare sector is one of the most vulnerable sectors to cyber security threats. The vulnerability of the global healthcare system to cyber attacks has increased over time. Cyber-security groups have observed a significant rise in cyber-attacks since the beginning of the modern era. The healthcare system, particularly nursing homes, has always been a primary target for cyber attacks. Recent incidents involving prominent physicians and health systems have exposed security issues in well-known healthcare companies.

Despite this, hackers have not adequately secured this critical sector. The healthcare industry remains vulnerable, and there has been a sharp increase in cyber-attacks from various perpetrators in recent times [16]. Figure 14.2 illustrates the smarter wellness data handling and protection monitoring chain.

Since they deal with confidential patient or recipient information and protect patient data, security and confidentiality are essential in the healthcare industry. The use of cutting-edge innovations by healthcare providers has increased over the past few decades. Automation systems, pattern recognition techniques to safeguard patient health records, cloud data storage, enhanced medical devices, etc., are some examples of these advancements. These technological developments have made healthcare providers' jobs much easier and resulted in a paperless workplace.

However, as a result, the risk of cyber attacks has increased. Patient care information is often not adequately protected by security measures, and medical professionals are frequently unaware of the potential vulnerabilities. Healthcare information technology (IT) is often so weak that it may take days or even weeks to detect a Cyberattack. Healthcare professionals may continue to use a compromised system without realizing the presence of attacks. This can result in significant financial costs and impact millions of people annually [17]. The healthcare sector has experienced numerous cyber attacks in recent years, with some of the most significant ones being:

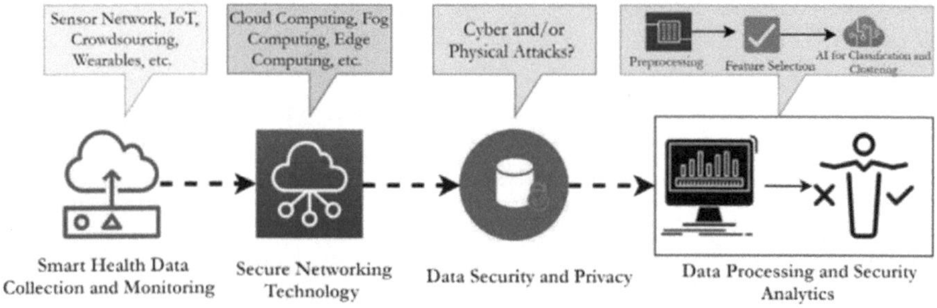

FIGURE 14.2 Advanced healthcare approach.

14.2.1.1 CYBER-ATTACK ON THE UNIVERSITY OF VERMONT HEALTH SERVICE SYSTEM

The UVM Medical Center was closed on October 28, 2020, following the discovery of a cyberattack. The institution was losing nearly $2 million every day, which included lost income from postponed services and costs associated with the attack's recovery. For approximately 40 days, primary care, including digital health records, was disrupted. More than 5,000 PCs were compromised since they were all connected to the same network. Around 300 employees were unable to work in November due to this interruption. Stephen Leffler, MD, President and COO of UVM Medical Center, stated that the healthcare institution expects the total cost of the incident to exceed $63 million by the time it is resolved [18].

14.2.1.2 ATTACK ON A NEBRASKA-BASED MEDICAL NETWORK

The Department of Health expects that used network devices will remain inaccessible. Nebraska Medicine first reported the outage in September 2020. As a result of the cyber attack, numerous patient appointments had to be canceled or rescheduled, impacting the Nebraska Medicine IT system. Since Nebraska Medicine supports other Regional Health Services' EHRs, the attack also affected other computers and EHRs. From February to May 2020, a security vulnerability at Machine Equipped, a company that handles donor identification for multiple institutions, including medical facilities, compromised patient information from over 46 hospitals and medical facilities [19].

14.2.1.3 HOSPITAL UNIVERSITY OF MONTPELLIER EXPERIENCES SECURITY BREACH

Since data breaches are one of the most frequent types of cyberattacks, the healthcare sector has experienced several incidents over the past few decades. Attackers commonly use phishing emails and deceptive web links to deceive users. When a person clicks on a suspicious web link received in an email, the attacker gains access to both the account and the remote server. In March 2019, the healthcare supplier for Montpellier University School of Medicine discovered that all job logins were accessible to outsiders. An employee of the health clinic accidentally clicked on a questionable attachment on a malicious website, giving the perpetrator access to both the victim's account and the hospital network [20]. The hospital later discovered that the accessed account contained private patient data, including identity, personal information, birthplace, clinical records, and more.

14.2.1.4 INTERNAL THREAT

Healthcare providers occasionally have to deal with internal cybersecurity concerns in addition to external attacks. These internal hazards to the companies are the result of either employee misconduct or human error. According to different studies, there are three major types of insider attacks: identity theft, dishonest or hostile insiders, and incompetence or ignorance from labor or suppliers.

14.2.2 CRACKERJACK

One of the three main categories of security breaches, which according to Crackerjack, is the act of assaulting and influencing a sensor device or other device with the intent to harm a victim. Any malfunction of a medical gadget at a medical facility is extremely unpleasant and could have fatal consequences. An improper prescription could be given as a result of inaccurate diagnostic information from any medical device. Incorrectly functioning medical equipment could harm patients and result in their death rather than helping them. Crackerjack is frequently targeted, particularly to harm powerful individuals and undermine the reputation of the healthcare provider. AI can support and enhance the security of medical device and instrument manipulation.

14.3 COMPUTERIZED COMPETENCE

14.3.1 LEADING COGNITIVE COMPUTING CONTRIBUTES TO CYBER SECURITY AND HEALTH CARE SYSTEM PROTECTION

A gadget or software program may be able to understand complex data thanks to human intelligence, such as visuals, videos, texts, voice, or other noises, and improve comprehension to accomplish tasks. With AI-powered computers designed to make decisions with minimal human involvement, some people wonder if machines will soon be capable of handling the complex decisions currently entrusted to doctors. According to David B. Agus, MD, a professor of medicine and technology at the University of Southern California University School of Medicine and Viterbi College of Engineering, computer vision is already making significant changes in the field of medicine, but it is crucial to distinguish reality from science fiction.

Thanks to modern advancements in processing capacity, ML algorithms can now be used to identify potential health outcomes through predictive analytics using various heterogeneous data characteristics. Cognitive computing algorithms are employed to make decisions with minimal human involvement, statistically spotting patterns and reducing risks by analyzing sensory receptors and network services.

14.3.2 INNOVATIVE UNIVERSAL HEALTH CARE USING AUTOMATION

Remote health is fundamental to the use of cellular devices and other hand-held platforms with wireless infrastructure to enhance healthcare (m-Health). M-health includes mobile devices, medical sensors, and telephones. Numerous research studies have proven that the use of AI technology in health services considerably increases the security of patient medical evaluation. For instance, in a study by [21], the author proposed an automated mobile application to alert consumers to impending heart failure. Currently, researchers and healthcare professionals are exploring straightforward methods for creating emergency notifications. However, the current system generates a large number of false alarms.

Another m-Health approach for voice recognition for individuals with dysarthria was presented in Ref. [22]. The researcher who conducted this study demonstrated how their method can assist with audio message generation. They utilized an embedding methodology tailored for each individual to analyze the general accessibility of a document employed in a learning algorithm. Markov chain modeling is used to create the AI that calculates the uncertainties for a moving target through predetermined surroundings. According to actual research involving nine experimental groups, their methodology's character recognition accuracy is only 69%.

14.3.3 COMPUTER CRIMES SYSTEMS AND THE WIRELESS SENSOR NETWORKS IN THE AGE OF AUTOMATION

One of the main targets for hackers in the Open Internet and computer criminal procedural computer hacking is healthcare systems in hospitals and clinics. From a hospital security perspective, the most important endpoints are devices such as infusion pumps, ventilation systems, anesthesia equipment, and patient health monitoring systems. The healthcare settings are utilizing the Internet of Things (IoT), which includes mobile devices, wearables, robotics, drones, and contactless gadgets. IoT is enabling the regulation of monitoring systems.

A group of sensors and devices that interact with each other and communicate with the physical world is known as a cyber-physical system (CPS). The medical equipment used in smart medical technology forms the foundation for various CPS implementations. By enabling capacity, adaptability, scalability, robustness, safety, and usability, advancements in CPS will also facilitate the use of cryptography in critical applications within the healthcare system. The emergence of automation is challenging the assumptions of CPS-based research, leading to new research theories. Integrating with CPS opens up new research opportunities with significant societal implications, particularly in secure healthcare operations. The connection between IoT and CPS workflows with AI and smart m-Health applications is demonstrated through the integration with wireless or Wi-Fi-enabled phones.

14.4 DIGITAL SECURITY

14.4.1 AI SECURITY STRATEGY

Cognitive technologies are frequently utilized in security, specifically in the field of AI. Digital computer security tools provide a framework for action and a clear picture of the security risks facing the global health sector, supporting them in life-or-death emergencies. By assisting human system managers with automation tracking, evaluation, and perceptions of aggressive attacks, advanced technologies are anticipated to improve protection. The results of the study on the interaction between intelligence and cyberspace may drastically change how we perceive the fundamentals of cybersecurity.

Additionally, these combined findings may contribute to the innovative motivation and education of healthcare professionals regarding cybersecurity in the AI era. Cybersecurity analytics and workforce development may benefit from the expansion of current AI possibilities and resources if fundamental AI research is combined with cybersecurity research. Computer vision, robotic process automation (RPA), relies on innovations like learning algorithms, supervised learning, speech recognition, and other discoveries to make it more difficult for hostile actors to obtain networks and other crucial data. Following a number of successes, AI is now concentrating on cybercrime. MIT estimates that AI can detect over 85% of intrusions, making it a valuable defense for IoT and CPS networks, especially those utilized by the health sector.

Because AI, Machine Learning (ML), and Deep Learning (DL) overlap, these terms are often misunderstood. AI technology can assist computers in modeling human behavior. Computers can automatically learn patterns and ways to characterize data sets through a type of AI known as computational learning. Deep learning, a subtype of ML, helps computers handle the complex multi-layer issues of neural networks. When gathering and evaluating sensed information from a connected sensor node, it is advisable to utilize AI, pattern recognition, and neural learning techniques.

In Ref. [23], researchers assess the performance of 11 well-known pattern recognition and fully convolutional algorithms using six primary data sources. When comparing the performance assessment matrices of different ML models, such as precision, recall, F1-score, correctness, processing time, area under roc curve (Receiver operating characteristic curve) rating, and misclassification rate, the authors of this paper showed that Random Forest is best. Moreover, they showed that ANN and CNN offer intriguing results in comparison to other deep learning models.

14.4.2 HOW AI CAN IMPROVE INFORMATION SECURITY?

By analyzing vast data sets to improve responsiveness and sustain constrained security operations, AI is transforming information security. Automation is a crucial element of cryptography because it helps identify potential risks. AI can clean up any transmissions or data sets by extracting features and unwanted information. Additionally, the majority of security professionals currently utilize AI to understand the digital platform.

14.4.2.1 MAINTAINING NETWORKS SECURE

When it comes to network protection within the healthcare industry and preventing unauthorized access to patient information through IoT sensors or public systems, AI is capable of managing data protection laws. The three major ways to utilize AI for network security are utilizing computer vision to identify cyber hazards based on AI, utilizing AI to enhance decision-making, and utilizing AI as a tool for overseeing access controls and network design. By detecting and analyzing risks, AI can identify new attacks before they exploit network vulnerabilities. However, it is important to note that humans can become complacent and overly reliant on cognitive computing to manage the network's cybersecurity.

14.4.2.2 WAY FASTER PROCESSING TIMES

An important advantage of AI in cybersecurity is its ability to quickly detect and identify any suspicious behavior or abnormalities, thereby protecting healthcare systems from potential attacks. By investing in AI, firms can strengthen their security systems and effectively respond to threats, safeguarding their financial investments and reputation. Three crucial tactics to enhance detection and response capabilities before attacks compromise critical healthcare services include maintaining security systems, developing AI technologies, and standardizing initial response procedures. Security monitoring service providers offer external supervision of security devices and networks.

The healthcare sector is now leveraging automation and faster vulnerability detection to identify trends and abnormalities resulting from cyber attacks and ransomware assaults. Centralization is crucial when dealing with cyber attacks, as it is currently lacking in the majority of the healthcare sector.

While human experts in digital security play a vital role in addressing these situations, AI frameworks can also be beneficial by suggesting methods to enhance response strategies.

14.4.2.3 SCAM IDENTIFICATION AND AVOIDANCE

Malicious actors attempt to deliver their payload through phishing attacks, making it one of the most common security issues for individuals and businesses to safeguard their information. AI and ML can play a significant role in preventing and thwarting phishing attacks. ML in computer-based intelligence can identify and investigate over 10,000 adaptable spoofing suppliers. AI-ML also effectively filters out phishing threats from around the world. Phishing attacks can be used for spreading malware, stealing money, or pilfering user credentials, among other purposes. The majority of phishing schemes aim to steal personal data. Its understanding of phishing attempts is not limited to a specific geographical region.

14.4.2.4 SECURE AUTHORIZATION

Secure administration or identification has become a significant issue since cellular modems play such important roles in providing a range of services. The main method for security verification might be the physically visible proof, which AI utilizes to investigate different security features to distinguish an individual. A Smartphone's scanner can be used to detect a user's distinct thumbprint and facial expression, enabling a secure login. The mobile internet software scans the fingerprinting and analyzes the user's body language to assess whether the login is legitimate. The usage of AI technology also allows users to view knowledge from any machine by verifying user credentials by looking at a wide range of characteristics.

14.4.2.5 PSYCHOLOGICAL BUSINESS INTELLIGENCE

The ability of AI to assess behavior is one of the key applications in cybersecurity. This implies that ML algorithms can analyze how you use your device and online platforms to learn about and create an example of your behavior. It is fascinating to read about how AI is being used in medicine for things like genetics and chromosome studies. The behavior aspect of cybersecurity involves people. Additionally, machine behavior has a significant impact on cyber incidents. Our way of life, including how we work, play, and live, is changing as a result of AI. Predictive and behavioral analytics enable the generation of insights and the taking of appropriate actions as more and more healthcare data is gathered and analyzed via multimodal systems and medical gadgets.

In conclusion, methodologies for AI have quickly developed from being trivially specialized. This will help ethical hackers manage operations related to the detection and defense against intrusions. AI can help workers discover security problems and guide them in the right direction. AI has the potential to significantly improve a number of data achievement processes, including product testing, information security, and criminality.

14.5 COMPLICATIONS WITH SMART CYBERSECURITY

The nation's entire economic and cyber-physical security interests are primarily focused on cybersecurity. Every organization's security analysts face a variety of cybersecurity-related difficulties, including protecting sensitive federal and state information. When developing the deep review, development, and deployment of AI in cybersecurity, it is important to distinguish between short-term and long-term goals. There are various direct applications for AI in cryptography.

Currently, there are several complex solutions to serious security vulnerabilities. Users will soon witness the promising results of implementing completely new data processing concepts. Managing data for cyber threats is an important area where AI can be applied. AI-based solutions are already being used in various applications, such as internal security features of software. However, as large datasets are created for healthcare systems, AI will find a broader range of applications. While many technologies are frequently mentioned in relation to AI for cybersecurity in healthcare databases, other technologies have the potential to advance to a highly sophisticated level and produce intelligence surpassing that of humans.

14.6 AI CYBERSECURITY BREAKTHROUGHS

Cybersecurity will be impacted by the development of AI and ML technology in various ways. Any network system can be targeted by cybercriminals at any time and from any location around the world. In recent years, there have been significant technological advancements in cybersecurity applications. There are several approaches to enhancing data security for AI, such as utilizing computer vision to effectively recognize cyber threats, employing intelligent automation to prevent phishing attempts, implementing automated network security, utilizing powerful behavioral analytics, and more. Business intelligence tools enable enhanced protection, and these cutting-edge approaches have a wide range of potential applications in sectors including hospitality, education, e-commerce, and others.

The use of intelligent machines to improve the security of digital communications and collected healthcare data raises various security-related concerns. The need to safeguard AI itself becomes more urgent when using AI security solutions to counter rapidly evolving threats. It is crucial to protect ML algorithms from manipulation, corruption, and abuse, especially when relying on them to identify and combat cyber attacks. As reliance on AI for critical tasks and services increases, attackers are not only motivated to target such processes but also the potential consequences of each successful attack become more severe.

One of the major challenges is enhancing the security and safety of AI. The use of AI technology has seen a recent surge across various industries, including healthcare. The importance of governments considering cybersecurity and AI is growing. Scientists have been working on reducing the risk of attackers gaining access to private AI training data or modeling methods in healthcare systems.

One of the primary security concerns for these systems is the potential for adversaries to compromise the reliability of their decision-making processes. This can be achieved by granting adversaries complete control over an information system, enabling them to manipulate the outputs and conclusions generated by the system. An intruder may attempt to influence these decisions directly by providing false classification models or inputs to an optimization technique.

14.7 ASPECTS OF THE FUTURE

Technology is constantly evolving, making it crucial to stay up to date. In the future, integrating composite programming would be a wise decision to secure clinical information. Specialists in cryptography should carefully supervise the system as business intelligence tools are still vulnerable to attacks. Data governance and compliance with relevant regulations should receive significant attention going forward, as more privacy and safety regulations are expected. ML techniques can simplify and enhance many cybersecurity applications. With this technology, future cybersecurity workloads will be lighter, reducing human error.

Medical diagnosis also significantly reduces human errors. According to [24], medication errors, including incorrect prescriptions, can lead to 3 million premature deaths. Additionally, many more people die each year due to delayed medical attention. Healthcare systems that incorporate AI into the diagnosis process and the smart health sector may experience a decrease in these fatalities, as it can accurately diagnose patients and identify issues earlier.

14.8 CONCLUSION

With several applications, AI is a rapidly growing field. The confidentiality, integrity, and accessibility (CIA) of data for affected businesses have all been compromised as a result of recent hacks on healthcare systems. These tragedies have also shown that, even when exercising caution, it just takes one error for a business or organization to lose millions of dollars and take years to correct the issue. The advancement of new technologies has also brought attention to the need for improvements in the medical field that can increase diagnostic efficacy and accuracy. One suggested tactic is to incorporate automation into protection and health services. Many different types of cancer and a wide range of other disorders are already diagnosed using AI in the medical industry. Faster treatment as well as more effective diagnosis could result from further AI integration in the smart health sector. In the world of cybersecurity, AI is already being used to either detect risks or assist experts in locating and neutralizing them. AI will continue to be incorporated into cybersecurity, resulting in more sophisticated and reliable systems that can withstand evolving cyber threats.

CONFLICTING IDEAS

Concerning this chapter, the authors state that they have no competing claims.

KEYWORDS

- **AI-driven internet of things**
- **artificial intelligence**
- **confidentiality, integrity, and accessibility**
- **cyber-physical system**
- **internet-of-things**
- **robotic process automation**

REFERENCES

1. McGee, T. M., (2016). Evaluating the Cyber Security in the Internet of Things: Smart Home Vulnerabilities. *West Point ETD*. 6. https://core.ac.uk/download/pdf/268449217.pdf (accessed on 27 January 2024).
2. Pramanik, M. I., et al., (2017). Smart health: Big data enabled health paradigm within smart cities. *Expert Systems with Applications, 87*, 370–383.
3. Harimoorthy, K., & Thangavelu, M., (2021). Cloud-assisted Parkinson disease identification system for remote patient monitoring and diagnosis in the smart healthcare applications. *Concurrency and Computation: Practice and Experience, 33*(21), e6419.
4. Kumar, N., (2017). IoT architecture and system design for healthcare systems. In: *2017 International Conference on Smart Technologies for Smart Nation (SmartTechCon)*. IEEE.
5. Malhotra, P., et al., (2021). Internet of things: Evolution, concerns, and security challenges. *Sensors, 21*(5), 1809.
6. Sadique, K. M., Rahmani, R., & Johannesson, P., (2018). Towards security on internet of things: Applications and challenges in technology. *Procedia Computer Science, 141*, 199–206.
7. Atlam, H. F., & Wills, G. B., (2020). *IoT Security, Privacy, Safety, and Ethics, in Digital Twin Technologies and Smart Cities* (pp. 123–149). Springer.
8. Yuan, S., Rao, R. H., & Upadhyaya, S., (2015). Emerging issues for education in E-discovery for electronic health records. *Security Informatics, 4*(1), 1–7.
9. Kute, S. S., Tyagi, A. K., & Aswathy, S., (2022). Security, privacy, and trust issues in internet of things and machine learning based e-healthcare. In: *Intelligent Interactive Multimedia Systems for e-Healthcare Applications* (pp. 291–317). Springer.
10. Atiyah, R. F., & Al-Mejibli, I., (2022). Security and privacy in IoT healthcare system: A systematic review. *Journal of Al-Qadisiyah for Computer Science and Mathematics, 14*(1), 15–23.
11. Institute of Medicine (US) Committee on Assuring the Health of the Public in the 21st Century (2002). The Future of the Public's Health in the 21st Century. Washington

(DC): National Academies Press (US); 5, The Health Care Delivery System. Available from: https://www.ncbi.nlm.nih.gov/books/NBK221239/books/NBK221227/ (accessed on 27 January 2024).
12. National Science Foundation (NSF), *Smart Health and Biomedical Research in the Era of AI and Advanced Data Science*. https://www.nsf.gov/pubs/2021/nsf21530/nsf21530.htm (accessed on 27 January 2024).
13. Lin, S. H., & Chen, M. Y., (2019). Artificial intelligence in smart health: Investigation of theory and practice. *Hu Li Za Zhi., 66*(2), 7–13. Chinese. doi: 10.6224/JN.201904_66(2).02. PMID: 30924509.
14. Gopal, G., Suter-Crazzolara, C., Toldo, L., & Eberhardt, W., (2019). Digital transformation in healthcare – architectures of present and future information technologies. *Clin. Chem. Lab. Med., 57*(3), 328–335. doi: 10.1515/cclm-2018-0658. PMID: 30530878.
15. Kamel, B. M. N., Peng, G., & VoPham, T., (2019). An overview of Geo AI applications in health and healthcare. *Int. J. Health Geogr., 18*(1), 7. doi: 10.1186/s12942-019-0171-2. PMID: 31043176; PMCID: PMC6495523.
16. Mashamba-Thompson, T. P., & Crayton, E. D., (2020). Blockchain and artificial intelligence technology for novel coronavirus disease-19 self-testing. *Diagnostics (Basel), 10*(4), 198.
17. Zeina, R., Marco, A., & Abdel-Badeeh, S., (2019). Machine learning approaches in smart health. In: *8th International Congress of Information and Communication Technology, ICICT 2019;Procedia Computer Science* (Vol. 154, pp. 361–368).
18. The University of Vermont (UVM) Health Network Cyberattack. https://www.uvmhealth.org/news/uvmhn/statement-uvm-health-network-cyberattack (accessed on 27 January 2024).
19. *Nebraska Medicine Reverts to Paper Records During Computer Network Outage: 4 Details* (2020). Laura Dyrda (Twitter).
20. Susan, A., (2020). Cybersecurity and Healthcare During COVID-19. GlobalSign Blog, Available from: https://www.globalsign.com/en-in/blog/cybersecurity-healthcare-during-covid-19 (accessed on 27 January 2024).
21. Larburu, N., Artetxe, A., Escolar, V., Lozano, A., & Kerexeta, J., (2018). Artificial intelligence to prevent mobile heart failure patients decompensation in real time: Monitoring based predictive model. *Mobile Information Systems, 2018*, 11. Article ID 1546210.
22. Hawley, M. S., Cunningham, S. P., Green, P. D., et al., (2013). A voice input voice output communication aid for people with severe speech impairment. *IEEE Transactions on Neural Systems and Rehabilitation Engineering, 21*(1), 23–31.
23. Vakili, M., Ghamsari, M., & Rezaei, M., (2020). *Performance Analysis and Comparison of Machine and Deep Learning Algorithms for IoT Data Classification.* arXiv preprint arXiv: 2001.09636.
24. Anderson, J. G., & Abrahamson, K., (2017). Your health care may kill you: Medical errors. *Stud. Health Technol. Inform., 234*, 13–17. PMID: 28186008.

CHAPTER 15

Delineating the Critical Cloud Infrastructure, Application, and Data for Business Agility and Reliability

M. RENUKA DEVI,[1] V. SINDHU,[2] and S. GAVASKAR[3]

[1]*Associate Professor, School of Information Science and Engineering, Presidency University, Bangalore, India*

[2]*School of Computer Applications, Lovely Professional University, Punjab, India*

[3]*Department of Computer Applications, Bharathiar University, Coimbatore, Tamil Nadu, India*

ABSTRACT

Cloud computing has recently evolved to become the most current IT paradigm. It is constantly changing the landscape of IT, allowing for the development and deployment of operations using the services of a cloud provider. Additionally, there is a vast warehouse structure for databases and data provided by the cloud provider. The cloud offers various unique infrastructures, with many others still evolving. The focus of this work is to identify, explore, and explain current trends and developments in cloud computing. The main components are SaaS, PaaS, and IaaS, which can be deployed on private, public, community, and hybrid clouds. This article examines the current development of cloud computing and provides guidance for new research. Contributions published in journals, conferences, and white papers were analyzed. Cloud consumers can access the resources, operations, and infrastructure provided by cloud providers on a pay-per-use basis. A similar approach can be seen in operations previously deployed by cloud providers for use by cloud users.

Leveraging Artificial Intelligence (AI) Competencies for Next-Generation Cybersecurity Solutions. Pethuru Raj, B. Sundaravadivazhagan, & V. Kavitha (Eds.)

15.1 INTRODUCTION

Beyond data centers, cloud computing has been a boon for all sectors, regardless of their size, in various industries, and has become an integral part of the ecosystem. Modern approaches and strategies for operational data integration are being implemented. Instead of investing in expensive hardware to manage and maintain their own data centers, businesses are turning to cloud service providers (CSPs) offered by major tech giants to have a flexible cloud framework for modernizing computing, networking, and warehouse systems. An organization should consider this when selecting and adopting a cloud computing architecture that best suits its business ecosystem, while respecting workflow conditions and delivering optimal results in terms of strength, flexibility, and scalability. Instead of setting up its own IT infrastructure on-premises, a business can pay to lease the cloud infrastructure and related capabilities and components from a trusted service provider. Cloud topology, also known as virtual computing topology, can be accessed by users over the Internet but lacks an extensive physical computing structure. The cloud architecture plays a crucial role in creating the systems and networks required for evolving cloud computing systems. The cloud network that stores data is always efficient and fast. By effectively understanding customer requirements using cloud technology and making changes to routers and software, powerful and creative cloud-focused results can be delivered. Additionally, designing, optimizing, and validating the functional aspects of the cloud platform are necessary. Cloud architectures support environmental proactiveness by providing virtual services instead of physical and hardware products, which reduces paper waste and improves energy efficiency. As a result, workers can access the cloud platform from anywhere with an Internet connection, reducing travel-related migration. According to a Peak Research report from a data center, power consumption is predicted to decrease by 31% between 2010 and 2020 due to the shift away from cloud computing and other virtual data options.

15.2 ELUDICATION OF CLOUD INFRASTRUCTURE

Cloud computing infrastructure is the compilation of hardware and software elements required to enable cloud computing. It includes computing power, networking, storage, and an interface for users to access their virtualized resources. The virtual resources mirror a physical structure, with components like servers, network switches, memory, and storage clusters.

A cloud-grounded structure has several crucial factors, including the below services.

- Servers;
- Software;
- Network devices; and
- Other storage resources.

These are the factors necessary for implementing cloud-based operations. These applications can be accessed over the internet through wide area networks (WANs), telecom services, and other networks. For example, an Electronic Data Interchange (EDI) provider can expand its services by utilizing a cloud-based EDI software model, allowing users to access the platform without the need for physical infrastructure on their premises. Cloud infrastructure types typically include a user interface (UI) for managing virtual resources through Infrastructure as a Service (IaaS), which is a prominent and convenient example of this model. With IaaS, a company or organization can acquire the necessary computing infrastructure over the Internet, including computing power (both physical and virtual machines), storage, and various requirements such as load balancers and firewalls. Instead of maintaining and managing their own physical infrastructure, they rent the resources they need from the IaaS provider. While this is a well-known example of cloud infrastructure or architecture, it encompasses a wider range of platforms and environments, including private and public clouds [1] (Figure 15.1).

Cloud infrastructure is built with all the sophisticated peripherals needed for cloud computing, including:

- Network
- Storage
- Hardware
- Virtualization

15.2.1 HARDWARE

The cloud, although presumed to be virtual, actually includes hardware as part of its infrastructure. A cloud network is composed of various physical hardware components that can be located in different geographical locations. This hardware includes networking devices such as routers, switches, firewalls, load balancers, backup devices, storage arrays, and servers. Virtualization connects the cloud servers together by abstracting and dividing resources to make them accessible to users. Any cloud infrastructure that requires physical hardware can be located anywhere geographically. A critical component of this hardware is the server, which can be computers or specialized programs that facilitate clients or users. Web servers provide Hypertext Markup Language (HTML) or Hypertext Pre-processor (PHP) pages and use the Hypertext Transfer Protocol (HTTP) to deliver server content with a vast amount of information, while mail servers authenticate and transmit messages over the internet. In the case of a private cloud, organizations can use dedicated servers that are responsible for storing data. The public cloud utilizes a multi-tenant model, which allows for the provision of services to multiple clients.

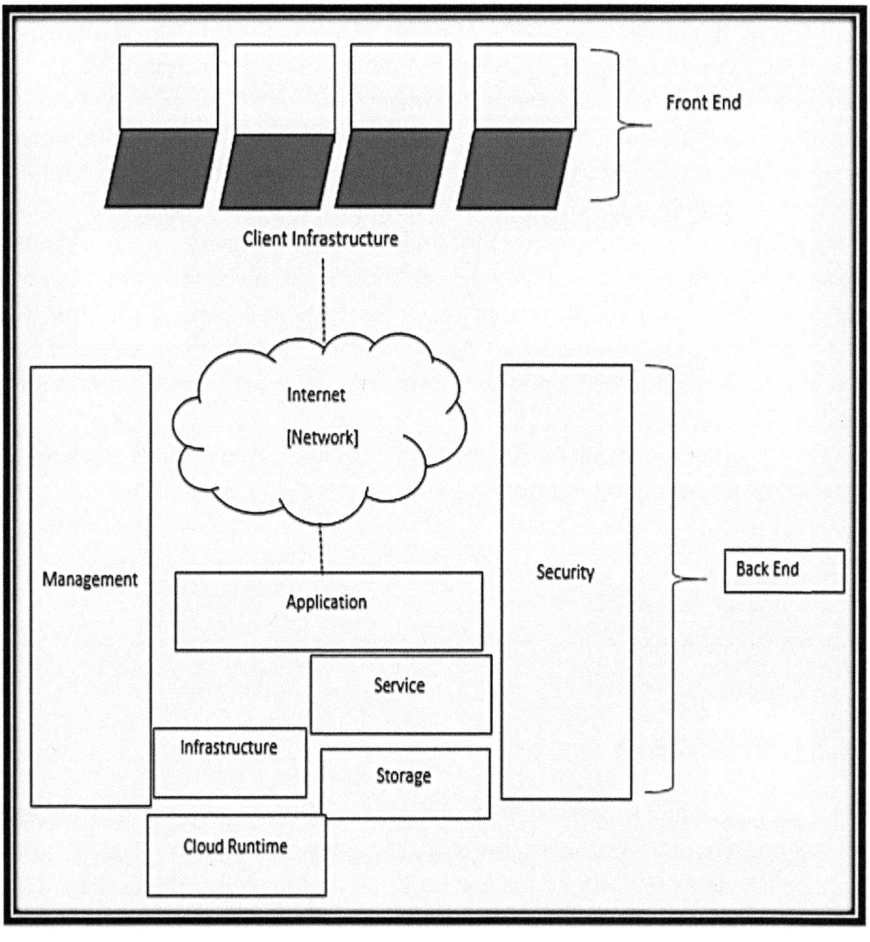

FIGURE 15.1 Cloud architecture.

15.2.2 NETWORKS

15.2.2.1 DIFFERENT ARCHITECTURES IN CLOUD INFRASTRUCTURES

Cloud structure encompasses the essential components required for cloud computing, including hardware, dedicated servers, storage, and grid servers. Think of cloud structure as the necessary tools to create a cloud. To host services and operations in the Cloud, a cloud structure is needed. Cloud structure consists of various components, all integrated into a unified framework that supports business operations. A typical setup may include hardware, virtualization, storage, and networking components. Cloud structure refers to the foundational layer of hardware and software components, such as server hardware, storage, networking equipment, and virtualization software, that form the underlying infrastructure for different Cloud deployment models like infrastructure as a service (IaaS), platform as a service (PaaS), containers as a service (CaaS), and software as a service (SaaS).

A cloud infrastructure is a collection of the factors and elements needed to provide cloud computing. This includes computing power, networking, storage, and an interface that allows users to access virtualized resources. Virtual resources mirror those of physical infrastructure, and they include elements like memory, network switches, servers, and storage clusters. They are necessary to perform operations that users can access through the cloud or retrieve through other technology. The cloud infrastructure approach offers benefits such as greater flexibility, scalability, and lower power costs.

A cloud infrastructure enables organizations to access data storage environments and computing capabilities as needed. Instead of building an on-premise IT architecture or leasing data center space, organizations can now rent cloud infrastructures and the necessary computing capabilities. Cloud infrastructure is available for private, public, and hybrid cloud systems. As a term, cloud infrastructure can be used to describe a complete cloud computing system — once all the pieces are put together—as well as the individual technologies themselves (Figure 15.2).

15.2.2.2 THREE CLOUD STRUCTURES

Cloud services are classified into four types: private cloud, public cloud, hybrid cloud, and multi-cloud. Cloud services are further divided into three types: Infrastructure-as-a-Service (IaaS), Platform-as-a-Service (PaaS), and Software-as-a-Service (SaaS). The choice of cloud selection is based on individual decision-making. The services provided by the cloud are not similar to one another, and they do not solve the same problems.

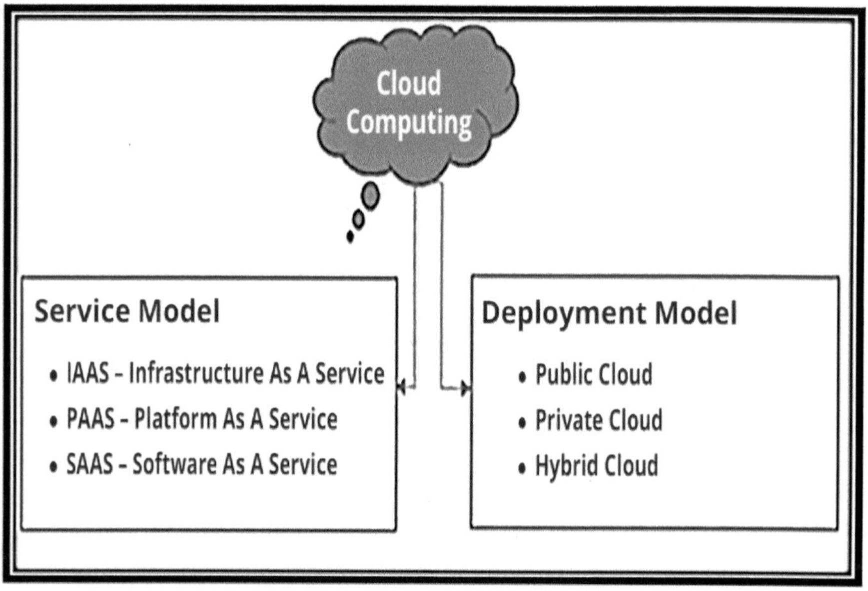

FIGURE 15.2 Cloud infrastructures.

Every cloud architecture pools and shares scalable computing resources across the network. Cloud computing is supported in all types of clouds as it helps manage workloads efficiently. Moreover, each cloud in the system is created using a unique combination of technologies, which mostly depend on the operating system, operation platform, and operation programming interfaces (APIs). Virtualization and robotization software are included in all types of clouds to provide additional capabilities and advantages.

15.2.2.2.1 Public Cloud

Public clouds are owned by tech giants in the market. They are not owned by any individual. The public cloud consists of third-party CSPs that offer cloud services to paying customers over the Internet. Public cloud providers use a multi-tenant model to achieve economies of scale and reduce computing power and data storage costs for their customers. While a multi-tenant environment is effective in reducing overall computing costs, it can also pose security concerns for companies dealing with sensitive data.

The public cloud infrastructure relies on the services of third-party CSPs that provide cloud services to multiple customers via the Internet. These providers operate a multi-tenant environment that reduces data storage and computing costs for customers. This approach is also highly effective in reducing the overall cost of computing. However, it can present security challenges for organizations handling sensitive data or personally identifiable information (PII).

The traditional public cloud has always been in high demand, but now public cloud providers have started offering cloud services in on-premise guest data centers. This has made positional and power differences irrelevant. Any cloud can become a public cloud when the neighborhood is divided and assigned to different users. Having physical structures is not currently a necessary characteristic of a public cloud, as some cloud providers (such as Massachusetts Open Cloud) allow users to utilize their cloud for free.

15.2.2.2.2 Private Cloud

Private clouds are generally defined as cloud environments reserved for only one individual or group, where the infrastructure typically operates behind the individual's or group's firewall. When the initial IT setup is dedicated to a single client with fully isolated access, any cloud can be considered a private cloud. In the private cloud architecture model, only one organization permeates the cloud structure. Private cloud hardware can be developed and maintained by on-site IT staff or provided by an external service provider. In the private cloud architecture approach, only one organization penetrates the cloud structure. Private cloud hardware can be built, developed, and maintained by a company's own IT team or outsourced.

However, private clouds no longer need to be sourced solely from on-premises IT infrastructure. Organizations are now building private clouds on leased, vendor-owned data centers located outside the country, rendering traditional hierarchies and power dynamics obsolete. This has also led to the emergence of various subtypes of private clouds, including managed private clouds. Customers create and utilize a private cloud that is hosted, configured, and managed by a third-party vendor. Managed private cloud is a cloud delivery option that helps understaffed or under-skilled IT businesses provide better private cloud services and infrastructure. It is essentially a cloud within another cloud. For example, an accounting department could have its dedicated cloud within the organization's private cloud.

15.2.2.2.3 Hybrid Cloud

A cold-blooded cloud is supposedly a single IT terrain created from multiple environments connected via native local area networks (LANs), wide area networks (WANs), virtual private networks (VPNs), and/or APIs. A hybrid cloud computing environment is defined as private and public cloud systems interacting in a separate but interconnected system. Organizations dealing with sensitive data may choose to maintain data sequestration by storing some sensitive information in on-point servers while hosting less sensitive operations and other cash registers in the public cloud, where costs may be lower (Figure 15.3).

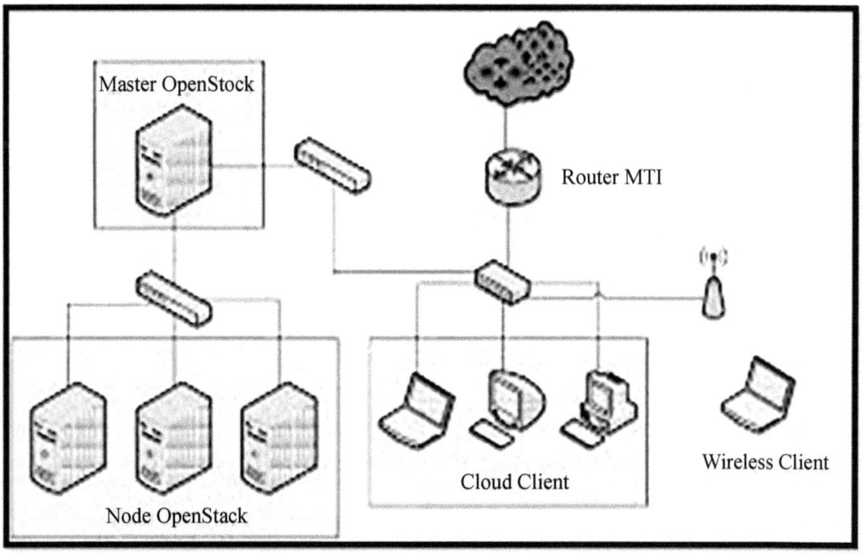

FIGURE 15.3 Hybrid cloud.

Organizations that use hybrid cloud maintain their private cloud environment but can utilize public cloud services for additional capacity or computing workloads on a flexible basis. A hybrid cloud architecture can be considered a best of both worlds approach, providing both private and public cloud infrastructures that interact within a connected but separate system.

This approach is ideal for organizations that handle sensitive information and personally identifiable information (PII), allowing them to store their most critical data in a private cloud and less sensitive data in a public cloud. With hybrid cloud deployments, organizations can maintain their private environment while leveraging public cloud services for other computing tasks and data storage capabilities.

The characteristics of a hybrid cloud are complex and can vary depending on who you ask. For example, it may be necessary to include a hybrid cloud.

- Should include a minimum of one private and public cloud
- More than one public clouds
- More than one private clouds
- The bare essence or virtual terrain is connected to at least one public or private cloud. It can also be connected to multiple clouds.

Multi-cloud is a cloud approach that provides services by incorporating multiple cloud services from multiple cloud vendors – public or private. All hybrid clouds are multi-clouds, but not all multi-clouds are cold-blooded clouds.

Multi-cloud terrain can exist intentionally (for better control of sensitive data or as spare storage for better disaster recovery) or unintentionally (generally due to shadow IT) [2].

15.3 EXISTING CRITICAL CLOUD INFRASTRUCTURE

Critical infrastructures offer important services that are essential to the economy, public safety, community well-being, and governmental efficiency. Governments and people rely heavily on these important service providers. The usage of ICT has significantly increased in several crucial infrastructure sectors, with automation playing an important role in manufacturing. ICT use has also started to grow in sectors such as agriculture, food, and water, where sensor equipment and management systems are helping to facilitate production and make it more sustainable to meet the increasing demands.

A long-standing trend that has seen a continuous rise in the sophistication and complexity of such services is the migration of services to the cloud. Even major infrastructure companies today are considering moving their services and data to the cloud, especially telecommunication companies that aim to operate their services as digital network services. These services are typically composed of multiple components, each with specific scalability and robustness requirements. Therefore, it is crucial to find a solution to the problem of describing the blueprint of how to build a service from its components, including the requirements of the components, and how to derive an actual deployment from such a design [26].

Given that nearly all establishments have already ICT infrastructures, cloud computing may have a massive effect on them [27]. Large ICT infrastructures that take benefit of economies of scale and may be replicated and dispersed across the world may be successfully accessed via cloud services. At least key desires of the United Kingdom electricity enterprise network are met via way of means of cloud computing (Figure 15.4).

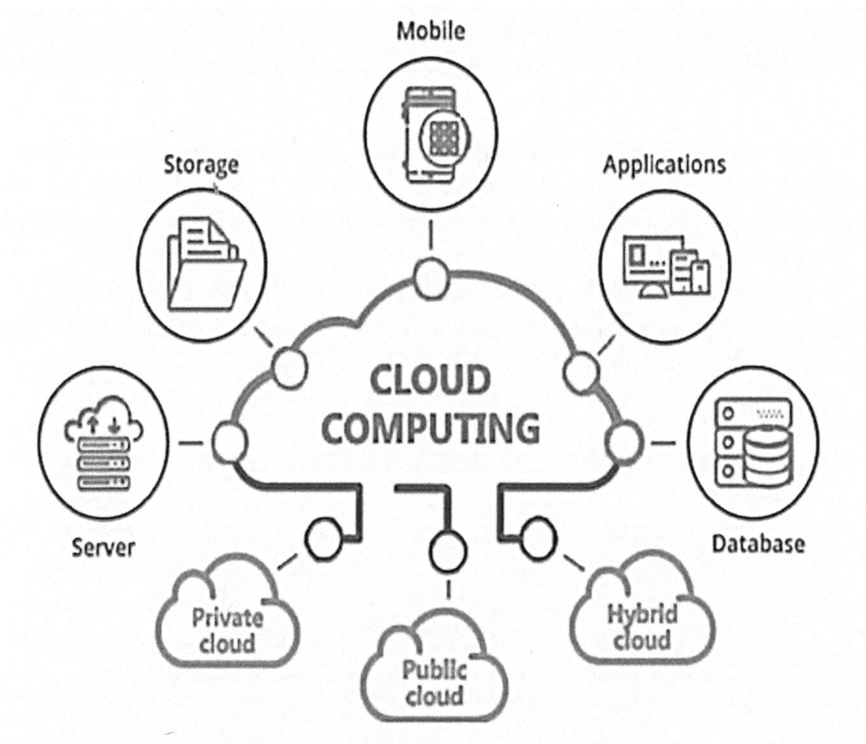

FIGURE 15.4 Structure of cloud computing.

First, depending on the type of energy being utilized or the current state of energy distribution, powerful network simulations require significantly varied amounts of computer resources. The output of renewable energy is often much less predictable than the consistent output provided by conventional energy sources such as coal, oil, gas, or nuclear power.

This is why the usage of the cloud to execute simulations permits dynamic scaling of important computing and information resources [28]. Not only the information, but also the remote terminal units (RTUs) and programmable logic controllers (PLCs) that operate the pumps and maintain the turbines, are crucial components of some critical infrastructures. Each infrastructure may have a different migration strategy for the cloud environment, but it is essential that the availability and integrity of the data are prioritized when it is in the cloud.

Corporate staff often use remote access to SCADA systems over the Internet to view and analyze the data collected by historians. The valuable data in this scenario comes from the infrastructure's records. Keeping the historian processes in an on-site private cloud can be another approach for infrastructure providers to take advantage of the cloud environment.

Due to the deployment of renewable energy resources and the ongoing development of smart grids, many operators struggle to cope with the increasing demand for specific historical and forecast simulations. These operators can reduce or prevent overspending on hardware resources and related maintenance costs thanks to cloud computing. Infrastructure providers will inevitably benefit from what cloud computing has to offer [29].

15.3.1 USE CASE

Critical infrastructure can also employ cloud environments to enhance performance and streamline automation processes. Many business plants utilize networked data servers to store process data and other business interfaces. The managed system's PLCs generate a significant amount of data and logs, which are stored in historian databases. This historical data is continuously recorded and can come from over 700 data points in some cases, making it easy for operators and engineers to retrieve [16, 30].

These front-end or management processors, which manage orders and query statistics from devices in network, provide the data to those central servers.

15.3.2 CRITICAL INFRASTRUCTURE COMPONENTS

The human-system interface (HMI) and various workstations, which run and manage machine software programs on conventional PC platforms, are commonly found within the control network. Physical processes such as electric power generation, transmission, and distribution can be directly monitored and controlled by the field network devices [31]. By transferring the historical operations from the historian database to a private cloud, it is possible to utilize the cloud environment for critical infrastructure. Utilizing a private cloud can facilitate a more efficient examination of the data and enable more skilled analysis [20].

This approach can address the challenges posed by processing the large volumes of data generated by the management systems. The cloud environment is well-suited for this purpose, as it offers substantial computing and storage capacity, is distributed and scalable, and provides higher processing speeds and efficiency compared to existing methods. The dataset can be used to simulate data flow and conduct behavioral analysis. By identifying trends and subtle changes in the data, it is possible to gain insights into the system's status without impacting its operation. The primary cyber-attack vector in control system topologies is the flow of network commands [32].

The behavior of the infrastructure is likely tested to distinguish between critical states delivered through faults or physical assaults and critical states delivered through cyberattacks by analyzing the sensor data from the historian in a private cloud environment.

15.3.3 REQUIREMENTS FOR CRITICAL INFRASTRUCTURE

The operational version of cloud computing is an enormous improvement within the ICT region that has exploded in popularity. Due to the viable price financial savings and benefits of more resilience as a result of elastic cloud behavior, this trend will likely also affect ICT services that support critical infrastructures. However, the use of cloud-based critical infrastructure services presents security and resilience needs that current solutions do not effectively address. For instance, due to the opacity of cloud environments, it can be difficult to assess the risks associated with deploying cloud-based critical infrastructure services, both from a technical and legal or financial standpoint [6].

15.3.4 CONTROL

A Service Level Agreement (SLA) outlines how both the customer and the provider will utilize the services. The responsibilities and procedures of each party should be clearly stated. The SLA cannot be modified or negotiated, which often discourages companies from making full commitments to the cloud. Additionally, there is no governing body, and no one is responsible for enforcing this compliance. As a result, there may be data breaches that the affected party is often unaware of. Control is a common issue as it depends on the selected deployment model and is not always in the owner's hands [21].

Private clouds give companies the ability to determine how their data is managed, stored, and secured. Where is the data stored when using public clouds? Who is responsible? Public clouds do not seem to be a good solution in this case [8].

15.3.5 DATA CENTRIC SECURITY APPROACH

An information-centric security approach should typically ensure that data protection measures are enforced across all offered security solutions, and that data owners have full control over who has access to and uses the data, as well as what they are allowed to do with it. Additionally, it is possible to specify and align institutional security policies and access regulations with the cloud environment. While requirement-based security challenges may vary for critical infrastructure applications and traditional IT applications, they must be considered together in the given context.

15.4 SERVICES FOR CRITICAL INFRASTRUCTURE PROVIDERS

To capitalize on the benefits of cloud computing, many critical infrastructure providers intend to utilize its unique advantages and migrate to the cloud computing paradigm. The UK's gas and electricity network, National Grid, has announced plans to replace its internal data centers with a CSC-hosted cloud. However, moving to the cloud without addressing all of the previously mentioned cloud security challenges is impossible to achieve in a short span of time. This section will investigate the security requirements of various critical infrastructure providers, including smart grids, telecommunications, transportation, and finance [7].

15.4.1 REQUIREMENT ANALYSIS

Critical infrastructure has different security requirements in unique environments, and providers operate using various kinds of infrastructures. Successful cloud migration of various critical infrastructure providers would have to meet all of their requirements. To find the common security requirements, we have inspected and evaluated the security requirements of various critical infrastructure services such as data security, compliance and audit, cryptography, and access control. Among them, one of the core requirements was the access control system [3].

Furthermore, the aggregation of data from multiple sources may unintentionally reveal sensitive information about consumers, and moving the aggregated data from one location to another may violate data privacy. To avoid data leakage issues, cloud consumers must know ahead of time where their data will be stored and how it will be separated. Additionally, the lack of visibility into how data is maintained and secured raises a number of concerns that must be addressed when the cloud moves. Data centers must deal with massive amounts of data collected from all over the cloud. Data centers cannot exist in isolation; they must be linked to other data centers. As a result, security and recess must be managed properly [9].

Virtualization is an important component of cloud computing because it brings well-known advantages to the cloud, but it raises several privacy concerns, such as virtualized performance and security concerns. Support for scalable multi-tenant billing and robust exclusion is critical for any system considering cloud deployment. In cloud computing, however, multiple networks may run on the same infrastructure. As a result, solid isolation is another requirement to ensure that cloud tenants do not interfere with each other's security or effectiveness. Metering and adjusting virtual resource consumption are necessary in cloud computing.

Customization of software and services, dealing with latency, removing technical barriers, and addressing the complex nature of integrating cloud services with existing legacy environments are all factors to consider. Highly configurable, secure, and easily customizable virtual machines are also required. Furthermore, because cloud computing is a shared environment with shared platforms, shared storage, and shared networks, its components must collaborate to achieve the intended mission, regardless of providers, storage, operating systems, and so on [17].

Encryption is commonly used to protect data in untrustworthy storage environments, such as cloud computing. However, if not handled properly, it can consume resources and become a source of extra storage and bandwidth usage. Key management is another complex problem that requires more attention [11].

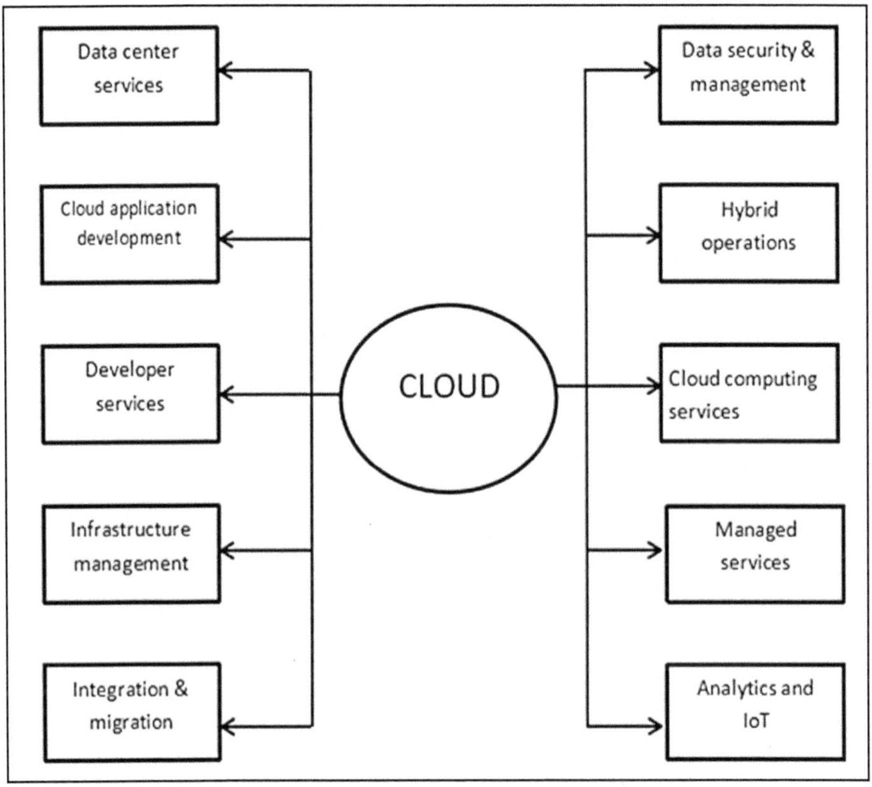

FIGURE 15.5 Requirements of cloud.

15.5 APPLICATIONS OF CLOUD INFRASTRUCTURE

15.5.1 ONLINE DATA STORAGE

Organizations need to store large amounts of data, and this data grows over time. This data can be presented in any format, including text, images, audio, and video. Organizations no longer need to install physical storage systems to store this massive amount of data. Their data can be stored in the cloud. The total amount of data that the company collects can be divided into two groups: historical data and current data. Current data is information that is often required to perform specific day-to-day tasks. Historical data, on the other hand, is information that has value even if it is not actively used and should be preserved [21].

15.5.2 BACKUP AND RECOVERY

Cloud service companies provide different methods for data recovery. They offer a wide variety of rehabilitation programs at various price points. Businesses can choose the type of package they need based on their requirements.

Data redundancy is an option provided by cloud service providers (CPS), which involves storing copies of data in multiple locations. This redundant storage system is implemented to safeguard data and provide alternative access to users. In the event that data becomes unavailable in the original location, it can still be easily accessed from alternate storage [13].

15.5.3 TESTING AND DEVELOPMENT

After development, testing is essential for a product to be completed and ready for release. Before distribution, the product must undergo extensive testing. This product must be tested on several devices with different infrastructures as the end user may be anywhere in the world. The load balancing test must also pass. Load Balancing: How the product works when multiple users are using it at the same time. Testing requires a variety of IT tools and computer systems to perform these tasks. These testing tools could now be available from one place thanks to the cloud. Businesses can easily evaluate how well their product performs online against a huge user base.

15.5.4 CLOUD COMPUTING IN MEDICAL FIELDS

The medical sector is also significantly affected by cloud computing. It is used to store patient data and allows for easy access to previous information without the need for physical storage or paperwork. Patient data can be accessed remotely from any location, reducing the wait time for emergency room staff to retrieve information from the hospital computer (Figure 15.6).

15.5.5 BIG DATA ANALYSIS

Big data analysis involves working with enormous amounts of data, with quantities ranging from terabytes to zettabytes (known as big data). Any normal database management system would find it difficult to handle this volume of data. Cloud computing can be utilized to store large data sets, including both structured and unstructured data from multiple sources, with sizes ranging from terabytes to zettabytes. In addition to storage, it provides a wide range of tools for analyzing this vast amount of data. The main purpose of storing such large amounts of data is to extract valuable insights from it [22].

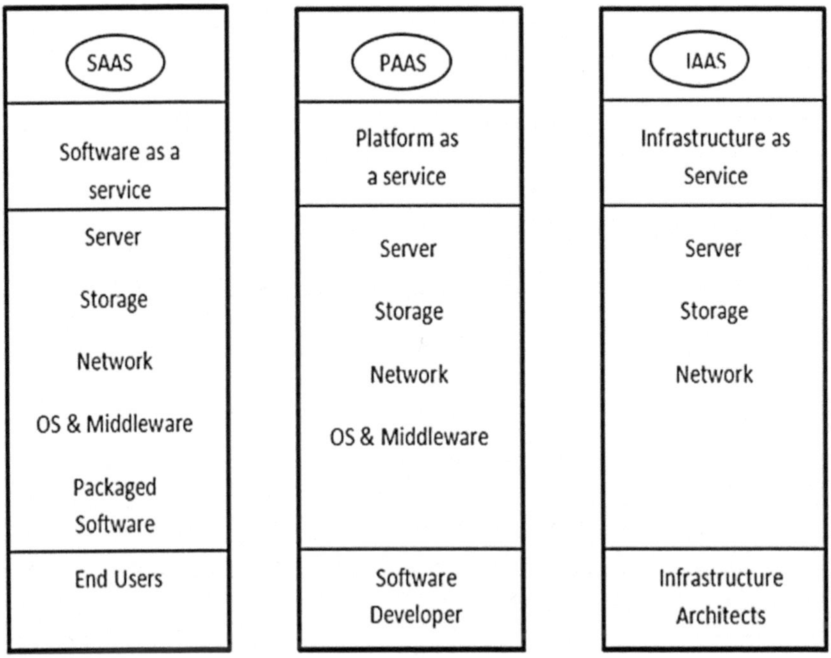

FIGURE 15.6 Application of the cloud.

15.5.6 *ENTERTAINMENT APPLICATION*

Today, the internet is a fantastic resource for entertainment, whether it is through online games, YouTube videos, or Netflix web series episodes. This data is heavily used by users from almost every nation on earth. It is also crucial to offer outstanding customer service so that you may access this information whenever you want. To achieve this, entertainment companies engage with their audience via a multi-cloud strategy. The entertainment industry is expanding to new heights thanks to the cloud.

15.5.7 SOCIAL NETWORK PLATFORM

Social networking sites are essential in modern life. They have changed how individuals interact and express themselves. These platforms have a sizable worldwide user base, making them great candidates for cloud computing adaptation. Social networking sites, which routinely upload copious quantities of multimedia content like photos and videos, can cause the entire network to run sluggishly. Cloud storage becomes useful in this situation. Cloud storage helps social networking programs run more effectively [12].

15.6 BUSINESS AGILITY AND RELIABILITY OF CLOUD INFRASTRUCTURE

15.6.1 BUSINESS AGILITY

Business agility means the ability of an organization to respond with creative business solutions to market changes and emerging possibilities quickly and cost-effectively.

15.6.2 BENEFITS OF BUSINESS AGILITY

- **Greater Business Continuity and Flexibility:** Cloud services can be expanded or contracted based on the business requirements, without the need to increase the amount of IT equipment that a firm must acquire and maintain. The pay-per-use flexibility of the cloud allows end customers to quickly scale up or down in response to the demands of the business [10].
- **Infrastructure Agility:** Utilizing the cloud enables businesses to drastically reduce the time needed to provide services. You can start small and scale up as necessary. If necessary, it is simple to make cuts. Establishing a cloud server is much simpler compared to physical server installation. Real-time augmentation of data processing capability is made possible by using Apache Hadoop, Spark, and Hive on the cloud [19] (Figure 15.7).
- **Automated Resource Allocation:** To achieve coherence and scalability, cloud computing relies on resource sharing and workload distribution. Automating the process and utilizing simple APIs and online terminals make it easier to provide, de-provision, and redeploy

resources. Compared to a traditional data center, the amount of time spent by an IT systems administrator on monitoring and supporting cloud infrastructure is significantly reduced [24].

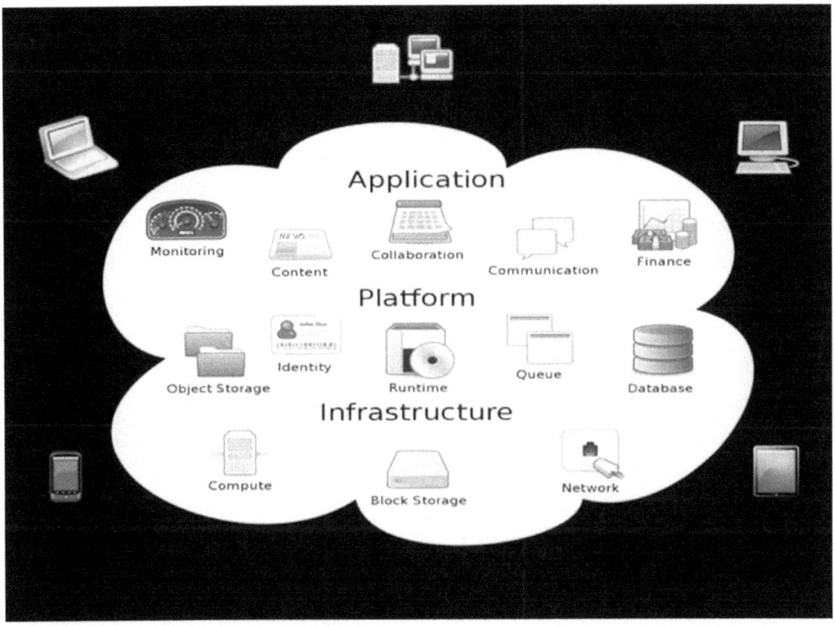

FIGURE 15.7 Applications of cloud.

- **Modern Technology Upgrades:** Because many dependencies must be considered, including infrastructure, operations, and software, the refresh cycles for an upgrade may be lengthy. However, the addition of services from vendors by cloud providers today makes it easier to update without impacting your overall solution. Environments using cloud data lake topologies can quickly update to a new Hadoop version.

15.6.3 ROLE OF CLOUD COMPUTING FOR ACHIEVING BUSINESS AGILITY

- **Scalability Upon Request:** All CPSs provide scale-up and scale-down based on the demands of the business, and these actions can be self-operating depending on several variables, such as the number of customers within the specified time. The business can grow and shrink as needed, which guarantees that they are only spending for their necessities and actual consumption. This is quite beneficial since it allows businesses to save money while also improving their ability to respond to client requests.

- **Data Availability:** We can now access more data than usual thanks to the cloud. Employees can now work remotely with ease, thanks to the cloud, which provides them with more flexibility and productivity.
- **Less Infrastructure maintenance:** Concerns can now focus on managing their applications instead of infrastructures, thanks to the rise in popularity of Platform as a Service (PaaS) offerings provided by the cloud. This has relieved them from higher costs [24].
- **Reduces Business Complexity:** By streamlining internal operations, enhancing the supply of internal resources, developing new modes of employee collaboration, hastening the implementation of new technologies or initiatives, and enhancing the capability to share, analyze, and act on data, the cloud works to make an organization less complex [14].

15.6.4 BUSINESS RELIABILITY

Reliability in cloud computing is essential for businesses of all sizes. Bugs in software can lead to decreased productivity, decreased sales, and diminished brand reputation. Before deploying your applications to the cloud, ensure they have been thoroughly tested against a variety of real-world scenarios. This will increase their dependability and enable them to meet client expectations.

Reliability in cloud computing can have different interpretations. Factors such as cost-effectiveness, performance, and security influence the level of reliability. Additionally, the frequency of component failures or unavailability of cloud services are also important factors to consider.

i. **Importance of Cloud Services in Business Reliability:** Any business utilizing cloud computing requires reliability in order to operate effectively. Faulty or poorly constructed apps, programs, and other software can significantly decrease productivity, leading to a cascading impact that may result in reduced profitability and a damaged reputation among clients [15].
ii. **Automatically Recover from Failure:** Any business utilizing cloud computing requires reliability in order to operate effectively. Faulty or poorly constructed apps, programs, and other software can significantly

decrease productivity, leading to a cascading impact that may result in reduced profitability and a damaged reputation among clients [15].

iii. **Increase Aggregate Workload Availability by Scaling Horizontally:** A huge single resource is replaced by multiple smaller ones through horizontal scaling. Individual workload failures become less significant as a result. Each system is more reliable when there is no single point of failure (Figure 15.8).

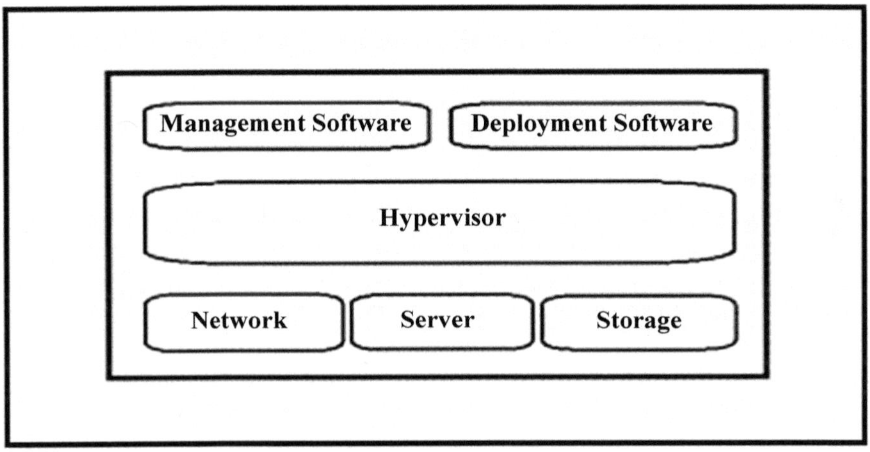

FIGURE 15.8 Reliability in cloud computing.

15.6.5 MEASURES TO IMPROVE CLOUD COMPUTING RELIABILITY

Businesses can improve the dependability of cloud computing by implementing cost-cutting measures, monitoring resource performance, and implementing advanced security features to protect data from vulnerabilities caused by improperly configured resources. While this may seem like a lengthy list of responsibilities, automation is commonly used to enhance the reliability of cloud computing.

Automation can be employed to enforce predetermined boundaries for services such as pricing, availability, and security requirements. Whenever an action violates these boundaries, the automated solution triggers a function to halt, pause, or reverse the action.

Organizations can, for instance, implement policies that:

- Notify corporate departments of anticipated budget overruns.
- After a certain amount of time, habitually erase unused resources.

- Prevent users from using resources that aren't authorized or compatible.
- Regularly check the resource configurations.
- Implement global tagging rules to make cost distribution and cloud governance simpler.
- Implement multi-factor authentication (MFA) and conditional access controls for privileged accounts.
- If questionable activity is found, revoke user access.

15.7 REQUIREMENTS FOR BUILDING A CLOUD INFRASTRUCTURE

15.7.1 SERVICE AND RESOURCE MANAGEMENT

All elements of a data center are virtualized through a cloud infrastructure. By adopting a public and/or private cloud provider, users can take advantage of utilizing the applications. In order for cloud administrators to effectively promote functionality, a streamlined framework for outlining and assessing services is crucial [18] (Figure 15.9).

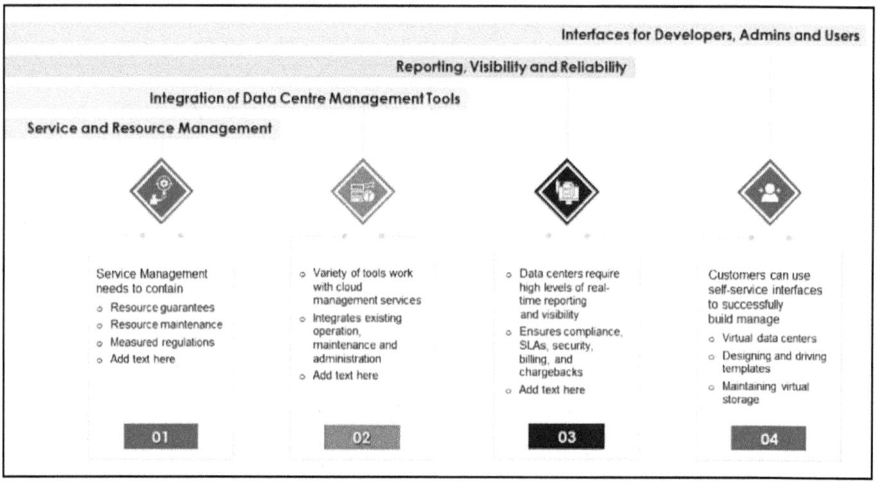

FIGURE 15.9 Requirements of cloud computing.

Services provided by the cloud include resource maintenance, guarantees, billing cycles, and measured regulations. All of these must be included in service management.

15.7.2 DATA CENTRE MANAGEMENT TOOLS INTEGRATION

For system management, security, provisioning, customer service, billing, and directories, among other things, the majority of data centers use a variety of IT technologies. Modern cloud services should leverage the latest software, physical devices, and various other technologies while supporting a data center's current architecture.

15.7.3 CLOUD REPORTING, VISIBILITY, RELIABILITY, A SECURITY

In order to maintain order, SLAs, data safety, product/service pricing, and chargeback's in cloud settings, data centers need to have live reporting and transparency. It is impossible to provide services, manage system performance, and perform other related activities without transparency in the system. The cloud environment should be able to provide service at any cost, even if multiple components fail.

15.7.4 PROVISION FOR ADMINISTRATION AND USERS

Self-service interfaces and automated deployment allow end users to more easily use complicated cloud services, which helps to reduce costs and increase acceptance. Customers can successfully deploy a cloud service using self-service interfaces by managing their virtual data centers, creating and modifying templates, managing virtual storage, managing networking resources, and utilizing libraries. The availability of information, VMs, and different cloud users is improved by administrator interfaces. Additionally, all of these structures integrate using developer-friendly APIs [25].

15.8 CONCLUSION

Cloud is an ecosystem that includes all the necessary hardware and software components for the smooth functioning of services. Customers can access a wide range of services online. IaaS, PaaS, and SaaS are the three main service model categories in cloud computing. Due to the rapid increase in demand for cloud services, the global market for cloud computing is growing at a significant rate. Many companies from various industries are now favoring cloud services due to the numerous benefits that cloud computing provides. As a result of the widespread use of cloud computing by organizations for various purposes, CPSs offer a wide range of applications in different industries.

KEYWORDS

- **cloud service providers**
- **electronic data interchange**
- **hypertext markup language**
- **hypertext transfer protocol**
- **infrastructure as a service**
- **platform as a service**
- **wide area networks**

REFERENCES

1. Jefry, W., & Sasi, S. (2017). *Does Cloud Computing Add Business Agility to Big Data Analytics. 2*(12).
2. Mell, P., & Grance, T., (2011). *The NIST Definition of Cloud Computing Recommendations of the National Institute of Standards and Technology* (Vol. 145, p. 7). Natl. Inst. Stand. Technol. Inf. Technol. Lab.
3. Hu, F., et al., (2011). A review on cloud computing: Design challenges in architecture and security. *Journal of Computing and Information Technology, 19*(1), 25–55.
4. Ali, M., Khan, S. U., & Vasilakos, A. V., (2015). Security in cloud computing: Opportunities and challenges. *Inf. Sci. (NY)., 305*, 357–383.
5. Davidovic, V., Ilijevic, D., Luk, V., & Pogarcic, I., (2015). Private cloud computing and delegation of control. *Energy Procedia, 100*, 196–205.
6. Jadeja, Y., & Modi, K., (2012). Cloud computing–concepts architecture and challenges. In: *2012 International Conference on Computing Electronics and Electrical Technologies ICCEET 2012* (pp. 877–880).
7. Bhardwaj, S., Jain, L., & Jain, S., (2010). Cloud computing: A study of infrastructure as a service (IAAS). *International Journal of Engineering., 2*(1), 60–63.
8. Lui, F., et al., (2011). *NIST Cloud Computing Reference Architecture: Recommendations of the National Institute of Standards and Technology*. NIST Spec. Publ. 500-292, 35.
9. Dhakar, R., Gupta, A., & Vijay, A., (2009). Cloud computing architecture. *China Cloud. Cn., 40*, 29–53.

10. Niharika, G., & Ritu, V., (2015). Cloud architecture for the logistics business. *Procedia Computer Science, 50*, 414–420.
11. Rani, B. K., Rani, B. P., & Babu, A. V., (2015). Cloud computing and inter-clouds – types topologies and research issues. *Procedia Comput. Sci., 50*, 24–29.
12. Verma, & Kaushal, S., (2011). *Cloud Computing Security Issues and Challenges: A Survey* (pp. 445–454).
13. Meena, M., Singh, A. R., & Bharadi, V. A., (2016). Architecture for software as a service (SaaS) model of CBIR on hybrid cloud of Microsoft azure. *Procedia Computer Science, 79*, 569–578.
14. Toy, M., (2015). Cloud services architectures. *Procedia Computer Science, 61*, 213–220.
15. Bou, A. J., & Demerjian, J., (2017). Evaluation of mobile cloud architectures. *Pervasive Mobile Computing, 39*, 284–303.
16. Ferrer, J., Pérez, D. G., & González, R. S., (2016). Multi-cloud platform-as-a-service model functionalities and approaches. *Procedia Computer Science, 97*, 63–72. [10 Things You Need to Know About Hybrid Cloud [online]].
17. Marissa, C., (2014). *Hybrid Cloud: What it is What it Does and Hybrid Cloud Use Cases.* [online] Available: http://searchservervirtualization.techtarget.com/tip/Hybrid-cloud-What-it-is-what-it-does-and-hybrid-cloud-use-cases (accessed on 27 January 2024).
18. Buyer, C. W., (2012). *Private Cloud: Get it Right and Exploit its Energy Economics*, 1–6.
19. Bohn, R. B., Messina, J., Liu, F., Tong, J., & Mao, J., (2011). NIST cloud computing reference architecture. *Proceedings–2011 IEEE World Congress on Services 2011* (pp. 594–596).
20. Amanatullah, Y., Lim, C., Ipung, H. P., & Juliandri, A., (2013). Toward cloud computing reference architecture: Cloud service management perspective. *Int. Conf. ICT Smart Soc.*, 1–4.
21. Gong, C., Liu, J., Zhang, Q., Chen, H., & Gong, Z., (2010). The characteristics of cloud computing. *Proceedings of the International Conference on Parallel Processing Workshops*, 275–279.
22. Miotto, R., & Weng, C., (2015). Case-based reasoning using electronic health records efficiently identifies eligible patients for clinical trials. *J. Am. Med. Inform. Assoc., 22*, e141–150. [PMC free article] [PubMed] [Google Scholar].
23. Li, L., Cheng, W. Y., Glicksberg, B. S., et al., (2015). Identification of type 2 diabetes subgroups through topological analysis of patient similarity. *Sci. Transl. Med., 7*, 311ra174. [PMC free article] [PubMed] [Google Scholar].
24. Bellazzi, R., & Zupan, B., (2008). Predictive data mining in clinical medicine: Current issues and guidelines. *Int. J. Med. Inform., 77*, 81–97. [PubMed] [Google Scholar].
25. *SOME CT*. https://www.nlm.nih.gov/healthit/snomedct/index.html (accessed on 27 January 2024).
26. Schöller, M., Stiemerling, M., Ripke, A., & Bless, R. (2013). Resilient deployment of virtual network functions. In *2013 5Th international congress on ultra-modern telecommunications and control systems and workshops (ICUMT)* (pp. 208-214). IEEE.
27. Li, S., Wang, R., & Zhou, G. (2012). Opinion target extraction using a shallow semantic parsing framework. In *Proceedings of the AAAI Conference on Artificial Intelligence* (Vol. 26, No. 1, pp. 1671–1677).
28. Wallom, D., Turilli, M., Martin, A., Raun, A., Taylor, G., Hargreaves, N., & McMoran, A. (2012). myTrustedCloud: trusted cloud infrastructure for security-critical computation

and data management. In *Proceedings of the WICSA/ECSA 2012 Companion Volume* (pp. 82–83).

29. Khorshed, M. T., Ali, A. S., & Wasimi, S. A. (2012). A survey on gaps, threat remediation challenges and some thoughts for proactive attack detection in cloud computing. *Future Generation Computer Systems, 28*(6), 833–851.
30. Mitchell, R., & Chen, I. R. (2014). A survey of intrusion detection techniques for cyber-physical systems. *ACM Computing Surveys (CSUR), 46*(4), 1–29.
31. Briesemeister, L., Cheung, S., Lindqvist, U., & Valdes, A. (2010, August). Detection, correlation, and visualization of attacks against critical infrastructure systems. In *2010 Eighth International Conference on Privacy, Security and Trust* (pp. 15–22). IEEE.
32. Muttoni, G., Carcano, C., Garzanti, E., Ghielmi, M., Piccin, A., Pini, R., ... & Sciunnach, D. (2003). Onset of major Pleistocene glaciations in the Alps. *Geology, 31*(11), 989–992.

CHAPTER 16

Machine Learning Approaches for Dynamic Cyber Security

M. KUNDALAKESI[1] and M. RENUKA DEVI[2]

[1]*Department of Computer Applications, Sri Krishna Arts and Science College, Coimbatore, Tamil Nadu, India*

[2]*Associate Professor, School of Information Science and Engineering, Presidency University, Bangalore,India*

ABSTRACT

In the context of computers, data science has recently led to significant changes in cybersecurity operations and technology. Making a security system automated and intelligent requires extracting patterns or insights related to security incidents from cybersecurity data and structuring a suitable data model. Data science, also known as diverse scientific approaches, machine learning (ML) techniques, processes, and systems, is the study of actual occurrences using data.

This paper aims to provide a quick overview of cybersecurity data science, which uses analytics to supplement the most recent data-driven patterns to deliver more efficient security solutions. The data is acquired from relevant cybersecurity sources. Compared to conventional ones, the idea of cybersecurity data science enables the computing process to be more intelligent. This paper offers a multi-layered framework for modeling cybersecurity based on ML. Additionally, this paper discusses related research problems and future approaches. Overall, the objective is to cover not only cybersecurity data science and appropriate methodologies but also to emphasize how data-driven intelligent decision-making can defend systems from cyber attacks.

Leveraging Artificial Intelligence (AI) Competencies for Next-Generation Cybersecurity Solutions. Pethuru Raj, B. Sundaravadivazhagan, & V. Kavitha (Eds.)

16.1 INTRODUCTION

Due to the increasing reliance on digitalization and the Internet of Things (IoT), there has been a significant rise in security incidents such as unauthorized access, malware attacks, zero-day attacks, breaches, denial of service (DoS), social engineering, phishing, and more in recent years. For example, in 2010, there were 50 million malware attacks reported globally. By 2012, this number had doubled to around 100 million, and in 2019, there were 900 million malicious attacks (AV-TEST, Germany).

Cybercrime and attacks can result in financial losses and have a significant impact on organizations and individuals. It is estimated that the annual cost of cybercrime is 8.19 million USD for the US and 3.9 million USD on average globally, with a total financial loss of 400 billion USD. According to Juniper Research, the number of data breaches is expected to increase significantly over the next five years. Therefore, it is crucial for organizations to adopt and enforce a robust digital security strategy to minimize the damage.

Why is it important for devices to be aware in order to develop essential cyber protection? There are several reasons. By making devices aware, digital security systems can analyze patterns and detect similar attacks, as well as adapt to changing circumstances. This can help digital protection teams proactively prevent threats and respond to attacks in real time. It can also reduce the time spent on repetitive tasks, allowing organizations to allocate their resources more effectively.

In short, awareness can make cyber security simpler, more proactive, cost-effective, and significantly more powerful. However, it can only be effective if the underlying data that supports the device's awareness provides a complete picture of the environment. As they say, garbage in, garbage out.

16.2 WHY USE MACHINE LEARNING (ML) IN CYBERSECURITY?

Legacy-based risk detection systems used heuristics and static signatures on various data logs to detect risks and anomalies. However, this meant that analysts had to be familiar with the appearance of daily data logs. The process involved data being ingested and processed through the traditional extraction, transformation, and load (ETL) phase. The transformed data is then analyzed by machines and reviewed by analysts who create signatures. These signatures are then evaluated by applying additional data. If a mistake is found, the rules are rewritten. Although signature-based change detection techniques are well understood, they are not robust because signatures need to be created on-the-fly for large volumes of data [3].

16.3 HOW MACHINE LEARNING CAN IMPROVE CYBERSECURITY

Over the safety of statistics, communications, commerce, and more, have never been more crucial. Despite the tremendous measures taken to protect against attacks-from extensive user authentication to firewalls to antivirus and malware software—hackers are still finding new ways to harm companies' networks [2].

A variety of network intrusion detection systems (NIDSs) have been deployed to thwart potential attackers. However, their development has led to an important paradox: as NIDSs acquire additional capabilities from training datasets—examples of past attack patterns and models of normal behavior that enable them to identify anomalous activity—their detection abilities can become overwhelmed by the extensive training process.

A recent machine learning (ML)-based scheme from a team of IEEE researchers is showing promise in solving this paradox. They propose a unique multi-level approach that, according to the team, can reduce computational complexity while maintaining detection capability—enabling more consistent performance [5].

"The proposed novel multi-level community intrusion detection framework reduces the general computational complexity of the deployed device mastering version—via decreasing the wide variety of capabilities in addition to the wide variety of education samples the version needs—whilst improving its overall detection performance in phrases of community assault detection accuracy," stated Abdallah Shami, senior member, IEEE, and professor with the ECE Department at Western University in London, Ontario, Canada. "Additionally, it money owed for a not unusual place hassle going through public cybersecurity datasets (i.e., the imbalanced nature of the datasets) via making use of an oversampling method to offer extra community assault samples to analyze assault styles/characteristics higher" [6] (Figure 16.1).

This paper extends the team's previous work on ML-based NIDS frameworks. The new framework is divided into three stages. The first stage preprocesses the data with Z-score normalization and utilizes the Synthetic Minority Oversampling Technique (SMOTE) to enhance the performance of the training model. The second stage employs a feature selection method to reduce the number of features in the ML classification model, thereby reducing its time complexity and training time while meeting performance metrics. The third stage of the framework optimizes the hyperparameters of the different ML classification models considered. This is achieved by exploring three different hyper parameter tuning/optimization models: random search, meta-heuristic optimization algorithms such as particle swarm optimization and genetic algorithms, and Bayesian Optimization algorithm [4].

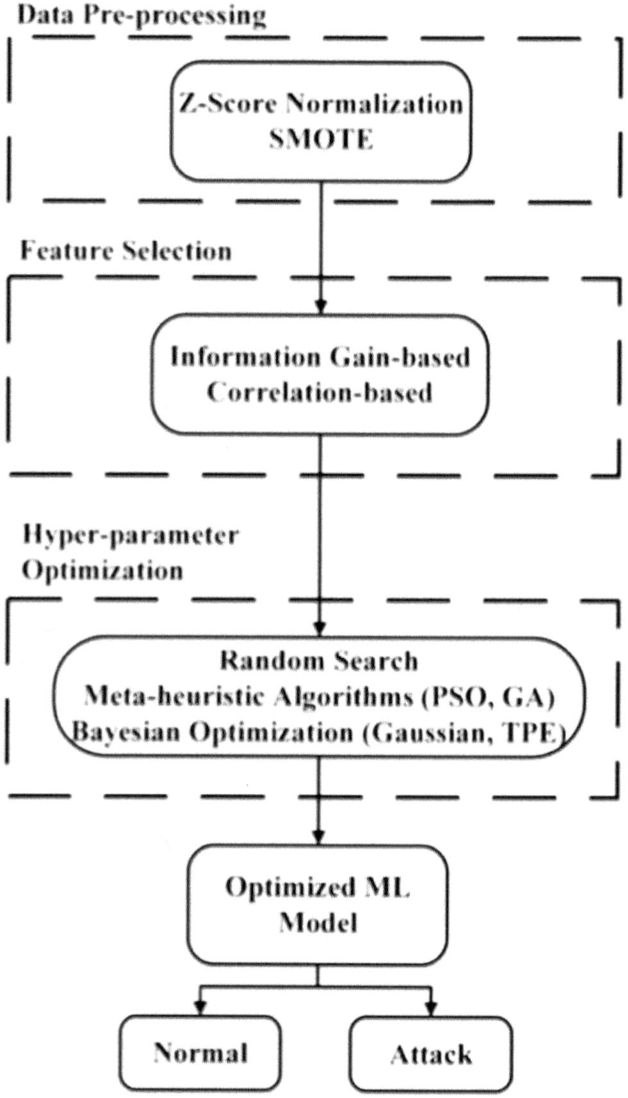

FIGURE 16.1 ML-Primarily based NIDS frameworks.

The results of these optimization stages are combined into the optimized ML classification model for the enhanced NIDS system, which can more effectively classify new instances as either normal behavior or attacks.

16.4 PROPOSED MULTI-STAGE OPTIMIZED ML-BASED NIDS FRAMEWORK

The paper presents an in-depth study of previous iterations of ML-based NIDS work and describes their shortcomings before exploring the mathematical backgrounds behind the different deployed techniques. The researchers then proceed to introduce their new framework, describe the datasets under consideration, and discuss the practical implications of the scheme [8].

According to the authors, their multi-level ML-based NIDS framework was able to reduce the required training pattern size by up to 74% and the feature set size by up to 50%. Additionally, the model's performance improves with hyper-parameter optimization, achieving detection accuracies of over 99% for each dataset. This outperforms the latest literature works by 1–2% in accuracy and a 1–2% decrease in false alarm rate.

"The proposed novel multi-level network intrusion detection framework is a comprehensive approach to optimize the entire machine learning pipeline, from feature selection to model training and testing," said Shami. "It demonstrates significant and promising results in terms of performance, with high network attack detection accuracy and low computational complexity. It also shows generality, as it is applicable to different data types/datasets, as demonstrated by testing on multiple cybersecurity datasets. Therefore, the proposed framework highlights the importance and benefits of considering/optimizing the entire pipeline rather than just the machine learning model" [10].

16.5 PROBLEMS THAT MACHINE LEARNING SOLVES

- **Face popularity:** Face recognition systems can identify individuals from digital images by analyzing facial features. These systems utilize 3-dimensional recognition and skin texture analysis to verify faces. They are similar to biometrics and are commonly used in security systems, such as using face recognition technology to unlock phones. Fake news has become widespread after the 2016 US presidential election. To combat this issue and distinguish between fake and real news, detectors have been developed. These detectors analyze the semantic and complex patterns of the article's content, as well as the credibility of the source, to differentiate between artificial and genuine information.

- **Sentiment evaluation:** Understanding the overall positivity or negativity of a document is essential for assessment, as it sets a clear boundary when making decisions. Sentiment evaluation structures perform sentiment mining to determine the sentiment.
- **Recommender structures:** These are structures that are capable of checking a purchaser's selection, primarily based on the personal records of previous selections made by the purchaser. This is another aspect that influences design determinations made by various equivalent clients. Such recommender systems are exceptionally popular and widely used by industries to promote movies, products, and services. Recommender systems, in a way, determine the go-to-market strategies for the organization based on collective likes or dislikes.
- **Fraud detection structures:** Fraud detection structures are created for danger mitigation and securities fraud in accordance with customer interest. Such structures detect outliers in transactions and raise flags by measuring anomaly coefficients.
- **Language translators:** Language translators are clever structures capable of translating not just phrase to phrase, but entire paragraphs at a time. Natural language translators utilize contextual data from multilingual files and are able to perform these translations [9].
- **Chatbots:** Intelligent chatbots are systems that enhance customer experience by providing automated responses when human customer support agents are unable to respond. They possess sentiment analysis capabilities and can also make recommendations. However, their utility extends beyond that of a mere digital assistant.

16.6 CURRENT CYBERSECURITY SOLUTIONS

Today, signature-based security systems are gradually being replaced by intelligent cybersecurity solutions. Machine learning products are effective in detecting new malware, zero-day attacks, and advanced persistent threats (Figure 16.2).

Insight from the enormous quantity of log data is aggregated through log correlation methods. Endpoint responses have been highly effective in identifying peripheral attacks. New machine learning-driven cybersecurity products have been proactive in strengthening infrastructure such as virtual machines. The following diagram provides a brief overview of some machine learning solutions in cybersecurity:

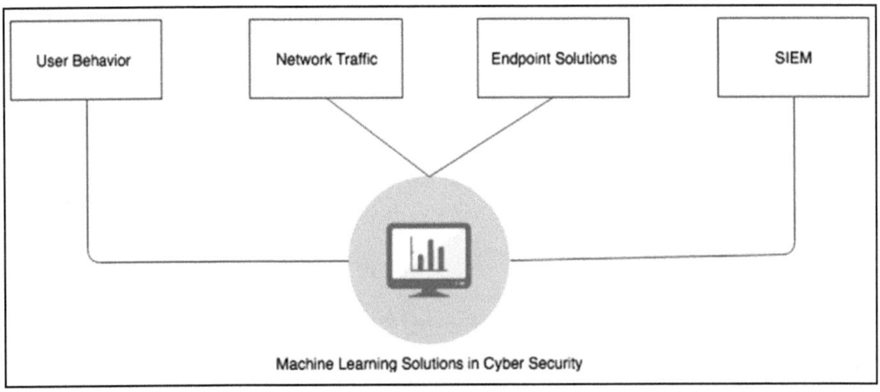

FIGURE 16.2 ML solutions in cyber security.

16.6.1 CYBERSECURITY TASKS AND MACHINE LEARNING

Instead of examining ML obligations and observing them in cybersecurity, let us explore the common role of cybersecurity obligations and machine learning opportunities. There are three dimensions (Why, What, and How).

The first dimension is the objective or task (e.g., detecting threats, anticipating attacks). According to Gartner's PPDR model, all security obligations can be divided into five categories:

i. Prediction
ii. Prevention
iii. Detection
iv. Response
v. Monitoring

The 2D measurement is a technical layer and a solution to the "What" query (e.g., at which stage to reveal issues). Here is the listing of layers for this measurement:

- Community (Community Visitors Evaluation and Intrusion Detection);
- Endpoint (Anti-Malware);
- Application (Waf or Database Firewalls);
- User (Uba);
- Technique (Anti-Fraud).

Each layer has standout subcategories. For instance, local area protection can be categorized as Wired, Wireless, or Cloud. Rest assured that equivalent calculations with equivalent hyperparameters for each location cannot be determined in the near future. The reason is the lack of data and algorithms to identify higher conditions of the three regions collectively, where substituting one set of rules for differences is appropriate.

The 0.33 measurement is a question of "How" (e.g., how to test the protection of a specific location):

- In Transit in Actual Time;
- At Rest; and
- Historically.

For example, in the case of approximate endpoint safety, one could investigate intrusion detection techniques, perform static binary analysis, and examine endpoint activity logs. Some obligations require resolution in three dimensions. Occasionally, certain obligations may not have values in certain dimensions. Approaches may be similar in one dimension, but each unique aspect of the three-dimensional cybersecurity landscape has its intricacies. It is challenging to address all of them, so let us focus on the most crucial dimension — the layers of technology. Consider the cybersecurity solution from this perspective [11].

16.7 MACHINE LEARNING FOR NETWORK PROTECTION

Network safety is not merely a physical location, but rather a collection of solutions that target various protocols, such as Ethernet, wireless, SCADA, or even digital networks like SDNs.

When it comes to network safety, famous Intrusion Detection System (IDS) solutions come to mind. Some of these solutions utilized Machine Learning (ML) years ago, primarily relying on signature-based approaches.

ML in network protection introduces new solutions known as Network Traffic Analytics (NTA), which aim to perform in-depth analysis of all the traffic at every layer, detecting attacks and anomalies [13] (Figure 16.3).

16.7.1 HOW CAN ML ASSIST HERE? THERE ARE A FEW EXAMPLES

- To regress, we need to consider the community packet parameters and evaluate them alongside the ordinary ones;

- Category to become aware of specific training on community assaults, including scanning and spoofing; and
- Clustering for forensic evaluation.

FIGURE 16.3 Machine learning for network protection.

16.8 MACHINE LEARNING TECHNIQUES FOR CYBER SECURITY

This proposal introduces the use of Bayesian groups to recognize attack signatures. Various popular ML techniques are discussed in this section. For each method, the effectiveness in cyber protection is highlighted. Three Bayesian networks are utilized, representing an interconnected acyclic graph of factors and their respective conditions. The nodes representing the variables depend on the determined nodes, and each node follows the conditions of the corresponding probability distribution and the random variable. Each node contributes to the overall outcome with different values. The specific probability tables are determined and illustrated in Figure 16.4.

Bayesian networks can be used to model unobserved variables and compare perceived attack patterns with streaming records of known attacks. They are also useful for detecting unseen intrusions. Jemili et al. improved the accuracy of an intrusion detection system using Bayesian networks. The KDD 1999 dataset was used, with nine attributes used to train the system. An average precision of 88% and 89% was achieved in normal and attack scenarios. The system achieved detection rates of nearly 100%, 21%, 89%, and 7% for Test, filter, DOS, and R2L attacks. However, due to limited training data for R2L attacks, the accuracy of the model suffered.

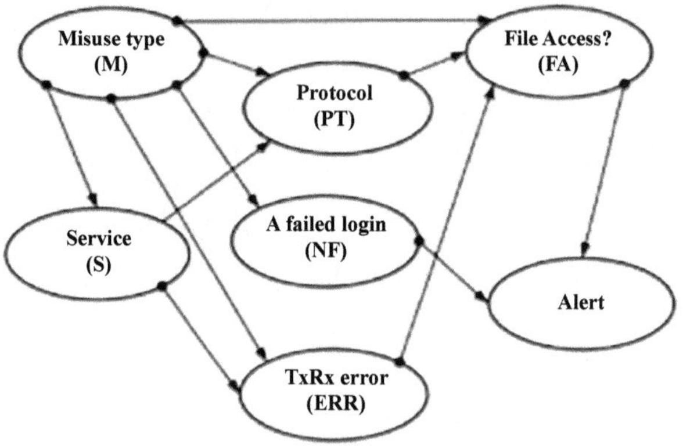

File Access state input variables and values	P(FA = True)	P(FA = False)
M=R2H, PT=NSF, ERR=0	0.95	0.05
M=R2H, PT=FTP, ERR=0	0.99	0.01
M=Probe, PT=none, ERR=50%	0.80	0.20
M=Probe, PT=PING, ERR=0	0.50	0.50
M=DoS, PT=POP, ERR=100%	0.80	0.20
M= DoS, PT=HTTP, ERR=50%	0.90	0.10

FIGURE 16.4 Machine learning techniques for cyber security.

Decision tree: A decision tree is a hierarchical structure resembling a tree. The leaves represent different classifications, and the branches are the links or features that guide the path to the classifications. ID3 and C4.5 are two well-known algorithms used to automatically construct decision trees from data. The approach of the Grunt towards the approaching visitors is steady due to the wide range of marks available. Kruegel and Toth et al. [15] modified 150 Grunt strategies by utilizing various ID3 algorithms. Their objective is to update the set of rules using a decision tree variant.

This approach may greatly enhance processing speed. Rule clustering is employed to update the GRUNT strategies, which reduces the number of necessary examinations. Additionally, it enables fair evaluation and accelerates the evaluation process. The clustered rule is applied to the DARPA 1999 dataset. The processing speed and performance of the enhanced version are compared to the baseline assessment.

Although Toth does not currently provide any quantitative figures, the study identified a significant speed improvement using the decision tree approach. The variant achieved a maximum improvement of 105%, with a minor speed increase of 5%. Furthermore, the number of strategies increased from 150 to 1,581 for further experimentation. Moreover, the processing time has significantly decreased. Three Grouping is an unmanaged approach that focuses on using a similarity degree to associate records together. Clustering algorithms can be studied from review records, and the explicit representation of different attacks prepared by the machine administrator is not necessary.

Hendry et al. [16] demonstrate that the continuous mark detection technique involves utilizing a clustering set of rules. The Straightforward Logfile Bunching Instrument (SLCT), a thickness-based thorough grouping plan, is employed to generate both conventional and abnormal local area visitors. Two grouping plans are utilized: Firstly, for identifying conventional and attack scenarios, the alternative plan can be used to select the expected visitors in a controlled manner. In this version, the M boundary is set to 97% to form the characteristic within the clustering. By setting the M boundary to 97%, 98% of attack records are detected with a 15% FAR. The marks are generated from the instances of excessive density. The KDD dataset is used to validate the developed version. Clustering is employed as the overall performance metrics enhance the accuracy of the model. An accuracy of 70% to 80% is achieved for attacks.

Considering the nature of attacks (new or zero-day), this level of accuracy is commendable. The ANN functions similarly to the human brain. The brain's local area has a layered structure. The input from the records stimulates the neuron in the subsequent layer of the area, which in turn leads to the next layer of classification. This process continues, and ultimately the output is generated through the final layer of the area. The internal area, which plays a crucial role within the local brain area, is isolated from external factors and is referred to as the hidden layer. One significant drawback of the local brain area is the strong focus on time due to the presence of local minima.

This method became nonexclusive in the mid-90s due to the emergence of SVMs. ANN started to decline. With the introduction of convolutional NN, the recognition of the brain's local area is once again gaining attention. Canady [17] describes an ANN version that utilizes a multi-class classifier to detect anomalies. The attack signatures have been incorporated into the system. Legitimate data is used to generate the records. Approximately 3,000 attacks have been simulated using tools such as Satan and Web Scanner out of the 10,000 recorded attacks.

The records are pre-processed using nine selected features: ICMP code, ICMP type, source address, destination address, protocol identifier, source port, destination port, raw data length, and raw data type. Then, the ANN is trained and evaluated using both normal and attack records. Canady et al. report a false positive rate of 0.058 and 0.070 during the training and testing phases. Therefore, an RMS of 0.070 translates to an average accuracy of 93% for the testing phase [14].

Five hereditary arrangements of rules and genetic programming are used in this study. Two of the most prevalent algorithm strategies based on the principle of natural selection are Genetic Algorithm (GA) and Genetic Programming (GP). These algorithms involve populations of chromosomes that are based on specific operators. The three basic operators used are selection, crossover, and mutation. The data is categorized as normal visitors or malicious visitors. The set of rules begins with randomly generated individuals, and a fitness cost is calculated for each individual. This indicates the ability of each individual to solve the current problem, and individuals with better chances are more likely to be selected in the mating pool.

Two successful individuals will undergo the next step called crossover, and eventually, everyone will undergo mutation. The best-fit chromosome will be preserved among the two mutated individuals over time. Three popular GP models have been used in this study: Linear Genetic Programming (LGP), Gene Expression Programming (GEP), and Multi Expression Programming (MEP). Abhram et al. [18] used a simplified GP version to construct a classifier for attacks. The model used various mathematical operators as feature sets. The DARPA 1998 dataset was used as the full dataset validated the developed model. The dataset consisted of four main types (U2R, R2L, DoS, and testing) of attacks with 24 different attack scenarios.

The False Alarms (FAR) of the above adaptation changed from 0% to 5% depending on the attack being researched. 3.6 Hidden Markov Models (Gee) This is a measurable Markov variant with an immovable state that can be interconnected with the utilization of change prospects that determine the geography of the rendition. The machine is assumed to be a Markov strategy with unseen boundaries. This form deals with the cost of a forward-in-reverse relationship that can be used to determine the concealed boundaries from the discernible boundaries. Since the open-door dissemination in each country is unprecedented, the machine can extrude states beyond normal time and address nonstationary arrangements. Joshi et al.'s AI used Well to build an intrusion detection machine. Five states are used, each having six articulation pictures consistent with the country.

The interconnection among the states is developed in the kind of way that any nation can change into any phenomenal country. In order to appraise the Well boundaries, the Baum-Welch strategy might be used. For the approval of the adaptation, the KDD 1999 dataset changed into utilized. Out of the colossal 41 abilities of the datasets, five changes into chosen for the assessment. The excellent recognition charge of the rendition added up to 79%, with a tremendous phony expense of 21%. If more prominent than five capacities are utilized inside the assessment, the precision of the form would conceivably improve no quantitative assessment changed into executed with the guide of utilizing the creators to direct this advancement guarantee. Inductive gaining the surmising of positive information from a dataset is called derivation.

On the contrary, the strategy for transitioning from precise articulation to expanding speculations and styles is referred to as inductive learning. These are the two primary methods for extracting information from the data. Inductive assessment facilitates the identification of several common patterns, which can be utilized to draw multiple theoretical conclusions. Fan et al. [20] introduced a structured anomaly generator to generate random events and unusual website visitors. Two prominent distribution methods were based on the overall anomaly period, and the filtered synthetic anomalies were used to generate these random anomalies. These generated data points were randomly integrated into the DARPA 1998 dataset. Fan et al. utilized this dataset to examine the overall performance of the developed inductive learning model. The model achieved a successful detection rate of 94% and a low false alarm rate (FAR) of 2%. This analysis specified the precise methodology for constructing the dataset that could be utilized for anomaly detection. It confirmed the effectiveness of inductive learning in the advanced dataset [19].

16.9 HOW IS MACHINE LEARNING BRINGING DOWN CYBERSECURITY COSTS?

Artificial intelligence (AI) and machine learning reduce the effort required to detect and respond to cyber threats, making them cost-effective solutions. According to Capgemini's research, the average cost reduction is 12%, with some companies saving more than 15%.

Machine learning can predict and mitigate risks before attackers can exploit them. By simulating scenarios and conducting high-level assessments, machine learning can anticipate potential vulnerabilities. For instance, an organization may utilize machine learning to identify social engineering attacks, user accounts with stolen credentials, and other risks [22].

Machine learning can also be used to prepare for and respond to attacks. For example, an IT department may use machine learning to create a security posture assessment to help organizations prepare for attacks [21].

In conjunction with data mining, machine learning can analyze large volumes of data in seconds, significantly faster than humans. Furthermore, they can identify fixes and mitigate risks in near real-time, greatly reducing the response time. The ability of attackers to quickly gain access to an organization's infrastructure makes sharp detection and response essential.

16.10 BENEFITS OF MACHINE LEARNING IN CYBERSECURITY

There are many advantages to using devices to study issues within the cybersecurity space. These include:

1. **Rapidly Synthesize Vast Volumes of Information:** One of the most demanding situations faced by analysts is the need to synthesize intelligence generated across their attack surface quickly, which has generally grown much faster than their teams can manually process. Machine learning can rapidly analyze large volumes of historical and dynamic intelligence, enabling teams to operationalize information from various sources in near real time.
2. **Activate Professional Intelligence at Scale:** Regular schooling cycles allow fashion to constantly examine their evolving pattern population, incorporating analyst-classified detections or analyst-reviewed indicators. This prevents false positives and permits fashion to examine and put professional-generated ground truth into effect.
3. **Automate Repetitively, Guide Responsibilities:** Applying machine learning to specific tasks can help alleviate security teams from mundane, repetitive tasks, acting as a force multiplier that enables them to scale their response to incoming signals and redirect time and resources towards complex, strategic projects.
4. **Augment Analyst Efficiency:** Machine learning can enhance analyst perception with real-time, updated intelligence, enabling analysts in risk assessment and security operations to effectively prioritize

resources to address their organization's critical vulnerabilities and investigate time-sensitive ML-alerted detections [20].

16.11 USE CASES OF MACHINE LEARNING IN CYBERSECURITY

Machine learning has a wide, ever-developing variety of use cases within the cybersecurity space. We can categorize these use cases into main groups:

1. Automated danger detection and reaction
2. Analyst-led operations assisted through device studying

16.11.1 AUTONOMOUS THREAT DETECTION AND RESPONSE

In the primary category, machine learning enables organizations to automate manual work, particularly in processes where it is crucial to maintain high levels of accuracy and respond with machine-level speed–such as automated threat detection and response or classifying new adversary patterns.

Applying machine learning in these scenarios enhances signature-based methods of threat detection with a generalized approach that learns the differences between benign and malicious samples and can quickly detect new in-the-wild threats.

16.11.2 DRIVING ANALYST EFFICIENCY WITH MACHINE LEARNING

Machine learning models can also assist in analyst-led investigations by alerting teams to investigate detections or by providing prioritized vulnerabilities for patching. Analyst evaluation can be particularly valuable in scenarios where there is insufficient data for models to predict outcomes with high levels of confidence or to analyze benign-performing behavior that may go unnoticed by malware classifiers [23].

16.12 EVALUATING THE EFFICACY OF MACHINE MODELS

16.12.1 MODEL EFFICACY FOR MALWARE CLASSIFIERS

One of the most common applications of machine learning in cybersecurity is malware classification. Malware classifiers provide a scored prediction on whether a given pattern is malicious or not, with "scored" referring to the confidence level associated with the classification. One way to evaluate the performance of these models is by representing predictions along two axes: accuracy (whether the results were correctly classified as "true" or "false") and output (the class assigned to a pattern by the model; "positive" or "negative"). It is important to note that the terms "positive" and "negative" in this context do not imply that a pattern is respectively "benign" or "malicious." If a malware classifier makes a "positive" detection, it means that the model predicts the pattern to be malicious based on the observed features it has learned from known malicious samples. To illustrate the meaning of these groupings, let's consider the example of models trained to analyze malicious files.

- **True Positive:** The version successfully expected a document to become malicious.
- **True Negative:** The version successfully expected that a document would no longer malicious.
- **False Positive:** The version incorrectly expected that a document became malicious (and it was not).
- **False Negative:** The version incorrectly expected that a document was not malicious (and it became).

16.13 CHALLENGES AND CONSTRAINTS OF MACHINE LEARNING

While studying fashion systems may be practical tools, each version operates within precise limitations:

- **Sufficient Exquisite Information:** Training high self-assurance models regularly requires access to substantial data sets, both to train and test machine learning models. In order to test models, a subset of the data is commonly set aside from the training set to evaluate the model's performance. This data should have minimal feature overlap with the training data; for example, representing a specific time span of data series or originating from a different data source. If there is insufficient quality data, a given problem area will not be suitable for implementing machine learning.
- **Tradeoffs among True and False positives:** As mentioned earlier, the sensitivity of each version needs to be calibrated in order to balance the detection threshold between true and false positives, thus maximizing detection efficacy.

- **Explain Ability:** Explainability refers to the capability to explain how and why a model behaves as it does. This enables information technology teams to understand which features in a model affect its performance and their relative importance. Explainability is crucial for establishing accountability, building trust, ensuring compliance with data regulations, and ultimately enabling continuous performance improvement in machine learning.
- **Repeatability:** Also referred to as reproducibility, this refers to the capability of machine learning experiments to be consistently reproduced. Repeatability drives transparency regarding how machine learning is used, what types of models are used, what data they are trained on, and what software environments or versions they operate in. Repeatability minimizes ambiguity and reduces potential errors as models move from testing to deployment and future update cycles.
- **Optimization for Target Environment:** Each version needs to be optimized for its goal manufacturing environment. Each environment will vary in its computational resources, memory, and connectivity availability. Therefore, every version has to be designed to perform in its deployment environment without burdening or interrupting the operations of the target host.
- **Hardening towards Adversarial Attacks:** Machine learning models have their attack surface that may be vulnerable to adversary attacks, in which adversaries may also attempt to exploit or manipulate model behavior (including causing the model to misclassify samples). To minimize the exploitable attack surface of models, data scientists "harden" models during training to ensure overall robust performance and resilience against attacks.

16.14 MISCONCEPTIONS MACHINE LEARNING

- **Misconception #1.** Machine learning is superior to traditional analytical or statistical techniques. However, machine learning may not be suitable for every problem area. Other analytics or statistical techniques may yield highly accurate and effective results, or even be more suitable for a given problem area, despite being less resource-intensive than a machine learning approach.
- **Misconception #2.** Machine learning should be used to automate as many tasks as possible. Machine learning can be a valuable and comprehensive resource, often requiring access to large amounts of

data, computational resources, and dedicated IT teams to build, train, and maintain models. To maximize the ROI of model maintenance, it is best implemented when target problems are high-value, frequently recurring, require speed and accuracy, and have sufficient quality datasets for ongoing training and testing.

16.15 HOW MACHINE LEARNING ENABLES THE FUTURE OF CYBERSECURITY

Machine learning helps provide cutting-edge cybersecurity solutions in various distinct ways. Individually, each one is valuable, and collectively they are game-changing for maintaining a strong defense posture in a dynamic threat landscape.

- **Identification and Profiling:** With new gadgets related to organizational networks, it is now imperative for an IT corporation to be aware of them. Machine learning can be used to identify and profile devices in a network. This profile can determine the distinct features and behaviors of a given device.
- **Automated Anomaly Detection:** Using system learning to detect recognized horrific behaviors unexpectedly is a first-rate use case for protection. After first profiling gadgets and knowledge of everyday activities, system learning knows what is daily and what is not.
- **Zero-Day Detection:** With conventional protection, a malicious actor must be visible at least once for it to be recognized as a malicious action. That is how legacy signature-based malware detection works. Machine learning can intelligently identify previously unknown types of malware and attacks to help protect organizations from potential zero-day attacks.
- **Insights at Scale:** With information and programs in many different locations, being able to identify trends across large volumes of devices is no longer humanly possible. Machine learning can do what humans cannot, enabling automation for insights at scale.
- **Policy Suggestions:** Constructing protection regulations is often a challenging endeavor with numerous obstacles. By utilizing machine learning, it becomes possible to provide policy recommendations for security devices, such as firewalls, based on an understanding of what constitutes normal behavior and what devices are present. Instead of manually navigating through conflicting access control lists for

different devices and network segments, machine learning can automate this process by making specific recommendations. With the increasing number of devices and threats emerging daily, and limited human security resources, machine learning is essential in handling complex situations and scenarios at scale, enabling organizations to address the cybersecurity challenge now and in the future [24].

16.16 MACHINE LEARNING IN THE DOMAIN OF CYBERSECURITY

Machines gaining knowledge in the cybersecurity field are now able to detect cyber-attacks, assisting humans in effectively managing and protecting their systems. This has led to the development of software programs that can handle complex tasks like mathematical calculations, which are challenging for humans. Consequently, the demand for human intervention has significantly increased. Subsequently, the next step involved enhancing the capabilities of these software programs through the implementation of artificial intelligence and machine learning techniques. As technology advanced, the amount of data being generated grew exponentially, minute by minute, hour by hour, and day by day. This led to the emergence of "big data," and as a result, systems became more intelligent in processing and gaining a deeper understanding of data. Numerous algorithms have been developed (and are still being developed) in response to technological advancements. Currently, these algorithms are utilized in various fields of study, such as image processing, speech recognition, biomedical research, and cybersecurity.

The purpose of device learning in cybersecurity is to provide a mechanism for software programs to mimic human behavior. The field of cybersecurity is a critical area of research. According to the Centre for Strategic and International Studies in 2014, annual costs to the global economy due to cybercrimes were estimated to be between $375 billion and $575 billion. Other sources may have different figures, but the average cost of a data breach incident for large companies exceeds $3 million. Researchers have developed intelligent systems to mitigate these costs in the cybersecurity field.

16.17 BIG FACTS AND MACHINE-GAINING KNOWLEDGE OF CYBERSECURITY

Machine learning is a part of artificial intelligence that enables computers to learn without being explicitly programmed. It involves the development of computer programs that can improve and adapt when provided with new data. Data generated by computers or sensors is collected and analyzed to extract valuable insights. The rise of machine learning can be attributed to the abundance of data available today. This phenomenon is commonly known as big data [25].

Big data is constantly being generated from various sources such as digital and social media platforms. It is transmitted through systems, sensors, and smartphones. Big data is often characterized by the three Vs:

- Volume;
- Variety; and
- Velocity.

Big data and machine learning are two complementary components that enhance each other. While significant data does not necessarily mean a specific quantity of data, the term is often used to describe terabytes, petabytes, or even exabytes of data accumulated over time. With the widespread adoption of IoT technology, the volume of data to be processed is expected to significantly increase in the future. To effectively handle big data, we need to employ machine learning techniques and develop intelligent frameworks that can leverage large datasets. Deep learning is currently one of the most prominent topics in machine learning, as it enables the creation of highly accurate intelligent systems by leveraging the depth of extensive data.

Real-time examples of cybersecurity utilizing machine learning include:

- Gmail utilizes device learning to filter emails and systems, providing protection against malicious emails.
- Google employs Deep Learning in its systems and applications. Deep Learning, a form of device learning, enables a deeper understanding of users and can proactively assist software based on data-driven insights.
- IBM introduced Watson cognitive learning, which expanded the scope of device learning for threat detection and cybersecurity purposes.
- Balbix is a platform that offers solutions for data breaches and security breaches. It utilizes AI-managed threat expectations to mitigate real-time security risks.

16.18 CONCLUSION

There are still significant areas to explore. They have even touched upon the fundamentals. Undoubtedly, there are challenges with interpretability, especially in deep learning algorithms. However, humans also struggle to interpret their own decisions accurately. On the other hand, machine learning is not a one-size-fits-all solution for security. Nevertheless, with the increasing amount of data and the utilization of various experts, ML is the most viable solution. It is effective now and will soon become essential. Despite its current limitations, considering this, hackers are also beginning to employ ML in their attacks.

KEYWORDS

- **artificial intelligence**
- **denial of service**
- **direct hereditary programming**
- **extraction, transformation, and load**
- **multi articulation programming**
- **quality articulation programming**
- **straightforward logfile bunching instrument**

REFERENCES

1. Handa, A., Sharma, A., & Shukla, S. K., (2019). Machine learning in cybersecurity: A review. *Wiley Interdisciplinary Reviews: Data Mining and Knowledge Discovery, 9*(4), e1306.
2. Dua, S., & Du, X., (2016). *Data Mining and Machine Learning in Cybersecurity*. CRC press.
3. Dasgupta, D., Akhtar, Z., & Sen, S., (2022). Machine learning in cybersecurity: A comprehensive survey. *The Journal of Defense Modeling and Simulation, 19*(1), 57–106.
4. Fraley, J. B., & Cannady, J., (2017). The promise of machine learning in cybersecurity. In: *Southeast Con 2017* (pp. 1–6). IEEE.
5. Thomas, T., Vijayaraghavan, A. P., & Emmanuel, S., (2020). Machine learning and cybersecurity. In: *Machine Learning Approaches in Cyber Security Analytics* (pp. 37–47). Springer, Singapore.
6. Sarker, I. H., Kayes, A. S. M., Badsha, S., Alqahtani, H., Watters, P., & Ng, A., (2020). Cybersecurity data science: An overview from machine e-learning perspective. *Journal of Big Data, 7*(1), 1–29.
7. Xin, Y., Kong, L., Liu, Z., Chen, Y., Li, Y., Zhu, H., & Wang, C., (2018). Machine learning and deep learning methods for cybersecurity. *IEEE Access, 6*, 35365–35381.
8. Apruzzese, G., Laskov, P., De Oca, E. M., Mallouli, W., Rapa, L. B., Grammatopoulos, A. V., & Franco, F. D., (2022). The role of machine learning in cybersecurity. *Digital Threats: Research and Practice*.

9. Martínez, T. J., Iglesias, C. C., & García-Nieto, P. J., (2019). Machine learning techniques applied to cybersecurity. *International Journal of Machine Learning and Cybernetics, 10*(10), 2823–2836.
10. Dasgupta, P., & Collins, J., (2019). A survey of game theoretic approaches for adversarial machine learning in cybersecurity tasks. *AI Magazine, 40*(2), 31–43.
11. Sarker, I. H., Abushark, Y. B., Alsolami, F., & Khan, A. I., (2020). IntruDTree: A machine learning-based cyber security intrusion detection model. *Symmetry, 12*(5), 754.
12. Bresniker, K., Gavrilovska, A., Holt, J., Milojicic, D., & Tran, T., (2019). Grand challenge: Applying artificial intelligence and machine learning to cybersecurity. *Computer, 52*(12), 45–52.
13. Shaukat, K., Luo, S., Varadharajan, V., Hameed, I. A., Chen, S., Liu, D., & Li, J., (2020). Performance comparison and current challenges of using machine learning techniques in cybersecurity. *Energies, 13*(10), 2509.
14. Das, R., & Morris, T. H., (2017). Machine learning and cyber security. In: *2017 International Conference on Computer, Electrical & Communication Engineering (ICCECE)* (pp. 1–7). IEEE.
15. Ford, V., & Siraj, A., (2014). Applications of machine learning in cyber security. In: *Proceedings of the 27th International Conference on Computer Applications in Industry and Engineering* (Vol. 118). Kota Kinabalu, Malaysia: IEEE Xplore.
16. Musser, M., & Garriott, A., (2021). *Machine Learning and Cybersecurity.* Center for Security and Emerging Technology: Washington, DC, USA.
17. Annamalai, C., (2022). Combinatorial and multinomial coefficients and its computing techniques for machine learning and cybersecurity. *The Journal of Engineering and Exact Sciences, 8*(8), 14713–01i.
18. Musser, M., & Garriott, A., (2021). *Machine Learning and Cybersecurity*. Center for Security and Emerging Technology: Washington, DC, USA.
19. Sornsuwit, P., & Jaiyen, S., (2019). A new hybrid machine learning for cybersecurity threat detection based on adaptive boosting. *Applied Artificial Intelligence, 33*(5), 462–482.
20. Apruzzese, G., Colajanni, M., Ferretti, L., Guido, A., & Marchetti, M., (2018). On the effectiveness of machine and deep learning for cyber security. In: *2018 10th International Conference on Cyber Conflict (CyCon)* (pp. 371–390). IEEE.
21. Boukerche, A., & Coutinho, R. W., (2020). Design guidelines for machine learning-based cybersecurity in the internet of things. *IEEE Network, 35*(1), 393–399.
22. Miao, Y., Chen, C., Pan, L., Han, Q. L., Zhang, J., & Xiang, Y., (2021). Machine learning-based cyber-attacks targeting on controlled information: A survey. *ACM Computing Surveys (CSUR), 54*(7), 1–36.
23. Kamhoua, C. A., Kiekintveld, C. D., Fang, F., & Zhu, Q., (2021). *Game Theory and Machine Learning for Cyber Security.* John Wiley & Sons.
24. Shaukat, K., Luo, S., Varadharajan, V., Hameed, I. A., & Xu, M., (2020). A survey on machine learning techniques for cyber security in the last decade. *IEEE Access, 8*, 222310–222354.
25. Calix, R. A., Singh, S. B., Chen, T., Zhang, D., & Tu, M., (2020). Cyber security tool kit (CyberSecTK): A Python library for machine learning and cyber security. *Information, 11*(2), 100.

CHAPTER 17

Cyber Security in Deep Learning: Challenges and Applications

M. JAITHOON BIBI and V. KRISHNAPRIYA

Department of Computer Science with Cognitive Systems, Sri Ramakrishna College of Arts and Science, Nava, Coimbatore, Tamil Nadu, India

ABSTRACT

This document presents an in-depth analysis of the potential use of deep learning in cyber-security. It includes a detailed overview of various cyber-attack detection methods and discusses the application of deep learning techniques such as generative adversarial networks (GANs). The datasets used to evaluate the effectiveness of cyber-attack detection methods are also discussed. The numerical study of documents published on cyber security along the function of "Deep learning" is controlled. Surviving viable cyber-security solutions established on "deep learning" to be labeled.

17.1 INTRODUCTION

Cyber-security appears to be essential for safeguarding dangerous infrastructures from attacks and illegal access. Numerous technologies and techniques are used in cyber-security. Some types of cyber-security are end-user learning, application–confidence, evidence–confidence, system–confidence disaster–rescue, and operative security. Although being a recent war without weapons, cyber-attacks are the most devastating and harmful because they expose individual and business information, disturb vital processes, exploit ongoing vulnerabilities, allow illegal and restricted access to hardware and

Leveraging Artificial Intelligence (AI) Competencies for Next-Generation Cybersecurity Solutions.
Pethuru Raj, B. Sundaravadivazhagan, & V. Kavitha (Eds.)

software, and have a substantial negative impact on the national economy. Some Attacking activities that aim to hack computer information-structures, organizations, computer networks, or laptop devices are referred to as cyber-attacks. Nation-states, persons, organizations, cultures, or organizations can all launch cyber-attacks. A cyber-attack could come from an unknown source. A specific target may be taken, changed, or eliminated through a cyber-attack.

Even if there are now enough attack detection systems provided, the rapid growth in criticism and the development of attacking skills require the creation of innovative funding types. Even though other ML techniques have been effective in recent years, their scalability over large networks and the capacity to identify cyber-attacks in highly scattered contexts remain major challenges. The fact that conventional ML algorithms use constructed features for the appreciation task is one of their difficulties. "Deep-learning" is the main AI recent research that is now receiving the most research, and it offers several potential ways to overcome the constraints of conventional "ML-ML" techniques. "Deep learning" prepares faster and more detailed analysis.

At present the survey conditions on cyber-security types are established on different "Deep-Learning" structures are evaluated through this paper, along with the fundamental concepts used in this paper, its benefits, risks, and a description of the open access datasets used in experimental tests of various types. The investigation of the DL technique for cyber-security attack detection is the main focus of our study.

Therefore, our primary contribution to this work is:

1. Analysis of current deep-learning techniques used to solve cyber-attack detection problems.
2. Classifying the Cyber-Security Deep-Learning methods that have succeeded.
3. Carrying out a statistical study of cyber-security deep-learning techniques that have succeeded.

17.2 PRIOR REVIEWS ON DEEP LEARNING IN THE CYBER-ATTACK DETECTION

The detection of cyber attacks is a major area of study, as evidenced by several surveys, analyses, and books. The examination of "CS-Deep-Learning" application issues in cyber-security is written quite frequently.

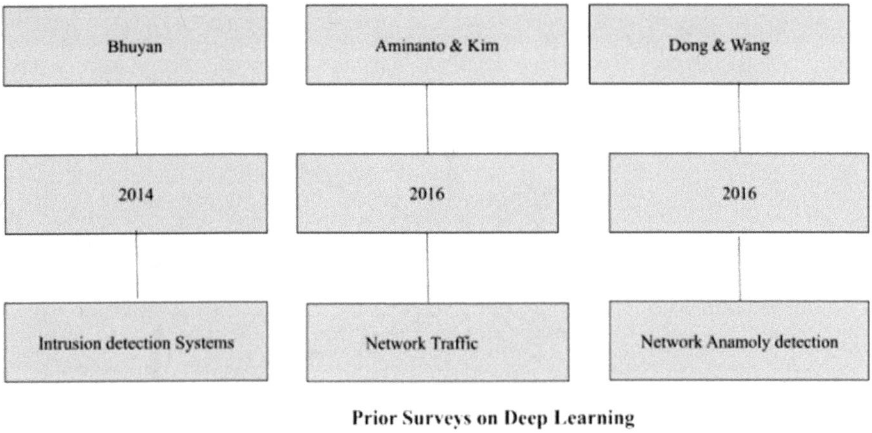

FIGURE 17.1 Prior surveys on deep learning.

17.3 "DEEP-LEARNING" USED IN CYBER HACKING DETECTION METHODS

The construction of a deep cyber-attack detection method classification scheme is shown in Figure 17.1. This plan was designed using the methods that were utilized. The strategies for detecting cyber-attacks in this study are divided into seven different types. Convolutional Neural Networks (CNN), "Restricted-Boltzmann-Machines" (RBM), Deep Belief Networks (DBN), Auto-Encoders, "Long Short-Term Memories" (LSTM), Deep Feedforward Neural Networks, and combination classifiers are examples of them. The methods built within the grouping learner classes are subdivided into sub-classes in Figure 17.2.

- **CNN Methods:** A CNN uses an image copy, provides it a load according to the various items in the image copy, and then differentiates from one another. As a combination of various "deep-learning" techniques, CNN-DL requires a very small initialization of the input. Most specific CNN's methods train its classifiers using antiquated techniques, which supports it to understand more about the properties of the particular object.

  CNN-DL's built using the same designs as that of "human brain neurons," other particular "Visual-Cortex." Each Neuron replies to a particular reward in a certain region of the visual field known as the accessible field. Those compilations extend beyond to capture the entire visual system (Figure 17.3).

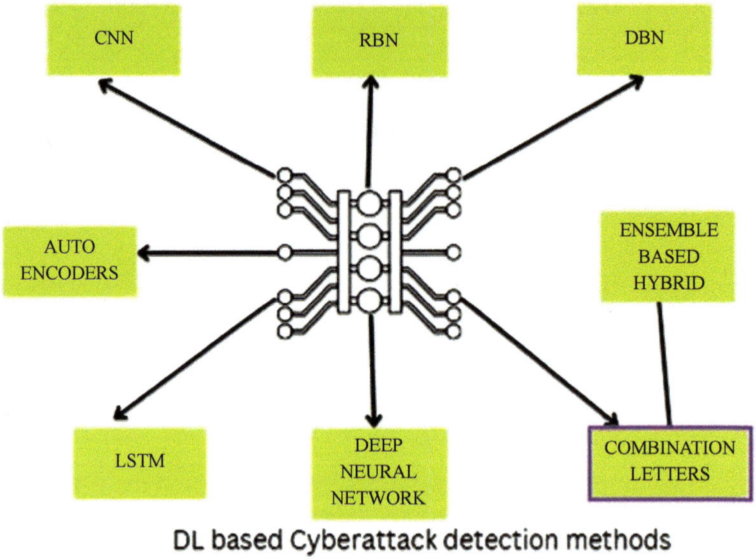

FIGURE 17.2 DL-based cyberattack detection methods.

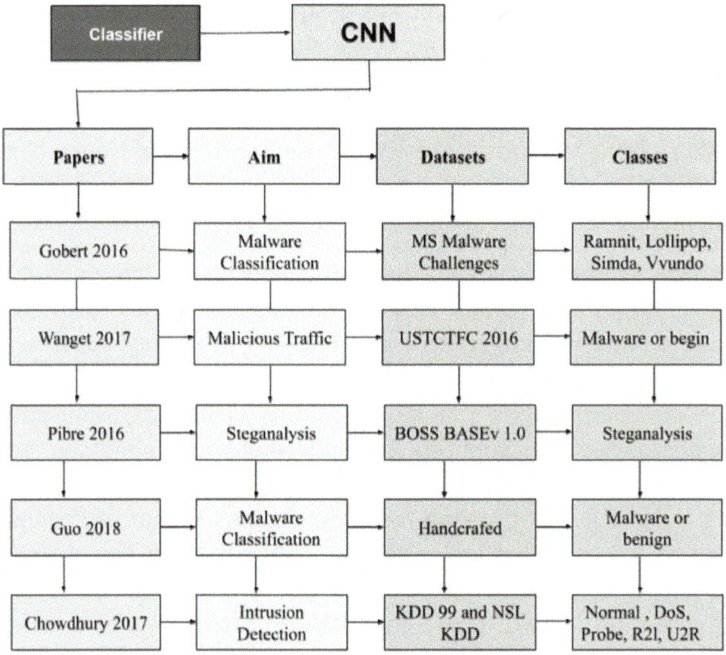

FIGURE 17.3 Review of surveys on CNN-based cyber-attack detection techniques.

- **RBN Techniques:** The "Gaussian-Bernoulli type" RBM technique is implemented, RBM and DBN are proposed, and this type for the discovery of network "DL-anomalies" based on a "Semi-Supervised algorithm" methodology is approached. The RBM is utilized in this work for feature reduction (Figure 17.4).

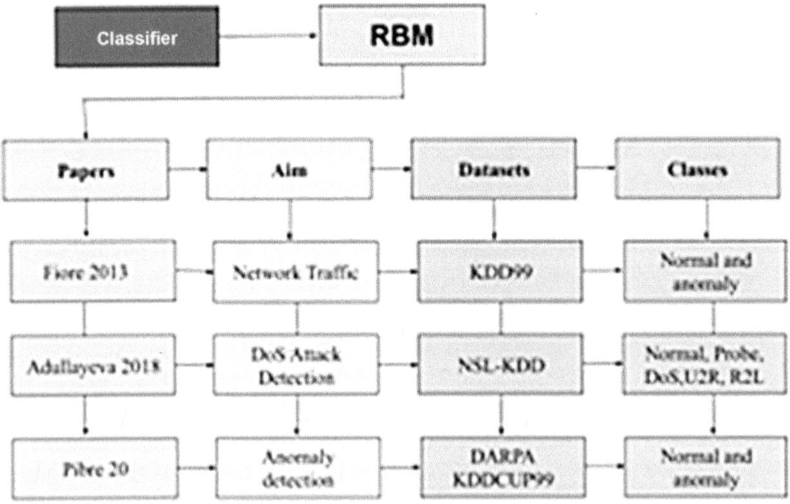

FIGURE 17.4 Review of surveys on RBN-based cyber-attack detection techniques.

- **DBN Method:** A "Cyber Security Deep Neural Network" (CS-DNN) based intrusion detection and prevention of (IDS), a D-B-N based IDS, and a malicious software detection technique types based on D-B-N are all suggested in a "DoS DL" hacking detection techniques based on Multi-layer D-B-Ns (Figure 17.5)

17.4 COMBINATION LEARNERS

Combinations of several approaches are studied in this chapter.

- **Ensemble Methods:** A decision is made by combining different perspectives using ensemble methods. There are three methods for constructing ensembles (Figure 17.6):

  1. Bagging;
  2. Boosting; and
  3. Stacking.

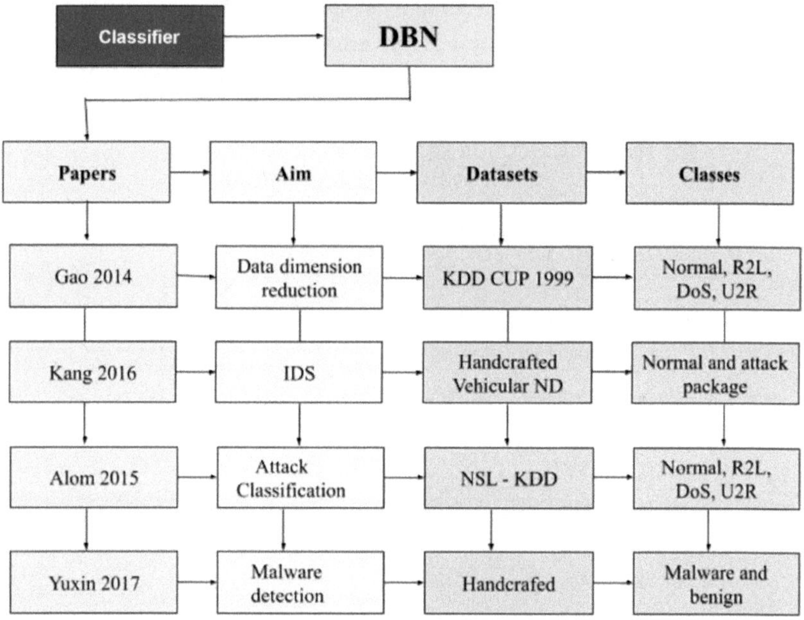

FIGURE 17.5 Review of surveys on DBN-based cyber-attack detection techniques.

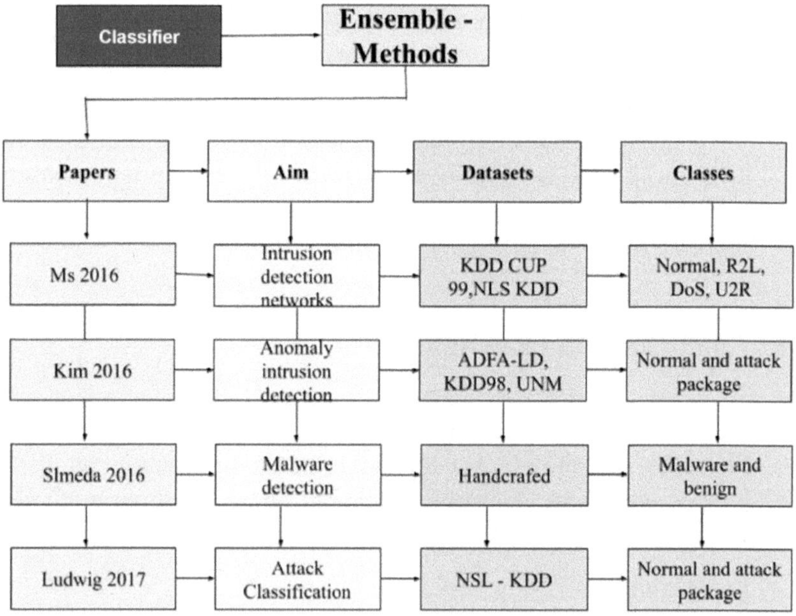

FIGURE 17.6 Review of surveys on ENSEMBLE-based cyber-attack detection techniques.

- **Hybrid Methods:** A hybrid intrusion detection approach built on a combination of D-B-N and S-V-M models is proposed, as is a fusion malicious software hacking approach based on hybridization of the Neural Networks and Deep Learning approaches, an unknown protocol detection technique, DL-based malware detection approach, IDS architecture, etc. The hybridization of classification models enhances the effectiveness of cyber-attack detection methods (Figure 17.7).
- **Auto-Encoder Methods:** This solution to network attack detection is dependent on the sparse auto-encoder and naïve deep learning techniques. A method of malware detection for programming interface requests is proposed. The "LSTM"-based decoder type is suggested for the acknowledgment of CAPTCHA images; a deep learning malicious detection method comprised of stacked-auto-encoders (SAEs) and a deep learning network interference detection model are also suggested; the threats and hacking focused on the 802.11 IEEE wireless network are evaluated; and an SAE-based approach is suggested for the classification of these hacks. A novel cyber-security model and a disseminated deep learning approach for detecting cyber-attacks in cloud-based computing are both proposed networks. An anomaly detection model based on autoencoder and feed forward NN architectures is presented and given the title Kitsune (Figure 17.8).
- **LSTM Methods:** An advanced architecture of Recurrent Neural Network (RNN) has been developed by humans and is widely used in the field of deep learning. Its most commonly used variant is the Long Short-Term Memory (LSTM) network, which is designed to eliminate the problem of long-term dependencies and includes feedback connections. Unlike other RNN models, LSTM can process entire sequences of knowledge along with individual data items, making it suitable for a wide range of applications such as detecting network traffic and IDS abnormalities, speech and pattern recognition, and intrusion detection systems.

LSTM units are composed of four major components, namely the cell, the input gate, the output gate, and the forget gate. These gates play a crucial role in normalizing the output from the cell by utilizing the cell's memory to store values over unexpected time intervals.

Since there are often intervals of unknown time duration between significant events in a time series, "DL-LSTM "networks are complementary to categorizing, handling, and building predictions based on "time series-data.""DL-LSTMs' 'were made to address the disappearance of

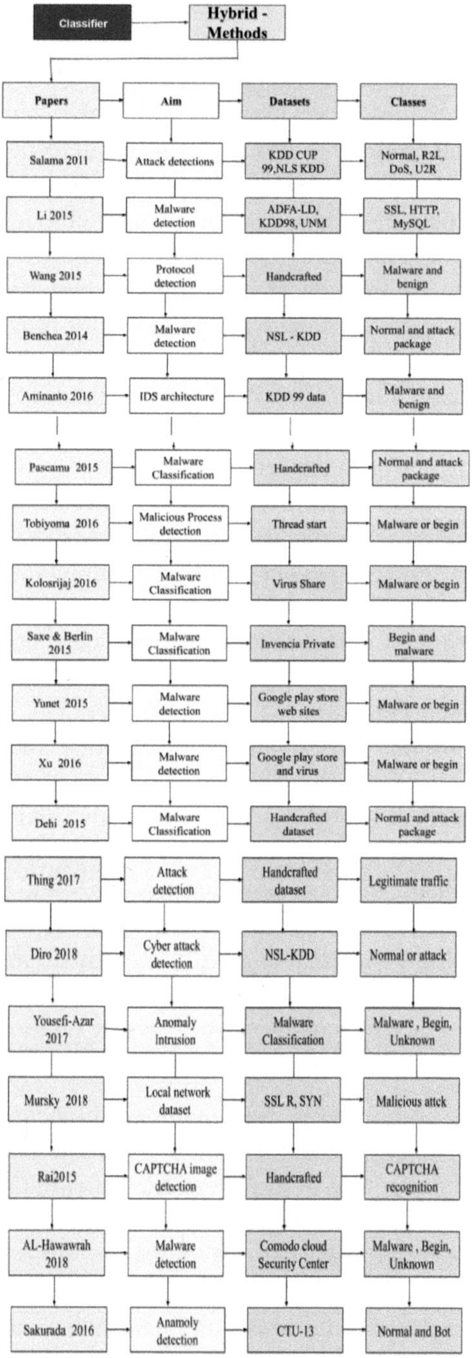

FIGURE 17.7 Review of surveys on hybrid-based cyber-attack detection techniques.

gradient issue capability when regular "DL-RNNs' are being trained. Often, "LSTM" has a benefit over RNNs, hidden "Markov- models," and some "sequence-learning" methods for finding "relative insensitivity" to breach length (Figure 17.9).

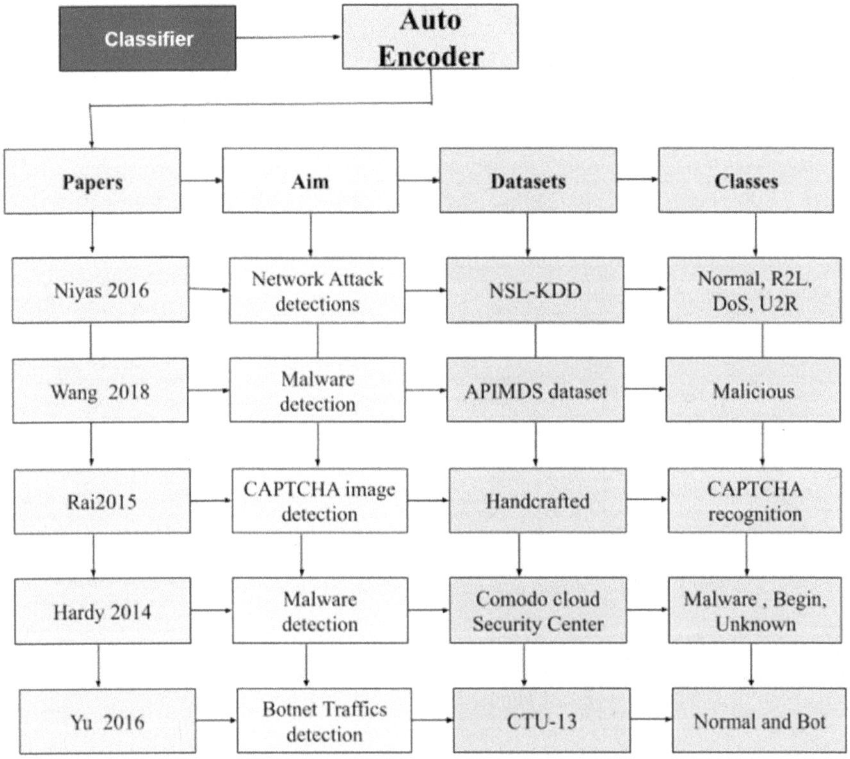

FIGURE 17.8 Review of surveys on auto-encoder-based cyber-attack detection techniques.

17.5 CYBER SECURITY–"DEEP NEURAL NETWORK (CS-DNN)"

CS-DNN should reveal the features of protected computer systems that affect the analysis used to detect network cyber-attacks by analyzing deviations from typical safety parameters. As a result, a neural network model that has been customized to detect network cyber-attacks on a particular computer system needs to be modified again to detect cyber-attacks on another computer system. In fact, the parameters of the "Neural Network Model" developed for signature identification are independent of the make-up and organization of a protected computer system. As a result, the obtained results will have a general pattern and enable measuring deep neural network (DNN) application potential for the security of various computer systems. As a result, efforts are focused on creating a "Deep-Neural-Network" model to identify the suggested network cyberattack (Figure 17.10).

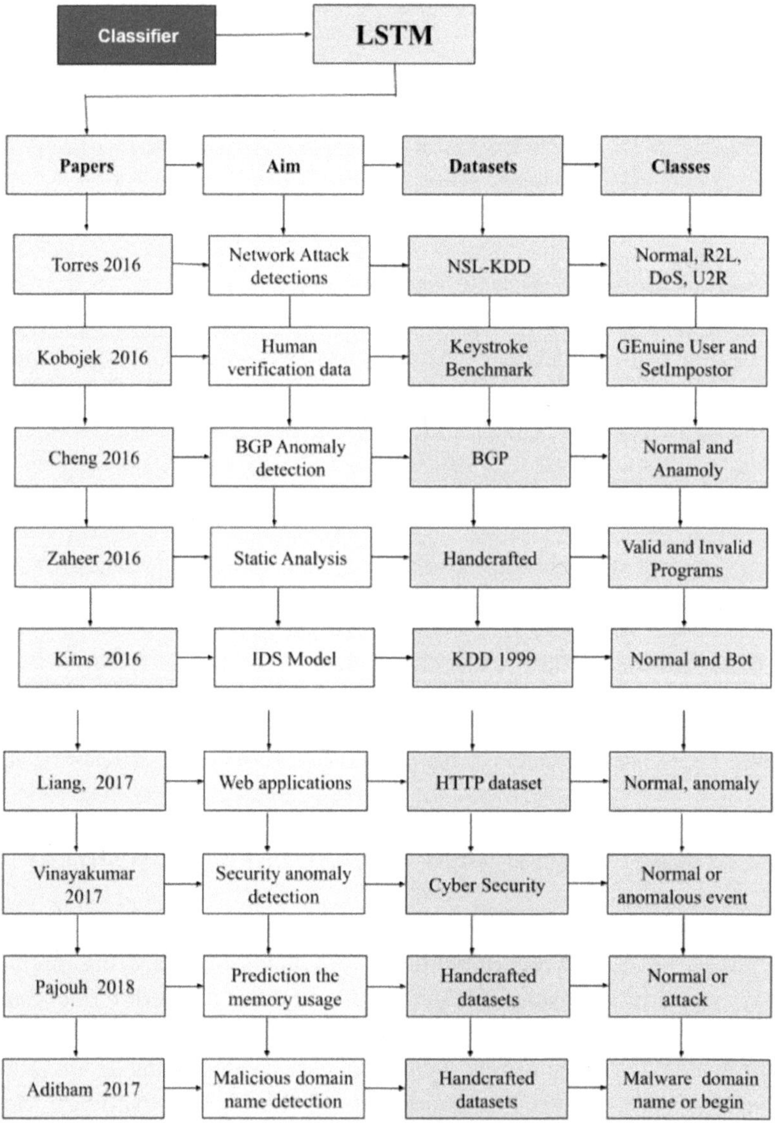

FIGURE 17.9 Review of surveys on LSTM-based cyber-attack detection techniques.

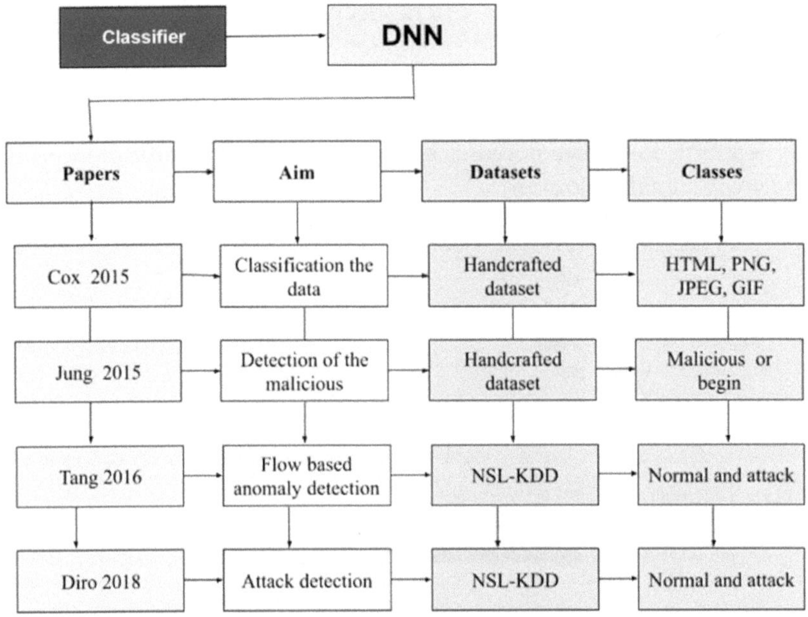

FIGURE 17.10 Review of surveys on DNN-based cyber-attack detection techniques.

17.6 ATTACK METHODS TO DEEP LEARNING

Despite the reality that deep learning is successfully used in numerous different fields, research shows that deep learning algorithms are highly susceptible to adversarial attacks. Despite the fact that DL is successfully used in many different sectors, research shows that DLCS algorithms are highly susceptible to DL adversarial attacks. The smallest amount of damage to an object causes it to be misidentified (Figures 17.11–17.13).

17.7 CYBERSECURITY SOLUTIONS

Only three businesses currently use CS-Deep-Learning for commercial "Cyber-Security" solutions:

- **A Sturdy Instinct:** The organization [https://www.deepinstinct.com] offers cross-enterprise security solutions and uses CS-deep learning for progressive anti-malicious software capabilities, stopping the execution of malicious software and files both at the endpoint and network levels.

- **Blu-Vector:** Blu-Vector offers deep learning-based solutions for detecting network traffic anomalies at [https://www.bluvectorcyber.com/].
- **Signal-Sense:** However, little is known about their plans to use deep learning for cyber defense, as they are still in stealth mode [http://www.signalsense.com].

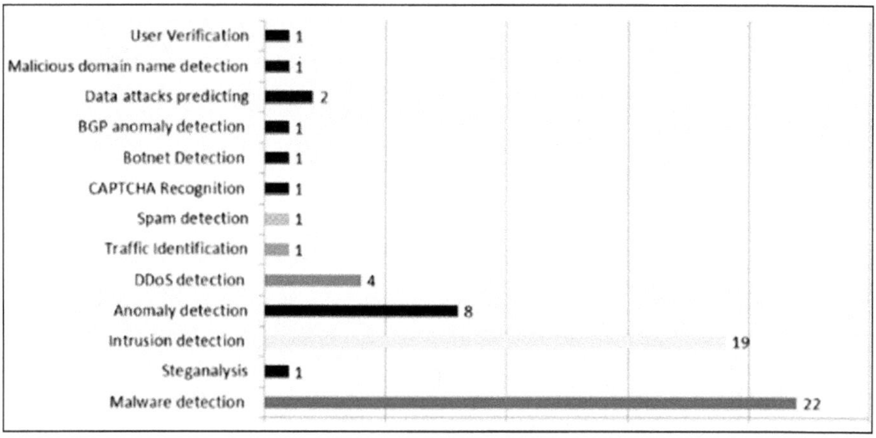

FIGURE 17.11 Statistics of the surveyed papers during the years 2011 to 2018 (April) by cyber security issues.

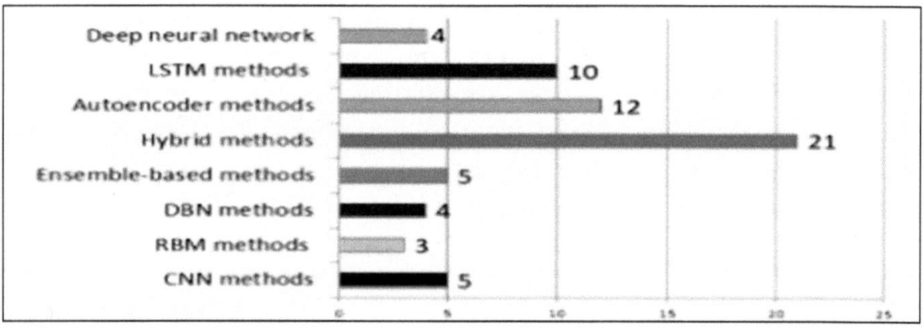

FIGURE 17.12 Number of published papers by method category from 2011 to 2018 (April).

17.8 CONCLUSION

An analysis of research examined into the current situation of deep learning-based cyber-attack detection methods. Over the past few years, DL has been used for a number of cyber-security-related activities. In DL-based cyber-attack detection systems, considerable human work and expert training are not required. The use of CS-DL models has no boundaries to certain fields and can be a generic approach to the challenges associated with cyber-attack detection.

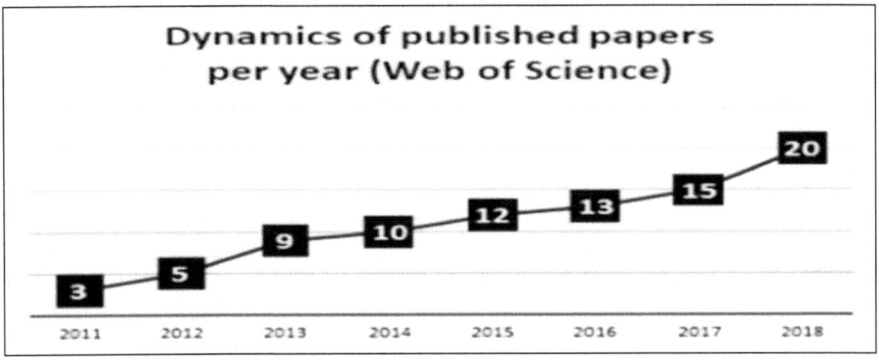

FIGURE 17.13 Dynamics of published papers on the application of the DL to cyber-attacks depicted by year.

KEYWORDS

- **convolutional neural networks**
- **cyber security deep neural network**
- **deep belief networks**
- **generative adversarial networks**
- **long short-term memories**
- **restricted-Boltzmann-machines**
- **stacked-auto-encoders**

REFERENCE

1. Aditham, S., Ranganathan, N., & Katkoori, S., (2017). LSTM-based memory profiling for predicting data attacks in distributed big data systems. In: *Proceedings of the IEEE International Parallel and Distributed Processing Symposium Workshops* (pp. 1259–1267). IEEE Press.
2. AL-Hawawreh, M., Moustafa, N., & Sitnikova, E., (2018). Identification of malicious activities in industrial internet of things based on deep learning models. *Journal of Information Security and Applications*, *41*, 1–11.

3. Alguliyev, R. M., Aliguliyev, R. M., & Abdullayeva, F. J., (2019). Hybridization of classifiers for anomaly detection in big data. *International Journal of Big Data Intelligence*, *6*(1), 11–19. doi: 10.1504/IJBDI.2019.097396.
4. Almeida, D. M., (2016). *Malware Classification on Time Series Data Through Machine Learning [thesis]*. Faculdade de Engenharia da Universidade Do Porto.
5. Alom, M. Z., Bontupalli, V. R., & Taha, T. M., (2015). Intrusion detection using deep belief networks. In: *Proceedings of the IEEE National Aerospace and Electronics Conference* (pp. 339–344). IEEE Press. doi: 10.1109/NAECON.2015.7443094.
6. Alrawashdeh, K., & Purdy, C., (2016). Toward an online anomaly intrusion detection system based on deep learning. In: *Proceedings of the 15th IEEE International Conference on Machine Learning and Applications* (pp. 195–200). IEEE Press. doi: 10.1109/ICMLA.2016.0040.
7. Aminanto, M. E., & Kim, K., (2016). Deep learning in intrusion detection system: An overview. In: *Proceedings of the International Research Conference on Engineering and Technology* (pp. 1–12). Academic Press.
8. Aminanto, M. E., & Kim, K., (2016). Deep learning-based feature selection for intrusion detection system in transport layer. In: *Proceedings of the Korea Institutes of Information Security and Cryptology Conference* (pp. 740–743). Academic Press.
9. Anderson, H. S., Woodbridge, J., & Filar, B., (2016). DeepDGA: Adversarially-tuned domain generation and detection. In: *Proceedings of the ACM Workshop on Artificial Intelligence and Security* (pp. 13–21). ACM.
10. Benchea, R., & Gavrilut, D. T., (2014). Combining restricted Boltzmann machine and one side perceptron for malware detection. In: *Proceedings of the International Conference on Conceptual Structures* (pp. 93–103). Academic Press. doi: 10.1007/978-3-319-08389-6_9.
11. Bhuyan, M. H., Bhattacharyya, D. K., & Kalita, J. K., (2014). Network anomaly detection: Methods systems and tools. *IEEE Communications Surveys and Tutorials*, *16*(1), 303–336. doi: 10.1109/SURV.2013.052213.00046.
12. Cakir, B., & Dogdu, E., (2018). Malware classification using deep learning methods. In: *Proceedings of the ACMSE Conference* (pp. 1–5). Academic Press. doi: 10.1145/3190645.3190692.
13. Chan, P. P., Lin, Z., Hu, X., Tsang, E. C., & Yeung, D. S., (2017). Sensitivity based robust learning for stacked autoencoder against evasion attack. *Neuro computing*, *267*, 572–580. doi: 10.1016/j.neucom.2017.06.032.
14. Chandy, S. E., Rasekh, A., Barker, Z. A., & Shafiee, M. E., (2018). Cyberattack detection using deep generative models with variational inference. *Journal of Water Resources Planning and Management*.
15. Cheng, M., Xu, Q., Lv, J., Liu, W., Li, Q., & Wang, J., (2016). MS-LSTM: A multi-scale LSTM model for BGP anomaly detection. In: *Proceedings of the IEEE 24th International Conference on Network Protocols Workshop on Machine Learning in Computer Networks* (pp. 1–6). IEEE Press. doi: 10.1109/ICNP.2016.7785326.
16. Chowdhury, M. U., Hammond, F., Konowicz, G., Xin, C., Wu, H., & Li, J., (2017). A few-shot deep learning approach for improved intrusion detection. In: *Proceedings of the IEEE 8th Annual Ubiquitous Computing, Electronics and Mobile Communication Conference* (pp. 456–462). IEEE Press. doi: 10.1109/UEMCON.2017.8249084.
17. Cox, J. A., James, C. D., & Aimone, J. B., (2015). A signal processing approach for cyber data classification with deep neural networks. *Procedia Computer Science*, *61*, 349–354. doi: 10.1016/j.procs.2015.09.156.

18. Cyberspace Policy Review, (2009). *Assuring a Trusted and Resilient Information and Communications Infrastructure*. Washington, DC: Executive Office of the President of the United States.
19. Dahl, G. E., Stokes, J. W., Deng, L., & Yu, D., (2013). Large-scale malware classification using random projections and neural networks. In: *Proceedings of the IEEE International Conference on Acoustics, Speech and Signal Processing* (pp. 1–5). IEEE Press. doi: 10.1109/ICASSP.2013.6638293.
20. David, O. E., & Netanyahu, N. S., (2015). Deep Sign: Deep Learning for automatic malware signature generation and classification. In: *Proceedings of the IEEE International Joint Conference on Neural Networks* (pp. 1–8). IEEE Press.
21. Diro, A. A., & Chilamkurti, N., (2018). Deep learning: The frontier for distributed attack detection in fog-to-things computing. *IEEE Communications Magazine*, *56*(2), 169–175. doi: 10.1109/MCOM.2018.1700332.
22. Diro, A. A., & Chilamkurti, N., (2018). Distributed attack detection scheme using deep neural network approach for internet of things. *Future Generation Computer Systems*, *82*, 761–768. doi: 10.1016/j.future.2017.08.043.
23. Dong, B., & Wang, X., (2016). Comparison deep learning method to traditional methods using for network intrusion detection. In: *Proceedings of the IEEE 8th International Conference on Communication Software and Networks* (pp. 581–585). IEEE Press. doi: 10.1109/ICCSN.2016.7586590.
24. Fiore, U., Palmieri, F., Castiglione, A., & Santis, A. D., (2013). Network anomaly detection with the restricted Boltzmann machine. *Neuro computing*, *122*, 13–23. doi: 10.1016/j.neucom.2012.11.050.
25. Gao, N., Gao, L., & Gao, Q., (2014). An intrusion detection model based on deep belief networks. In: *Proceedings of the Second International Conference on Advanced Cloud and Big Data* (pp. 247–252). Academic Press. doi: 10.1109/CBD.2014.41.
26. Gibert, D., (2016). *Convolutional Neural Networks for Malware Classification.* Master's thesis. Universitat Politècnica de Catalunya.
27. Goodfellow, I., Pouget-Abadie, J., Mirza, M., Hu, B., Warde-Farley, D., Ozair, S., & Bengio, Y., (2014). Generative adversarial nets. In: *Proceedings of the 27th International Conference on Neural Information Processing Systems* (pp. 2672–2680). Academic Press.
28. Guo, W., Wang, T., & Wei, J., (2018). Malware detection with convolutional neural network using hardware events. In: *Proceedings of the CCF National Conference on Computer Engineering and Technology* (pp. 104–115). Academic Press. doi: 10.1007/978-981-10-7844-6_11.
29. Hardy, W., Chen, L., Hou, S., Ye, Y., & Li, X., (2016). DL4MD: A deep learning framework for intelligent malware detection. In: *Proceedings of the International Conference on Data Mining* (pp. 61–67). Academic Press.
30. Hu, W., & Tan, Y., (2017). Black-Box Attacks Against RNN Based Malware Detection Algorithms, 1–10. doi: https://doi.org/10.48550/arXiv.1705.08131.
31. Huang, S., Zhou, C. J., Yang, S. H., & Qin, Y. Q., (2015). Cyber-physical system security for networked industrial processes. *International Journal of Automation and Computing*, *12*(6), 567–578. doi:10.1007/s11633-015-0923-9.
32. Huang, W., & Stokes, J. W., (2016). MtNet: A multi-task neural network for dynamic malware classification. In: *Proceedings of the 13th International Conference on Detection of Intrusions and Malware, and Vulnerability Assessment* (pp. 399–418). Academic Press.

33. Imamverdiyev, Y. N., & Abdullayeva, F. J., (2018). Deep learning method for denial-of-service attack detection based on restricted Boltzmann machine. *Big Data*, *6*(2), 159–169. doi: 10.1089/big.2018.0023 PMID:29924649.
34. Jung, W., Kim, S., & Choi, S., (2015). Poster: Deep learning for zero-day flash malware detection. In: *Proceedings of the 36th IEEE Symposium on Security and Privacy* (pp. 1, 2). IEEE Press.
35. Kang, M. J., & Kang, J. W., (2016). Intrusion detection system using deep neural network for in-vehicle network security. *PLoS One*, *11*(6), 1–17. doi: 10.1371/journal.pone.0155781PMID:27271802.
36. Kim, G., Yi, H., Lee, J., Paek, Y., & Yoon, S., (2016). LSTM-Based System-Call Language Modelling and Robust Ensemble Method for Designing Host-Based Intrusion Detection Systems. 1–12. doi: DOI: https://doi.org/10.48550/arXiv.1611.01726.
37. Kim, J., Kim, J., Thu, H. T., & Kim, H., (2016). Long short-term memory recurrent neural network classifier for intrusion detection. In: *Proceedings of the International Conference on Platform Technology and Service* (pp. 1–5). Academic Press. doi: 10.1109/PlatCon.2016.7456805.
38. Kim, T. Y., & Cho, S., (2018). Web traffic anomaly detection using C-LSTM neural networks. *Expert Systems with Applications*, *106*, 66–76. doi: 10.1016/j.eswa.2018.04.004.
39. Kobojek, P., & Saeed, K., (2016). Application of recurrent neural networks for user verification based on keystroke dynamics. *Journal of Telecommunications and Information Technology*, (3), 60–70.
40. Kolosnjaji, B., Zarras, A., Webster, G., & Eckert, C., (2016). Deep learning for classification of malware system call sequences. In: *Proceedings of the Australasian Joint Conference on Artificial Intelligence* (pp. 137–149). Academic Press. doi: 10.1007/978-3-319-50127-7_11.
41. Li, Y., Ma, R., & Jiao, R., (2015). A hybrid malicious code detection method based on deep learning. *International Journal of Security and Its Applications*, *9*(5), 205–216. doi: 10.14257/ijsia.2015.9.5.21.
42. Liang, J., Zhao, W., & Ye, W., (2017). Anomaly-based web attack detection: A deep learning approach. In: *Proceedings of the VI International Conference on Network, Communication, and Computing* (pp. 80–85). Academic Press. doi: 10.1145/3171592.3171594.
43. Lin, T. C., (2016). Financial weapons of war. *Minnesota Law Review*, *102*(3), 1377–1440.
44. Ludwig, S. A., (2017). Intrusion detection of multiple attack classes using a deep neural net ensemble I. In: *Proceedings of the IEEE Symposium Series on Computational Intelligence* (pp. 1–7). IEEE Press. doi: 10.1109/ SSCI.2017.8280825.
45. Ma, T., Wang, F., Cheng, J., Yu, Y., & Chen, X., (2016). A hybrid spectral clustering and deep neural network ensemble algorithm for intrusion detection in sensor networks. *Sensors (Basel)*, *16*(10), 1–23. doi: 10.3390/s16101701PMID:27754380.
46. *Microsoft Malware Classification Challenge (BIG 2015)*, (n.d.). Retrieved from: https://www.kaggle.com/c/malware-classification (accessed on 27 January 2024).
47. Mirsky, Y., Doitshman, T., Elovici, Y., & Shabtai, A., (2018). Kitsune: An Ensemble of Autoencoders for Online Network Intrusion Detection. arXiv:1802.09089v2 [cs.CR] 27 May 2018. doi: https://doi.org/10.48550/arXiv.1802.09089.
48. Netzer, Y., Wang, T., Coates, A., Bissacco, A., Wu, B., & Ng, A. Y., (2011). Reading digits in natural images with unsupervised feature learning. In: *Proceedings of the NIPS Workshop on Deep Learning and Unsupervised Feature Learning* (pp. 1–9). Academic Press.

49. Niyaz, Q., Sun, W., Javaid, A. Y., & Alam, M., (2016). A deep learning approach for network intrusion detection system. In: *Proceedings of the 9th EAI International Conference on Bio-Inspired Information and Communications Technologies* (pp. 21–26). Academic Press.
50. Pajouh, H. H., Dehghantanha, A., Khayami, R., & Choo, K. R., (2018). A deep recurrent neural network-based approach for internet of things malware threat hunting. *Future Generation Computer Systems, 85*, 88–96. doi: 10.1016/j.future.2018.03.007.
51. Papernot, N., McDaniel, P., Swami, A., & Harang, R., (2016). Crafting adversarial input sequences for recurrent neural networks. In: *Proceedings of the IEEE Military Communications Conference* (pp. 49–54). IEEE Press. doi: 10.1109/MILCOM.2016.7795300.
52. Pascanu, R., Stokest, J. W., Sanossian, H., Marinescu, M., & Thomas, A., (2015). Malware classification with recurrent networks. In: *Proceedings of the IEEE International Conference on Acoustics, Speech, and Signal Processing* (pp.1916–1920). IEEE Press. doi: 10.1109/ICASSP.2015.7178304.
53. Pibre, L., Pasquet, J., Ienco, D., & Chaumont, M., (2016). Deep learning is a good steganalysis tool when embedding key is reused for different images, even if there is a cover source mismatch, Media watermarking, security, and forensics. In: *Proceedings of the IS&T International Symposium on Electronic Imaging* (pp. 1–12). Academic Press.
54. Rhode, M., Burnap, P., & Jones, K., (2017). Early-Stage Malware Prediction Using Recurrent Neural Networks. arXiv:1708.03513 [cs.CR], doi: 10.1016/j.cose.2018.05.010.
55. Rui, C., Jing, Y., Rong, H., & Shu-guang, H., (2013). A novel LSTM-RNN decoding algorithm in CAPTCHA recognition. In: *Proceedings of the Third International Conference on Instrumentation, Measurement, Computer, Communication, and Control* (pp. 766–771). Academic Press. doi: 10.1109/IMCCC.2013.171.
56. Sakurada, M., & Takehisa, Y., (2014). Anomaly detection using autoencoders with nonlinear dimensionality reduction. In: *Proceedings of the MLSDA 2014 2nd Workshop on Machine Learning for Sensory Data Analysis* (pp. 1–8). Academic Press. doi: 10.1145/2689746.2689747.
57. Salama, M. A., Eid, H. F., Ramadan, R. A., Darwish, A., & Hassanien, A. E., (2011). Hybrid intelligent intrusion detection scheme. In: *Soft Computing in Industrial Applications* (pp. 293–303). Springer.doi: 10.1007/978-3-642-20505-7_26.
58. Saxe, J., & Berlin, K., (2015). Deep neural network based malware detection using two dimensional binary program features. In: *Proceedings of the 10th International Conference on Malicious and Unwanted Software* (pp. 11–20). Academic Press. doi: 10.1109/MALWARE.2015.7413680.
59. Silva, L. A., Costa, K. A., Ribeiro, P. B., Rosa, G. H., & Papa, J. P., (2016). Learning spam features using restricted Boltzmann machines. *IADIS International Journal on Computer Science and Information Systems, 11*(1), 99–114.
60. Tang, T. A., (2016). Deep learning approach for network intrusion detection in software defined networking. In: *Proceedings of the International Conference on Wireless Networks and Mobile Communications* (pp. 1–12). Academic Press. doi: 10.1109/WINCOM. 2016.7777224.
61. Tang, T. A., Mhamdi, L., McLernon, D., Zaidi, S. A., & Ghogho, M., (2016). Deep learning approach for network intrusion detection in software defined networking. In: *Proceedings of the IEEE International Conference on Wireless Networks and Mobile Communications* (pp. 258–263). IEEE Press. doi:10.1109/WINCOM.2016.7777224.

62. Teyou, G. K., & Ziazet, J., (2019). Convolutional Neural Network for Intrusion Detection System in Cyber Physical Systems. Available from: https://arxiv.org/vc/arxiv/papers/1905/1905.03168v1.pdf.
63. *The CTU-13 Dataset*, (n.d.). Retrieved from: https://stratosphereips.org/category/dataset.html (accessed on 27 January 2024).
64. *The UNB ISCX 2012 Intrusion Detection Evaluation Dataset,* (n.d.). Retrieved from: http://www.unb.ca/cic/research/datasets/ids.html (accessed on 27 January 2024).
65. Thing, V. L., (2017). IEEE 802.11 network anomaly detection and attack classification: A deep learning approach. In: *Proceedings of the IEEE Wireless Communications and Networking Conference* (pp. 1–6). IEEE Press. doi:.10.1109/WCNC.2017.7925567.
66. Tobiyama, S., Yamaguchi, Y., Shimada, H., Ikuse, T., & Yagi, T., (2016). Malware detection with deep neural network using process behavior. In: *Proceedings of the IEEE 40th Annual Computer Software and Applications Conference* (pp. 577–582). IEEE Press. doi: 10.1109/COMPSAC.2016.151.
67. Torres, P., Catania, C., Garcia, S., & Garino, C. G., (2016). An analysis of recurrent neural networks for botnet detection behavior. In: *Proceedings of the IEEE Biennial Congress of Argentina* (pp. 1–6). IEEE Press. doi:10.1109/ARGENCON.2016.7585247.
68. Vinayakumar, R., Soman, K. P., & Poornachandran, P., (2017). Long short-term memory-based operation log anomaly detection. In: *Proceedings of the IEEE International Conference on Advances in Computing, Communications, and Informatics* (pp. 236–242). IEEE Press. doi: 10.1109/ICACCI.2017.8125846.
69. Vinayakumara, R., Somana, K. P., & Poornachandran, P., (2018). Detecting malicious domain names using deep learning approaches at scale. *Journal of Intelligent & Fuzzy Systems, 34*(3), 1355–1367.doi:10.3233/JIFS-169431.
70. Wang, W., Sheng, Y., Wang, J., Zeng, X., Ye, X., Huang, Y., & Zhu, M., (2017). HAST-IDS: Learning hierarchical spatial-temporal features using deep neural networks to improve intrusion detection. *IEEE Access, 6*, 1792–1806. doi: 10.1109/ACCESS.2017.2780250.
71. Wang, W., Zhu, M., Zeng, X., Ye, X., & Sheng, Y., (2017). Malware traffic classification using convolutional neural network for representation learning. In: *Proceedings of the International Conference on Information Networking* (pp. 712–717). Academic Press.
72. Wang, X., & Yiu, S. M., (2018). A Multi-Task Learning Model for Malware Classification with Useful File Access Pattern from API Call Sequence. arXiv:1610.05945 [cs.SD], doi: https://doi.org/10.48550/arXiv.1610.05945.
73. Wang, Z., (2015). *The Applications of Deep Learning on Traffic Identification.* Black hat.
74. Xu, L., Zhang, D., Jayasena, N., & Cavazos, J., (2016). HADM: Hybrid analysis for detection of malware. In: *Proceedings of SAI Intelligent Systems Conference* (pp. 1–23). Academic Press. doi: 10.1007/978-3-319-56991-8_51.
75. Yadav, S., & Selvakumar, S., (2016). Detection of application layer DDoS attack by feature learning using stacked autoencoder. In: *Proceedings of the IEEE International Conference on Computational Techniques in Information and Communication Technologies* (pp. 1–6). IEEE Press. doi: 10.1109/ICCTICT.2016.7514608.
76. Yan, R., Xiao, X., Hu, G., Peng, S., & Jiang, Y., (2018). New deep learning method to detect code injection attacks on hybrid applications. *Journal of Systems and Software, 137*, 67–77. doi: 10.1016/j.jss.2017.11.001.
77. Yang, Z., Qin, X., Li, W. R., & Yang, Y. J., (2014). A DDoS detection approach based on CNN in cloud computing. *Applied Mechanics and Materials, 513*(517), 579–584. doi: 10.4028/www.scientific.net/AMM.513–517.579.

78. Yousefi-Azar, M., Varadharajan, V., Hamey, L., & Tupakula, U., (2017). Autoencoder-based feature learning for cyber security applications. In: *Proceedings of the IEEE International Joint Conference on Neural Networks* (pp. 3854–3861). IEEE Press.
79. Yu, Y., Long, J., & Cai, Z., (2017). Network intrusion detection through stacking dilated convolutional autoencoders. *Security and Communication Networks*, 1–10. doi: 10.1155/2017/4184196.
80. Yu, Y., Long, J., & Cai, Z., (2017). Session-based network intrusion detection using a deep learning architecture. In: *Proceedings of the International Conference on Modeling Decisions for Artificial Intelligence* (pp. 144–155). Academic Press. doi: 10.1007/978-3-319-67422-3_13.
81. Yuan, Z., Lu, Y., Wang, Z., & Xue, Y., (2014). Droid-sec: Deep learning in android malware detection. *Computer Communication Review*, *44*(4), 371, 372. doi: 10.1145/2740070.2631434.
82. Yuxin, D., & Siyi, Z., (2017). Malware detection based on deep learning algorithm. *Neural Computing & Applications*, *31*(2), 1–12.
83. Zaheer, M., Tristan, J. B., Wick, M., Guy, L., & Steele, G. L. Jr., (2016). Learning a static analyzer: A case study on a toy language. In: *Proceedings of the International Conference on Learning Representations* (pp. 1–10). Academic Press.
84. Zhao, B., Feng, J., Wu, X., & Yan, S., (2017). A survey on deep learning-based fine-grained object classification and semantic segmentation. *International Journal of Automation and Computing*, *14*(2), 119–135. doi: 10.1007/s11633-017-1053-3.

CHAPTER 18

Revolution of Digital Transformation Technology: Artificial Intelligence and Blockchain Technology in Cyber Security

S. SRIVIDHYA,[1] D. SOWMYADEVI,[2] and S. R. LAVANYA[3]

[1]*Department of Computer Science, KPR College of Arts Science and Research, Coimbatore, Tamil Nadu, India*

[2]*Department of Computer Science, PSG College of Arts and Science, Coimbatore, Tamil Nadu, India*

[3]*Department of Computer Science, KPR College of Arts Science and Research, Coimbatore, Tamil Nadu, India*

ABSTRACT

The influence and rapid advancements in digital transformation technology have reshaped the world. In the late 1920s, the digitization of society and its fast acceleration drove the need for digital transformation across all sectors, including healthcare, hospitality, insurance, retail, and more. The collaboration of digital technologies can have a significant business impact. Artificial Intelligence (AI) and blockchain technology, being buzzing tools of digital transformation, have made their mark in cybersecurity. The COVID pandemic has highlighted the need for ubiquitous connectivity and digitalization, which has led to an increase in cyber-attacks. With the growing scale of these attacks, AI offers many advantages and applications in cybersecurity. AI, in collaboration with blockchain technology, can automate threat detection in an efficient manner. Additionally, AI techniques can detect intrusions, malware attacks, phishing attempts, and cloud attackers. Companies like Microsoft, Google, and FireEye have utilized AI approaches and techniques

Leveraging Artificial Intelligence (AI) Competencies for Next-Generation Cybersecurity Solutions.
Pethuru Raj, B. Sundaravadivazhagan, & V. Kavitha (Eds.)

to detect malware, phishing, and the spread of miscellaneous information. Microsoft, for example, has analyzed trillions of security signals, nation-state groups, and hacker groups using AI technologies. AI-based facial recognition software also provides security in social media platforms. Blockchain technology is rapidly emerging in various areas, particularly in finance with the help of cryptocurrencies. It is an innovative solution that integrates with other buzzing technologies like AI and Blockchain with cybersecurity.

18.1 INTRODUCTION

Technology drives towards digital transformation to enable innovation. The process of incorporating technology into the traditional business process aims to create new trends to meet the challenges in the evolving market and satisfy customer expectations. Digital Transformation utilizes digital technologies to improve the current business culture and its linked processes. It also aims to satisfy the changing demands of customers. Digital transformation is the revival of business in this digital era. It becomes an absolute necessity to transform from paper to smart applications to meet the demands of Industry 4.0 and engage our customers. Every digital transformation begins and ends with customers, says Chairman and Co-CEO of Salesforce, Mr. Marc Benioff. It is essential to know the dissimilarities of digital transformation and its other verticals of digitalization and digitization. Digitization is a transition or converting data from analog to digital. This involves converting handwritten documents and ledgers into digital computer files. Digital records make it easy to search and share information. Work is made easier by using digital data in digitalization. Due to digitalization, records can be easily retrieved. Due to the enormous growth of digital technology, new strategies have evolved for the growth of business. These technologies create new things and new methods of doing things. Digital transformation gives utmost importance to customer interaction and entirely changes the new methods of conducting business. Digitalization has advantages like time optimization, increased return on investment, improved internal and external communication, enhanced employee experience, improved customer retention, and new sales.

Digital transformations can be successfully enabled by focusing on key concepts such as Digital Twins, Privacy, Culture, Augmented Intelligence, and Digital Product Management. A digital twin is a virtual model designed to simulate a physical object, implemented as a software model that reflects a physical object, process, or person. The second important aspect is privacy, as digital transformation may be rejected if it is perceived as violating privacy and personal data security. Digital transformation has the potential to greatly improve the healthcare domain by enabling early disease detection and diagnosis with the assistance of Artificial Intelligence (AI). Robotic assistance allows for precise surgeries. The automotive industry has also seen significant advancements through digital transformation, with self-driving cars reducing accidents, carbon emissions, and insurance premiums. Digital transformation has also made an impact in areas such as 3D printing, manufacturing, resource management, and sustainability. According to an article on Consultancy.eu [1], most organizations have embraced digital transformation and reported being digitally ready (Figure 18.1).

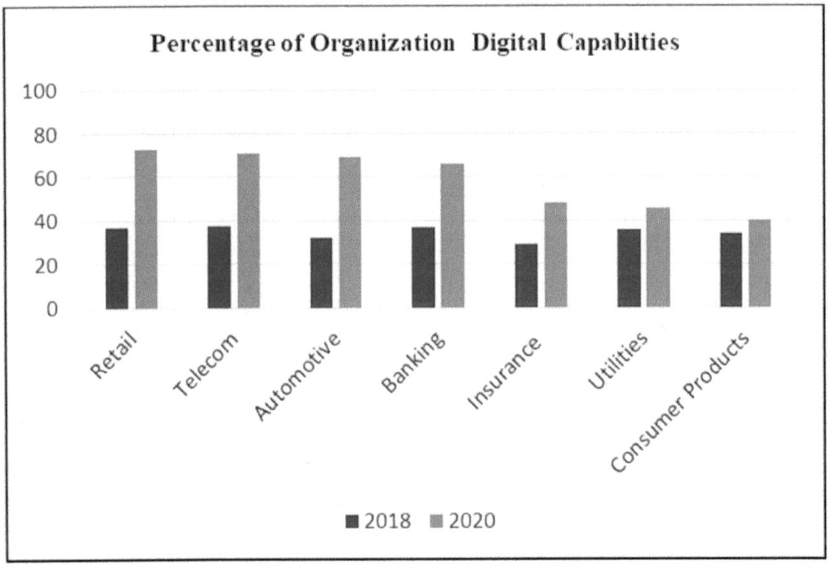

FIGURE 18.1 Digital readiness.

The main objective of digital transformation is to develop your current business processes by integrating digital transformation tools. This transformation is crucial to maintain a high level of competitiveness in the industry. Many organizations digitize their work by converting data into a technology-compatible format. They also automate procurement processes, utilize algorithms for decision-making, and employ software to gain insights from their data. However, success cannot be achieved solely through these measures. Instead, the entire business must be involved in a radical redesign, deploying new business models, activities, and processes to fully leverage the technological opportunities for a comprehensive transformation.

18.2 CYBER THREATS

Cybernetics is the study of movement and control of systems. A cyber threat refers to a malicious attack that utilizes unauthorized data for unofficial purposes. Common types of threats include ransomware, DoS attacks, malware attacks, SQL injection attacks, zero-day exploits, social engineering, and man-in-the-middle attacks [2]. These malicious activities can involve unauthorized access to data, misuse of sensitive information, digital theft of money, or disruption of ongoing processes. Cyber threats come in various forms, as shown in Figure 18.2:

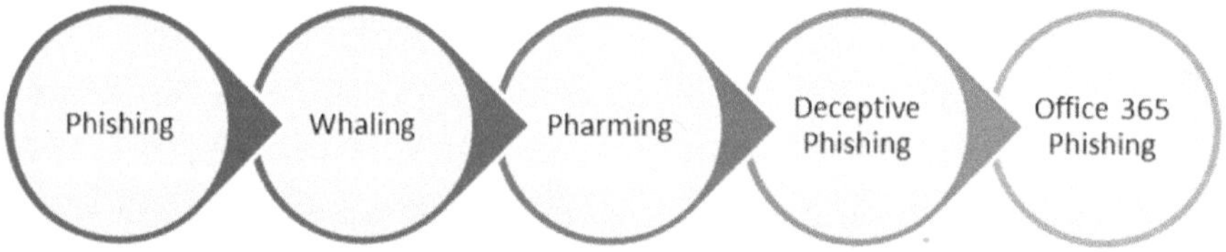

FIGURE 18.2 Types of cyber threats.

- **Phishing:** The method of sending false communications that appear to come from a trusted source through email is known as phishing. The main objective of this attack is to steal credit card data or login credentials. These attacks can gain access to online accounts and obtain personal data. Cybercriminals are well-versed in deceptive phishing techniques, which involve tricking others into clicking on a malicious link in an email. Phishing relies on social engineering techniques through email or SMS messages. Phishers gather information from social networks such as WhatsApp, Facebook, Twitter, Instagram, etc., including work history, interests, and personal details of their victims. They also obtain names, job titles, and email addresses of their victims. Common types of phishing attacks include Spear Phishing, whaling, Pharming, Deceptive Phishing, and Office 365. Spear Phishing targets specific individuals instead of a group of people. Phishers search for victims using social media websites to identify more vulnerable targets. According to the SANA Institute, 95% of attacks on enterprise networks are a result of spear phishing.
- **Whaling:** Phishers target high-profile individuals, such as a company's CEO, in a type of attack known as whaling. This attack is aimed at senior members of an organization with the goal of stealing money,

obtaining sensitive information, or gaining unauthorized access to computers for illegal purposes. The attackers often create a sense of urgency, coercing the victim into making a wire transfer. This whaling attack is also referred to as CEO fraud, as it involves spoofing emails and websites to deceive victims into revealing sensitive data or making unauthorized money transfers. The primary objectives of a whaling attack may include financial gain, control over a network, supply chain disruption, corporate espionage, malware distribution, or personal vendettas. In cases of corporate espionage, attackers may steal intellectual property or other personal data to benefit competitors. Victims of whaling attacks may suffer a loss of reputation and prominence.

- **Pharming:** Pharming is an attack that causes users to disclose personal information. This attack involves executing malicious code on the user's machine, redirecting them to a machine controlled by the attacker. Since the code runs on the user's machine, the attacker does not rely on the user replying to an email or clicking a link. Instead, the attacker utilizes code that targets the victim on their own website. Pharming can be accomplished by either passing through individual computers or by poisoning a server. Both methods result in redirecting to a server, but their execution may differ. The different types of pharming attacks include hacking individual computers and poisoning a DNS or DNS cache. Detecting a pharming attack can be challenging, but symptoms may include subtle changes to a link or website, unsafe connections, unusual banking activity, unauthorized password changes, and unfamiliar application downloads. Some techniques to prevent pharming attacks are:
    - o Clear DNS Cache
    - o Run an antivirus program
    - o Suspicion server is attacked which is then to contact Internet Service Provider
    - o Install a VPN for secured browsing
- **Deceptive Phishing:** This is online identity theft that steals the identity of users and financial account details. Techniques used in deceptive phishing include legitimate links, blending malicious and benign code, redirecting and shortening links, and modifying brand logos. Deceptive phishing is a common type of phishing. An example of deceptive phishing is a mail from a bank asking the user to click a link to verify their account.

- **Office 365 Phishing:** Attackers gaining access to Office 365 email accounts is called Office 365 Phishing. This type of phishing takes the form of an email from Microsoft. The email will ask the user to log in and reset their password, claiming that there is an issue with their account that requires attention. The email will also include a URL link as a solution to the problem.

18.3 CYBER SECURITY

The mechanism of providing security to computers, mobile devices, servers, and networks against malicious attacks is called cybersecurity. Cyber in cybersecurity refers to the systems, data, network, and servers. Security in cybersecurity refers to the security provided to the cyber. Cybersecurity is otherwise termed as electronic means of information security and information technological security. Today, cybersecurity is very significant since all the systems are interconnected and with the revolution of IoT, cybersecurity plays a vital role which has to be implemented in all the systems that can be attacked to prevent loss of data, misuse of information, and resource prevention. Cybersecurity is defined as a method to protect networks, attacks on software and data, unauthorized data, and data damage as defined by Craigen et al. [3]. Cybersecurity is defined by Aftergood et al. [4] as a method to protect networks, programs, data, and systems from vulnerable attacks, access without permission, modification, and destroying the data.

Confidentiality, Integrity, and Availability (CIA) are important components of cybersecurity, popularly known as the CIA triad. Confidentiality protects data by providing access only to authorized individuals. Encryption, access control, authentication, authorization, and physical security are some tools used for confidentiality. Data encryption is a good example of confidentiality.

Integrity ensures that data remains unchanged and accurate, without unauthorized modifications from external sources. It also guarantees the authenticity of the source. Common integrity tools include backups, checksums, data correcting codes, and availability.

Availability refers to the accessibility of data for authorized users. Physical protection and computational redundancies are availability tools.

There are various types of cyber threats, such as phishing, ransomware, malware, SQL injection, etc. (Figure 18.3).

Most enterprises adopted ICT technologies to adapt to the digital transformation by compromising security attacks. The WannaCry ransomware attacks had a significant impact on many organizations. In addition, the theft of $81 million from Bangladesh's central bank was a cyber-attack targeting SWIFT international transactions. These attacks draw attention to the importance of cybersecurity in this era of digital transformation. The essence of digital transformation is closely tied to key technologies such as AI and the Internet of Things (IoT). It also encompasses Big Data and Cloud Computing as common storage areas. All of these technologies have a significant impact and create a new paradigm where cybersecurity must be prioritized. The scope of cybersecurity relies on factors such as Technology Evolution, Regulatory Compliance, Sophistication of Attacks, and limited resources. Emerging cybersecurity solutions exhibit the following characteristics: Security Automation and Intelligence, Cybersecurity datasets and Data-driven approaches, standards, Integrated policies, Security as a service, decentralization, and blockchains. Organizations have begun adopting digital technologies to enhance customer experiences and develop new business models, but this has also resulted in an increase in cyber-attacks and data breaches.

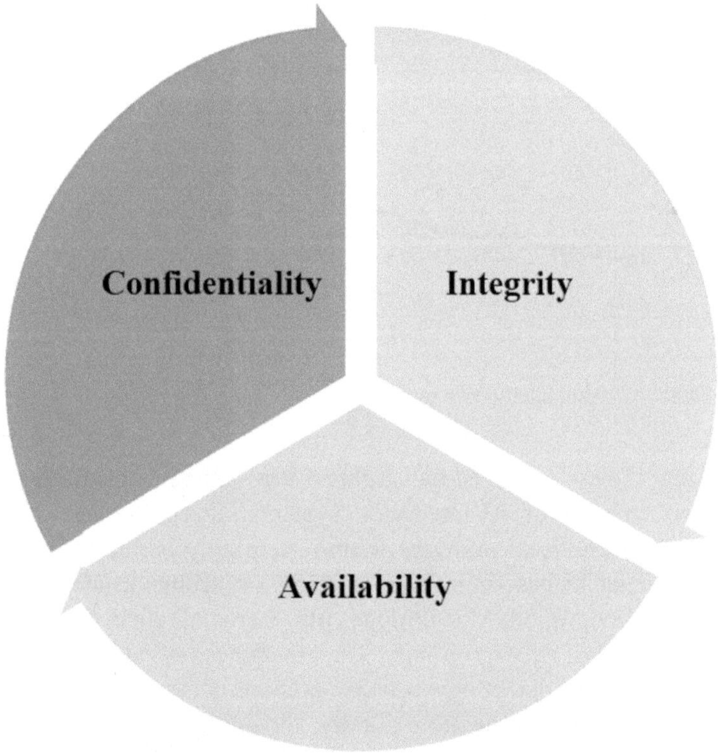

FIGURE 18.3 Cyber security goals.

18.3.1 ARTIFICIAL INTELLIGENCE DRIVEN CYBER SECURITY

AI plays a vital role in digital transformation services and Industry 4.0. The power of digital transformation lies in incorporating AI into business processes. AI in cybersecurity incorporates AI features along with the required software to simulate human expertise to detect different attacks. The revenue of AI in the world market from 2016–2025 has been tremendously increasing from \$5,000 to \$40,000. This has been depicted in Figure 18.4.

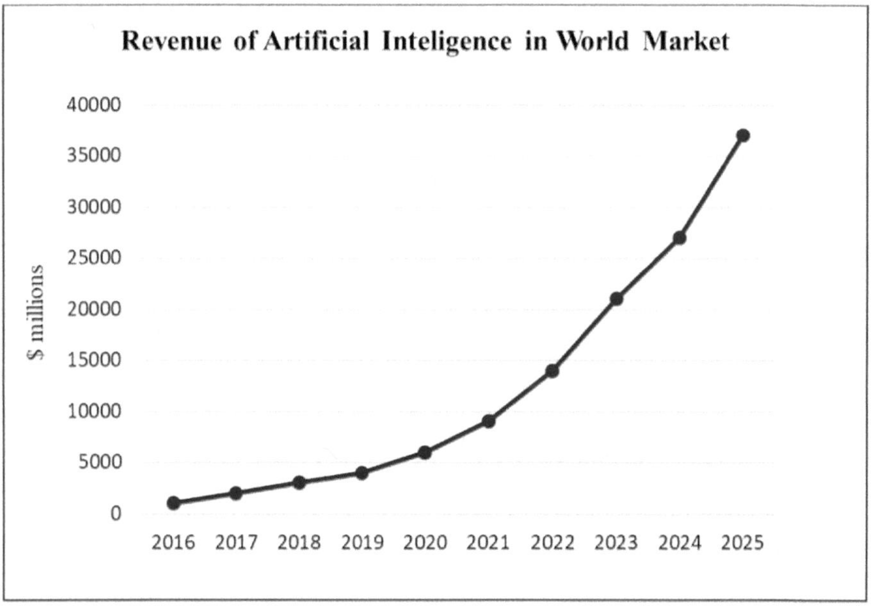

FIGURE 18.4 Artificial intelligence revenue.

Changing the code in even the slightest way can have a negative impact on the performance of AI-integrated systems. Deep Locker, developed by IBM, aims to conceal malware within seemingly harmless images and videos. It utilizes AI-based machine learning algorithms to identify targets through facial recognition. AI techniques play a crucial role in preventing and safeguarding against cybercriminals. The advantages of AI in cybersecurity include the ability to detect new attacks, combat bots, predict breach risks, and enhance endpoint protection. Firstly, the implementation of chatbots to address level 1 queries has become commonplace on most websites, reducing customer waiting time to virtually meet with an agent. Secondly,

AI-driven sales techniques have shown remarkable improvements, resulting in increased sales and revenue. This is achieved by analyzing effective sales strategies over time and identifying promotional techniques that can generate long-term revenue growth. AI-enabled systems can analyze vast amounts of data from external sources to identify patterns, correlations, and anomalies, enabling organizations to address issues and mitigate risks. AI in cybersecurity plays a crucial role in identifying threats by analyzing millions and trillions of relevant data points specific to an organization's security.

Cybersecurity processes can be automated and made more intelligent than traditional security systems by incorporating AI methods in the security intelligence model. AI-based security can make intelligent decisions in the field of cybersecurity. The following are the benefits of integrating AI in cybersecurity:

1. **AI Learning Power is High:** The intelligence of this technology improves the security power by using ML and deep learning techniques to recognize patterns and identify anomalies. The ability of Artificial Neural Networks (ANN) to improve security in the future is by blocking similar threats that have already been recorded.
2. **Unknown Threats Identification:** Conventional security techniques identify the discovered threats, which may not be very effective for the undiscovered threats. Traditional security techniques with AI can improve the detection rate. AI models can be implemented in organizations to enhance threat detection processes.
3. **Handling Massive Data Empowerment:** A large amount of data is being transferred in a business every day, which needs to be protected from malicious software and hackers. The threats masked by any activity can be detected by AI, which is automated to skim through huge data. AI-based digital data transfer technologies help to transfer data and will detect and identify hidden threats.
4. **Vulnerability Management:** Conventional vulnerability management tools are ineffective in dealing with the increasing threats and the vast amount of data. Many teams rely on vulnerability tools that are not very effective in distributed and hybrid environments. AI-enabled security solutions analyze real-time data to assess the severity of vulnerabilities. A robust vulnerability strategy can be achieved by incorporating a knowledge base that includes asset prioritization. Additionally, mapping should be performed to align the inventory with the vulnerability strategy. By utilizing the appropriate AI tools and implementing a well-planned strategy, vulnerabilities can be identified, prioritized, and remediated efficiently.

5. **Improves Detection and Response Time:** Integration of AI with cybersecurity identifies threats and responds to them more quickly, protecting the network from damage. ML algorithms can also be utilized in AI-based cybersecurity platforms to detect attackers by identifying outliers. AI and ML-based cybersecurity solutions enhance system efficiency and enable software to safeguard organizations against threats. The following outlines the cybersecurity challenges that can be addressed with AI and ML-based solutions.
    - High rate of detecting threat
    - Malware and Intrusion Detection
    - Discover Code Vulnerabilities
    - Network Security Enhancement
    - Phishing Detection
    - data breach risk prediction
    - Predictive Analytics
    - Prescriptive Analytics

There are a few challenges faced when implementing AI in providing solutions to cybersecurity. First and foremost, the challenge is the cost of implementation. Since AI is a complex technology, more time and money need to be invested in implementing AI-powered solutions. Therefore, it is highly necessary to study the product demand before developing AI-based cybersecurity solutions. The second challenge comes with the requirement of an extensive dataset for developing AI-driven solutions. Thirdly, it is necessary to be cautious as hackers also use AI techniques to attack. To prevent such attacks, ML-based detection systems are created to identify AI and human-generated attacks. Finally, AI-driven systems are also vulnerable to attacks that try to confuse the model developed as a cybersecurity solution. Five aspects are to be considered when developing AI-based cybersecurity solutions, as depicted in Figure 18.5.

18.3.2 ADVANTAGES AND LIMITATIONS OF AI IN CYBER SECURITY

There are remarkable advantages to applying AI in cybersecurity, along with some limitations. The advantages and limitations are discussed below:

Pros of AI in Cyber Security:

- **Accuracy:** AI finds patterns and derives insights from large sets of data points that even humans have missed.

- **Speed:** AI-enabled systems appear to be faster, and in the event of any issues, notification will be provided promptly.
- **Tireless Work:** AI-powered systems tend to operate effortlessly without rest, maintaining the same level of efficiency.
- **No Interruption:** AI systems work without any interruption and distractions.
- **Risk Assessment:** Systems with AI techniques can interpret data more easily, enabling faster risk assessment.
- **Change Detection:** AI monitors the entire network systems for any physical or digital modifications.

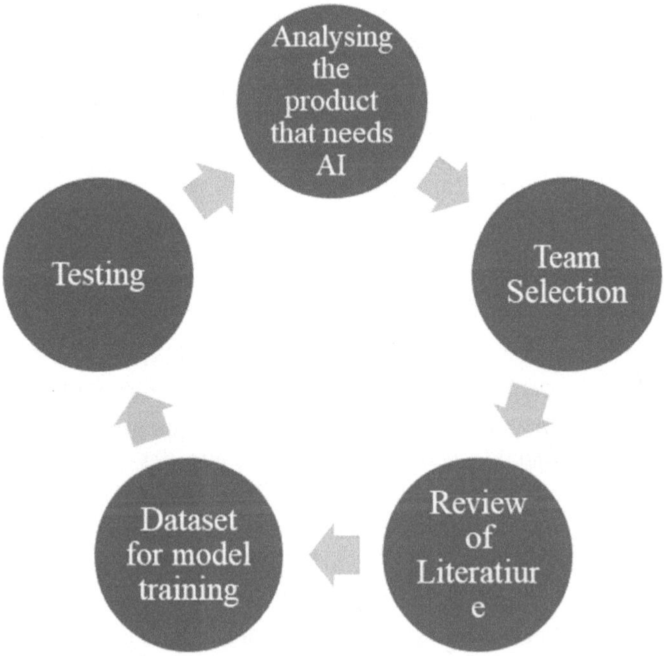

FIGURE 18.5 Considerations for AI-based cyber security solutions.

18.3.3 LIMITATIONS OF AI IN CYBER SECURITY

AI can be manipulated. Hackers or cyber attackers can easily attack by providing false information to the system:

- **Hacking:** As AI is used to prevent and detect attacks, it can also be used for attacking too.

- **Biased:** AI systems tend to discriminate against people as they can be biased toward people.
- **Compromises Privacy:** AI can be used to track one's activity, which could have an impact on the individual's privacy.
- **Infancy:** It is highly important to understand AI's vulnerabilities in order to utilize cyber security, which is considered to be very critical.

18.4 AI IN CYBER SECURITY – ITS FUTURE

It is very clear that AI will be predominantly used in Cyber Security for the following reasons:

- Quicker identification of threats
- Attack prevention by auto blockage of activity which is suspicious.
- Resilience of network improvement
- Quicker recovery from cyber attack
- Improvisation of digital system security.

18.5 AI WITH CYBER SECURITY IN BUSINESS

Increasing business competitiveness and raising awareness to protect against cyber-attacks in emerging technologies using AI is highlighted by Narcisa Roxana et al. [5]. The authors also point out the vulnerability of information technology companies to cyber-attacks, as the internet is a shared resource. All data stored in a system connected to the local network or internet is susceptible to attack. The authors also demonstrate how cyber protection can safeguard a company's data. The article discusses the current state of AI, provides case studies and applications, and aims to make the public more aware of the problems and issues in cyber security. This includes leaders, engineers, academicians, researchers, students, and industrialists, as well as presentations on business implications, government implications, and policy recommendations [4].

18.6 BLOCK CHAIN FOR CYBERSECURITY

Bitcoin, as the first application in cryptocurrency, has brought about tremendous changes in industries and the global economy. It operates as a decentralized and transparent system, eliminating the need for a central server to protect against attacks. Participants update the ledger with each transaction, allowing all participants to view transaction details and trace assets or products through the blockchain. Blockchain technology is being implemented across various domains, offering protection against data theft and transaction malpractices. The encryption, transparency, and immutability of blockchain technology also enable the detection of data falsification and improvement of cyber defense. Additionally, blockchain can be used to manage SSL certificates in organizations, making it difficult for hackers to forge digital certificates. Transactions recorded on the ledger, embedded with blockchain technology, are digitally signed and time stamped, enabling companies to track and locate their transactions using public addresses.

18.7 CYBER SECURITY THREATS AND IMPACT ON BLOCKCHAIN NETWORK

It has been found that there were over 60 threats that occurred in the past few years, which had a negative impact on blockchain systems. These threats are classified as hacks, scams, and smart contract flaws. The Ponzi Scam resulted in the highest loss in 2018, totaling US $660 million [7]. More than 32,000 victims in Vietnam suffered losses from investing in Ifan and Pincoin.

The next highest loss is $530 million in cryptocurrency [8]. This is due to the Coincheck incident in 2018, where the BTC wallet's contract security feature stored blocks in a single wallet. Hybrid hot and cold wallet systems were used for storage, employing multi-sig contract security. The third loss occurred in 2014, amounting to $480 million, known as the Mt. Gox hack. Hackers breached the exchange and stole large amounts of cryptocurrency [9]. Mt. Gox experienced three attacks: one in June 2011, totaling $8.75 million; the next in October 2011, with a sum of $8.35 million; and the third in February, with a sum of $480 million. These failures were a result of inadequate security controls and subpar software. The BTC blockchain network also faces transaction malleability, which alters the transaction identifier (TXID) without invalidating the transaction. Modifying the TXID deceives individuals into believing that the transaction failed, only to be confirmed at a later time, causing delays. The altered TXID replaces the original transaction ID, making it undetectable by the currency exchange, which then confirms the transaction as failed. Attackers exploit this vulnerability in Mt. Gox attacks to withdraw coins from the exchange, freezing users' accounts and halting withdrawals [10]. Hacking incidents account for 48 cases, resulting in a loss of $1.6 billion. These compromises involve computer systems, email accounts, private keys, servers, and SIM swapping attacks on cold and hot wallets. Various flaws in blockchain user software include memory, security, performance, Graphical User Interface, configuration, and compatibility hard fork flaws. The blockchain network flaws mentioned above can potentially expose users' private keys. In 2014, blockchain.info made a mistake while updating their software, causing users who generated their key pairs using the infected software to have their private keys leaked. This issue affected 0.002% of users, resulting in the theft of some bitcoins. The problem was resolved within two and a half hours.

18.8 IMPROVEMENT OF CYBERSECURITY THROUGH BLOCKCHAIN

The various characteristics of blockchain technology, such as the blockchain chaining process, its distributed architecture, validation mechanism, and encryption process, enhance the cybersecurity process.

- **Chaining Process:** The arrangement of blocks in the blockchain mechanism executes transactions in a secure manner. A block in the blockchain consists of a header and a body. The blocks in the blockchain are signed by creators and time stamped. The chain of blocks as shown in Figure 18.6 has a pointer in each block pointing to the previous block. Immutability is also ensured by the block by having the previous block's hash code.

18.9 ROLE OF BLOCK CHAIN IN CYBER SECURITY

The development of technology has led to an increase in cyber threats. Nowadays, organizations should concentrate on data protection. The available data protection measures are not sufficient and, at times, are inefficient. Cyberattack seem to grow in a sophisticated manner, creating a demand for effective cybersecurity. With the demand for effective and efficient cybersecurity methods, there is a need for new technology to meet the demands of cybersecurity. The emerging blockchain technology, with its nature of immutability, decentralization, and transparency, can improve the security of data. Most organizations have turned to adopting blockchain technology to combat cyber threats by introducing faster detection and mitigation (Figure 18.7).

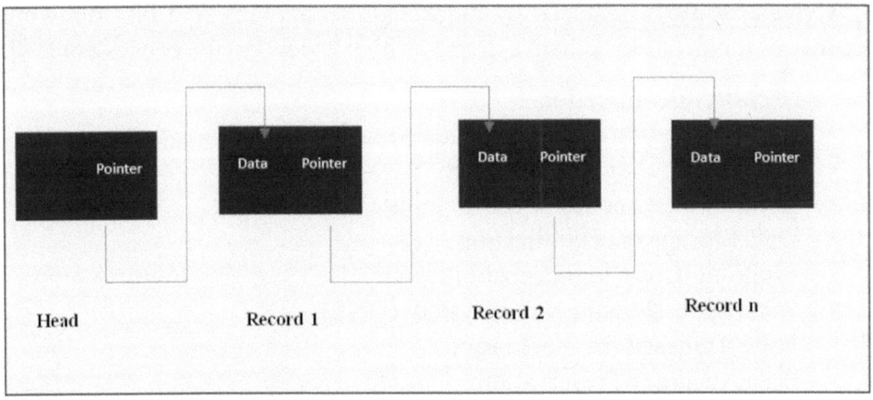

FIGURE 18.6 Blocks in blockchain.

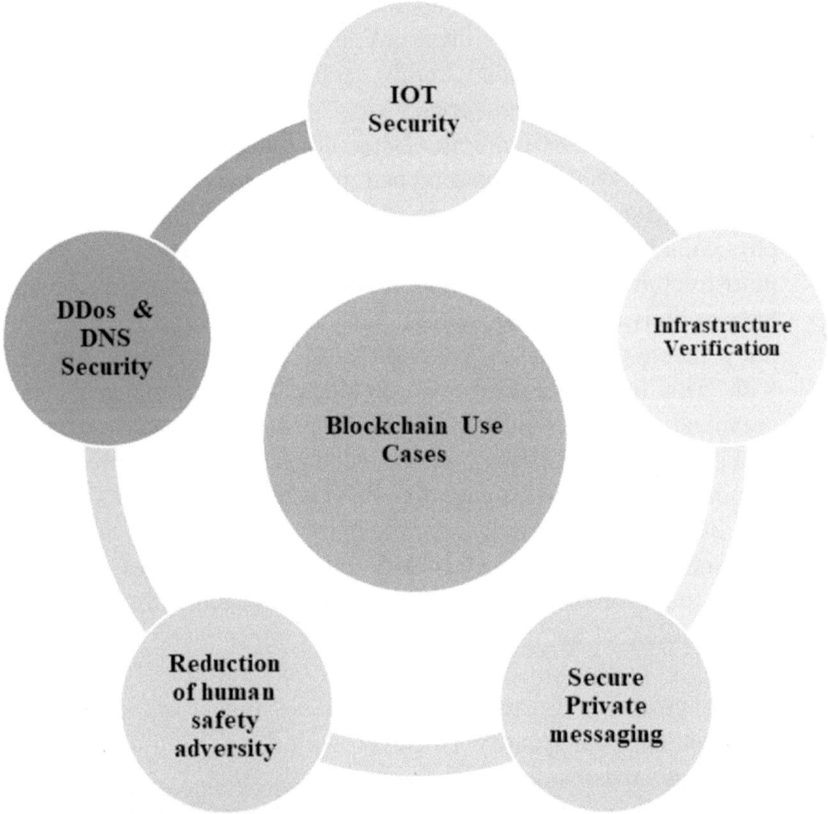

FIGURE 18.7 Blockchain use cases for cyber security.

Cybersecurity is reshaped by blockchain technology and the following are the solutions of blockchain cybersecurity provided to the ecosystem [10].

- Security Identification
- Communication and Transaction Security
- Data Manipulations Prevention
- Compromised nodes or server failure protection
- Data Storage decentralization
- Multi-signature login capabilities
- Security in Domain Name System (DNS)
- False Transactions blocking
- Data Accuracy Preservation

- **IoT Security:** The decentralized nature of blockchain technology secures vulnerable systems and devices. The incorporation of AI and IoT with blockchain technology has enabled the system to make wise decisions without human intervention. The central authority system makes the overall system highly secure and detects threats from known and unknown networks.
- **Infrastructure Verification:** Data integrity is compromised by methods such as data tampering and misconfiguration. The configuration of blockchain technology assists in verifying and authenticating the physical infrastructure of the network. Data generated by the blockchain infrastructure appears to be more reliable throughout the entire chain.
- **Secure Private Messaging:** In most cases, data and services are not protected with strong passwords, which can be easily tampered with. This leads to a shift towards blockchain technology by many organizations to provide encryption and protect data. Blockchain's API framework offers cross-messenger capabilities. Data breaches are common in social media networks like Twitter, WhatsApp, Facebook, etc., but can be prevented and protected by blockchain technology.
- **Reduction of Human Safety Adversity:** In the case of AV, data is transmitted from sensors to databases. In such a situation, hackers can gain access to Car Area Networks and compromise security. When security is compromised, attackers can take control of the automation process, which can have a significant impact on human safety and even national security. However, with the support of blockchain technology, these vulnerabilities can be prevented.
- **DDoS, DNS Security:** A Domain Name System (DNS) is convenient for attackers since it is a centralized system that facilitates the filtering of connections between websites and IP addresses. DNS attacks can

render the system inaccessible and redirect users to other websites. When a website denies access, a Distributed Denial of Service (DDoS) attack can occur, which can either shut down or slow down the system. To prevent these attacks, implementing blockchain technology can decentralize the DNS entries.

- **Confidentiality:** Confidentiality ensures that only those who need the data and have permission can access it. The encryption of blockchain technology ensures the prevention of unauthorized individuals accessing the data, as the data moves across both reliable and unreliable networks.
- **Integrity:** Organizations can ensure data integrity by leveraging the immutability and traceability features inherent in blockchain technology. In the event of a 51% cyber control attack, ledger splitting can be managed and prevented by establishing consensus-based procedures within businesses.
- **Availability:** DDoS attacks are the most frequent types of cyberattacks nowadays, aiming to reduce the availability of technical services. However, DDoS attacks are costly in blockchain-based systems because the attacker attempts to overwhelm the network with numerous small transactions. Due to the absence of a single point of failure, IP-based DDoS attacks are less likely to disrupt business as usual when utilizing blockchains. The platforms and systems are resilient thanks to their distributed operation and multiple nodes.

18.9.1 HANDLING COMMON FORMS OF CYBER ATTACKS USING BLOCKCHAIN

A different aspect of storing and performing information and functions for applications against cyberattacks is to enhance cybersecurity. This includes user identity security, transaction and communication security, and protection of critical infrastructure of organizations. The main characteristics of blockchain are immutability, hash function, and cryptography. These are combined with a decentralized structure, making it impossible to change the data in the ledger. It is also difficult to access the data through cyberattacks. Some common forms of cyberattacks and how they are handled through blockchain cybersecurity are discussed below:

- **Malware:** Malware is malicious software introduced in various ways into networks, such as phishing, infected files, OS loopholes, connected peripherals, and malicious websites, in order to compromise an organization's data and damage the information system. Through these methods, hackers gain access to login credentials and other sensitive information. Enhancing cybersecurity with blockchain can assist law enforcement in tracking transactions, recovering assets, and identifying hackers through the use of tools in blockchain technologies.

- **Web Application Attack / Credential Stuffing:** Mining user data is considered a web application attack. The data obtained through this attack is used in a brute force attack, where login credentials are systematically guessed to gain unauthorized access to users' accounts and steal sensitive information from individuals, organizations, and networks. This type of attack is commonly known as credential stuffing. Hackers use stolen data to attempt unauthorized access to users' financial accounts and their associated data.
- **Distributed Denial of Service (DDoS) Attack:** DDoS attacks transmit a continuous flow of data traffic to a targeted node to weigh down the system and interrupt system operations permanently or temporarily. Existing DNS makes it difficult to prevent DDoS attacks since it is only partially decentralized, which means hackers can still target the centralized part of DNS and continue crashing the websites. By using multilevel protection strategies like firewalls, VPN, anti-spam, and content filtering in networks and applications, we can identify and prevent DDoS attacks. Additionally, merging blockchain with cybersecurity will create one more layer of security that contributes to the defense mechanism in depth. DNS is completely decentralized using blockchain technology by distributing the contents to more nodes, which makes the hacker's work very difficult.
- **Man-in-the-Middle Attack (MITM):** MITM is a type of cyberattack in which hackers secretly intercept and relay information between users and trusted parties. There are two types: a standard attack, which occurs in an unsecured network like a WIFI router, and a man-in-the-browser attack, which injects malware into the user's system and tracks the data being sent to a trusted party website without the user's knowledge. Any system that relies on key agreement protocols is vulnerable to Man-In-the-Middle attacks. The best way to prevent a Man-In-Middle attack is to implement systems that eliminate the need for secrets.
- **Ransomware Attack:** Denying users access to organization files available on their system is a type of malware called Ransomware.

It encrypts the files and demands payment to decrypt them. Many financial institutions, such as cryptocurrency, are facing these attacks. Solutions based on blockchain technology are decentralized, making it difficult to find even a single copy that could be held for ransom. Only privileged customers can safely access the data.

Placing the data in the blockchain prevents even the administrator from swapping the data, as the blockchain is immutable. This technique uses sequential hashing, which applies a mathematical function to generate a unique hash code. This process is used to test and secure the integrity of data transactions in the blockchain. Additionally, blockchain makes it infeasible for anyone to alter the ledger without detection, thanks to data encryption.

18.10 BLOCKCHAIN TECHNOLOGIES TO IMPROVE CYBER SECURITY

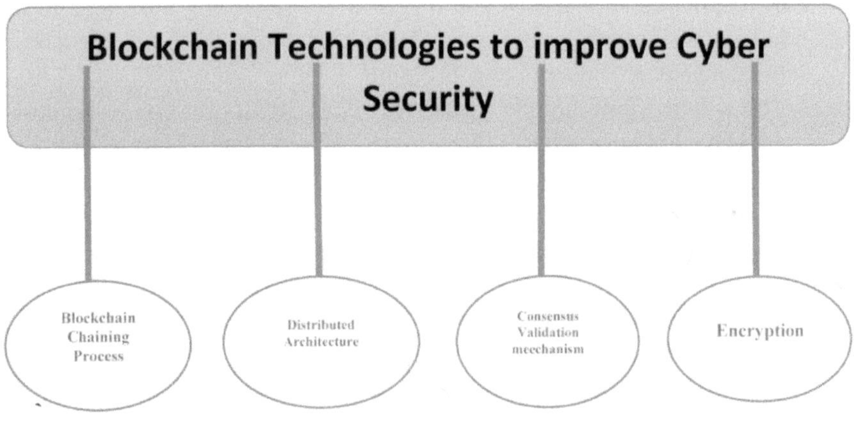

FIGURE 18.8 Blockchain technologies to improve cyber security.

- **Block Chaining Process:** In Blockchain, each block consists of records that are divided into multiple blocks, and all blocks are linked together by hash values. The hash value of the previous block can be found in each block of the blockchain. In addition to hash values, it also includes timestamps and transaction data. If a block has been altered by any of the transactions, then the block's hash value will be non-identical to the previous one. Due to this cryptographic link, transaction data in the block cannot be modified. Additionally, it is possible to verify the entire blockchain by issuing the first block's hash value and checking whether it is present in the second block, and this process will be repeated until the entire block has been verified.

- **Distributed Ledger Technology:** All transactions are recorded in a distributed immutable ledger using blockchain. The backbone of blockchain ensures that all nodes in the network have a copy of the blockchain and it is updated when transactions occur. Blockchain eliminates the need for a central authority by incorporating distributed computing with a peer-to-peer architecture, making the blockchain secure, transparent, and immutable. The risk is distributed among nodes, so any attack on one or more nodes will not compromise the ledger. As a result, ransomware attacks have less impact on the blockchain network, as compromising the entire network of nodes is required.
- **Consensus Mechanism:** Blockchain is a self-regulatory system that does not operate under a single authority. In this system, shared ledgers require reliable and secure mechanisms to ensure the legitimacy of all transactions occurring on the blockchain network, as well as to establish consensus among all nodes regarding the status of the ledger. These verification and authentication processes are carried out by consensus mechanisms utilizing various consensus algorithms. The system continuously verifies the integrity of past transactions and new block transactions. Additionally, network rules and each block, digitally signed with a hash value, prevent hackers from compromising the nodes in the blockchain.
- **Cryptographic Hash Function:** Encryption of blockchain technology prevents sensitive information from being hacked and misused by providing multi-layered protection. It makes use of a cryptographic algorithm which includes an asymmetric key algorithm and SHA-256 hashing algorithm. The main role of a hash function is to link blocks to one another and maintain data integrity for each block. Different hash values will be noticeable if there is a slight change in the data, which may lead to inconsistency and break the blockchain, making it invalid.

18.11 CHALLENGES FACED BY BLOCK CHAIN A PART OF CYBER SECURITY

- **Data Privacy:** Blockchain comes in four types-public, private, hybrid, and consortium. Public blockchain allows unrestricted access to anyone with an internet connection. Users can simply sign on to the platform. The only consideration is that the owner should closely monitor the publicly available information. Permissioned blockchain is a hybrid of public and private blockchain, where multiple users are granted access to the platform. It addresses privacy concerns by providing data access or viewing only to trusted parties.

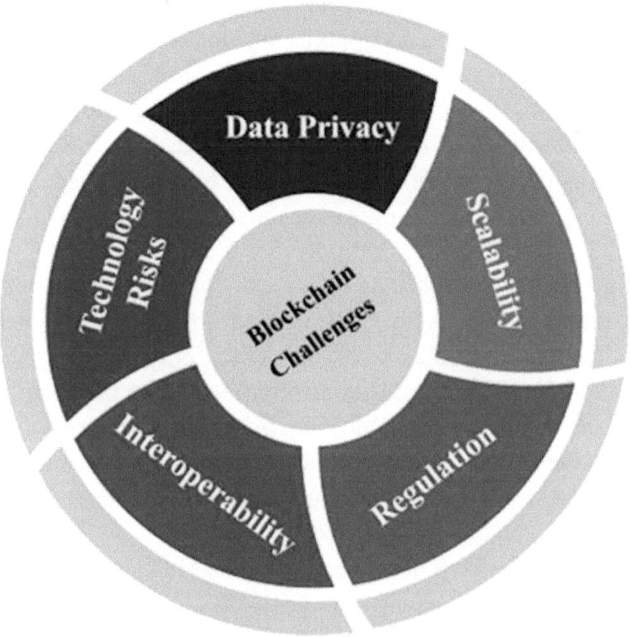

FIGURE 18.9 Challenges of blockchain a part of cyber security.

- **Scalability:** Limitation is the primary issue in blockchain scalability. The transaction details are added to every node in the ledger whenever a new transaction is processed. This growing transaction history can cause the entire system to become unstable. Additionally, blockchain faces hardware limitations when scaling the network to a larger extent. It becomes challenging to establish and maintain the necessary hardware for the expanded network and operating nodes in the blockchain network.
- Another factor is transaction fees, which highlight the scalability problem in blockchain. The number of transactions is directly proportional to the complexity of validating them. This may be attributed to the increased demand for computational power in mining. As the blockchain network continues to grow, users are required to pay

higher transaction fees for the validation process. All transactions in the network must undergo a validation process, which can result in longer validation times due to the number of transactions in the queue. Response time is another factor contributing to the blockchain scalability problem, and it is directly linked to high transaction fees.

- **Regulation:** Even though blockchain is used in many businesses and organizations, members are facing difficulties in understanding how the structure and complexity of blockchain fit into the evolving landscape of data protection, privacy compliance, and regulations. General Data Protection Regulations and laws allow all members to request the deletion of their data from the ledger. In certain cases, "Right to be forgotten" laws can also be established. However, the technology risks violating government rules by preventing network members from deleting or modifying their data [12].
- **Interoperability:** Various types of environments are used by blockchain networks for their transactions and consensus models. Scalability is limited by weak interoperability. Some of the obstacles may be created by the misconfiguration of the platform, mistrust in communication, errors, and flaws in the development of the application, as well as cross-chain smart contract logic problems. The pace of awareness, exposure, and implementation depends on the extent to which these hurdles are resolved.
- **Technology Risks:** There are several benefits available in blockchain networks, such as security, integrity, efficiency, and more. The rapid and drastic growth in technology paves the way for new risks and problems in the systems if they are not managed properly. Some of the risks are [12].
    - **Key Management:** One of the main challenges of cryptographic systems is efficient and secure key management. If hackers can find the key using mechanisms like side-channel attacks or weak encryption, it may lead to complete or irreversible loss. If the keys are not secure, nothing in the network will be secure.
    - **Unintended Hard Forks and Chain Split Attacks:** Smart contracts are programs stored in the blockchain that will run when predetermined criteria have been met without any human intervention. While upgrading these smart contracts, some nodes in the blockchain may not support these new standards at the time of the consensus phase. This can result in the blockchain splitting and creating new chains. It may also introduce new risks such as

replay attacks, double spending, 51% attacks, and unauthorized users being able to block, repeat, or reverse transactions.

- **Adequacy Issues in Scheme Selection:** Different cryptographic algorithms are used in different blockchains. Encryption schemes used for storing or transmitting data may not provide sufficient protection. Intrusions of this kind, along with improper verification of signatures, weak encryption, and key management errors, can make a blockchain vulnerable.
- **Application Programming Interface (API) Integration:** There are several APIs available in blockchain for business wallets, payment processing, simple text queries, and blockchain exploration. To implement blockchain with the above-specified processes, third-party API integration occurs either in a public or private blockchain. This can lead to issues of trust and accidental leakage of sensitive data from the blockchain network.
- **Blockchain Vendors:** The vendors in the network will go through primary vendors of blockchain who tend to provide the Blockchain-as-a-Service product. Nevertheless, all organizations and industries, whether public or private, that are willing to deploy applications, utilize blockchain services, and develop solutions [13].
  - i. **Amazon Web Services (AWS):** AWS is a cloud hosting platform by Amazon that also provides an open-source blockchain framework that can be easily deployed to enhance cyber security. Threats are identified and monitored in the network activity within the cloud environment.
  - ii. **Microsoft Azure:** With the introduction of cloud computing and the fast-growing progress of the Internet, there is a growth of Microsoft Azure, which is a cloud computing platform that provides a variety of services without the need to buy hardware. With the availability of cloud platforms, organizations have the opportunity to build efficient and effective blockchain solutions to improve cybersecurity.
  - iii. **IBM Blockchain:** The IBM Blockchain platform provides organizations with a Hyperledger Fabric. Since the Hyperledgers are modular, they provide control over cryptography, smart contracts, consensus algorithms, and identity. The BaaS service of IBM can operate, model, and create networks with better performance and security.

18.12 PROS OF BLOCK CHAIN BASED CYBER SECURITY

Blockchain's decentralized peer-to-peer network helps users to view network transactions, as there is no need for third-party verification. Transactions in blockchains are digitally signed and time-stamped, allowing network users to easily track transaction history and account balances. This feature also provides businesses with accurate information about assets or product distribution. Network members enjoy a high level of anonymity due to the use of public-key cryptography to authenticate users and encrypt their transactions. However, some blockchain-based startups take this technology a step further. For instance, Guard Time has developed a Keyless Signature Infrastructure (KSI) that allows users to validate their signatures without revealing their keys.The right to be forgotten is crucial for data privacy, even in the case of unalterable information. Blockchain technology ensures data privacy even if users forget their keys, as the data cannot be decrypted by anyone. The peer-to-peer connections and distributed consensus make it easy to identify malicious behavior in the event of a hack. Currently, blockchains are considered 'unhackable' since attackers can only have an impact on the network if they control 51% of the network nodes. Furthermore, the system continues to function normally due to the presence of multiple copies of the ledger, as blockchain technology does not have a single point of failure.

Distributed ledger safeguards data against modification or destruction. Furthermore, the technology ensures the authenticity and irreversibility of transactions. Encrypted blocks contain immutable data that cannot be hacked. The network can operate around the clock, even if some nodes are offline or under attack. Although blockchain technology cannot improve the quality of the data, it can ensure the accuracy and quality of data once it has been encrypted in the blockchain. To avoid risks associated with lost keys or human error, employees may require continuous access to a blockchain from multiple devices, putting a company's control over their private keys at risk. REMME's blockchain assigns each user and device a unique Secure Sockets Layer certificate, eliminating the need for passwords. Unauthorized network access is impossible with this approach.

Business correspondence contains sensitive data that can be effectively protected if cybersecurity is implemented using a blockchain. Many startups are encrypting business communication. Obsidian, for example, utilizes blockchain-based networks to mitigate vulnerabilities end-to-end. Blockchain technology can significantly improve smart contract security by reducing the risks of cyber-attacks and bugs. Smart contracts are ledger-based software programs. These programs ensure contract terms are followed and parties are verified. There is no need to keep sensitive data in a single location because blockchain technology enables the user to maintain multiple copies of their data that are always accessible to network users. Furthermore, blockchain technology enables users to instantly provide their clients with details about their products and services. Clients will have more faith in each other if they can ensure a high level of data security.

18.13 DRAWBACKS OF BLOCKCHAIN TECHNOLOGY FOR CYBERSECURITY

Blockchains mainly rely on private keys to encrypt data. However, if lost, these private keys are irretrievably lost, resulting in a permanent loss of access to encrypted data. Checking the network's scalability is crucial since blockchain networks have fixed block volumes and transaction restrictions. Companies may encounter challenges when integrating Blockchain technology, as it involves a complete overhaul of existing systems. Blockchain requires high levels of computing and storage capacity, resulting in higher costs compared to non-Blockchain applications.

Blockchain concepts are still unregulated on a global scale. The development of rules and frameworks is necessary to maintain governance in Blockchain systems. Learning Blockchain technology requires in-depth familiarity with various development, programming, and other tools. Due to this, there is currently a shortage of available Blockchain developers. Despite these drawbacks, blockchain technology also offers many advantages.

18.14 IMPLICATIONS OF BLOCKCHAIN TECHNOLOGY IN CYBERSECURITY

- **No Failures at a Single Point:** Due to the decentralized nature of blockchain systems, when there is a failure at a single node, it does not affect the entire network. Because several copies of the ledgers are kept, the system remains secure even during DDoS attacks. Private blockchains cannot provide this benefit.
- **User Privacy:** The public key cryptography used in a blockchain network helps to keep users' information private.
- **Protected Data Transfers:** Blockchain's Public Key Infrastructure (PKI) upholds authentication during data transfers. The automatic

implementation of agreements between two parties during a transfer is made easier with the aid of smart contracts.

- **Transparency and Traceability of Data:** A history of each of these transactions is kept and is thus always accessible. Members of the Blockchain network digitally sign the transaction data to preserve transparency.
- **Safe Data Processing and Storage:** Data can be stored safely and securely due to the main feature of blockchain, which is immutability and the recording of any changes.

18.15 CONCLUSION

It is evident that there is an increase in cyberattacks in terms of two factors: volume and complexity. AI provides hand-to-hand support in securing operations, enabling analysts to stay away from threats. ML and NLP, which are part of AI technology, reduce daily alert noise and response time. Blockchain, being the buzzing technology in this era, also has non-traditional features. Even though blockchain technology is gaining attention, scalability, security, and sustainability need to be concentrated on. To protect organizations from attacks, they need to adopt cyber security controls and standards.

KEYWORDS

- **artificial intelligence**
- **cyber threats**
- **cyber security**
- **confidentiality, integrity, and availability**
- **Internet of things**
- **artificial neural networks**
- **distributed denial of service**

REFERENCES

1. https://www.consultancy.eu/news/5628/how-digital-transformation-has-fast-tracked-in-two-years (accessed on 27 January 2024).

2. Sarker, I. H., Kayes, A. S. M., Badsha, S., Alqahtani, H., Watters, P., & Ng, A., (2020). Cybersecurity data science: An overview from machine learning perspective. *J. Big Data, 7*(1), 1–29.
3. Craigen, D., Diakun-Thibault, N., & Purse, R., (2014). Defining cybersecurity. *Technol. Innov. Manag. Rev., 4*(10):13–21.
4. Aftergood, S., (2017). Cybersecurity: The cold war online. *Nature, 547*(7661), 30.
5. Yaga, D., Mell, P., Roby, N., & Scarfone, K., (2018). *NISTIR 8202 Blockchain Technology Overview.* Retrieved from: National Institute of Standards and Technology, US Department of Commerce.
6. Dannen, C., (2017). *Introducing Ethereum and Solidity* (Vol. 1). Berkeley, A press.
7. Curran, B., (2018). *The History of the Coin check Hack: One of the Largest Heists Ever.* https://blockonomi.com/coincheck-hack/ (accessed on 27 January 2024).
8. Conti, M., Sandeep, K. E., Lal, C., & Ruj, S., (2018). A survey on security and privacy issues of bitcoin. *IEEE Communications Surveys Tutorials, 20*(4), 3416–3452.
9. https://www.techfunnel.com/information-technology/how-to-use-blockchain-technology-in-cyber-security/ (accessed on 27 January 2024).
10. https://www.blockchain-council.org/blockchain/the-state-of-blockchain-applications-in-cybersecurity/ (accessed on 27 January 2024).
11. https://www.infosys.com/insights/cyber-security/cybersecurity-blockchain.html (accessed on 27 January 2024).
12. https://www.blockchain-council.org/blockchain/the-state-of-blockchain-applications-in-cybersecurity/ (accessed on 27 January 2024).

CHAPTER 19

A Survey of Deep Learning Algorithms and Methods for Automated Cyber Security in Healthcare

M. PRANEESH[1] and N. NIVETHA[2]

[1]*Department of Computer Science, Sri Ramakrishna College of Arts and Science, Coimbatore, Tamil Nadu, India*

[2]*Department of Computer Science, Adithya College of Arts and Science, Coimbatore, Tamil Nadu, India*

ABSTRACT

In order to defend the CIA triangle of computer resources, networks, software programs, and data against attacks, a variety of policies, tactics, technologies, and processes are combined to form what is known as cybersecurity. This survey article provides a literature review of deep learning algorithms and techniques for automated cybersecurity. It discusses and analyzes the suitability of deep learning techniques for security applications. Additionally, the article reviews and emphasizes various cybersecurity datasets. Lastly, it analyzes the prevention of cybersecurity vulnerabilities in COVID-19 pandemic situations.

19.1 INTRODUCTION

Cybercriminals attack healthcare institutions of all sizes. The increasing number of healthcare-related hacks suggests that smaller healthcare providers are becoming more targeted by cybercriminals. Large healthcare

Leveraging Artificial Intelligence (AI) Competencies for Next-Generation Cybersecurity Solutions.
Pethuru Raj, B. Sundaravadivazhagan, & V. Kavitha (Eds.)

organizations typically have the resources to establish a strong cyber defense strategy, including a chief security officer, a security operations center, and access to the best threat intelligence services.

In the past decade, notable cyber-attacks have been directed at healthcare institutions. Smaller hospitals, doctors, and dentists often lack the budget for expensive cyber security measures. They face the same cyber threats and criminal opportunities as larger organizations. According to the World Medical Association, most medical offices in the U.S. have 10 or fewer physicians, with around 10% being solo practitioners. The inability or unwillingness to pay hefty ransoms often leads many smaller healthcare professionals to close their practices. However, even if a ransom is paid, there is no guarantee that the hacker will return the data or equipment, or refrain from selling patients' information on the dark net (Figure 19.1).

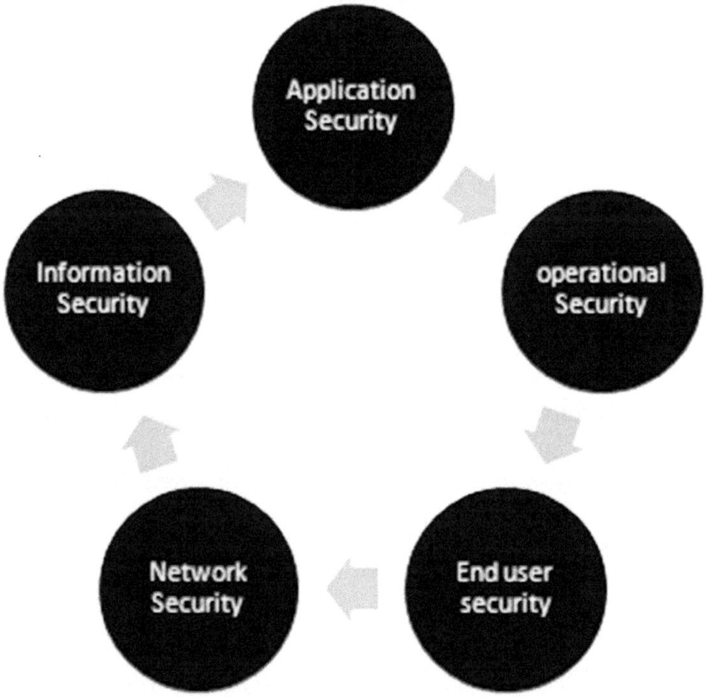

FIGURE 19.1 Cyber security levels.

The article is structured as follows: Section II provides an explanation of the importance of healthcare in cyber security. Section III describes the various healthcare sectors. Section IV discusses the consequences and suggestions regarding cyber security vulnerabilities in cyber-attacks. Section V provides an overview of Deep learning methods and architecture. Finally, section VI concludes the paper.

19.2 IMPORTANCE OF CYBER SECURITY IN HEALTH CARE

Cyber threats can hinder doctors from accessing patient data, leading to potentially catastrophic medical errors or gaps in treatment. If healthcare providers are unable to operate their businesses due to IT challenges, they face the risk of financial loss and damage to their reputation. For instance, during a ransomware attack, MedStar Health, a large healthcare organization in Maryland, had to shut down its email and website for several days. According to TechCrunch, critical services like cancer radiation were disrupted [14].

The increasing adoption of electronic health records (EHRs) has further amplified the IT risks in the industry. Inadequate spending on cyber security and the high value of compromised patient information make healthcare providers an attractive target for hackers. Common healthcare cyber security threats include email phishing, outdated systems, software vulnerabilities from manufacturers, insufficient IT resources, and lax security procedures [14].

Interest in digital security has significantly grown over the past decade. However, compared to other countries like the United States, there has been a general delay in addressing cyber security issues in various countries. This delay can be attributed to the perception that the healthcare sector in the United States is undeniably an industry, not just in terms of visibility but also in terms of its practices. The response to the crisis in the country has primarily focused on trade and consumption. Only recently has the issue started to receive the attention it deserves (Figure 19.2).

19.3 SPECIFIC HEALTH SECTORS WILL BE SUBJECTED TO HEIGHTENED SCRUTINY IN THE CONTEXT OF CYBERSECURITY

19.3.1 WEARABLE MEDICAL DEVICE

Implantable medical devices, in particular, are wearable healthcare devices [2–6] that are part of a hybrid approach. The ability of a heterogeneous system to communicate with each other over wireless networking creates a setting that may be vulnerable to cyber-attacks. An attacker could potentially send purposefully false data to the control algorithm.

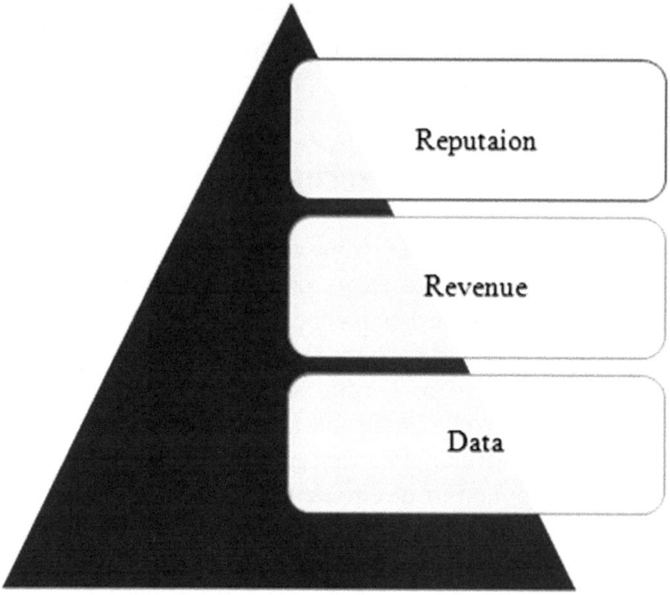

FIGURE 19.2 Cyber security and health care.

19.3.2 IMAGE ARCHIVING AND COMMUNICATION SYSTEM

A PACS consists of numerous components, including elaborators, workstations, databases, data stores, and applications. These systems are composed of specialized archive systems. Since a PACS serves as a repository for patient data, it is clear that these behaviors raise concerns about data security and integrity. This is particularly true when considering the implications of issues related to data privacy and protection (Figure 19.3).

19.3.3 HEALTH CARE NETWORKS

This is a well-known fact that hospital corporations nowadays heavily rely on technologies to run their operations. According to references [2, 3, 7], the number of cyber-attacks is rapidly increasing (Figure 19.4).

i. First and foremost, there has been a "dematerialization" of administrative operations.
ii. As previously stated, the increasing dependence on computerized medical, biological, and non-biomedical technologies is a concern.

iii. In addition, there is a vast volume of information held throughout the Hospital Information Systems (HIS).

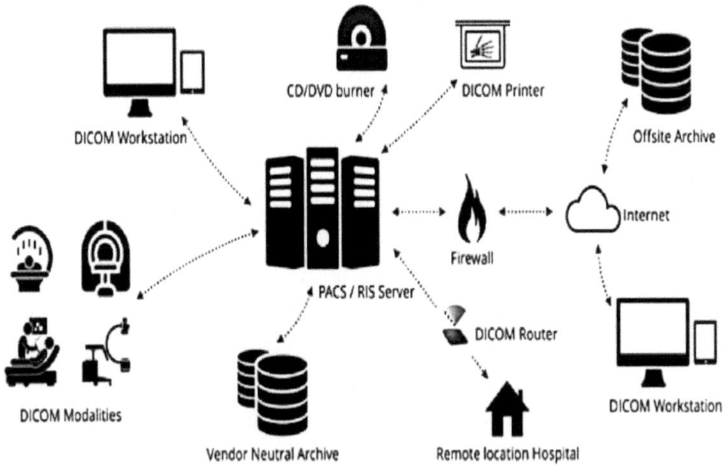

FIGURE 19.3 Picture archiving and communication system.

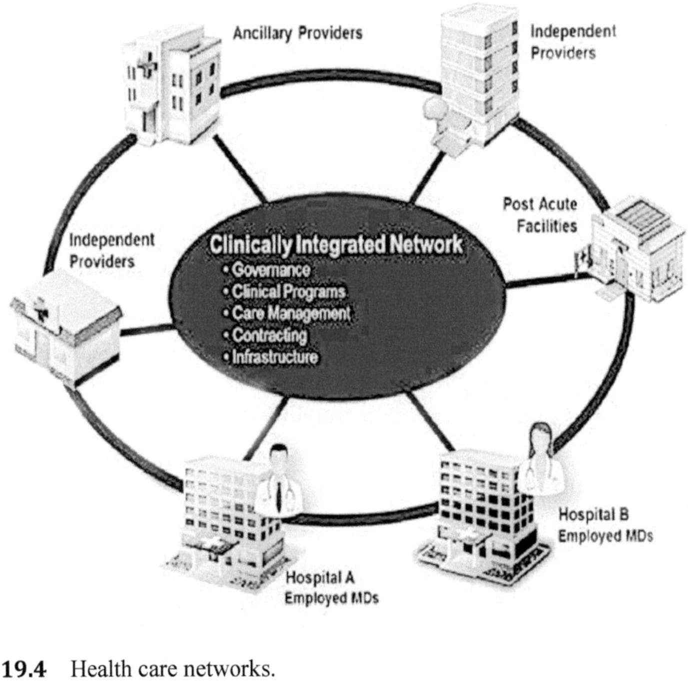

FIGURE 19.4 Health care networks.

The HIS can therefore be targeted and hacked with respect to both privacy and activity [7], implying that it is vulnerable. When it comes to health information systems, it is important to remember that the activity (which is based on specially designed software) is directly related to the health of people. Regarding the HIS, cybersecurity plays a crucial part in the defense of information technology infrastructures as well as in the eventual analysis of the individual citizen.

19.4 VULNERABILITY ANALYSIS OF CYBER ATTACKS IN HEALTHCARE

The healthcare sector is particularly vulnerable to a variety of cyber threats. Healthcare institutions must be aware of the changing threat landscape and put in place efficient preventive measures, even while legislation is in place to address these problems and deal with such cyber-attack situations. Therefore, it is important to be aware of some of the most significant ways that organizations in the healthcare sector may be particularly vulnerable to cyber-attacks.

In general, phishing awareness training informs IT personnel and employees about the likelihood of phishing assaults through social media. Healthcare IT systems in Asian countries that have been impacted must implement phishing awareness training using various contemporary training platforms, considering the data on phishing attacks targeting Asian healthcare institutions. During ransomware attacks, data is encrypted, leading to exorbitant ransom demands.

19.5 DEEP LEARNING FRAMEWORK AND ALGORITHMS

The "Network in Network" (NIN) suggested by Lin et al. [7] is a version of the regular CNN in which the 1x1 convolution filter is a Multi-Layer Perceptron (MLP) instead of typical linear filters, and the fully connected layers are replaced with a global average pooling layer (Figure 19.5). Because the micro network is composed of a stack of MLPCONV layers, the resulting structure is referred to as the MLPCONV layer. Unlike CNN, NIN can enhance the capacity to abstract latent concepts. They successfully demonstrated that the final MLPCONV layers of NIN were confidence maps of the categories, enabling them to perform object recognition using NIN. Figure 19.6 illustrates the convolution operation.

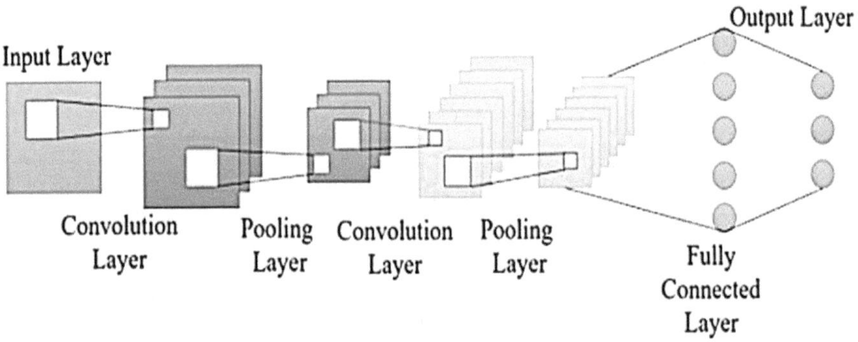

FIGURE 19.5 The architecture of CNN shows alternating convolution and pooling layers.

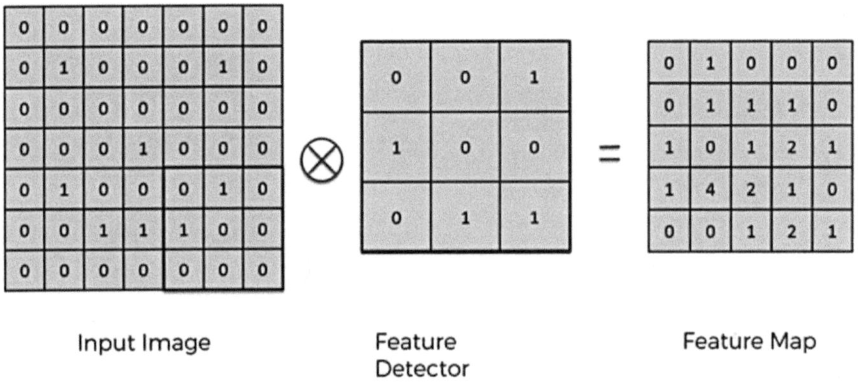

FIGURE 19.6 Convolution operation.

19.5.1 CLUSTERING LEVEL

This addresses the issue of over sizing. Some of the common grouping operations include maximum grouping, average grouping, stochastic grouping, spectral grouping, pyramidal spatial grouping, and grouping without order on multiple scales [8]. Figure 19.7 illustrates the functionality of maximum grouping.

i. **Layer Fully Connected:** In a typical neural network, the neurons in this layer are fully connected to all neurons in the preceding level. It is not possible to have a convolutional layer following a fully connected layer because the neurons are not spatially organized in one dimension. Recently, some architectures have replaced the fully connected layer, as in Lin et al.'s "NIN" [10], where a global average pooling layer is used instead of the fully connected layer.

$$f(x)=\max(0, x) \quad (1)$$

ii. **Loss Layer:** As a penalty for the discrepancy between the desired and actual outputs, the final fully connected layer serves as a loss layer that calculates the loss or error. The SoftMax loss is used to predict a single class out of K mutually exclusive classes. This loss function is widely popular and is essentially a multinomial logistic regression. It normalizes and associates predictions with non-negative values to obtain a probability distribution across classes. Support Vector Machine (SVM), a classifier with a large margin of error, is trained by estimating hinge loss. Euclidean loss can be applied to regress back to real-value labels.

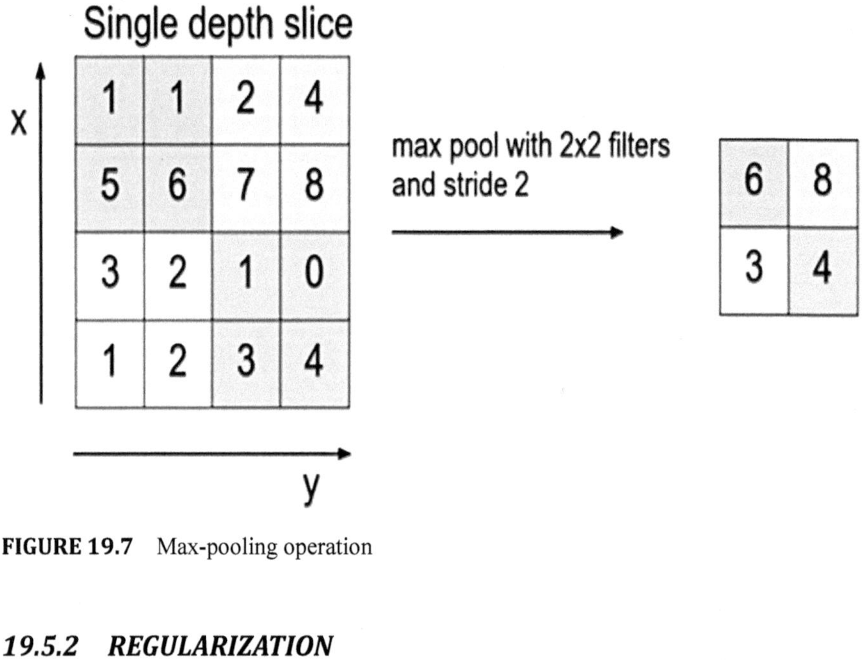

FIGURE 19.7 Max-pooling operation

19.5.2 REGULARIZATION

The problem of over sizing is one of the key challenges of deep convolutional networks. The approach used to address this problem is regularization. The two primary regularization approaches are Dropout and Drop-Connect. Figure 8(a) shows a conventional neural network. The dropout method, illustrated in Figure 8(b), is similar to sampling a neural network. During the training phases, some of the neurons' activations are set to zero. On the other hand, Drop-Connect randomly sets the connection weights to zero, as shown in Figure 8(c).

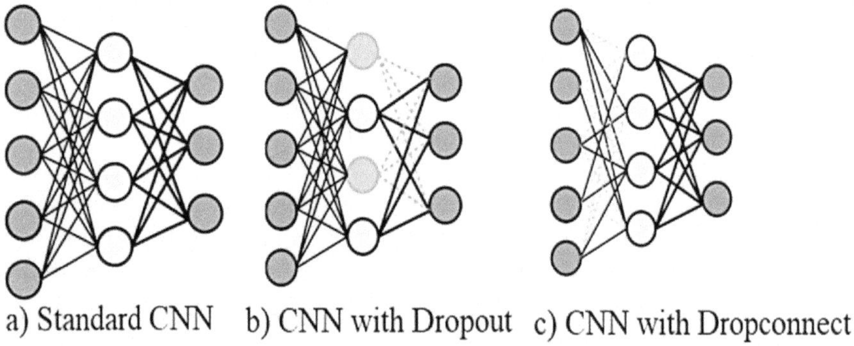

FIGURE 19.8 CNN with many layers.

19.5.3 CNN TRAINING AND TESTING

Due to the non-convex structure of the loss function [11], constructing a multi-tiered CNN can be a laborious task, especially when dealing with large datasets that require a high-processing GPU. It is uncommon for a CNN to be randomly initialized and trained from scratch. With proper network initialization, convergence can be accelerated. If the training dataset is small or if there is significant weight initialization, the transfer learning technique can be employed. In this scenario, only the classifier or the last layers need to be retrained, reducing the training time by half. By carefully selecting parameters such as learning rate, number of iterations, training and test sample sizes, batch size, etc., a good model can be created.

When training the same network with different parameter settings, an ensemble model can be created to achieve more accurate results. Back-propagation and gradient descent are used to train the network and update the weights. The vanishing gradient problem, also known as the issue of reduced gradient flow or prolonged delay, can affect back-propagation training [12]. Training the lower layers of a multilayer neural network can be challenging due to the exponential increase or decrease in error rates at these levels.

The solution to this issue is to use a sparse activation function, such as sigmoid or tanh. However, due to their dispersion characteristics and reduced vulnerability to the reduced gradient flow, recently rectified linear units (ReLU) have been shown to be more efficient for ConvNets [14]. The classifier is the top layer of a ConvNet. The SoftMax classifier, which generates probabilities for each class that sum up to 1, is one of the most popular choices. However, alternative classifiers like SVM, probabilistic classifiers, etc., are also used depending on the requirements. SVM is employed for structured forecasting. Classifier sets are also widely used.

19.5.4 COMMON CNN ARCHITECTURES

- **LeNet:** It was one of LeCun et al.'s groundbreaking works in CNNs. Improvements were made in 1990 [2] and 1998 [15]. ConvNets were used in their research to solve the problem of handwritten digit recognition, discover uses for deciphering postal codes, numerals, etc. The use of CNN was halted at the time due to a lack of high-processing machines.
- **AlexNet:** Alex Krizhevsky et al. [16] are recognized as the first to promote Convolutional Networks in the field of artificial vision using this design. AlexNet was identical to LeNet, but instead of rotating between convolutional and pooling layers, it had all of them. Additionally, compared to LeNet, this network is significantly larger. AlexNet was able to achieve impressive results in the ILSVRC-2012 competitions, with error rates between the top 1 and top 5 in the test dataset.
- **GoogleNet:** The ConvNet architecture proposed by Szegedy et al. [18] won the 2014 ILSVRC competition. Google introduced a new design called Inception (v1), which allows for better utilization of the network's computer resources. Google Net is one version of the Inception module with 22 layers, but with fewer parameters than AlexNet. Subsequently, Inception-v1 underwent several changes, the most significant of which was the addition of batch normalization, resulting in Inception-v2 by Ioffe et al. [19]. The Inception-v3 architecture is mentioned, and additional features have been incorporated into this version.
- **VGGNet:** Holding all other variables constant, Karen Simonyan and Andrew Zisserman examined the impact of the depth factor in a ConvNet. This research led to the development of the VGGNet ConvNet architecture, which is more accurate. In the 2014 ILSVRC tournament, it achieved second place.

19.6 CONCLUSION AND FUTURE WORK

Cybersecurity, application security, information security, operational security, recovery plans and operational continuity, and end-user training are all examples of generic cybersecurity actions that may be found in both the consumer and industrial sectors. Because the behavior is linked to people's health, it is important to recognize that healthcare systems are extremely crucial. As a result, ensuring that cybersecurity is implemented correctly and effectively is of paramount importance. Every aspect of traditional healthcare as well as those coming from Health and mobile healthcare must be handled with the utmost care, and all of these areas may and should be explored by academics and researchers. Training and information sharing must be essential components of cybersecurity in the healthcare industry.

CONFLICT OF INTEREST

No conflict in this work.

KEYWORDS

- **electronic health records**
- **multi-layer perceptron**
- **network in network**
- **rectified linear units**
- **support vector machine**

REFERENCES

1. Buczak, L., & Guven, E., (2016). A survey of data mining and machine learning methods for cyber security. *IEEE Commun. Surv. Tutor., 18*, 1153–1176. [CrossRef].
2. Nguyen, T. T. T., & Armitage, G., (2008). A survey of techniques for internet traffic classification using machine learning. *IEEE Commun. Surv. Tutor., 10*, 56–76. [CrossRef].
3. Garcia-Teodoro, P., Diaz-Verdejo, J., Maciá-Fernández, G., & Vázquez, E., (2009). Anomaly-based network intrusion detection: Techniques, systems, and challenges. *Comput. Secur., 28*, 18–28. [CrossRef].

4. Sperotto, A., Schaffrath, G., Sadre, R., Morariu, C., Pras, A., & Stiller, B., (2010). An overview of IP flow-based intrusion detection. *IEEE Commun. Surv. Tutor., 12*, 343–356. [CrossRef].
5. Wu, S. X., & Banzhaf, W., (2010). The use of computational intelligence in intrusion detection systems: A review. *Appl. Soft Comput., 10*, 1–35. [CrossRef].
6. Torres, J. M., Comesaña, C. I., & García-Nieto, P. J., (2019). Machine learning techniques applied to cybersecurity. *Int. J. Mach. Learn. Cybern.*, 1–14.
7. Xin, Y., Kong, L., Liu, Z., Chen, Y., Li, Y., Zhu, H., Gao, M., Hou, H., & Wang, C., (2018). Machine learning and deep learning methods for cybersecurity. *IEEE Access, 6*, 35365–35381. [CrossRef].
8. Apruzzese, G., Colajanni, M., Ferretti, L., Guido, A., & Marchetti, M., (2018). On the effectiveness of machine and deep learning for cyber security. In: *Proceedings of the 2018 10th IEEE International Conference on Cyber Conflict (CyCon)* (pp. 371–390). Tallinn, Estonia.
9. Wickramasinghe, C. S., Marino, D. L., Amarasinghe, K., & Manic, M., (2018). Generalization of deep learning for cyber-physical system security: A survey. In: *Proceedings of the IECON 2018-44th Annual Conference of the IEEE Industrial Electronics Society* (pp. 745–751). Washington, DC, USA.
10. Al-Garadi, M. A., Mohamed, A., Al-Ali, A., Du, X., & Guizani, M., (2018). *A Survey of Machine and Deep Learning Methods for Internet of Things (IoT) Security*. arXiv, arXiv:1807.11023.
11. McCulloch, W. S., & Pitts, W., (1943). A logical calculus of the ideas immanent in nervous activity. *Bull. Math. Biophys., 5*, 115–133. [CrossRef].
12. Hebb, D. O., (1949). *The Organization of Behavior*. John Wiley Sons, Inc.: New York, NY, USA.
13. Rosenblatt, F., (1958). The perceptron: A probabilistic model for information storage and organization in the brain. *Psychol. Rev., 65*, 386. [CrossRef].
14. Rumelhart, D. E., Hinton, G. E., & Williams, R. J., (1986). Learning representations by back-propagating errors. *Nature, 323*, 533. [CrossRef].
15. Deng, L., & Yu, D., (2014). Deep learning: Methods and applications. *Found. Trends Signal Process., 7*, 197–387. [CrossRef].
16. Hinton, G., Osindero, S., & Teh, Y. W., (2006). A fast-learning algorithm for deep belief nets. *Neural Comput., 18*, 1527–1554. [CrossRef].
17. Fukushima, K., (1975). Cognitron: A self-organizing multilayered neural network. *Biol. Cybern., 20*, 121–136 [CrossRef] [PubMed].
18. Jarrett, K., Kavukcuoglu, K., & LeCun, Y., (2009). What is the best multi-stage architecture for object recognition? In: *Proceedings of the 2009 IEEE 12th International Conference on Computer Vision* (pp. 2146–2153). Kyoto, Japan.
19. Vincent, P., Larochelle, H., Lajoie, I., Bengio, Y., & Manzagol, P. A., (2010). Stacked denoising autoencoders: Learning useful representations in a deep network with a local denoising criterion. *J. Mach. Learn. Res., 11*, 3371–3408.
20. Ranzato, M., Boureau, Y. L., & LeCun, Y., (2008). Sparse feature learning for deep belief networks. In: *Advances in Neural Information Processing Systems* (pp. 1185–1192). MIT Press: Cambridge, MA, USA.
21. Ranzato, M., Huang, F. J., Boureau, Y. L., & LeCun, Y., (2007). Unsupervised learning of invariant feature hierarchies with applications to object recognition. In: *Proceedings of the CVPR'07 IEEE Conference on Computer Vision and Pattern Recognition* (pp. 1–8). Minneapolis, MN, USA.

22. Benigo, Y., Lamblin, P., Popovici, D., & Larochelle, H., (2007). Greedy layer-wise training of deep networks. In: *Advances in Neural Information Processing Systems* (pp. 153–160). MIT Press: Cambridge, MA, USA.
23. Mohamed, A. R., Dahl, G. E., & Hinton, G., (2012). Acoustic modeling using deep belief networks. *IEEE Trans Audio Speech Lang. Process., 20*, 14–22. [CrossRef].
24. Sarikaya, R., Hinton, G. E., & Deoras, A., (2014). Application of deep belief networks for natural language understanding. *IEEE/ACM Trans. Audio Speech Lang. Process., 22*, 778–784. [CrossRef].
25. Lee, H., Grosse, R., Ranganath, R., & Ng, A. Y., (2009). Convolutional deep belief networks for scalable unsupervised learning of hierarchical representations. In: *Proceedings of the 26th Annual International Conference on Machine Learning* (pp. 609–616). Montreal, QC, Canada.
26. Bengio, Y., Simard, P., & Frasconi, P., (1994). Learning long-term dependencies with gradient descent is difficult. *IEEE Trans. Neural Netw., 5*, 157–166. [CrossRef].
27. Hochreiter, S., & Schmidhuber, J., (1997). Long short-term memory. *Neural Comput., 9*, 1735–1780. [CrossRef].
28. El Hihi, S., & Bengio, Y., (1996). Hierarchical recurrent neural networks for long-term dependencies. In: *Advances in Neural Information Processing Systems* (pp. 493–499). MIT Press: Cambridge, MA, USA.
29. Sutskever, I., (2013). *Training Recurrent Neural Networks*. University of Toronto: Toronto, ON, Canada.
30. Pascanu, R., Mikolov, T., & Bengio, Y., (2013). On the difficulty of training recurrent neural networks. In: *Proceedings of the International Conference Machine Learning* (pp. 1310–1318). Atlanta, GA, USA.
31. Sutskever, I., Vinyals, O., & Le, Q. V., (2014). Sequence to sequence learning with neural networks. In: *Advances in Neural Information Processing Systems* (pp. 3104–3112). MIT Press: Cambridge, MA, USA.
32. Cho, K., Van, M. B., Gulcehre, C., Bahdanau, D., Bougares, F., Schwenk, H., & Bengio, Y., (2014). *Learning Phrase Representations Using RNN Encoder-Decoder for Statistical Machine Translation*. arXiv:1406.1078.
33. Bahdanau, D., Cho, K., & Bengio, Y., (2014). *Neural Machine Translation by Jointly Learning to Align and Translate*. arXiv:1409.0473.
34. Graves, A., Mohamed, A. R., & Hinton, G., (2013). Speech recognition with deep recurrent neural networks. In: *Proceedings of the 2013 IEEE International Conference Acoustics, Speech, and Signal Processing (ICASSP)* (pp. 6645–6649). Vancouver, BC, Canada.
35. LeCun, Y., Boser, B. E., Denker, J. S., Henderson, D., Howard, R. E., Hubbard, W. E., & Jackel, L. D., (1990). Handwritten digit recognition with a back-propagation network. In: *Advances in Neural Information Processing Systems* (pp. 396–404). MIT Press: Cambridge, MA, USA.
36. LeCun, Y., Bottou, L., Bengio, Y., & Haffner, P., (1998). Gradient-based learning applied to document recognition. *Proc. IEEE, 86*, 2278–2324. [CrossRef].
37. Srivastava, N., Hinton, G., Krizhevsky, A., Sutskever, I., & Salakhutdinov, R., (2014). Dropout: A simple way to prevent neural networks from overfitting. *J. Mach. Learn. Res., 15*, 1929–1958.
38. Long, J., Shelhamer, E., & Darrell, T., (2015). Fully convolutional networks for semantic segmentation. In: *Proceedings of the IEEE Conference on Computer Vision and Pattern Recognition* (pp. 3431–3440). Boston, MA, USA.

39. Ning, F., Delhomme, D., LeCun, Y., Piano, F., Bottou, L., & Barbano, P. E., (2005). Toward automatic phenotyping of developing embryos from videos. *IEEE Trans. Image Process., 14*, 1360–1371. [CrossRef].
40. Taigman, Y., Yang, M., Ranzato, M. A., & Wolf, L., (2014). Deep Face: Closing the gap to human-level performance in face verification. In: *Proceedings of the IEEE Conference on Computer Vision and Pattern Recognition* (pp. 1701–1708). Columbus, OH, USA.
41. Krizhevsky, A., Sutskever, I., & Hinton, G. E., (2012). Image Net classification with deep convolutional neural networks. In: *Advances in Neural Information Processing Systems* (pp. 1097–1105). MIT Press: Cambridge, MA, USA.
42. Waibel, A., Hanazawa, T., Hinton, G., Shikano, K., & Lang, K. J., (1990). Phoneme recognition using time-delay neural networks. In: *Readings in Speech Recognition* (pp. 393–404). Elsevier: Amsterdam, The Netherlands.
43. Sainath, T. N., Mohamed, A. R., Kingsbury, B., & Ramabhadran, B., (2013). Deep convolutional neural networks for LVCSR. In: *Proceedings of the 2013 IEEE International Conference Acoustics, Speech, and Signal Processing (ICASSP)* (pp. 8614–8618). Vancouver, BC, Canada.
44. Collobert, R., Weston, J., Bottou, L., Karlen, M., Kavukcuoglu, K., & Kuksa, P., (2011). Natural language processing (almost) from scratch. *J. Mach. Learn. Res., 12*, 2493–2537.
45. Zhang, X., & LeCun, Y., (2015). *Text Understanding from Scratch.* arXiv, arXiv: 1502.01710.
46. Goodfellow, I., Pouget-Abadie, J., Mirza, M., Xu, B., Warde-Farley, D., Ozair, S., Courville, A., & Bengio, Y., (2014). Generative adversarial nets. In: *Advances in Neural Information Processing Systems* (pp. 2672–2680). MIT Press: Cambridge, MA, USA.

CHAPTER 20

An Intrusion Detection System Using the NSL-KDD Dataset's Convolutional Neural Network and Data Analytics

D. KALAIVANI

Department of Computer Technology, Dr. SNS Rajalakshmi College of Arts and Science, Coimbatore, Tamil Nadu, India

ABSTRACT

The rapid expansion of intrusion attacks as a result of the internet's fast evolution has made them a major concern in the present day. Cyber-attacks can target any of the millions of internet users, as well as global businesses and governmental organizations. Consequently, developing advanced algorithms to detect these network intrusions is one of the most crucial tasks in the field of cyber-security research. Internet traffic log data is utilized to train intrusion detection systems (IDS) in recognizing malicious traffic inputs. These methods enable the identification of malicious traffic inputs. The Network Security Laboratory's Knowledge Discovery and Data Mining (NSL-KDD) team maintains the most commonly used database for internet traffic records. It also serves as the benchmark for current internet traffic. This framework aims to distinguish between normal and abnormal (denial of service (DoS), probe, user to root (U2R), and remote to local (R2L)) categories in the NSL-KDD database with high detection accuracy and low false alarm rates. Several classifiers, including Naive Bayes (NB), K-Nearest Neighbor (KNN), Support Vector Machine (SVM), linear discriminant analysis (LDA), and Convolution Neural Network (CNN), will be employed to achieve this. In this study, unique and cutting-edge supervised detection techniques will be utilized as the fundamental approaches

Leveraging Artificial Intelligence (AI) Competencies for Next-Generation Cybersecurity Solutions.
Pethuru Raj, B. Sundaravadivazhagan, & V. Kavitha (Eds.)

to address the issue of the need for more labeled data collected during IDS training. The trial results demonstrate that the CNN classifier outperforms both recently proposed approaches and other currently used methods in terms of classification performance.

20.1 INTRODUCTION

20.1.1 A SUMMARY OF IDS

Over the past few years, there has been a rise in the demand for online services. In fact, it is anticipated that there will be 50 billion gadgets connected to the internet by the year 2020. Cyber-attacks are a persistent threat to information and communication technology systems, but there are benefits as well. Malware attacks have gotten more complex, challenging to detect, and dangerous to society and the economy. Billions of dollars are lost each year as a result of IT service breaches, and in the coming years, this number is expected to increase. Consequently, the main concern in modern civilization is cyber security (Figure 20.1).

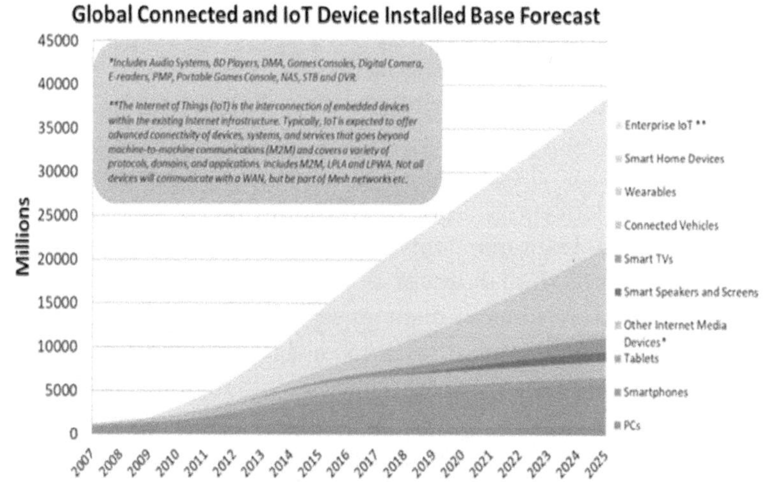

FIGURE 20.1 Count of internet-connected devices.

The most recent survey conducted by Strategy Analytics found that as of the fourth quarter of 2018, there were 22 billion devices throughout the world that were linked to the internet (www.businesswire.com). By 2030, there will be 50 billion connected devices, and by 2025, there will be 38.6 billion. At this moment, network traffic data monitoring and analysis is essential for identifying potential attack trends. To address the dangers and security challenges, several academics are currently focusing on creating IDSs. IDSs are primarily responsible for monitoring hosts and networks, analyzing computer behavior, providing alerts, reacting to suspicious behavior, and sending warnings when they detect suspicious activity and unknown dangers. In addition, IDSs provide alerts and responses to suspicious behavior.

An intrusion is any action that jeopardizes an information system. Computer systems are monitored by intrusion detection systems (IDSs), which check for odd behavior that a conventional packet filter might miss. To achieve optimal cyber flexibility and counter destructive operations, it is crucial to be able to detect unauthorized access to an information system by analyzing network packets for indications of malicious activity. Developing sophisticated and intelligent Intrusion Detection Systems (IDS) is essential for enhancing cyber security in such scenarios and mitigating threats. Many IDSs employ various methods to initially detect intrusions. One such method is Signature-Based IDS (SIDS), also known as Knowledge-Based Detection, which generates a signature identifier for known malware, enabling future identification of similar malware. If an exact signature match is found, the traffic can be classified as malicious. SIDS often demonstrates exceptional detection accuracy, particularly for reported intrusions.

Because SIDS may be as simple as updating the signature-based information, three issues arise. First, malware's polymorphic nature makes it easy to deceive security mechanisms that rely on signatures. This method fails the similarity test as it does not match any of the signatures in the IDS database, providing attackers with a way into the computer system. Second, as more signatures are stored in the database, it takes more time to analyze and refine the large amount of data. This is due to the exponential growth of the database. The limitation of SIDS has been addressed thanks to the development of anomaly-IDS, or AIDS, systems. These systems are increasingly used to identify hostile attacks on computer systems. This approach is based on the idea that profiles of potentially dangerous behavior are distinct from those of typical user activity. AIDS creates a statistical model of average user activity, detecting any abnormal actions that deviate from the standard model. To classify anomalous occurrences based on their deviation from expected behavior, the AIDS technique monitors, records, and visualizes the typical behavior profile. Anomaly detection systems are capable of identifying new attacks and have a high level of generalizability. However, due to the dynamic nature of cyber-attacks, they may also generate a significant number of false positives. Unknown user behaviors are classified as intrusions since they differ from normal behavior. The HIV/AIDS framework consists of two phases: training and testing.

After the model has learned the regular traffic profile through the use of data sets that duplicate typical activity, it is then evaluated based on a set of data that was not shown during the training process. Primarily, AIDS could be fought using machine learning (ML)-based methods. However, due to the enormous amount of heterogeneous data generated by multiple sources, ML techniques are ineffective and unsuitable for effectively addressing such security concerns [4]. Even with its low computing cost, it is unable to comprehend the intricate non-linear relationships that are present in the sizable dataset. By utilizing DL technology to create a more complicated IDS, the constraints mentioned above are eliminated, and intrusion detection performance is enhanced [5]. In fact, through the use of a hierarchical learning process, DL is able to automatically build exceedingly abstract levels based on the representations that are fed into it.

20.1.2 EXISTING TECHNIQUES

For the purpose of detecting intrusions, a wide variety of methods are utilized, including host-based, network-based, rule-based, and signature-based systems. The methods discussed above require a significant amount of calculation time and have large overheads. Currently, both network-based approaches and SIDS are the only types of IDS in use, and both types can only detect known threats with incredibly low rates of false positive detection. ML and deep learning technologies have proven to be some of the most successful methods for developing safe systems.

20.1.3 PROPOSED METHOD

Delivering a strong IDS architecture that would boost network security is the aim of the proposed study. Attacks can be quickly detected by such a system. We investigate, discuss, and assess a variety of cutting-edge ML and DL algorithms that can be applied to enhance network security on the NSL-KDD dataset. The proposed system includes feature extraction, pre-processing, and classification.

Stage 1: Data values are normalized and the basic data format is adjusted during the data pre-processing step. Normalized data must be transformed into an image data format for the CNN model to function more effectively.

STEP 2: The process of extracting features is one of the most important steps involved in the development of ML and DL models. The relevance of the features in the data has an effect on the accuracy of the model, but it also makes the training process more time-consuming. It is necessary to use the feature extraction technique to develop an IDS.

STEP 3: By continuously modifying the parameters, the AI model is trained to perform better. To enhance the model's performance, several parameters are also adjusted during the training process.

STEP 4: After training, the accuracy of the ML/DL model should be evaluated using the provided test data (STEP 3). For instance, if the accuracy rate reaches the goal level, the training would be considered complete. If not, the model would continue the training operation.

Following the training, in Stage 5's assessment step, the performance of the model is evaluated. Common evaluation measures include the accuracy rate, detection rate, and false alarm rate.

The graphic below illustrates the four steps of the IDS mode using the established data analytics algorithm and DL model (Figure 20.2).

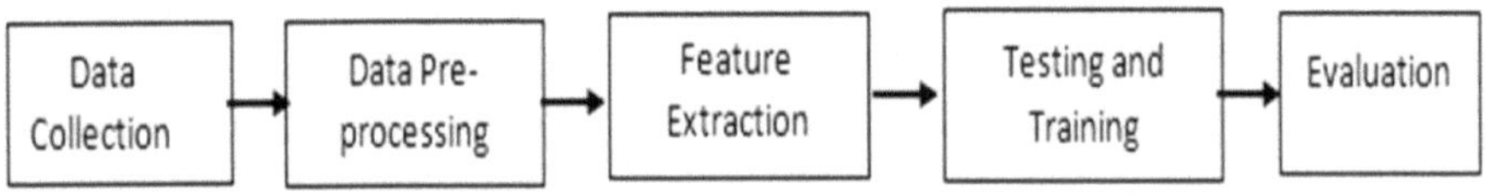

FIGURE 20.2 IDS framework.

The DL algorithm and DL model provided in this work are then used to demonstrate the 5 steps of the IDS in the accompanying image.

20.1.4 LIMITATION OF EXISTING TECHNIQUES

The NSL-KDD database has many redundant and null records, which is one of its key drawbacks. In contrast to unique recordings, which are often more challenging for R2L and U2R networks to handle, this characteristic compels learning algorithms to train on regular recordings. Additionally, the presence of these repeated records in the test set will impact the evaluation outcomes. Here are a few more drawbacks:

- The premise that underpinned the awarding of a high score was that the activity was suitable for the system.
- The extensive and difficult accessibility to a large amount of data.

- Ability to receive training from attackers.
- Setting metrics and parameters is challenging.

20.1.5 PROBLEM DEFINITION

The amount of network traffic in the web-enabled world has significantly increased as a result of the rapid expansion of online users and their online communication. As the number of internet users increases, so do the security risks. The effectiveness of the network is significantly impacted by the magnitude and frequency of these attacks. Therefore, it is reasonable to assert that despite the various security measures that have been recently established, such as peripheral protection mechanisms and multiple authentication and access control systems, achieving comprehensive protection against intrusions is not possible. This assertion is supported by recent developments in each of these areas. One potential solution to this problem is the implementation of IDS. The primary mission of IDS is to recognize and anticipate normal as well as deviant patterns of behavior. An IDS is a type of software that observes and collects data from various networks and computers to detect suspicious activities intended to gain unauthorized access to, modify, or disable computer systems. IDSs, which demonstrate typical system or network traffic behavior, are able to identify any system or network traffic behavior that deviates from the predefined and trained data as abnormal. Anomaly-based IDSs face significant issues such as how to teach these systems, how to identify relevant attributes that constitute typical behavior of a system or network environment, and how to represent this behavior computationally. All of these questions must be answered before these systems can be effectively used.

The purpose of this research was to find solutions to these problems by utilizing data analytics and AI (ML/DL) techniques. Due to their effectiveness, one of the most frequently used tactics employed by the IDS is the application of AI algorithms. In recent years, ML and DL-based IDS have become more accurate and better equipped to recognize new types of attacks.

20.1.6 CONTRIBUTION

This research takes advantage of the feature extraction method that is a part of the NSL-KDD dataset in order to overcome fundamental constraints such as data repetition and the presence of null values. Specifically, these issues were addressed by incorporating this method. To improve the detection rate of the proposed work and extend the training time required to develop the IDS technique, this method is used to eliminate irrelevant and null features that hinder the detection rate.

- An architectural design is offered for an efficient IDS.
- A comprehensive review of various types of attacks, available attack datasets, and research-ready attack detection systems.
- Data analytics-based AI techniques for IDS aim to enhance classification accuracy and improve the quality of the obtained results.
- The creation and use of supervised IDS to handle the vast amount of labeled data.

20.2 LITERATURE REVIEW

20.2.1 BACKGROUND RESEARCH

ML, deep learning, and ensemble techniques are just a few of the cutting-edge technologies that have been developed in recent years by various researchers to detect intrusions. Other advanced technologies include anomalies, which are data points that deviate from the normal distribution of a network [6]. For instance, ML techniques like SVM, DT, and KNN are applied to labeled datasets to create classifiers capable of identifying intrusions. According to the research conducted by R. Doshi and colleagues in 2019, deep learning strategies such as Auto Encoder, Deep Neural Network (DNN), Deep Belief Network (DBN), and Recurrent Neural Network (RNN) enable automatic extraction and categorization of features. The last group enhances detection performance by utilizing a wide range of ensemble and hybrid tactics, including bagging, boosting, stacking, and combination classifier methods. Several IDSs evaluating the effectiveness and performance of our proposed models have been published in the research literature, using the KDD99 and NSL-KDD datasets. These IDSs can be found here.

20.2.2 AI TECHNIQUES FOR IDSS

20.2.2.1 ML METHODS

Binhan Xu and colleagues came up with the idea for the incremental k-NN SVM methodology in 2017. This method combines the benefits associated with SVM and k-NN. It utilizes an R*-tree to locate the k-NN, which efficiently improves the total quantity of training data. The expansion k-NN, SVM IDS methodology can train and update with new data in a reasonable amount of time, according to studies that made use of the free database KDDCUP 99. Additionally, the amount of time required for forecasting does not significantly increase throughout the expansion learning process.

Anish Halimaa et al. compared the dataset's post-preprocessing detection rate and error rate in 2019. According to the execution data, SVM accuracy is 93.85% for 19,000 instances, while Naive Bayes accuracy is 71.01%. When compared with SVM, Naive Bayes incorrectly classifies more data for 19,000 cases. Over 19,000 instances, the accuracy rate provided by Naive Bayes was lower than expected.

Al-Yaseena et al. suggested a hybrid IDS that combines ELM and SVM to identify known and unknown attacks. On the KDD Cup 1999 dataset, the system had a false alert rate of only 1.87%, indicating that it worked quite well.

E. D. Alalade et al. [9] developed an IDS that combines ELM with an AI system to detect anomalies in the smart home system (AIS-ELM). By optimizing AIS using the clonal method, the input parameters are analyzed by ELM to achieve improved convergence in detecting abnormal activity.

Rohan Doshi et al. [10] introduced a packet-level ML approach for DoS detection, capable of identifying DoS attacks from consumer IoT traffic. They evaluated five ML classifiers, including KNN, LSVM, DT, RF, and NN, using a database of cyber-attack (DoS) movements after testing a network design for consumer IoT devices. With any of these techniques, the test set of data can be predicted with an accuracy level of 99.9% or higher.

Ingre et al. [11] constructed a three-layer MLP to classify different types of attacks in the NSL-KDD database. By dividing the data into binary classes and five different classes, the following conclusions (type of attack) were reached. Various performance metrics were utilized during the data analysis, indicating an increase in accuracy. The detection rates for attack classifications range from 79.9% to 81.2%, depending on whether we are referring to intrusions or attacks. The average detection rate for intrusions is 79.9%. Compared to the current classification tasks of binary classes and five classes, the new method outperforms the existing system, achieving a higher detection rate.

Feature selection had a negative impact on the operational performance of replicas produced by NSL-KDD rather than having a positive impact on the amount of time needed to compute models based on the UNSW-NB15 dataset. On the other hand, normalization had a greater impact on model performance than feature selection, regardless of the dataset. This implies that the NSL-KDD dataset is less developed and unsuitable for constructing a reliable current IDS compared to the UNSW-NB15 database. The Random Forest model outperforms all other models on both datasets, achieving an accuracy of 99.75% and 98.51% for the NSL-KDD data and 98.51% for the UNSW-NB15 data, respectively.

Cremer et al. [13] share their findings from analyzing several open datasets, including KDD-Cup'99, the Center for Applied Internet Data Analysis (CAIDA), NSL-KDD, and CICIDS 2017. These datasets were used for their research. Additionally, they provide a comparison of different IDS datasets, focusing on the types of computer threats covered by each dataset and the aspects prioritized by each dataset. The authors present their classification findings for the selected datasets, based on their previous research. Their model, consisting of an MLP neural network and a payload classifier, achieved an accuracy rate of 95.2% when tested on the CICIDS 2017 dataset.

K-means is a method for recognizing DDoS attacks that took place during CICIDS 2017. In addition to this, to overcome what they referred to as "the curse of feature dimensionality," they developed a fusion-based feature optimization technique. This method eliminated irrelevant attributes from the data fed into the model. At this stage of the process, every available quality and trait is carefully considered. Before the features can be processed, they need to undergo several preliminary steps, including data preprocessing, feature scaling, and feature subset selection. After this step, the application will output some of the features it has been processing thus far. Ultimately, they implemented the suggested optimization strategy, resulting in an accuracy rate of 96.50% with an error rate of 30.5%.

20.2.2.2 DL METHODS FOR IDSS

Using unsupervised feature learning, a two-stage classification DL model method in order to categorize the traffic in the NSL-KDD Cup dataset as either attack or normal. A sparse autoencoder assists the deep learning model in learning features, while a sparsity penalty is employed for classification. Although the model reduces false positives, the accuracy can be further enhanced by utilizing a signature-based IDS.

CNN was utilized in order to automatically recognize traffic features from the raw data set and alter the cost function weight coefficient for each class based on the numbers associated with it. This approach reduces false alerts and improves class accuracy with small integers (FAR). To minimize computing expenses, the traffic vector data is converted into an image. The

CNN model was evaluated using NSL-KDD data. The proposed model outperforms traditional standard approaches in terms of accuracy, FAR, and processing cost. It is a reliable method for detecting network breaches.

A CNN-based IDS was proposed to boost internet security. To identify anomalies and potential threats, an Intrusion Detection System (IDS) classifies network packet traffic as either normal or abnormal. Using the CICIDS 2017 dataset, the suggested model was trained and validated. We have evaluated the system's accuracy, attack true positive rate, false alarm rate, and training cost.

A CNN model using DL on the CICIDS 2018 dataset, which included more samples but the same feature set as CICIDS 2017. The CICIDS 2018 database was used to construct and assess study models. Simulating multi-class classification models required only some of the dataset's classes at once. The experimental results imply that the RNN, a general DL model for time sequence data, may not perform as well as the CNN-based IDS. Using normal and DoS samples from CICIDS 2018, the CNN technique reported in this research achieved 96.7% accuracy. Compared to CNN and RNN models, only 82.8% of the dataset's predictions were accurate.

A study that utilized anomaly analysis for intrusion detection with KNN and DL was presented in 2019 by K. Atefi and colleagues. The purpose of the study was to assess the capabilities of ML and DL classification. CICIDS 2017 was utilized to replicate the model's research performance. DNN outperformed KNN significantly, they found. Compared to the KNN's 90% accuracy, their DNN is 96.427. Also compared were the running time overheads. DNN's 110-second CPU time is shorter than KNN's 130 seconds.

DL models for IoT network cyber security were published in 2019. The authors examined the MLP model, LSTM, CNN, and LSTM-CNN fusion using CICIDS 2017 DDoS assault samples. LSTM achieved the highest precision (98.44%), followed by CNN and the hybrid model. The MLP simulation accuracy was 88.47%. The authors compared these findings to ML models and found that all studied DL models outperformed ML models such as SVM, Bayes, and RF, except MLP.

DL-MAFID was developed by Louati, Ktata, and others in 2020 to defend against multi-class cyber-attacks using multiagent systems and auto-encoders. This method applies an autoencoder to KDDCup'99 to reduce the dimensions of the data. Shallow classifiers like MLP and KNN are utilized to identify between the five different dataset classes. DL-MAFID speeds up detection while maintaining 99.95% accuracy.

In 2020, Rashid et al. compared NSL-KDD and CIDDS-001. They optimized fusion features before employing self-learning (ML/DL) classification algorithms such as SVM, NB, k-NN, Neural Networks, DNN, and DAE. The IDS was evaluated using well-known performance criteria. k-NN and NB techniques achieved around 99% accuracy on the CIDDS-001 dataset, while SVM, NN, and DNN classifiers achieved around 100% accuracy on the NSL-KDD dataset. Table 20.1 summarizes the literature review and lists the authors, techniques, and reliability.

TABLE 20.1 Summary of Literature Review

Author Names (Years)	Proposed Method	Accuracy (%)
Halimaa et al. [8]	SVM	93.85
Al-Yaseena et al.	SVM and Extreme Learning Machines	98.13
Rohan Doshi et al.	Packet-level Machine Learning	99.9
Ingre et al. [11]	3-layer MLP	81.2
Umaret et al. [12]	Random Forest	98.51
A. Khraisat et al.	MLP	95.2
Yonghao Gu et al.	Semi-supervised K-means algorithm	96.50
K. Atefi et al.	DNN	90.913
Monika Roopak et al.	LSTM	98.44
Louati and Ktata et al.	Shallow classifiers (MLP and KNN)	99.95
A. Rashid et al.	DNN and DAE	95
Jiyeon Kim et al.	CNN model	96.77

This has aided DNNs in coping with intrusions by allowing for intrusion detection and the categorization of hazardous actions in this portion of the research review. To accomplish this, it begins by classifying ML-based IDSs into various categories based on the methods of feature selection that they use. Next, it outlines how each system aims to use ML techniques to recognize the various types of intrusions. In the aforementioned IDS systems, the shallow learning techniques used in conjunction with the ML techniques are also investigated. To provide a complete knowledge of the analyzed IDS frameworks, the primary contributions, advantages, and shortcomings are also listed. Additionally, the datasets and evaluation measures of each area are compared. Finally, it can be argued that DL is a fascinating method that presents the IDS with both opportunities and difficulties. According to the aforementioned literature research, our work focuses on DL approaches on the NSL-KDD database for intrusion detection framework.

20.3 RESEARCH METHODOLOGY

20.3.1 INTRODUCTION

Recent events, along with an ever-increasing demand for the use of the Internet in all applications and fields, have led to a growth in both the quantity of moving packets and the burden placed on the network. Therefore, even though there are many cyber protection systems, such as the effective IDS, which is an effective system of protection and prevention, the most sensitive information is still at risk. This section provides an overview of the NSL-KDD database, which was used for this investigation.

20.3.2 CLASSIFICATION

20.3.2.1 ML TECHNIQUES

In the interdisciplinary field of study known as ML, the main emphasis has been on developing algorithms for machine learning. Learning is nothing more than creating databases of features for learning purposes. The expert system can commonly solve diagnostic problems using historical data, thanks to the development and application of ML systems. Reinforcement learning, unsupervised classification, and supervised classification are just a few of the various learning techniques that are now available for classification tasks. One of the techniques for supervised learning is known as classification, and the classification tool is used to make predictions regarding the target class [15] (Figure 20.3).

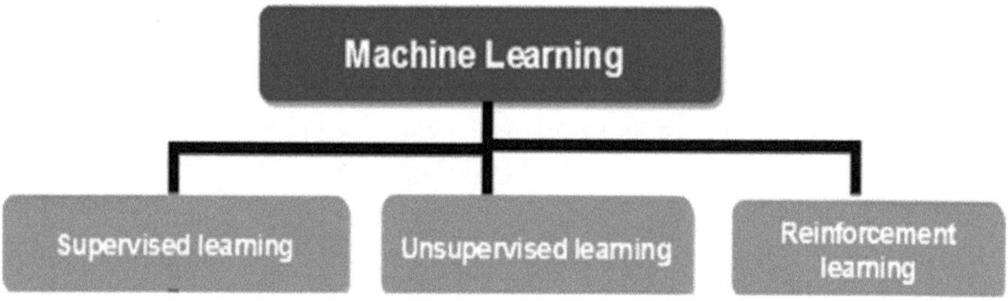

FIGURE 20.3 ML methods.

The usage of supervised learning can be found in the vast majority of real-world applications of ML. The fact that all of the data is labeled makes it possible for the algorithms to learn how to make predictions about the results based on the data that they are given as input. Neural networks (Multilayer Perceptron), SVM, logistic regression, naive Bayes, and the KNN Algorithm are a few examples of these types of algorithms. In unsupervised learning, none of the data is labeled; rather, the algorithms make conclusions about the underlying structure based on the data that is being input. This type of learning is considered to be more exploratory than supervised learning. Some examples of these kinds of models include neural networks, c-means, fuzzy c-means, hidden Markov models, hierarchical, Gaussian, and c-means mixes. The fundamental idea behind reinforcement learning is called "decision-making in sequential order," and it is at the center of the process. To put it another way, the result of the input that came before it determines the result of the input that comes after it, and the condition of the input that comes after it is determined by the result of the input that came before it. Decision trees, linear regression, ensemble techniques, and neural networks are just a few examples of the kinds of methods that fall under this category. One of the most significant actions that must be taken in order to successfully implement this system is to choose a suitable algorithm. In recent years, a large variety of supervised and unsupervised ML algorithms have been developed, and each one applies a different learning strategy in order to maximize efficiency [16].

The classification challenge in cyber-attacks is a well-known classification model, and this study addresses it using supervised machine learning techniques. The study utilizes scientific data to solve the problem, employing the classification model. We possess a collection of training records that are labeled with the letters "A" and "B" for each distinct category, denoted by the notation $D = X_1, X_2$, and X_n. Based on the classification model, the characteristics of the testing record are assigned to one of the many class labels. The model then predicts the class label for a particular instance or set of features associated with an unknown class. This prediction is made based on the information input into the model. The supervised learning techniques are built upon the foundation provided by the annotated training properties of the input image. The process of supervised binary classification is depicted in the following image [16].

According to the technique, the NSL-KDD dataset was divided into normal and malignant categories using various supervised learning strategies. This division was done to gain a better understanding of the data. The supervised learning process relies heavily on the labeled instances in the training dataset as a reference point. This facilitates the efficient training of learning models, enabling accurate classifications at a high level. Therefore, it is crucial to utilize learning algorithms that strictly adhere to established principles.

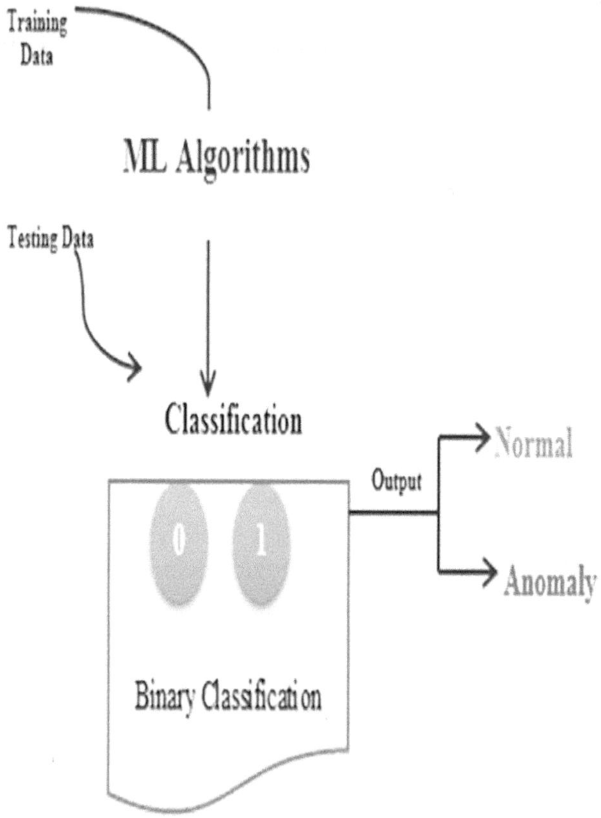

FIGURE 20.4

During supervised learning, a connection is established between a set of input variables (X) and a single output variable (Y). This connection is then utilized for subsequent learning. The next step involves using this connection to predict outcomes based on hypothetical data. In fact, the majority of machine learning algorithms are built upon the foundation of supervised learning. Once all the data is labeled, the findings are analyzed using algorithms based on the obtained information. Within the scope of this study, various machine learning classification algorithms such as neural networks, SVMs, LDA, and convolutional neural networks (CNNs) are examined and compared.

20.3.2.1.1 Naïve Bayes (NB) Algorithm

NB is one of the most significant ML classification algorithms. This method is enabled by the "NB" classifier, which is based on the concept of independence. It is used for various purposes, including spam filtering and text classification. The combined probabilities of characteristics or features and grades are used in the calculation to determine the probability score of grades given an assumed feature subset. This classifier utilizes a simple probabilistic classifier to partition a set of data (dr) into classes (C (i=1) m = c1, c2, etc.). According to A. Ali and colleagues (2020), the maximum posterior (MAP) class returns are the best for the "NB" approach.

$$\underset{C1\epsilon C}{\text{Cmap}} = \arg\max P(C_i)\,P\,(d_r \mid C_i)$$

In this scenario, the class "P (ci)" can be determined by dividing the total number of features in class "ci" by the total number of features. The value of P (dr | ci) indicates the number of times the characteristic appears in the data "dr," which is a member of the ci class. It is going to be determined what the probability "P (ci | dr)" is for each latent class, but "P (dr)" is going to remain the same. Given this, it's possible that the denominator can be removed. This is done by calculating the likelihood of each class and then using that information to find the classes that are most likely given the facts. This process is referred to as "cmap" (called "d") (Figure 20.5).

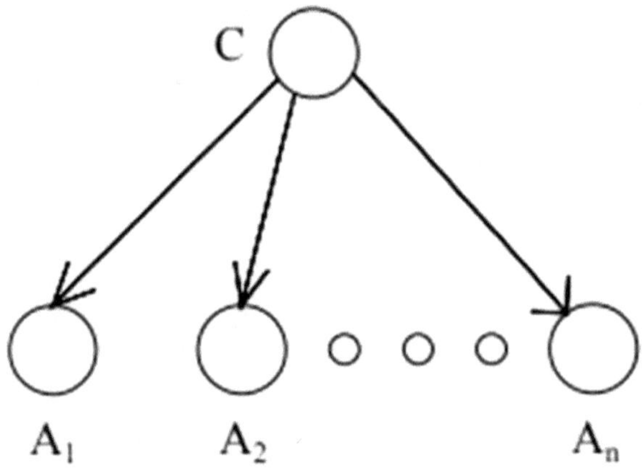

FIGURE 20.5 Functioning of NB.

The following picture provides an illustration of the fundamental structure of the NB classifier, in which the classification node serves as the parent node for all other nodes. In an NB structure, no additional connections are permitted [17].

20.3.2.1.2 K-Nearest Neighbor (KNN) Algorithm

ML systems heavily rely on the KNN technique. It is a component of supervised learning and is widely utilized in various industries, such as pattern recognition and intrusion detection. KNNs are applied in practical scenarios where non-parametric approaches are required.

These techniques do not establish any patterns for the distribution of data. The KNN method splits the dataset's correlatives into clusters that can be recognized by a certain trait. The key advantage of this strategy is that it generates comparable results for comparable training data. It finds the most appropriate classification for all or a portion of the samples in the input population [18].

Examine the sample populations $X_i = \{x_1, x_2, \ldots, x_{iN}\}$ $X_j = \{x_1, x_2, \ldots, x_{jN}\}$ in order to calculate the distance between them and determine how similar they are.

$$\text{Dist}(X_i, X_i) = \sqrt{\sum_{m=1}^{N} (x_{im} - x_{jm})^2}$$

The equation above represents the Euclidean distance, which measures the similarity between two pixel locations. As a result, the pixels are assigned to the category that the majority of instances of those pixels belong to. In the KNN algorithm, there are a total of K nearest neighbors at any given time. The primary consideration in this decision is the number of people living nearby, and in settings with two classes, K is typically chosen as an odd number. When K is equal to 1, this calculation is known as "the computation of the closest neighbor," named after its namesake. This outcome presents the fewest obstacles to overcome [18].

20.3.2.1.3 SVM Algorithm

It is possible to construct a supervised machine learning model known as an SVM by employing either a linear or nonlinear methodology. Because datasets are frequently nonlinearly separable, the primary objective of the SVM approach is to capture the best surface that can discriminate between positive and negative training feature samples based on the principle of reducing experimental threat (training set in addition to test set error). This is done because datasets are frequently separable in a nonlinear fashion. By utilizing hyperplanes, this method can attempt to specify a decision boundary in a high-dimensional feature space. The hyperplane makes a choice based on the support vector, which is obtained by factoring and splitting the input into two groups. The decision is made based on the support vector. The following scenario, which serves as an example and is offered for your reference, can be used to demonstrate how the SVM works [19].

To supply a training set consisting of "N" linearly separable objects as well as a feature vector denoted by "x" that has dimensions. There is a need for two optimizations.

Here, $\alpha \in R^N$ then $y \in \{1, -1\}$

The result of SVMs may then be summarized as below:

$$\vec{\alpha}^* = argmin\left\{-\sum_{i=1}^{n}\alpha_i + \sum_{i=1}^{n}\sum_{j=1}^{n}\alpha_i\alpha_j y_i y_j \left\langle \vec{x_i}, \vec{x_j} \right\rangle\right\}$$

Where;

$$\sum_{i=1}^{n}\alpha_i\, y_i = 0; \qquad 0 \le \alpha_i \le C$$

A linear dataset is divided into two categories by an SVM classifier, which utilizes a single hyperplane and a limited number of features. In this scenario, kernel functions are employed to map the data to a higher dimensional space that is linearly separable, enabling the management of nonlinear datasets with more than two classes. This approach allows for more effective handling of the data.

The following diagram illustrates the objective of SVM, which is to distinguish between positive and negative classes in feature space. This objective can be understood as finding a hyperplane that accurately separates these classes with maximum margin, which is the most crucial challenge.

20.3.2.1.4 Linear Discriminant Analysis (LDA)

For the purpose of carrying out LDA, an approach called supervised classification may be utilized. Within the context of a general classification problem, an arbitrary variable with the name X is assigned to one of the K classes, and several class-specific probability densities are presented (x). A discriminant rule will attempt to partition the data space into K distinct parts so that it can accurately represent all of the classes (visualize the cases on a chessboard).

The discriminant analysis classification method simply determines that x belongs to class j if any one of these regions contains the variable of interest [20]. Using these two allocation criteria, the region of the data x has been logically located as follows:

Applying the maximization of likelihood criteria, place x in class j if there is an equal chance of occurrence across all classes.

$$j = \arg\max_i f_i(x)$$

Bayesian rule: In the event that we are aware of the class prior probability, we should place x in class j.

$$j = \arg\max_i \pi_i f_i(x)$$

Explicit variants of the aforementioned allocation rules can be developed if it is assumed that the data come from a multivariate Gaussian distribution and that the distribution of X can be described by its mean and covariance. This will allow the production of explicit variants of the rules. According to the Bayesian rule, a piece of data is assigned to a certain category, such as class j of K, if it has the highest likelihood among all of K's categories, and we set I = 1.

$$\delta y(X) = \log f_i(x) + \log \pi_i$$

The term "discriminant role" is used to characterize the aforementioned function. It is important to note how the log-likelihood was utilized in this situation. The discriminant function, in essence, provides us with information about the likelihood that data x belongs to each class. Consequently, the set of x that contains the points at which the two discriminant functions have the same value serves as the decision boundary between any two classes, k and l. Since we are unable to make a decision, any information regarding the decision boundary is equally likely to come from either class.

When we assume that the covariance in each of the K classes is equal, Linear Discriminant Analysis (LDA) is employed. In other words, instead of having a covariance matrix for each class, all classes share a single covariance matrix. As a result, we can derive the discriminant function as stated below:

$$\delta_k(X) = X^T \sum^{-1} \mu_k - \frac{1}{2} \mu_k^T \sum^{-1} \mu_k + \log \pi_k$$

Everyone can see, without a doubt, that this is a linear function in x. This method is called LDA because each decision boundary between two classes is also a linear function in x. The quadratic term of the likelihood does not cancel out without the equal covariance assumption, hence the resulting discriminant function is a quadratic function in x.

$$\delta_k(X) = -\frac{1}{2}\log|\Sigma_k| - \frac{1}{2}(x-\mu_k)^T \sum_k^{-1}(X-\mu_k) + \log \pi_k$$

The decision boundary in this instance is quadratic in x. LDA is the term used for this [20].

20.3.2.2 DL TECHNIQUES

Deep learning is a crucial concept in machine learning. In deep learning, data representation is achieved through a layered hierarchy technique that mimics the functioning of the human brain. By analyzing enormous volumes of data, deep learning can produce conclusions with higher accuracy rates compared to traditional categorization techniques. Many different fields are utilizing deep learning techniques to address various problems. Numerous prestigious companies are actively working on deep learning-related initiatives, applying them to a range of big data applications. However, these triumphs also come with numerous challenges. The future of deep learning looks promising as computer scientists collaborate with specialists from various industries to develop algorithms that can effectively manage these challenges [20].

Recent research studies have employed a wide variety of machine learning approaches in both domains for anomaly identification. Conventional machine learning techniques often struggle to scale up to large platforms due to the lack of labeled training datasets and their heavy reliance on manually extracted features. Artificial neural networks (ANNs) have played a significant role in the development of the cutting-edge machine learning paradigm known as deep learning (DL), which surpasses the capabilities of more traditional machine learning approaches [20].

Simply a few of the networks used in DL include CNNs, DBNs, Restricted Boltzmann Machines (RBMs), RNNs, and DNNs. The learning processes for these networks can be supervised, semi-supervised, or unsupervised. The usage of hierarchical layers, which extract suitable high-level properties from the raw input data instead of relying on manual features, is also advantageous. Since DL has garnered a lot of interest in the field of IDS, the research community has developed a number of DL-based anomaly detection models that can handle a wide variety of cyber-attacks and other forms of security risks.

20.3.2.2.1 Convolutional Neural Network

By utilizing the convolutional features, CNNs are able to significantly reduce the amount of weight and computation required during training on multi-dimensional input. The backpropagation technique requires training with a CNN, which is a multi-layer network, in order to recognize two-dimensional figures and minimize the number of parameters needed. Additionally, various attributes can be derived from the unlabeled data [21].

Xiao et al. introduced CNN-IDS, a network IDS based on a CNN paradigm. The CNN is used to extract the characteristics of the network traffic, and supervised learning is employed to gather the necessary information for intrusion detection. This technique reduces the execution cost by transforming the traffic vector into an image. The authors demonstrate, using the KDDCup99 performance evaluation tool, that the CNN and IDS model achieves a good FAR, accuracy, and timeliness rating, which measures the model's performance against several criteria. However, this technique does not address the low detection rate of attacks such as R2L and U2R.

20.4 MODEL IMPLEMENTATION

In this section, the various applications of the model will be discussed. This chapter provides an explanation of how to implement the concept, including topics such as software and hardware configuration, as well as the decision-making process. The complete test is conducted on a computer running a 64-bit version of macOS, equipped with a Radeon Pro 560X 4GB GPU, 16 gigabytes of DDR4 RAM operating at 2,400 MHz, and an Intel 2.6 GHz 8-core i7 processor. All programs were written in Python 3.8 and executed within the Anaconda environment (Table 20.2).

TABLE 20.2 Libraries for Implementation

Libraries	Versions
Keras	2.3.1
Scikit-learn	0.21.3
Numpy	1.21.5
Matplotlib	3.1.1
Pandas	0.25.1
Psutil	5.6.3
Joblib	0.13.2

20.4.1 DETAILS OF THE DATABASE

Numerous IDSs frequently refer to the NSL-KDD dataset as a benchmark in their evaluations. It is a fragment of the very first dataset that was generated by KDD99. The prior version of KDD99, which was criticized for favoring more frequent samples, received criticism since the test and train sets contained redundant and duplicate data. The NSL-KDD protocol authentically addresses these issues. The NSL-KDD dataset can be obtained free of charge from the Canadian Institute of Cyber security. KDDTrain+ and KDDTest+ are the names of the two datasets used for testing and training, respectively. In particular, the KDDTest+ dataset had 17 new attack types that were not available in the KDDTrain+ dataset for proper categorization; as a consequence, cases that met these categories were removed from the dataset. More information regarding the KDDTrain+ and KDDTest+ sets can be found in Table 20.3.

TABLE 20.3 Feature Details of NSL-KDD Dataset

Duration is guest login protocol type count service srv count flag error rate source bytes srv error rate.	Destination bytes error rate land srv error rate wrong fragment same srv rate urgent diff srv rate hot srv diff host rate.	Num failed logins dst host count logged in dst host srv count cont num compromised dst host same srv rate cont root shell dst host diff srv rate su attempted dst host same src port rate.	Num root dst host srv diff host rate num file creations dst host error rate num shells dst host srv error rate num access files dst host error rate Num outbound cmds dst-host srv error rate is host login.

20.4.2 EXPERIMENTAL DESIGN

20.4.2.1 DATA PREPROCESSING

Following the min-max pre-processing strategy, the numeric feature values z_i2 were mapped into the numeric range [0-1], as demonstrated by the following equation:

The following equation demonstrates how the min-max pre-processing method was utilized to map the numeric feature values z_i2 into the numeric range [0-1].

$$z_{i2} = \frac{z_{i1} - z_{min}}{z_{max} - z_{min}}$$

20.4.3 FEATURE EXTRACTION

The technique known as one-hot encoding was applied to transform the three categorical features, which include protocol type, service, and flag, into numerical values. Each categorical property is uniquely represented by a binary value. For example, the protocol type has three properties: tcp, udp, and icmp. The one-hot encoding method was used to convert each of them into a binary vector, resulting in the values [1,0,0], [0,1,0], and [0,0,1], respectively. Similarly, the vectors representing service and flag were transformed using one-hot encoding.

Following the selection of the categorical data characteristics, a data frame consisting entirely of categorical attributes is subsequently constructed. One-hot encoding of category properties can be accomplished in Pandas through the usage of the acquire dummies () function. This processing module is responsible for extracting the characteristics that have a high level of correlation. The fraction of zeros for each continuous feature is analyzed using the KDDTrain+ and KDDTest+ sets, respectively. Figure 20.13 presents a representation of the distribution of null values for each numerical variable that is part of the KDDTrain+ set (Figure 20.6).

Attack labels are separated into "normal" and "abnormal" groups during the labeling procedure. The numeric attributes of the binary class dataset and the encoded label attribute are then used to generate a data frame. The next step is to identify characteristics that have a correlation of at least 0.5 with the encoded attack label property. The combined encoded, one-hot-encoded, and original attack label attribute file is then saved to the disk after the encoding process is complete. Attributes are chosen using the Pearson correlation coefficient approach.

20.4.4 CLASSIFICATION

In Figure 20.7, the 20 characteristics that are deemed undesirable according to this research are presented in red font. After merging the remaining 18 continuous features with 84 one-hot-encoding vectors, a feature vector with 102 dimensions is generated as a result of this operation. When comparing IDS, you can utilize the various ML and DL algorithms that were discussed in the previous section. Only a few selected intrusion detection datasets are covered in the conducted literature review. Since the NSL-KDD dataset is the preferred choice for the majority of researchers in the field, we will compare and contrast various ML and deep learning IDS methods using that dataset. Import multiple ML and deep learning models using the acquired dataset to evaluate the accuracy of each model.

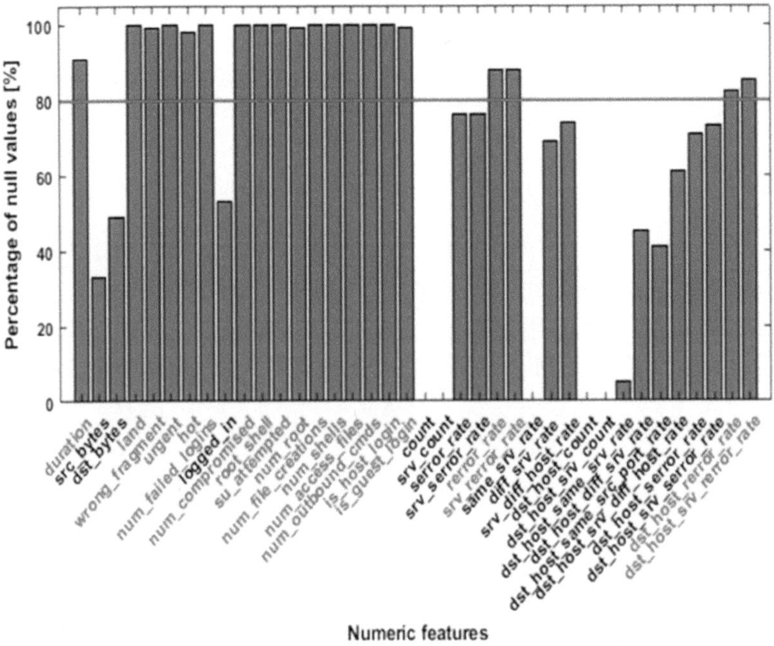

FIGURE 20.6 The graph shows the 38 numerical properties of the KDDTrain+ set's null values.

20.4.4.1 PERFORMANCE EVALUATION METRICS

The IDS is often assessed using the confusion matrix measurement in the following ways (Figure 20.7).

- **Accuracy or Classification Rate (CR):** The CR is calculated as the IDS's overall classification accuracy for both routine and intrusion assaults:

$$Accuracy = \frac{TP + N}{TP + TN + FP + FN}$$

- **Precision (P):** P is the ratio of the total number of true positive (TP) and false positive (FP) instances divided by the total number of true positive (TP) instances:

$$Precision = \frac{TP}{TP + FP} \times 100\%$$

		Predicted class	
Actual class		Normal	Intrusion
	Normal	**True Negative (TN)**	**False Positive (FP)**
	Intrusion	**False Negative (FN)**	**True Positive (TP)**

FIGURE 20.7 Confusion matrix.

- **Recall (R):** The ratio of the total number of true positives (TP) and false negatives (FN) instances to the total number of correctly identified relevant outcomes (TR) is:

$$Recall = \frac{TP}{TP + FN} \times 100\%$$

- **F1-Score:** It represents the average of recall and precision. When only one precision metric is required as an evaluation measurement, F-Measure is preferred.

$$F - measure = \frac{2 * Recall * Precision}{Recall + Precision}$$

For the classification process, several effective classification algorithms such as NB, KNN, SVM, LDA, and CNN are imported. The network weights of a CNN model can be iteratively adjusted by the Adam optimizer based on training data to improve accuracy. The dropout function is utilized in the CNN model to filter the data and prevent overfitting. Built-in functions were employed to train the data in this study. Approximately 25% of the data were reserved for testing during the evaluation, with a random state value of 42. The various ML models were trained using built-in libraries [21] (Table 20.4 and 20.5).

TABLE 20.4 Accuracy Values of ML Classifier on KDDTest+ Dataset

Algorithms	Accuracy
NB	82.6
KNN	85.6
SVM	88.57
LDA	89.7
CNN	92.23

TABLE 20.5 Shows How Well the NB, KNN, LDA, and CNN Classifiers

Algorithms	NB	KNN	SVM	LDA	CNN
Precision	82.5	80.5	82.3	86.3	92.7
Recall	85.67	85.89	85.61	87.37	89
F1-score	87.45	86.74	86.01	87.46	89.78

20.5 CONCLUSION

Cybercriminals have started employing sophisticated methods, such as polymorphism, to generate new attacks and constantly modify the signature to launch a large number of attacks. For Intrusion Detection Systems (IDSs) to be considered effective, they must be able to reliably detect both known and zero-day attacks. In this research, we propose a novel approach to develop an intelligent IDS that surpasses currently available IDSs. This will be achieved by applying algorithms to identify both known and unknown threats, with a particular focus on the latter. The main subcategories within IDSs include abuse detection tactics and anomaly detection methods. Abuse detection algorithms utilize pre-determined signatures of security threats and behaviors that compromise system integrity to detect intrusions. When it comes to handling intrusions, a topic extensively explored in security literature, researchers investigate various machine learning (ML) and deep learning (DL) methods. The dynamic field of deep learning has recently been adopted by many IDS systems to enhance the efficiency and effectiveness of misuse detection in different contexts. We have demonstrated that a Convolutional Neural Network (CNN) classifier can achieve higher detection rates compared to other ML algorithms, considering factors such as detection rate, accuracy, specificity, sensitivity, and the F1 score. In this specific investigation, the intrusion was classified using ML and DL classifiers. In terms of performance, the CNN algorithm exhibited significantly better results than ML approaches.

ML and deep learning techniques can detect network breaches by predicting risks using training data. Over the years, numerous ML and deep learning algorithms have been introduced, proving to be more accurate than earlier versions of cyber security systems.

KEYWORDS

- **deep belief network**
- **denial of service**
- **intrusion detection systems**
- **linear discriminant analysis**
- **network security laboratory's knowledge discovery and data mining**
- **support vector machine**
- **user to root**

REFERENCES

1. https://www.businesswire.com/news/home/20190516005700/en/Strategy-Analytics-Internet-of-Things-Now-Numbers-22-Billion-Devices-But-Where-Is-The-Revenue (accessed on 27 January 2024).
2. Wang, A., (2020). Internet of things computer network security and remote control technology application. In: *2020 5th International Conference on Mechanical, Control and Computer Engineering (ICMCCE)* (pp. 1814–1817).
3. Nie, L., et al., (2022). Intrusion detection for secure social internet of things based on collaborative edge computing: A generative adversarial network-based approach. In: *IEEE Transactions on Computational Social Systems* (Vol. 9, No. 1, pp. 134–145).
4. Kotenko, I., Saenko, I., Lauta, O., & Karpov, M., (2021). Situational control of a computer network security system in conditions of cyber-attacks. In: *2021 14th International Conference on Security of Information and Networks (SIN)* (pp. 1–8).
5. Ge, B., & Xu, J., (2020). Analysis of computer network security technology and preventive measures under the information environment. In: *2020 5th International Conference on Mechanical, Control and Computer Engineering (ICMCCE)* (pp. 1978–1981).
6. Musa, U. S., Chhabra, M., Ali, A., & Kaur, M., (2020). Intrusion detection system using machine learning techniques: A review. In: *2020 International Conference on Smart Electronics and Communication (ICOSEC)* (pp. 149–155).
7. Xu, B., Chen, S., Zhang, H., & Wu, T., (2017). Incremental k-NN SVM method in intrusion detection. In: *2017 8th IEEE International Conference on Software Engineering and Service Science (ICSESS)* (pp. 712–717).

8. Anish, H. A., & Sundarakantham, K., (2019). Machine learning based intrusion detection system. In: *Proceedings of the Third International Conference on Trends in Electronics and Informatics* (pp. 916–920). IEEE Xplore, Tirunelveli, India.
9. Alalade, E. D., (2020). Intrusion detection system in smart home network using artificial immune system and extreme learning machine hybrid approach. In: *2020 IEEE 6th World Forum on Internet of Things (WF-IoT)* (pp. 1, 2).
10. Doshi, R., Apthorpe, N., & Feamster, N., (2018). Machine learning DDoS detection for consumer internet of things devices. In: *2018 IEEE Security and Privacy Workshops (SPW)* (pp. 29–35).
11. Ingre, B., & Yadav, A., (2015). Performance analysis of NSL-KDD dataset using ANN. In: *2015 International Conference on Signal Processing and Communication Engineering Systems* (pp. 92–96).
12. Mubarak, A. U., & Chen, Z., (2020). Effects of feature selection and normalization on network intrusion detection. *Communication, Networking and Broadcast Technologies*. 10.36227/techrxiv.12480425.v2.
13. Cremer, F., Sheehan, B., Fortmann, M., et al., (2022). Cyber risk and cyber security: A systematic review of data availability. *Geneva Pap Risk Insur. Issues Pract., 47*, 698–736.
14. Gu, Y., Li, K., Guo, Z., & Wang, Y., (2019). Semi-supervised K-means DDoS detection method using hybrid feature selection algorithm. *IEEE Access, 7*, 64351–64365.
15. Abrar, I., Ayub, Z., Masoodi, F., & Bamhdi, A. M., (2020). A machine learning approach for intrusion detection system on NSL-KDD dataset. In: *2020 International Conference on Smart Electronics and Communication (ICOSEC)* (pp. 919–924).
16. Dwibedi, S., Pujari, M., & Sun, W., (2020). A comparative study on contemporary intrusion detection datasets for machine learning research. In: *2020 IEEE International Conference on Intelligence and Security Informatics (ISI)* (pp. 1–6).
17. Ali, A., et al., (2020). Network intrusion detection leveraging machine learning and feature selection. In: *2020 IEEE 17th International Conference on Smart Communities: Improving Quality of Life Using ICT, IoT and AI (HONET)* (pp. 49–53).
18. Belgrana, F. Z., Benamrane, N., Hamaida, M. A., Mohamed, C. A., & Taleb-Ahmed, A., (2021). Network intrusion detection system using neural network and condensed nearest neighbors with selection of NSL-KDD influencing features. In: *2020 IEEE International Conference on Internet of Things and Intelligence System (IoTaIS)* (pp. 23–29).
19. Wang, W., Du, X., Shan, D., Qin, R., & Wang, N., (2020). Cloud intrusion detection method based on stacked contractive auto-encoder and support vector machine. In: *IEEE Transactions on Cloud Computing.*
20. Subbiah, S., Anbananthen, K. S. M., Thangaraj, S., Kannan, S., & Chelliah, D., (2022). Intrusion detection technique in wireless sensor network using grid search random forest with Boruta feature selection algorithm. In: *Journal of Communications and Networks* (Vol. 24, No. 2, pp. 264–273).
21. Zhang, X., Ran, J., & Mi, J., (2019). An intrusion detection system based on convolutional neural network for imbalanced network traffic. In: *2019 IEEE 7th International Conference on Computer Science and Network Technology (ICCSNT)* (pp. 456–460).

CHAPTER 21

Cyber Security Through Big Data Analytics

S. LEELALAKSHMI,[1] K. TAMILSELVI,[2] and T. BHUVANESWARI[3]

[1]*Department of Computer Science, Queen Mary's College, Chennai, Tamil Nadu, India*

[2]*Department of Computer Science, Bharathi Women's College, Chennai, Tamil Nadu, India*

[3]*Department of Computer Applications, Queen Mary's College, Chennai, Tamil Nadu, India*

ABSTRACT

Analyzing data for cyber security is a challenging task that requires the use of specialized tools. Although security analysts can examine a large amount of information quickly, there are limitations. The file generated by Wire shark during a network capture is called a .pcap file, which can be analyzed using software such as libpcap or WinPcap libraries to understand the network data's characteristics. To manage network traffic efficiently, new approaches are necessary for network design and management. Recently, Big Data Analytics (BDA) algorithms and machine learning (ML) advancements have emerged as valuable problem-solving techniques for detecting network anomalies. However, transforming network data into Big Data presents additional challenges. Open-Source Analytics Tools such as R, Mongo DB, MySQL, Apache Drill, Python, etc., can help manage the heterogeneous data from the current Cloud Data Environment. Analytic tools can also assist in selecting appropriate algorithms to analyze the data. ML models can effectively handle classification effectiveness, outlier detection, and anomaly detection. The combination of ML algorithms and Data Analytics tools can

Leveraging Artificial Intelligence (AI) Competencies for Next-Generation Cybersecurity Solutions. Pethuru Raj, B. Sundaravadivazhagan, & V. Kavitha (Eds.)

provide deeper insights into network data. Reinforcement learning is also a promising choice for learning about cyber security threats.

21.1 INTRODUCTION

Big data refers to the generation and management of large amounts of data, which is a common occurrence in our daily lives. The volume of data we generate through our mobile phones alone is staggering, let alone the data generated by businesses, MIS, and EIS. It is essential to have a good methodology in place to manage this data to ensure it doesn't become overwhelming. Big Data Analytics (BDA) is a powerful tool that allows us to leverage and make sense of this data. The amount of data generated every moment is doubling, and every household has become a centre for data, with all family members using mobile devices and connected to a network. The COVID-19 pandemic has only increased the amount of data generated, as work, education, and extracurricular activities have shifted to online platforms. While this has generated an enormous amount of data, security has become a concern. BDA can help identify potential security threats and weaknesses in the system. As we generate more and more data, the importance of managing and analyzing it will only continue to grow.

21.1.1 BIG DATA AND CYBER SECURITY

What is the connection between big data and cyber security? The amount of data generated in cyberspace is enormous, which cannot be efficiently managed with traditional tools. To address this problem, it is necessary to have tools and techniques that can handle big data analytics (BDA) and machine learning (ML).

21.1.1.1 CYBER SECURITY

Network visualization tools have evolved to handle large-scale data sets. Despite the challenges of rendering voluminous data and numerous log entries, these tools provide a comprehensive view of the data. In the field of cyber security, analyzing PCAP files to identify network traces is crucial. By extracting files from full packet captures, security analysts can save time. Web traffic consists of various types of transfers, some of which may be malicious and fall into different categories. Identifying network information requires multiple approaches, and traditional methods can pose difficulties. Analyzing and extracting features and applying machine learning algorithms can be more useful in identifying network information. In order to manage network traffic effectively, new approaches are needed in the design and management of networks. As networking advances from handling normal data to handling big data, managing it manually becomes increasingly difficult. By using machine-learning-based methods, we can make more accurate predictions and gain deeper insights into the data. Big data analytics (BDA) can be particularly useful in areas such as malware detection, intrusion detection, multi-factor authentication (MFA), and so on.

New approaches in the field of cyber security or cyber defense are being developed, taking into consideration various aspects:

- Unification of data from various sources,
- Attack detection before software vulnerability (zero-day)
- Detection of various malware
- Identification and Detection of Intrusion
- Provision of authentication for multifactor
- Distribution of Data sharing across different threat detection systems
- Resource-constrained data processing,
- Anomaly detections on data with the help of Time series Analysis
- Rule-based heuristics

21.1.2 TECHNIQUES AND TOOLS FOR BIG DATA

Most of the techniques that are involved in BDA are:

- Machine learning;
- Predictive analytics;
- Stream analytics;
- Distributed storage;
- No SQL databases;
- Apache drill;
- Apache spark; and
- Mongo DB.

Cyber Analytics and Security can present various challenging problems in the field of Network Analysis, among others. This field presents several research challenges for different interdisciplinary fields.

i. Intrusion Management - Managing Intrusion Detection, Systems for Removing Detection, Containing Intrusion Detection
ii. Anomaly Detection System for mining large-scale networks, predictive analytics for the anomalies and consequences.
iii. Ensuring the quality-related factors for trust: infusing integrity, availability, confidentiality, and privacy in the system.

In the Indian scenario, we have a large amount of data. Dealing with big data from a big country can be challenging. TCS DigiGOV™ FACTS Biometrics for Identification, Registration, and Verification assists the Police Department. The fingerprint matching solution helps maintain law and order by accurately and quickly identifying criminals, ensuring they are apprehended. FACTS aids the police in identifying repeat offenders after an arrest and helps expedite convictions in court. Fingerprint biometrics collected at crime scenes are compared against the existing fingerprint data in FACTS, enabling fingerprint experts to solve cases accurately and efficiently.

PCAP files are commonly used to analyze various aspects of network data. These files are generated when traffic data is captured from the network. The generated files contain different fields (e.g., source, destination) to facilitate understanding for end users. PCAP files have the extension .pcap and can be utilized for network analysis. They provide valuable network information, allowing for a better understanding of data traffic status. The Wireshark application is used to capture and analyze data, helping identify different aspects of network data and communication. Wireshark files can be accessed on various operating systems such as Windows, MAC, and Linux. Applications like Wireshark and Packet Square can process and manage Wireshark files.

Packet Capture or PCAP (also known as libpcap) captures live network data from any source and destination. Users with Wireshark can capture and analyze the data. The captured data includes OSI Layers 2–7. Various network analysis tools are available to record files with the .pcap extension. PCAP files come in different formats such as Libpcap, WinPcap, and PCAPng. PCAP files can handle TCP/IP and UDP packets. To understand how packet data capturing works, we can capture network data using available tools. Some popular networking tools include:

- SolarWinds;
- Network Packet Sniffer;
- Wireshark;
- Tcpdump; and

- WinDump.

The data generated during the network analysis has an enormous volume of log details. It also consists of a large number of events (audio, video, etc.). To record all of these, a reliable network visualization tool is required. This tool should be fast and provide a clear visualization view. When the tool presents data in the correct format, it can save a significant amount of time. The field of cybersecurity faces numerous challenges, and analyzing PCAP files to identify network traces provides crucial information. Nowadays, networking infrastructure requires various approaches. Time series analysis and rule-based heuristics are commonly used to handle them. As networking evolves from data to big data, it becomes increasingly difficult for humans to manage. ML-based methods can offer more accurate predictions compared to other methods.

- **Deep Reinforcement:** Deep reinforcement is a combination of machine learning (ML) and artificial intelligence (AI). Machines can learn from their actions and store the actions, which is called deep reinforcement. Humans learning from their actions is very common, and deep reinforcement algorithms also follow the same principle. In this algorithm, learning is based on either reward or penalty. Rewards are planned accordingly when the system reaches the target outcome, while penalties are given for undesired actions.

This algorithm learns from a set of actions performed, continuously adapting to dynamic situations. It is based on artificial neural networks (ANNs) that facilitate reinforcement learning (Figure 21.1).

Deep reinforcement learning has proven its power in gaming areas. This technology can be applied to:

- Manufacturing;
- Automotive;
- Finance;
- Healthcare; and
- Bots and cyber security.

21.1.3 Q-LEARNING

Q-learning is based on the policy of Reinforcement learning. This algorithm works by finding the next possible state when given the current state. It aims to move towards the desired output and maximize the reward (Figure 21.2).

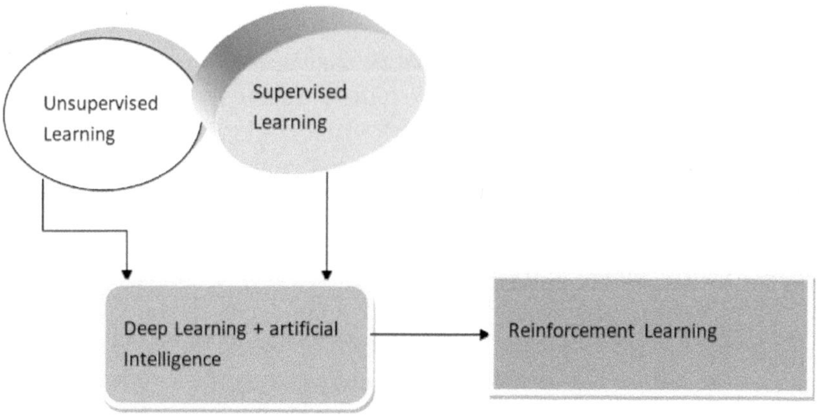

FIGURE 21.1 Deep-reinforcement learning.

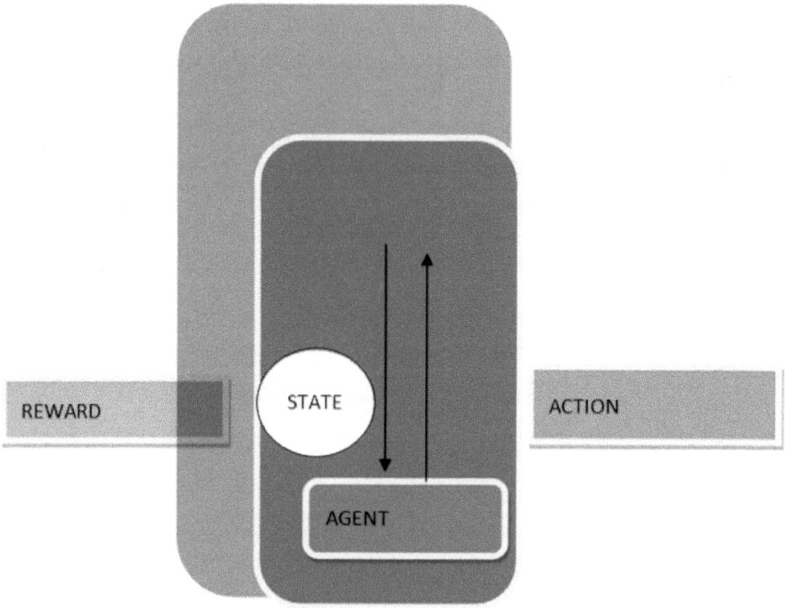

FIGURE 21.2 State diagram of Q-learning.

Given the current state of the system, this reinforcement learning model will find the best action suited for the particular system based on the system environment, parameters, and attributes.

The agent of the particular system is identified, and then the necessary course of action will be followed.

This learning methodology will have its own set of rules and regulations to determine the next step of action.

21.1.4 SEQUENTIAL MINIMAL OPTIMIZATION (SMO) ALGORITHM

The SMO is based on the decomposition method. It considers the optimization of two points at each iteration. The requirement that the condition P` i=1 yii = 0 is enforced throughout the iterations implies that the smallest number of multipliers that can be optimized at each step is 2: whenever one multiplier is updated, at least one other multiplier needs to be adjusted to keep the condition true.

21.1.5 FATT

FATT is a scripting language used for extracting network data. This script also includes fingerprint management, with the ability to trace JA3 and HASSH fingerprints. FATT is commonly used in forensic analysis and can detect anomalies in honey pots. It is compatible with any operating system.

FATT can be used to manage the following protocols and render its service support.

SSL/TLS, SSH, RDP, HTTP, gQUIC.

- Fingerprinting
- JA3: TLS client/server fingerprint
- HASSH: SSH client/server fingerprint
- RDFP
- HTTP header fingerprint
- gQUIC/iQUIC fingerprint in the future can be added
- JSON output

21.2 RELATED WORKS

General features of PCAP files, when taken into consideration, consume a significant amount of time, which can impact efficiency and generate false alarms. This, in turn, may affect the accuracy of the Network System. The intrusion characteristic-specific features are derived and utilized for detection. In this paper, only 16 optimized predictors are employed for analysis. Intrusion detection can be analyzed with the assistance of these predictors. Here, the focus is on considering the important features. The PCAP files are directed to the preprocessor for further processing. A classification algorithm is utilized to extract the required features and generate a classifier [2]. This approach proves useful in monitoring the network and preventing further contiguous attacks. Network intrusion is detected using the Snort tool. This article explains the management, tracking, and detection of attacks by analyzing the traffic pattern [3]. Packet capture is the commonly and repeatedly used method by network analysts to identify any abnormal differences in traffic. This is done to detect any suspicious activity in the network. An intrusion detection system (IDS) is employed to analyze any abnormal differences in traffic activity.

IDS tools are used to classify different attacks or analyze traffic. Open-source products and IDS tools are also available. In addition to these, there are several sets of IDS tools available. These IDS tools can classify the 32 attacks or traffic analysis based on data gathered from PCAP (packet capture) files managed from honeypots or servers. When considering analysis for the PCAP files, Wireshark [4] is the best tool of choice. When compared with other tools, Wireshark offers compatible facilities and is effective. Wireshark not only provides a good idea of network analysis but also has an excellent GUI. It not only lists the information of individual packets but also includes analysis and statistics tools.

The Wireshark has the entire individual TCP conversations that allow it to track within a PCAP file. Wireshark helps to retrieve related metrics regarding the TCP conversation. Processing a PCAP file programmatically can provide more insights into the network traffic. The entire network data or network communication details can be stored in a single file and used for further investigation. This paper [5] proposes a methodology for PCAP processing. A new approach was used to process the PCAP files in this methodology. The time for investigating the PCAP files is reduced more effectively. The investigation of PCAP files was done in different ways, using tools that are available for free and open source. The open-source packet analyzer is used to manage this.

A better evaluation of the PCAP Funnel prototype was discussed here. Nine domain experts were involved in the experiment, covering a wide range of different scenarios. Various domains, including educational purposes and exploratory network data analysis, were identified. The experiment recorded the initial orientation and the improvement in the network dataset. The application was developed in collaboration with industry partner Flowmon Networks [6]. This company specializes...

In analyzing network traffic in different domain areas, deep learning and reinforcement learning algorithms together provide a new methodology called deep reinforcement learning. This technology, known as DRL, finds its application in finding solutions for cyber-attacks. For example, DRL methods can effectively handle falsified data injection to cyber systems [7], deception attacks to systems [8], and distributed denial-of-service attacks [9]. Using benchmark data from the KDD (knowledge discovery and data mining) competition managed by DARPA, efficient, qualitative, and accurate classifiers [10] can identify data to detect intrusions. The comparison is based on the performance of the data in neural networks and support vector machines (SVMs), and systems for intrusion detection were discussed. This paper identifies datasets that contain reliable benign data and seven common attack network flow data. This dataset is publicly accessible and caters to real-world problems in the network arena. The paper compares the performance of network traffic features with ML algorithms to identify the best features for detecting attacks on network traffic and determining the best method for identifying intrusions.

21.2.1 DEEP LEARNING, LSTM, FNN

Deep learning methodologies can be efficiently used to detect abnormalities in traffic. By detecting abnormalities at the packet level, the time involved in identifying malicious traffic can be significantly reduced. Deep learning technologies are applied in real-time online detection of malicious traffic, making them promising for cyber security. In this study, a detection process using packet classification is proposed, providing insights into packet classification [11]. Word embedding is utilized for malicious classification systems. Two deep learning-based anomaly detection models, using autoencoder and denoising autoencoder, are discussed for anomaly detection. The threshold value is determined using a stochastic approach [12]. In comparison to other models, this proposed model demonstrates higher accuracy. The model is tested with KDD datasets.

The ML-based intrusion detection models for traffic analysis were implemented using different classification techniques. These classification techniques include extreme learning machine (ELM), k-nearest neighbor, decision-tree algorithm, etc. The classification metrics such as area under the curve, precision-recall curve, etc., can be used to evaluate deep and conventional ML models [13]. DNN applications for cyber security problems will provide more research ideas. Anomaly detection can be performed with the help of DNN features like Auto-encoders (AE), Deep Belief Networks (DBNs), and LSTM. A survey of deep networks IDSs is discussed based on previous and current works. This paper reviews various ML techniques and their ability to detect anomalies [14]. The features that will be selected for network analysis and the selected features that influence the behavioral methodology of ML-based IDS are discussed [14]. This highlights the importance of feature selection in IDS classification, as it plays a major role. This paper suggests an algorithm called ELM [15] for single-hidden layer feed-forward neural networks (SLFNs).

Taking into account the advantages of using neural networks, this paper focuses on the analytical determination of the output weights of SLFNs. In this approach, the hidden nodes are randomly selected.

According to theory, this algorithm demonstrates good generalization performance. The management of learning capabilities of neurons in this method is remarkably fast.

Real-time experiments are also proved to be the same. Here the inputs for the experiments were given some artificial and benchmark function approximation. The very large data which are also complex produced effective generalization performance. This proves that this algorithm learns more quickly than the other cases. This algorithm as per the experiments learns faster in terms of feed-forward neural networks.

Wang et al. [16] proposed an ensemble learning method to manage abnormal changes and record information from the datasets. The first step in this methodology involves using the CNN source domain as input. Then, the CNN is trained. This well-trained CNN can be used to identify necessary information from different datasets. Transfer learning was used in this methodology.

This method utilizes the knowledge learned by the CNN during the training sessions with the source data. It helps reduce the time required for modeling and finds application in mechanical fault diagnosis.

Ma, J., et al. [17] proposed a multivariate analysis to identify the important features of air quality factors. In this paper, XGBoost is compared with other existing models in the first step, and its parameters are checked using the Bayesian Optimization algorithm. This method effectively determines the important features that influence air quality when given input of the air quality. The experiments showed that six different factors have the largest impact on air quality.

A Network Analyst or Cyber security Analyst's main task is to find intrusion detection for any computer network-based systems. Intrusion detection can occur in various places such as government data centers, important network centers of organizations, and other locations where data transmission takes place. This must be addressed carefully, as it can result in the compromise of sensitive information. Traditional detection methods may not be able to identify new or unknown strategies. Anomaly detection in network systems differentiates between illegal or malicious events and normal patterns in a network system. Anomaly detection is considered a classification problem. This approach can be used to build models of network transmission based on normal behavior, which can then be used for detection. While anomaly detection systems typically record normal and abnormal activity, previous network behavior can also be taken into consideration for detection. In this paper, the authors propose a real-time collective anomaly detection model based on neural network learning and feature operating factors. A Long Short-Term Memory Recurrent Neural Network (LSTM RNN) is trained using basic data, enabling the methodology to predict future details based on input data. In this approach, the first step involves training an LSTM RNN using normal time series data. Once trained, it can be used for live prediction at each corresponding time step.

Malicious detection can now be effectively managed using the deep learning methodology application. Character-level embedding was utilized for malicious detection. When comparing this method with traditional machine learning algorithms, this system excels in detecting URL phishing.

Deep learning architectures and conventional algorithms, when compared, the performance of deep learning architectures proved to be extraordinary [19]. Sequence learning can be achieved with the use of Recurrent Neural Networks (RNNs). RNNs build directed cycle connections between units, which sets them apart from normal neural networks and multilayer perceptrons.

Unlike basic neural networks and multilayer perceptrons, RNNs can map input data to source vectors. By utilizing RNNs, the entire input from previous data can be used. RNNs with memory can effectively capture previous data, making them suitable for analyzing network data.

In this paper, the author proposes a combination of RNN and Long Short-Term Memory (LSTM) for detecting phishing attacks in the data. This combination has proven to be effective in identifying phishing attacks in network data. Additionally, the paper compares the analysis of statistical URLs with lexical features using a random forest classifier. The results of this analysis are used for detection. As suggested by the paper, both models have shown successful outcomes.

The registry key, path, and URL that are abnormal are detected by CNN with character-level Keras embedding used for network anomaly detection in Ref. [20].

Cyber security-based problems can be addressed in conjunction with the assistance of deep learning networks. There are deep learning evaluation benchmarks for performance available in the existing system. In this paper, a C-LSTM neural network was proposed to obtain results for spatial and temporal information. This can be done in real-time and with new data to detect traffic differences or deviations [21].

In this paper [22], a row-based DNN methodology was introduced for intrusion detection. The proposal for a C-LSTM neural network was designed based on the traffic data. The modeling of this traffic data takes into consideration the spatial and temporal information. These data are defined as one-dimensional data with a time series signal. The model extracts features of both spatial and temporal information from the data, automatically extracting this information from the raw data.

This paper discusses the identification of different attacks based on payload. Various identifying methodologies for attacks planned on payload are discussed. The input in this case is raw payload. CNN and LSTM neural network methodologies [23] were applied to network intrusion detection using different payload datasets.

The LSTM neural networks were trained to learn the latent properties of the data requests [24]. This paper also identifies the shortfalls and difficulties faced in Android malware detection. A model summarizing different threats that need to be handled by Android developers and researchers is discussed.

In the deep learning algorithm methodology used for detection, raw payload data is utilized. This implementation is end-to-end, providing a good and effective detection model [25]. The details discussed in this research paper pertain to an experiment involving the steps taken during the evolution of a malware detection system. This system identifies the detective performance of various data set features for malware detection through thorough investigation. In the selection of effective features from the already collected data set, a 10-feature selection was taken into consideration. This approach was implemented for malware detection.

The proposed system takes into consideration hierarchical, spatial, and temporal features. These features, combined with an IDS, were designed [26]. For understanding and training the network data features, different levels of high and low levels were selected. Low-level spatial data features utilize deep CNN, while LSTM is used for understanding high-level temporal features.

In this [27] paper, an application was developed for multiple neural networks. The development focused on network anomaly detection, utilizing RNN, CNN, and autoencoders. The implementation of a spam detection system using DBN was discussed in Ref. [28]. DBN was constructed based on a feed-forward neural network (FFN) with multiple hidden layers. It was noted that at least one or two layers of this FFN differed from the traditional conventional method.

Three public datasets were utilized in this study. The performance evaluation of this method was discussed, employing an unsupervised greedy algorithm as the methodology to address low generalization problems.

Issues related to email, such as indifferent patterns, fraudulent emails, and spam, have been causing significant problems for various internet operators.

Nowadays, neural networks are used in conjunction with nonlinear data. This is utilized for modeling tools that manage RNN (Recurrent Neural Networks). The RNN algorithm was introduced to untangle emails that were entangled into [29] threads. Following this, two or five zones of email classification were prepared. Each zone maintains different information, including signatures, headers, body, and greetings. The proposed algorithm in this case is the Artificial Neural Network (ANN). The ANN model for email classification is discussed here. The feed-forward back propagation algorithm is used for training in this model. ANN possesses the property of adaptiveness, allowing it to adjust its structure over time. During the network learning process, it can adapt to both internal and external information.

There are many detection and classification algorithms available. In order to make it even more efficient, a further reduced scheme is proposed here. This scheme significantly reduces the number of n-grams and training samples. The information received is managed for radical and affinity propagation is managed for sample selection. In this work, a step-by-step study on the investigation of permission-induced risk in Android apps was conducted [30]. This study involved large-scale data analysis, where three levels were discussed.

To detect malicious applications [58] on specific subsets of risky permissions, a discussion was held here. The algorithms utilized in this context are Decision Tree and Random Forest.

This paper presents BAdDroIds [31], a machine learning methodology for mobile application analysis. This methodology was introduced to identify malware on resource-constrained devices. BAdDroIds operates in the background and detects applications as they are installed. The analysis is conducted transparently, preventing infection from infiltrating the device. BAdDroIds is designed using static analysis techniques and built upon Android OD features.

This combination provides a robust model for Android applications, incorporating application program invocations and permission-related features. Additionally, it encompasses an ad-hoc supervised classification method, enabling resource-efficient malware detection.

Most neural networks developed for network anomaly detection employ a combination of RNN, CNN, and autoencoders. In this paper, a filter and wrapper feature selection method is utilized to maximize classifier accuracy.

VANET should have good detection accuracy. This can be achieved through the experience of communication between vehicles (nodes). Distributed ML can be considered as a good framework for this purpose. This framework can be used to design an architecture that is scalable and supports collaborative detection algorithms. The main challenge lies in data exchange. Malicious nodes have a higher chance of extracting important information from other nodes, especially by inferring already recorded data. In this paper [33], a privacy-preserving machine-learning-based collaborative IDS (PML-CIDS) for VANETs is proposed.

21.3 SYSTEM MODEL

The PCAP file contains a vast amount of data, making it highly valuable for network analysis. Information such as frame number, transaction timestamp, byte count, and protocol can be extremely useful for network administrators. For instance, even for someone with limited knowledge, details like source and destination IP addresses can provide valuable insights. Now, let's discuss the four divisions of what can be done with the data generated by the PCAP:

1. Making data in a more readable format;
2. Machine learning and PCAP;
3. Reinforcement learning and big data; and
4. Bridging fingerprint with OS fingerprint.

21.3.1 EXAMPLE OF MAKING PCAP DATA INTO A MORE READABLE FORMAT

The availability of PCAP files: There are many public PCAP files available on the internet. However, there is no need to worry if we have downloaded

Wireshark and Tshark. Capturing packet captures is not a problem on all operating system platforms. If we have internet access, capturing and analyzing the data should not be an issue.

21.3.1.1 SYSTEM ARCHITECTURE

FIGURE 21.3 System architecture.

- Data Capture for Any Cyber Threats.
- Data Capturing can be done with Wireshark or Tshark.

Parsing and process can be done with any Big data analytical tools like R, Python, Weka, Apache Drill, etc.

After executing the code below, the R program provides the following output and establishes a connection with the Apache Drill Big Data Analytics tool.

```
filter(my_pcaps, type=="TCP") %>% + count(dst_port, sort=TRUE)
# Source: lazy query [?? x 2]
# Database: DrillConnection #
Ordered by: desc(n) dst_port n
443 37
80 8
52302 8
52301 8
52304 7
52303 4
2048 2
52292 2
52253 1
52246 1 #.
More rows will be listed afterward
Apache Drill Display of PCAP
```

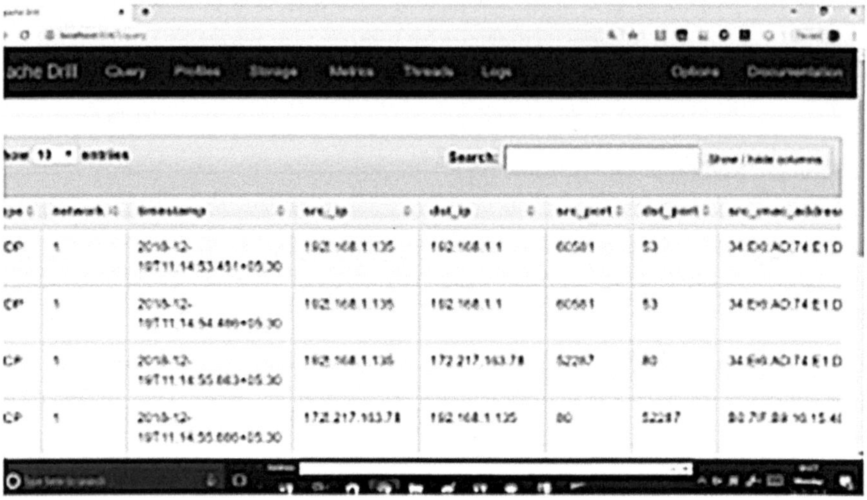

FIGURE 21.4 Apache drill display of pcap.

Identifying the protocols that fall under the destination machines, as well as expert information.

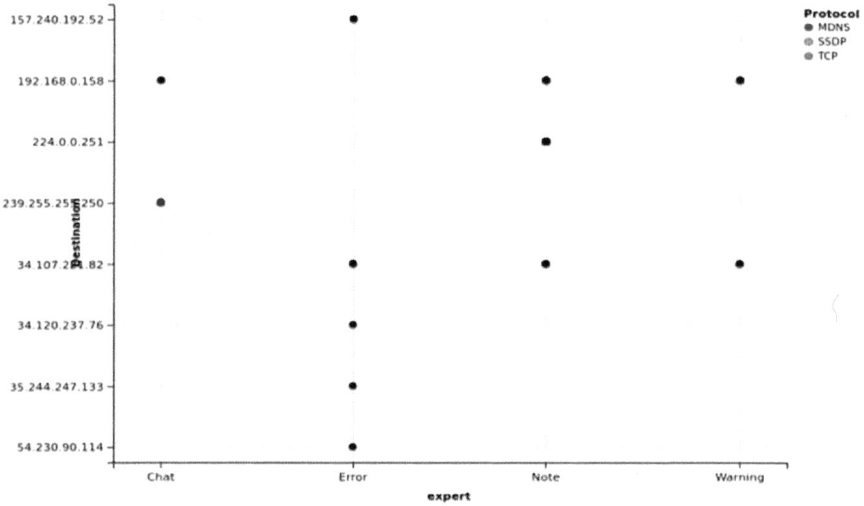

FIGURE 21.5 Protocol identification.

21.3.2 MACHINE LEARNING AND CYBER SECURITY

ML problems are generally used for predicting, categorizing data, and finding patterns within the data. Cybercrimes can exhibit patterns. Therefore, a well-designed ML algorithm can be helpful in (Figure 21.6).

a. Grouping; and
b. Categorizing.

FIGURE 21.6 Training data.

The data needs to be identified first, and it can be from any network. After identifying the data source and data Collection:

a. Features can be extracted from where some meaningful data can be categorized
b. A suitable methodology needs to be determined from a variety of ML algorithms that are somewhat suited to the current scenario. The data should then be trained accordingly.
c. Most of the studies now focus on deep learning, LSTM, and RNN algorithms, which can assist in identifying malicious activity in traffic data and monitoring the traffic. The proposed architecture for Malicious Detection is shown in Figure 21.7.

Malicious detection is performed using various network classification and machine learning algorithms.

Python is the software utilized for the experiment. The evaluation is based on parameters such as Precision, Recall, False Positive Rate, Accuracy, and F1-score. The obtained values are stored and compared using the two-algorithm methodology: DNN and Long-Short-Term Memory (LSTM).

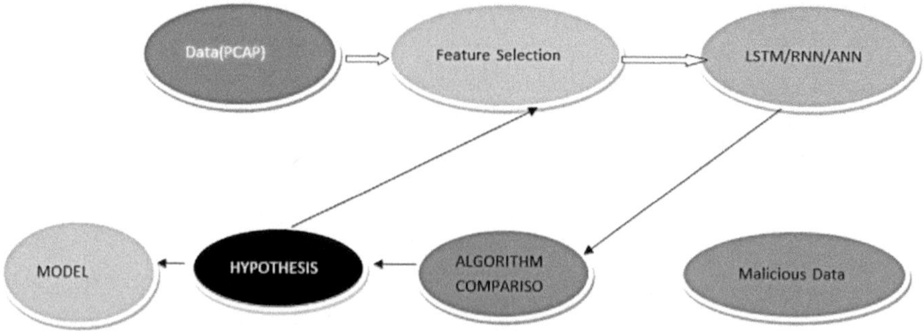

FIGURE 21.7 Training data learning.

The discussion includes the utilization of multi-dimensional features [34] and the optimization approach SMO. The malicious detection method is processed using the LSTM-RNN algorithm. The SMO-LSTM-RNN test cases involve malicious traffic types such as dos, probe, r21, and u2r. The results of the test cases are presented in Table 21.1.

TABLE 21.1 Malicious Traffic Parameter Comparison Details

Class	Accuracy	Precision	Recall	False Positive Rate	F-Score
Normal	0.8788	0.8251	0.9714	0.2201	0.8923
Dos	0.9813	0.9684	0.9704	0.0139	0.9694
Probe	0.9847	0.8919	0.8427	0.0064	0.8666
r21	0.9015	0.8085	0.2074	0.0065	0.3301
u2r	0.998	0	0	0	0
Overall	0.8722	0	0.9912	0.0494	0.6117

Figure 21.8 represents the confusion matrix of the multi-class test. In this matrix, the rows indicate the predicted class-output class, while the columns indicate the actual class-target class. The data being examined here is the malicious detection data. The classification of the data is based on the diagonal dark and light blue cells. The column on the right side indicates the predicted class, while the bottom row represents the behavior of the actual class.

21.3.3 REINFORCEMENT LEARNING

Incorporating back propagation algorithms into reinforcement learning can be a better choice for cyber security. Algorithms like Q-learning can be used.

Q-learning can be considered as a special case of the temporal difference algorithm that learns by reducing the difference among the estimations. This estimation is made fixed by the agent at different periods of time. Using Q-learning with PCAP files can provide better results. Fixing the back propagation algorithm by binding a neural network for the look-up table can be done. Each Q (state, action) update can be used for the training data.

There are a number of successful reinforcement learning systems

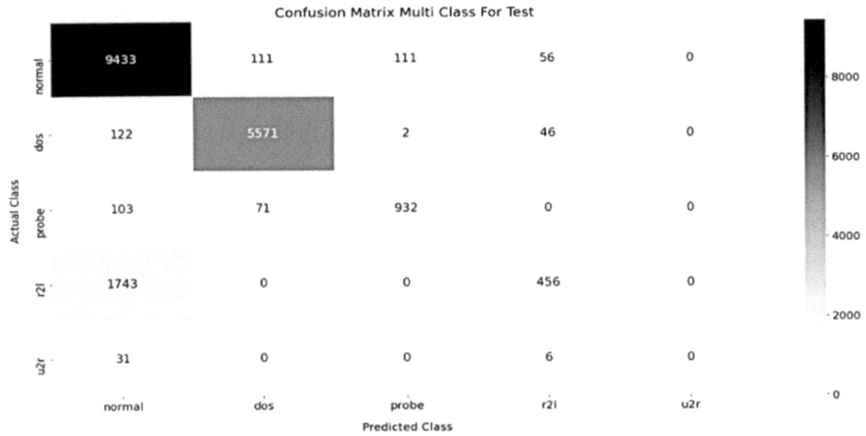

FIGURE 21.8 Architecture for the proposed malicious detection.

21.3.4 PROPOSED ARCHITECTURAL DESIGN OF REINFORCEMENT LEARNING IN PCAP

Suppose, for example, an attacker sends a different set of data that is not in the given pretext. When detected, it can activate the Q-learning algorithm (Figure 21.9).

The TD is more efficient when there are more distant-looking heads. Reinforcement learning aligns with dynamic programming approaches to Markov decision processes. The proposed architectural design of Fingerprint.

21.3.5 OS FINGERPRINT WITH THE HELP OF PCAP FILES

The goal of the research done here [35] was to understand if it was viable to check with fingerprints for malicious detection. Fingerprinting that can detect malicious data has to be considered. Packet fingerprinting is divided into two types: active and passive fingerprinting (Figure 21.10).

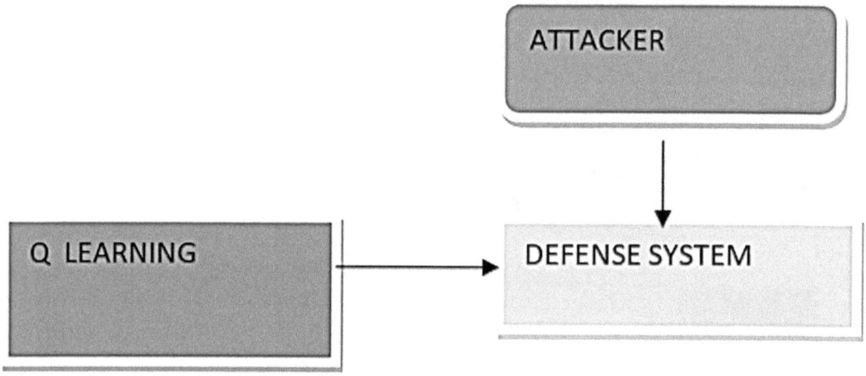

FIGURE 21.9 Q Learning defense system.

- **Active-Fingerprinting:** When an attacker wants to target a system, there are various processes that can be employed. One such process is active fingerprinting, where the system is flooded with packets. By analyzing the system's response, certain aspects of the system can be learned, such as its operating system (OS). However, these attacks can also create new opportunities for further exploitation. It is important to note that repeated communication may raise suspicion and alert the host.
- **Passive Fingerprinting:** This involves sniffing packets sent from the machines where an attack has to be done. Inspection of the IP header, TCP header, UDP header, or ICMP header will help the attacker check for the data. The properties of the headers can be used to identify the OS machines of the target system.
  Tools used for fingerprinting

  Ettercap

  NMap

  Wireshark(Can debug finger-printing Issues)

  pOf

  prads

  Packet

  PAcketFence

  Satori etc.

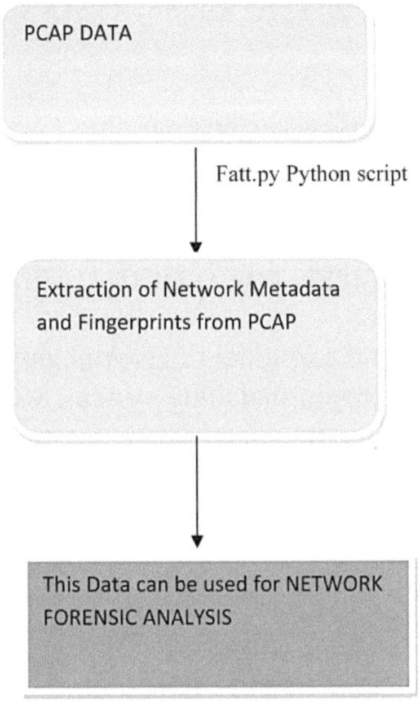

FIGURE 21.10 Architecture for extraction of fingerprints.

21.3.6 FRAMEWORK FOR FUTURE CYBER SECURITY ATTACKS FOR INDIAN SCENARIO

All the biometric data has to be collected. This collection of data will result in a huge storage. As TCS is providing software [37] for criminal tracing systems, we need to analyze the data if there is any data from other systems that have the same data.

Steps for Checking

1. If a crime is occurring at any location (or the fingerprint of a terrorist), it must first be examined through fingerprint analysis.
2. If any fingerprint is available, it should be matched with other location data, such as ATM machines or local banks, if their fingerprints are available. Classification and cluster analysis methods can be used.
3. Once this is done, we need to check if the corresponding fingerprint analysis leads to any computer (in case any international terrorism is linked).

After this step, we can proceed with the PCAP analysis, as discussed in this paper.

For example:

Network metadata Analysis can be done.

Reinforcement learning can be used to identify malicious data etc.

21.4 BRIDGING BETWEEN THE ORIGINAL FINGERPRINT

Bridging the gap between the original fingerprint and attempting to trace the system used by the fingerprint matching system, while identifying the OS fingerprint.

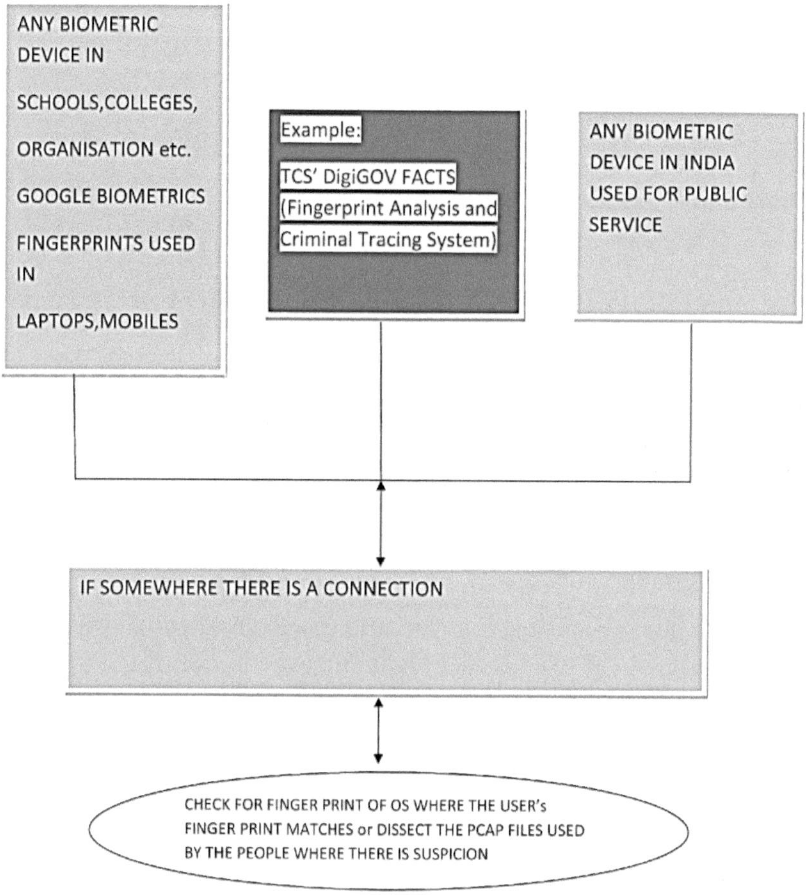

FIGURE 21.11 Architecture.

- **Big Data and Cyber Security:** Whenever we discuss Cyber Security, big data has to come into existence. This huge amount of data can only be managed by big data technology. If we implement a combination of real-time applications discussed in this paper, it can provide good results. By analyzing all the fingerprint data at our disposal, we can identify cyber threats from these individuals, which can provide us with a more realistic solution.

21.5 CONCLUSION

BDA holds enormous data. It requires huge flat data storages, servers, and heterogeneous data handling. With the use of Big Data technologies, data can be protected against attackers. Attackers are also likely to employ sophisticated tools and technologies. The show will be won by better tools that can reinforce, learn, and discover improved methods for combating attacks. Multiagent DRL can be explored as a solution to tackle this problem. Additionally, incorporating more reinforcement, LSTM, and RNN can yield better results.

KEYWORDS

- **artificial intelligence**
- **auto-encoders**
- **big data analytics**
- **cyber security**
- **forward neural network**
- **long short-term memory recurrent neural network**
- **machine learning**

REFERENCES

1. Leelalakshmi, S., Tamilselvi, K., & Rameshkumar, K., (2019). Examine and visualizing packet capture files. *International Journal of Recent Technology and Engineering, 8* (2 Special Issue 11), 3930–3933.

2. Chandramohan, A., & Srinivasa, K. S., (2020). Predictor selection and attack classification using random forest for intrusion detection. *Journal of Scientific & Industrial Research, 79*, 365–368.
3. https://www.opensourceforu.com/2015/04/make-your-network-secure-with-pcap-and-snort/ (accessed on 27 January 2024).
4. https://vnetman.github.io/pcap/python/pyshark/scapy/libpcap/2018/10/25/analyzing-packet-captures-with-python-part-1.html (accessed on 27 January 2024).
5. Pradeepini, G., Muni, S. G., & Arun, V., (2018). Hybrid PCAP analyzer using t-shark a tool that makes use of open-source analyzer that can meet industrial standards. *International Journal of Engineering & Technology, 7*(4.17), 85–88.
6. Uhlár, J., Martin, H., & Vít, R., (2021). PCAP Funnel: A tool for rapid exploration of packet capture files. In: *2021 25th International Conference Information Visualization (IV). AI & visual analytics & data science*. Sydney: IEEE Biometric Council, 2021, pp. 69–76. ISBN 978-1-6654-3827-8. ISSN 2375-0138.
7. Akazaki, T., Liu, S., Yamagata, Y., Duan, Y., & Hao, J., (2018). Falsification of cyber-physical systems using deep reinforcement learning. In*: International Symposium on Formal Methods* (pp. 456–465).
8. Gupta, A., & Yang, Z., (2018). *Adversarial Reinforcement Learning for Observer Design in Autonomous Systems Under Cyber Attacks*. arXiv preprint arXiv:1809.06784.
9. Malialis, K., & Kudenko, D., (2015). Distributed response to network intrusions using multiagent reinforcement learning. *Engineering Applications of Artificial Intelligence, 41*, 270–284.
10. Iman, S., Arash, H. L., & Ali, A. G., (2018). Toward generating a new intrusion detection dataset and intrusion traffic characterization. In: *4th Int. Conf. on Inf. Sys. Sec. and Priv.* Portugal.
11. Hwang, R. H., Peng, M. C., Nguyen, V. L., & Chang, Y. L., (2019). An LSTM-based deep learning approach for classifying malicious traffic at the packet level. *Applied Sciences, 9*(16), 3414. https://doi.org/10.3390/app9163414.
12. Aygun, R. C., & Yavuz, A. G., (2017). Network anomaly detection with stochastically improved autoencoder based models. In: *Proc. IEEE 4th Int. Conf. Cyber Secur. Cloud Comput. (CSCloud)* (pp. 193–198).
13. Naseer, S., et al., (2018). Enhanced network anomaly detection based on deep neural networks. In: *IEEE Access* (Vol. 6, pp. 48231–48246). doi: 10.1109/ACCESS.2018.2863036.
14. Hodo, E., Bellekens, X., Hamilton, A., Tachtatzis, C., & Atkinson, R., (2017). *Shallow and Deep Networks Intrusion Detection System: A Taxonomy and Survey*. [Online]. Available: https://arxiv.org/abs/1701.02.
15. Huang, G. B., Zhu, Q. Y., & Siew, C. K., (2006). Extreme learning machine: Theory and applications. *Neurocomputing, 70*(1–3), pp. 489–501. [Online]. Available: http://linkinghub.elsevier.com/retrieve/pii/S0925231206000385 (accessed on 27 January 2024).
16. Jianyu, W., Zhenling, M., Heng, Z., & Qiang, M., (2020). Ensemble diagnosis method based on transfer learning and incremental learning towards mechanical big data. *Measurement. 155*, 107517.
17. Jun, M., Yuexiong, D., Jack, C.P. C., Feifeng, J., Yi, T., Vincent, J. L. G., & Zhiwei, W., (2020). Identification of high impact factors of air quality on a national scale using big data and machine learning techniques. *Journal of Cleaner Production, 244*, 118955.

18. Loic, B., Van, L. C., McDermott, J., & Nhien-An Le-Khac, (2017). *Collective Anomaly Detection Based on Long Short Term Memory Recurrent Neural Network.* https://paperswithcode.com/paper/collective-anomaly-detection-based-on-long (accessed on 27 January 2024).
19. Zhao, R., Ruqiang, Y., Jinjiang, W., & Kezhi, M., (2017). Learning to monitor machine health with convolutional bi-directional LSTM networks. *Sensors, 17*(2), 273.
20. Vinayakumar, R., Soman, K. P., & Poornachandran, P., (2018). Evaluating deep learning approaches to characterize and classify malicious URL's. *Journal of Intelligent & FuzzySystems, 34*(3), 1333–1343.
21. Zhang, X., Zhao, J., & LeCun, Y., (2015). Character-level convolutional networks for text classification. In: *Advances in Neural Information Processing Systems* (pp. 649–657).
22. Tae-Young, K., & Sung-Bae, C., (2018). Web traffic anomaly detection using C-LSTM neural networks. *Expert Systems with Applications, 106*, 66–76.
23. Perdisci, R., Davide, A., Prahlad, F., Giorgio, G., & Wenke, L., (2009). McPAD: A multiple classifier system for accurate payload-based anomaly detection. *Computer Networks, 53*(6), 864–881.
24. Chen, T., Mao, Q., Yang, Y., Lv, M., & Zhu, J., (2018). TinyDroid: A lightweight and efficient model for Android malware detection and classification. *Mob. Inf. Syst., 2018*, 4157156.
25. Mahindru, A., & Sangal, A., (2021). FSDroid: A feature selection technique to detect malware from android using machine learning techniques. *Multimed. Tools Appl.*, 1–53.
26. Wang, W., Xing, W., Dawei, F., Jiqiang, L., Zhen, H., & Xiangliang, Z., (2014). Exploring permission-induced risk in android applications for malicious application detection. *IEEE Transactions on Information Forensics and Security, 9*(11), 1869–1882.
27. Aonzo, S., Alessio, M., Mauro, M., Luca, O., & Francesco, P., (2017). Low-resource footprint, data-driven malware detection on android. *IEEE Transactions on Sustainable Computing, 5*(2), 213–222.
28. Wah, Y. B., Nurain, I., Hamzah, A. H., Abdul-Rahman, S., & Simon, F., (2018). Feature selection methods: Case of filter and wrapper approaches for maximizing classification accuracy. *Pertanika Journal of Science & Technology, 26*(1).
29. Alghoul, A., Sara, Al. A., Ghada, Al. J., Ghayda, H., & Abu-Naser, S. S., (2018). *Email Classification Using Artificial Neural Network.*
30. Chen, T., Mao, Q., Yang, Y., Lv, M., & Zhu, J., (2018). TinyDroid: A lightweight and efficient model for android malware detection and classification. *Mobile Information Systems, 2018*, 1–9. 10.1155/2018/4157156.
31. Wang, W., Li, Y., Wang, X., Liu, J., & Zhang, X., (2017). Detecting android malicious apps and categorizing benign apps with ensemble of classifiers. *Future Generation Computer Systems, 78.* 10.1016/ j.future.2017.01.019.
32. Aonzo, S., Merlo, A., Migliardi, M., Oneto, L., & Palmieri, F., (2020). Low-resource footprint, data-driven malware detection on android. In: *IEEE Transactions on Sustainable Computing* (Vol. 5, No. 2, pp. 213–222). doi: 10.1109/TSUSC.2017.2774184.
33. Cai, J., Luo, J., Wang, S., & Yang, S., (2018). Feature selection in machine learning: A new perspective. *Neurocomputing, 300*, 70–79. ISSN 0925-2312.
34. Leelalakshmi, S., & Rameshkumar, K., (2022). Traffic monitoring and malicious detection multidimensional PCAP data using optimized LSTM RNN. *International Journal of Information Security and Privacy (IJISP), 16*(2), 1–22.

35. https://medium.com/thg-tech-blog/fingerprinting-network-packets-53ee32ddf07a (accessed on 27 January 2024).
36. GitHub–0x4D31/fatt: *FATT/finger print All The Things – A Pyshark Based Script for Extracting Network Metadata and Finger prints from PCAPFiles and Live Network Traffic*.
37. https://www.tcs.com/facts-fingerprint-biometric-identity-authenication (accessed on 27 January 2024).

Index

A

B

C

D

E

F

J

K

O

Q

R

S

T

U